BEHAVIOR OF MARINE ANIMALS

Current Perspectives in Research

Volume 4: Marine Birds

BEHAVIOR OF MARINE ANIMALS

Current Perspectives in Research

Series Editors: Howard E. Winn and Bori L. Olla

BEHAVIOR OF MARINE ANIMALS

Current Perspectives in Research

Volume 4 : Marine Birds

Edited by

Joanna Burger
Rutgers University
New Brunswick, New Jersey

Bori L. Olla
National Marine Fisheries Service
Highlands, New Jersey

and

Howard E. Winn
University of Rhode Island
Kingston, Rhode Island

PLENUM PRESS • NEW YORK-LONDON

Library of Congress Cataloging in Publication Data

Winn, Howard Elliott, 1926-
  Behavior of marine animals.

  Vol. 4 edited by J. Burger, B. L. Olla, and H. E. Winn.
  Includes bibliographies and index.
  CONTENTS: v. 1. Invertebrates. — v. 4. Marine birds.
  1. Marine fauna — Behavior. I. Olla, Bori L., joint author.
II. Burger, Joanna, joint author. III. Title.
AL121.W5   591.52'636   79-167675
ISBN 0-306-37574-5(v.4)

© 1980 Plenum Press, New York
A Division of Plenum Publishing Corporation
227 West 17th Street, New York, N.Y. 10011

Printed in the United States of America

CONTRIBUTORS

C. G. Beer Institute of Animal Behavior, Rutgers University, Newark, New Jersey

Richard G. B. Brown Canadian Wildlife Service, Bedford Institute of Oceanography, Dartmouth, Nova Scotia, Canada

Francine G. Buckley North Atlantic Regional Office, National Park Service, Boston, Massachusetts

Paul A. Buckley North Atlantic Regional Office, National Park Service, Boston, Massachusetts

Joanna Burger Department of Biology, Livingston College, and Center for Coastal and Environmental Studies, Rutgers University, New Brunswick, New Jersey

Roger M. Evans Department of Zoology, University of Manitoba, Winnipeg, Canada

Michael Gochfeld Department of Health, Trenton, New Jersey; and Division of Environmental Health, Columbia University School of Public Health, New York, New York

George L. Hunt, Jr. Department of Ecology and Evolutionary Biology, University of California, Irvine, California

W. A. Montevecchi Department of Psychology, Memorial University of Newfoundland, St. John's, Newfoundland, Canada

J. M. Porter Department of Biology, Acadia University, Wolfville, Nova Scotia, Canada. *Present address*: Department of Zoology, University of Manitoba, Winnipeg, Manitoba, Canada

John P. Ryder Department of Biology, Lakehead University, Thunder Bay, Ontario, Canada

William E. Southern Department of Biological Sciences, Northern Illinois University, DeKalb, Illinois

Bernice M. Wenzel   Department of Physiology and Brain Research
Institute, UCLA School of Medicine, Los Angeles,
California

PREFACE

The majority of the chapters in this volume are structured to include a balance between literature review, original data, and synthesis. The research approaches taken by the authors are generally of two kinds. One centers on the long-term, in-depth study of a single species in which many aspects of its natural history are examined in detail. The other is a comparative one which involves investigating particular questions by examining species or by comparing groups of species that may include taxonomic and/or ecological affinities.

Most of the chapters concern obvious aspects of breeding behavior including habitat selection, the effects of age on breeding, communication, mating systems, synchrony of breeding activities, development of behavior, prefledging parental care, and postfledging parental care. Of these topics, many relate directly to the advantages and disadvantages of coloniality—a conspicuous behavior pattern in marine birds. As such, they provide parapicuou for the further study of coloniality and the social behavior of many other animals. Other important areas of marine bird breeding behavior (such as courtship behavior, antipredator behavior, information transfer) have not been included because of space limitations.

Since man's encroachment on the seashore and continental shelf poses certain threats to marine birds, a volume elucidating various aspects of their biology has multiple uses. As well as being of value to ornithologists, the volume should be useful to managers involved with coastal planning. Behaviorists and ecologists whose studies do not usually encompass marine birds, but who are interested in basic mechanisms, should also find useful information and ideas in this volume.

We are especially grateful to Deborah J. Gochfeld, Michael Gochfeld, Betty Green, Jill Grover, and Joseph Luczkovich for their help in indexing, and to Paul Buckley, Michael Gochfeld and Guy Tudor for their expertise in seabird taxonomy.

J. Burger
B. Olla
H. Winn

INTRODUCTION

Among birds, the colonial marine species provide unusual opportunities to explore basic problems in behavior, ecology, and evolution. The overall aim of this volume is to provide a representative selection of the current research being performed on this group, especially concentrating on various aspects of their behavior and ecology.

Traditionally, marine birds have included those species that breed in large colonies on offshore islands or along the coasts. Based on this, the following have usually been recognized as constituting marine families: Spheniscidae (penguins), Diomedeidae (albatrosses), Procellariidae (fulmars, petrels, shearwaters), Hydrobatidae (storm petrels), Pelecanoididae (diving petrels), Phaethontidae (tropicbirds), Pelecanidae (pelicans), Sulidae (gannets, boobies), Phalacrocoracidae (cormorants), Anhingidae (anhingas), Fregatidae (frigatebirds), Chionidae (Sheathbills), Stercorariidae (skuas), Laridae (gulls, terns), Rynchopidae (skimmers), and Alcidae (alcids). In compiling the present volume we have tended to adhere to this view and have limited the articles accordingly.

However, it should be noted that the categorization of a species as marine is not as straightforward as it would appear. For example, species such as Franklin's Gull, *Larus pipixcan*, breed inland away from the sea but depend upon it during other times of the year. The Herring Gull, *Larus argentatus*, although usually found along the coast, is able to reside throughout the year (including the breeding season) in a nonmarine environment. Other species, simply because of taxonomic similarities, have been included with marine bird families but are entirely independent of the sea [e.g., some anhinga (*Anhinga* sp.)]. Then there are others that, because of the traditional categorization, are not considered marine but do, in fact, depend upon the sea for at least part of their yearly cycle (e.g., shorebirds, some ducks, and loons). From a perspective of environmental impacts, it is perhaps time to broaden our perspectives of what constitutes marine bird populations by considering behavioral–ecological relationships and include those species that depend upon the sea for any or part of their life cycle.

Marine birds comprise only 3% of the known avian species in the world, yet the literature on this group far exceeds their relative numbers.

This has not always been the case. Analyzing articles written in *Auk* and *Ibis* from 1930 to 1979, we found that in 1930 only 4% of the articles were on marine birds. This rose gradually, and, in the period from the early 1940s to 1965, marine bird studies averaged about 7 to 8%. Beginning in 1965, the percentage again increased, reaching 13% in 1970 and 15% in 1979. In 1979, the percentage of all titles that concerned marine birds in European and other major U.S. ornithological journals ranged from 12% (*Wilson Bulletin*) to 24% (*Bird-Banding*). The increase in interest is shown not only by an increase in articles in the ornithology journals but by the recent formation of groups in the United States, Australia, Europe, New Zealand, and South Africa solely devoted to the study of marine birds.

The steady increase in the study of marine birds has been stimulated in large measure by the fact that they are especially accessible during the breeding season when they primarily occur in large colonies. At this time they lend themselves to examination of individual variation and adaptation. Many species can be handled and individually marked without obvious ill effect or more than transient behavioral changes. This makes it possible to follow the behavior and reproductive success of individuals under different environmental conditions. Further, because marine birds are long-lived, they can be followed over a number of years, allowing for examination of population dynamics as well as behavioral and physiological effects. Thus, most research on marine birds has concentrated on breeding biology with large gaps in our knowledge during the times when they are not on the breeding grounds. For example, during the nonbreeding season many species generally disperse over wide areas. Even during the breeding season, many species forage away from the colony and are difficult to follow. In addition, there are those species that do not even come to land until they reach maturity at three years or older.

Future studies on marine birds will continue to center on examining breeding biology and behavior. The life cycle and ecological factors when marine birds are away from the breeding colonies present a myriad of research possiblities that up until the present time have hardly been touched upon. However, if we are ever to understand fully the selective factors that operate on marine birds, we can no longer neglect their biology when they are not breeding. For birds that spend their first three to eight years away from the nesting colonies and are away much of the year thereafter, this period is critical.

J. Burger
B. Olla
J. Winn

CONTENTS

Chapter 3
Habitat Selection and Marine Birds

Francine G. Buckley and Paul A. Buckley

Chapter 4
Mate Selection and Mating Systems in Seabirds

George L. Hunt, Jr.

Chapter 8
Development of Behavior in Seabirds: An Ecological Perspective

Roger M. Evans

Chapter 9
Parental Investments by Seabirds at the Breeding Area with Emphasis on Northern Gannets (*Morus bassanus*)

W. A. Montevecchi and J. M. Porter

Volume 4: Marine Birds

Chapter 1

SEABIRDS AS MARINE ANIMALS

Richard G. B. Brown

Canadian Wildlife Service
Bedford Institute of Oceanography
P. O. Box 1006
Dartmouth, Nova Scotia, Canada B2Y 4A2

I. INTRODUCTION

Seabirds are marine animals. A great deal is known about their ecology, yet paradoxically almost all this information has been collected at the birds' breeding sites on land (see, e.g., Bédard, 1969; Belopol'skii, 1961; Cramp *et al.*, 1974; Fisher, 1952; Fisher and Lockley, 1954; Nelson, 1970; Nettleship, 1972; Richdale, 1963; Salomonsen, 1955; L. M. Tuck, 1961). Yet remarkably little is known about the 50% or more of their lives that seabirds spend at sea. This is partly because of the technical difficulties in studying seabirds at sea. But it is unfortunate, too, that the pelagic aspects of seabird ecology fall between the disciplines of ornithology and oceanography, and neither side has been very eager to bridge the gap. Ornithologists tend to think of the sea as something flat, wet, and relatively uniform, over which their birds happen to fly. Oceanographers and marine biologists know that it is far from uniform, but seem reluctant to concede that these airborne, highly mobile animals, which cannot even breed at sea, could ever be associated with specific and relatively localized marine habitats or species communities.

This chapter attempts to bridge the gap. It describes, in a preliminary and often speculative way, some of the links between seabirds and oceanography that are beginning to appear now that the basic question of the birds' distributions at sea has been more or less answered. Its emphasis is on the seabird as a marine animal and as a member of a marine community, and virtually ignores the land-based part of the birds' lives. If this approach seems

1

perverse, it is not wholly unrealistic. After all, most seabirds have long subadult periods, which many of the more pelagic species spend away from land. During those years they are indeed truly, if temporarily, marine animals.

A review of this kind inevitably leans heavily on the important reviews of Bourne (1963, 1976a) and Ashmole (1971), and on Murphy's (1936) *Oceanic Birds of South America*—the best book on seabirds, perhaps on any group of birds, yet written. Wynne-Edwards's (1935) classic paper on seabird distributions in the North Atlantic is also essential reading. Other pelagic surveys are reviewed in the course of the chapter.

II. HISTORICAL PERSPECTIVES

A. Identification and Distribution

The first seabird observers were seamen. A working knowledge of seabird identification and the distances the various species would travel from land was an important part of the early navigator's art. On September 17, 1492, in mid-Atlantic at 36°W, Christopher Columbus saw a tropicbird *Phaethon* sp. and concluded, wrongly, that he must be near land because the species "is not accustomed to sleep on the sea" (Jane, 1960, p. 11). Later mariners' observations were more accurate and more detailed. The *English Pilot* of 1767 [facsimile in Lysaght (1971, Fig. 4)], in describing "Some Directions which ought to be taken Notice of by those sailing to *Newfoundland*," notes:

> There is also another thing to be taken Notice of by which you may know when you are upon the [Grand] Bank, I have read an Author that says, in treating of this Coast, that you may know this by the great quantities of Fowls upon the Bank, viz. *Sheer-waters* [Greater and Sooty Shearwaters *Puffinus gravis* and *P. griseus*], *Willocks* [alcids: probably Common and Thick-billed Murres *Uria aalge* and *U. lomvia*], *Noddies* [Northern Fulmar *Fulmarus glacialis*], *Gulls* [probably Black-legged Kittiwake *Rissa tridactyla*] and *Pengwins* [Great Auk *Pinguinus impennis*], &c. without making any Exceptions; which is a mistake, for I have seen these Fowls 100 Leag. off this Bank, the *Pengwins* excepted. It's true that all these Fowls are seen there in great quantities, but none are to be minded so much as the *Pengwin*, for these never go without [i.e., away from] the Bank as the others do, for they are always on it, or within it, several of them together, sometimes more, other times less, but never less than 2 together. . . .

Such detailed instructions were obviously the result of long, careful observation. They also tell us more than we can ever now discover of the pelagic distribution of the now-extinct Great Auk in the western Atlantic.

All this should have given a promising start to the study of birds at sea, and the great marine explorers of the eighteenth and early nineteenth centuries followed it by including seabirds in their collections (e.g., Beaglehole, 1963). On land this collecting phase of ornithology evolved into distribution studies, with increasing emphasis on field rather than museum work, and these, in turn, laid the foundations of our modern ecological approach. This did not occur in marine ornithology until very recently. The results of the early collections were never properly published, and so gave little impetus to further research (Bourne, 1963). The lack of adequate identification techniques and the enormous logistical difficulties in taking observers to areas where they could collect information on pelagic distributions were even more important obstacles.

Identification was a problem because it has always been difficult to collect birds from ships. This was a serious enough drawback in the days when respectable ornithology was strictly a museum science. But even in our own era the difficulty of checking field identifications against museum specimens has hindered the development of field guides to seabirds. The breakthrough came relatively recently with Alexander's (1928) *Birds of the Ocean*, followed by the Smithsonian Institution's series of regional identification manuals (Watson, 1963, 1966, 1975; W. B. King, 1967), and by G. S. Tuck's (1978) new all-ocean field guide. Even now, many field identification problems remain; "the various white-rumped and dark-rumped storm petrels in the Pacific present as great a problem to the field worker as any two groups of sibling species known" (Crossin, 1974, p. 155). Nor is the museum taxonomy yet settled—the well-known Giant Petrel was recently split into two siblings, *Macronectes giganteus* and *M. halli* (Bourne and Warham, 1966), and subspecies of such birds as the Gentoo Penguin *Pygoscelis papua* and Cory's Shearwater *Calonectris diomedea*, marginally identifiable in the field, may yet be raised to specific rank (Murphy, 1924, 1964).

Up to a point the logistical difficulties can be avoided, because it is not absolutely essential to go to sea in order to study aspects of the pelagic ecology of seabirds. Seabird movements can be observed from the shore, as Lockley (1953) showed for Manx Shearwaters *Puffinus puffinus* in the English Channel and as Ainley (1976) showed for various seabirds along the California coast. Counts of seabirds found dead on beaches after storms can be used as indirect indices of movements (e.g., Cooper, 1978; Veitch, 1978), although this technique is used mainly to assess the impacts of oil and other pollutants (e.g., Bourne, 1976b). Much can also be discovered about seabird distributions and movements at sea by combining museum data with published and unpublished sight records (e.g., Phillips, 1963; Roberts, 1940; Stresemann and Stresemann, 1970; Voous and Wattel, 1963; Watson *et al.*,

1971). In practice, however, there is no real substitute for direct observation of birds out at sea.

Unfortunately, the observer almost always has to sail on a ship whose primary purpose is trade, fishery, or oceanographic research, which restricts observations to where the ship happens to be going. This can be a serious limitation. For example, the well-traveled shipping route between Britain and South Africa permits extensive coverage of the barren subtropical South Atlantic, but of virtually none of the oceanographically and ornithologically important Benguela Current (e.g., Stanford, 1953). Casual travelers on commercial carriers are further limited by their inability to make repeated series of observations—the essential basis for any assessment of a bird's "normal" distribution and of how it may vary seasonally—except on coastal ferries (Finch et al., 1978). Here, however, the observations of ships' deck officers have made an important contribution, from the early reports of Swinburne (1886) and Paessler (e.g., 1909), through those from observers on World War II convoy duty (Baker, 1947; Rankin and Duffey, 1948), to Mörzer Bruijns (e.g., 1965), Chapman (1969), Dorval (1969), and the observers whose reports are summarized in the Royal Naval Bird-Watching Society's journal, Sea Swallow.

The first observers to work away from the regular shipping lanes were on whaling and fishing ships, from Collins (1884) and Murphy (1914) to such recent workers such as Bagenal (1951), Bierman and Voous (1950), Garcia Rodriguez (1972), Lambert (1971), Lockley and Marchant (1951), Martin and Myres (1969), Rees (1963), and van Oordt and Kruijt (1953, 1954). But nowadays oceanographic research cruises provide more opportunities. Jespersen (1924, 1929, 1930) and Hentschel (1933) were pioneers in this work, followed by Moore's (1941, 1951; see also Butcher et al., 1968) summaries of observations made on Woods Hole Oceanographic Institute ships. Recently ornithologists have taken full advantage of the boom in oceanographic research that began in the 1950s. Bourne (1976a), Brown (1977, 1979), Brown et al. (1975b), Joiris (1976), and Harris and Hansen (1974) in the North Atlantic; Brown et al. (1975a), Harris and Hansen (1974), Jehl (1974a), Rumboll and Jehl (1977), Summerhayes et al. (1974), and Tickell and Woods (1972) in the South Atlantic; Bartonek and Gibson (1972), Jehl (1974b), W. B. King (1970, 1974), J. E. King and Pyle (1957), Kuroda (1955, 1960), Sanger (1970), Shuntov (1974), Swartz (1967), Wahl (1975), and Watson and Divoky (1972) in the North Pacific; Brown et al. (1975a), Devillers and Terschuren (1978), Jehl (1973), Shuntov (1974), Summerhayes (1969), Szijj (1967), and Webb (1973) in the South Pacific; and Bailey (1966, 1968, 1971), Johnstone and Kerry (1976), Ozawa (1967), and Shuntov (1974) in the Indian Ocean are examples of the extent of recent coverage, obtained for the most part on oceanographic cruises. We now know, in broad outline, the pelagic distributions of the commoner seabirds, in summer at least, in almost all the world's oceans. The main gap [surprisingly, since its seabird *colonies* have been so extensively studied by such researchers as Belopol'skii (1961) and Uspenski (1958)] is in the Barents Sea. There is still room for further detail, of course, and some important work is in progress. Surveys in the Canadian Arctic and Alaska, led by the Canadian Wildlife Service and the U.S. Fish and Wildlife Service, and off the coast of New England by Manomet Bird Observatory, will soon provide valuable new information on seabird pelagic distributions (e.g., Bartonek and Nettleship, 1979).

B. Development of Quantitative Survey Techniques

Along with this increasing amount of ocean coverage of seabird distributions there has been a corresponding development in quantitative recording methods. The first attempts were made before World War I, when Paessler (1909), a ship's captain on the route between Hamburg and the western coast of South America, kept regular daily lists of all the species he saw—a rudimentary system, but a significant advance on the casual records kept by Swinburne (1886) a quarter of a century earlier. Murphy (1914) improved on this by adding approximate numbers to his lists, as did Jespersen (1924, 1930), Hentschel (1933), Redfield (1941), Moore (1951), and, most recently, Tickell and Woods (1972). The difficulty, however, is that quantitative lists lack a proper statistical base. At best they give estimates of the numbers of birds seen per day, but without stating such important information as the number of hours of observation during which these numbers were seen; nor can one always be sure that the *absence* as well as the presence of a species is being recorded.

Wynne-Edwards (1935) carried out the first properly quantified pelagic seabird survey. He counted the numbers of birds seen per hour in the course of eight Atlantic crossings he made between Canada and England in 1933. This systematic technique, along with the repeated recensusing of a standard route, provided for the first time not only quantified distribution data, but also showed how, where, and when these distributions changed in the course of the summer. Several subsequent surveys used these 1-hr watch periods (e.g., Bailey, 1966; J. E. King and Pyle, 1957; Kuroda, 1960). The main drawback was that so long a watch is too coarse-grained for use in areas where seabird numbers and species change very rapidly (e.g., at boundaries between water types) (see Section IV). The Smithsonian Institution's Pacific Ocean Biological Survey Program's (POBSP) seabird scheme avoided this by recording *all* the birds seen during the day (W. B. King, 1970; W. B. King *et al.*, 1967). This worked well in tropical seas, where seabirds were scarce, but not with the large seabird populations of

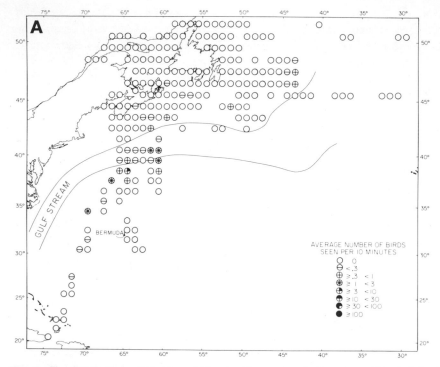

Fig. 1. The distributions of Cory's and Greater Shearwaters in the western North Atlantic, June–September. (A) Greater Shearwater, *Puffinus gravis*. (B) Cory's Shearwater, *Calonectris diomedea*. Averages are of the numbers of birds seen per 10-min watch from a moving ship in good visibility, and refer to 1°N × 1°W "blocks." Data were collected between 1969 and 1976. The absence of a symbol indicates absence of coverage, not absence of birds. The Greater Shearwaters seen south of the Gulf Stream were all June birds, presumably migrant stragglers. [After Brown (1977), maps 2 and 3.]

temperate and polar seas. As a compromise, the Programme Intégré des Recherches sur les Oiseaux Pélagiques (PIROP) introduced 10-min watch periods for use in the northwest Atlantic (Brown *et al.*, 1975*b*), and this unit seems to have been generally adopted for use in shipboard work. The POBSP scheme also devised a computer program to deal with the mass of collected observations, and this adapted by PIROP and other seabird reporting schemes.

Figures 1–4 show the kinds of information on seabird distributions which these quantitative surveys can produce. Figures 1 and 2 show the quantitative distributions of Cory's, Greater, and Sooty Shearwaters in the boreal summer in the northwest Atlantic. Cory's and Greater Shearwaters show very little overlap in distribution (Fig. 1); they are, respectively, birds of the Gulf Stream and adjacent waters, and of the cooler waters farther

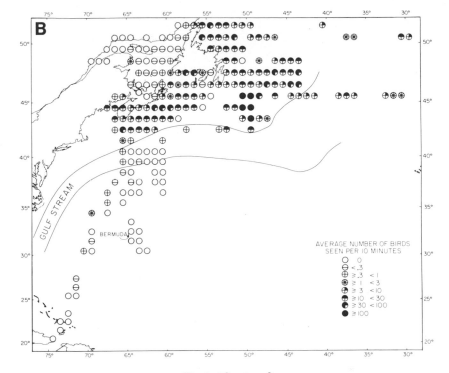

Fig. 1. (*Continued*)

north. Greater and Sooty Shearwaters, both migrants from the South Atlantic, do overlap, and a nonquantitative assessment of their distributions (Palmer, 1962, pp. 169 and 176) suggests that they are identical. But quantitative data show that Sooty Shearwaters pass rapidly through eastern North American waters and that by September are mostly in the northeast Atlantic (Fig. 2; Phillips, 1963). By contrast, the bulk of the Greater Shearwaters seem to stay in the western Atlantic (Fig. 1; Brown *et al.*, 1975*b*; Voous and Wattel 1963). Taken together, Figs. 1 and 2 provide the kind of detailed information on the comparative distributions of the three species, as well as the route and timing of the migration of one of them, which has long been available for many land birds, but not for seabirds. Figures 3 and 4 show, on a finer grid, the distributions of Dovekies *Alle alle* and Red Phalaropes *Phalaropus fulicarius* in August off southeast Baffin Island and eastern Labrador, respectively. Both cases take quantitative surveys to the stage where one can begin to ask *why* the birds are where they are. The Dovekies are clearly associated with pack ice, while the main Phalarope concentrations are at the eastern boundary of the Labrador Current (Fig. 5). These points are discussed further in Sections IV and V.

However, the figures also illustrate some of the basic limitations of sur-

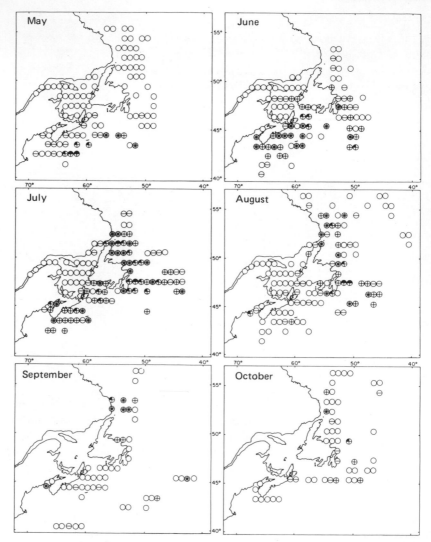

Fig. 2. The migration of Sooty Shearwaters, *Puffinus griseus*, through the northwest Atlantic, May–October. Data were collected between 1969 and 1973. Symbols and other details as in Fig. 1. [After Brown *et al.* (1975*b*), map 5.]

veys done from ships. Figures 1–3 combine data from several years in order to give more complete coverage. (Figure 4 shows that the area that can be covered in a single cruise is usually quite small.) But by doing this, year-to-year comparisons of distribution patterns are ruled out. In the surveys shown in Fig. 4 the ship visited different areas from one year to the next, so one cannot tell whether concentrations A and B occurred again in 1978, or even how long they lasted in 1977. Moreover, as the ship spent little time close inshore, there is no indication of the use which the birds made of that habitat. Finally, since ships rarely visit the Arctic in winter, it would be hard to obtain Dovekie distribution data in, say, January that would be comparable to those shown for August (Fig. 3).

Beyond these limitations there are also difficulties in the actual counting techniques (Bailey and Bourne, 1972). Counts are biased toward large, conspicuous species especially if, like fulmars and many gulls, they regularly follow ships. It is also important to allow for the ship's speed, because the number of ship followers increases sharply as soon as the ship stops, and even more sharply if it does so as part of some fishing activity. As a result, the surveys will produce maps like those in Figs. 1–4, which show where within its range a seabird is *relatively* most abundant. However, it is difficult to convert these figures into *absolute* estimates of population density. Up to a point these difficulties can be resolved by, for example, eliminating the conspicuousness bias by confining counts to birds seen within a relatively small, fixed distance from the ship (e.g., Bartonek and Gibson, 1972; Cline *et al.*, 1969). But it seems unlikely that a completely satisfactory solution will ever be found to the problems of shipboard censusing.

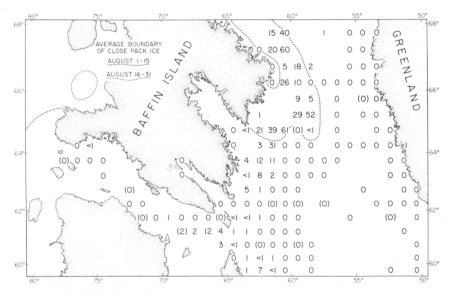

Fig. 3. The distribution of Dovekies *Alle alle* off southeast Baffin Island in August, in relationship to the average boundary of the pack ice. Averages refer to 30'N × 1°W "blocks." The number of 10-min watches on which each average was based was rarely more than 5. Numbers in parentheses represent averages that are the result of only a single watch. Data were collected between 1970 and 1978. [Pack-ice boundaries are from Anon. (1958). Other data from R. G. B. Brown, unpublished manuscript.]

An alternative method is to count from the air. Rankin and Duffey (1948), on antisubmarine patrols in World War II, were the first to appreciate the greater scope of observation from an aircraft. Recently De Havilland Twin-Otters, high-winged twin-engined aircraft, have been used extensively for quantitative surveys in the Canadian Arctic (Bradstreet and Finley, 1977; Johnson *et al.*, 1976; Nettleship and Gaston, 1978; see also Fig. 6). These flights are made at a standard height and speed (30 m; 100 kt = 185 km/hr), and all species seen within a 200-m band on each side of the aircraft are counted; correction factors for the positions of the

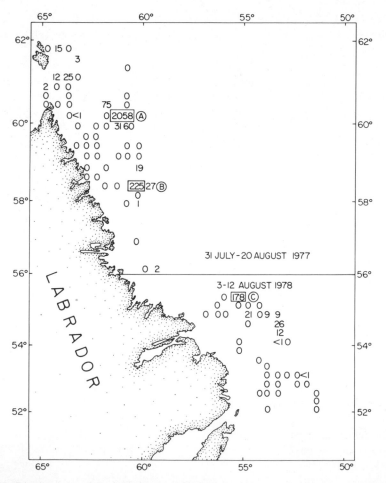

Fig. 4. Distributions of Red Phalaropes, *Phalaropus fulicarius*, on two August cruises off Labrador. A, B, and C are concentrations per 10-min watch shown Fig. 5. For further details, see Fig. 3. [R. G. B. Brown, unpublished manuscript.]

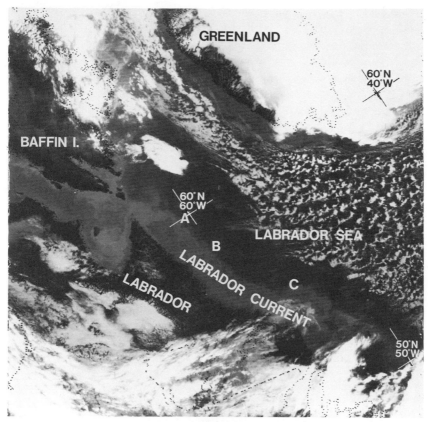

Fig. 5. The Labrador Current photographed by satellite on September 3, 1977. The cold current shows up as paler than the marginally warmer Labrador Sea. In terms of surface sea temperature the current was 1°–2°C colder, both in August and October 1977. [Unpublished Bedford Institute of Oceanography data, J. R. Lazier, personal communication] A, B, and C mark the approximate positions of the concentrations of Red Phalaropes shown in Fig. 4. [Photograph by U.S. N.O.A.A Weather Satellite, courtesy of Atmospheric Environment Service, Toronto.]

observers in different parts of the aircraft are an integral part of the analyses. These surveys have considerable advantages over those from ships. They can operate in the Arctic far earlier in the year than any ship can, they are well suited to the great distances involved, can sample all habitats and can be repeated many times in a season, and are free of such behavioral biases as ship following.

Figure 6 shows Thick-billed Murre distributions in western Lancaster Sound as shown by two aerial surveys flown in early August 1976 (Nettleship and Gaston, 1978). These showed, among other things, a change in murre foraging areas over a period of only 5 days—a fact that would almost

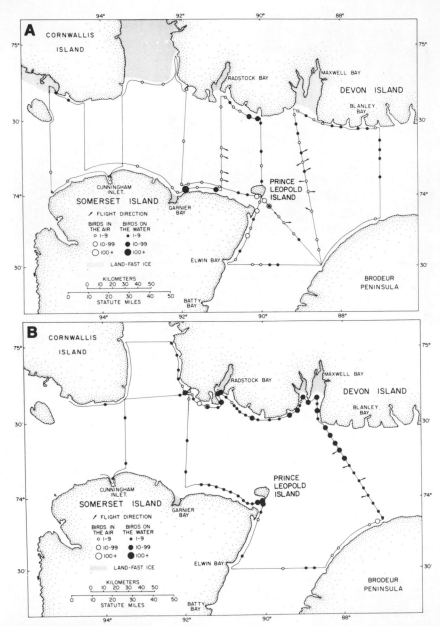

Fig. 6. Distributions of Thick-billed Murres *Uria lomvia* in Barrow Strait and western Lancaster Sound, N.W.T., as shown by aerial surveys. (A) August 2, 1976. (B) August 6, 1976. The birds are from the large colony on Prince Leopold Island, as the flight directions show, and many are flying at least 60 km to feed. Note changes in foraging areas that took place between August 2 and 6. [After Nettleship and Gaston (1978), Figs. 14 and 15.]

certainly have gone undetected in the more limited scope of a shipboard survey. Survey lines such as these can be repeated from year to year, and many times during a season. Those of Johnson *et al.* (1976) in eastern Lancaster Sound provided a picture of seasonal population fluctuations of seabirds from early May to late September 1976 that, in shipboard terms, could only have been obtained by combining several seasons' data (e.g., Fig. 2). Nonetheless there are drawbacks to aerial studies. The principal drawback, apart from considerations of cost and range, is identification—closely related species such as alcids are very difficult to separate from the air. This is not a serious problem in the relatively simple seabird communities of the Canadian Arctic, but it could be a considerable limitation if the technique were applied to surveys of the more complex communities of Alaska and the North Atlantic.

III. SEABIRDS AND THEIR MARINE HABITATS

One of the advantages of the boom in oceanographic research has been that it has provided ornithologists not only with platforms from which to observe birds, but also with oceanographic information to help them interpret their results. It is now possible to integrate the ornithological and oceanographic research on a cruise, as Bailey (1966), Brown *et al.* (1975*a*), J. E. King and Pyle (1957), Pocklington (1979), and Summerhayes *et al.* (1974) have tried to do. Even when direct integration is not possible—as, for example, on geological and geophysical cruises—the ornithologist can often fall back on oceanographic data on file for the cruise area at the appropriate time of year.

In interpreting seabird distributions it is important to understand how a marine habitat is defined, as well as how it resembles and differs from a terrestrial one. Ornithologists have tended to use habitat definitions based on the depth of water and/or distance from land, such as "inshore" (from the high-water mark out to about 5 mi), "offshore" (from there to the edge of the continental shelf), and "pelagic" (farther out still) (Wynne-Edwards, 1935). This is valid as far as it goes, but it is much too sweeping. The oceanographic practice of dividing the sea into water types is more precise and therefore more likely to be useful in integrating seabirds with the other members of their marine community. In physical terms a water type is defined by a fairly limited range of temperature–salinity relationships (as opposed to temperature or salinity taken separately), which set it apart from other water types. Figure 7, based on Pickard (1971), shows this for the Chilean fjords. On Wynne-Edwards's classification these fjords would

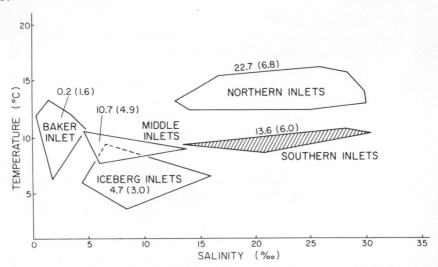

Fig. 7. Water types in the Chilean fjords, defined in terms of their temperature–salinity relationships at the surface (see Pickard, 1971). The Blue-eyed Shag *Phalacrocorax atriceps* was dominant in all the water types except the Southern Inlets (*shaded area*), where it was replaced by the sibling King Shag *P. albiventer*. Average numbers of the two siblings combined per 10-min watch in each water type are followed by parenthetical Secchi disk readings (m); the lower the number, the higher the silt content and, presumably, the difficulty with which shags can pursue their prey underwater. [After Brown *et al.* (1975a), Fig. 4.]

all be grouped as "inshore," but a temperature–salinity plot shows that they include at least five water types. On a broader scale, Pocklington (1979) has recognized nine water types over the Indian Ocean as a whole, and the subdivisions in other oceans are summarized by Pickard (1964). Each water type has its own, more-or-less distinctive community of fish, plankton, and other biota (e.g., Bary, 1963; Fager and Longhurst, 1968; Fager and McGowan, 1963). (It should be emphasized that the animals in these communities are not necessarily reacting directly to temperature or salinity as such; these parameters should be thought of merely as labels.) Clearly, the term "sea" includes many habitats—perhaps as many as "grassland" does in a terrestrial context.

The production cycles in terrestrial and marine habitats are also comparable. In both, nitrates, phosphates, and other nutrients fertilize the growth of plants, which in turn leads to the production of first herbivorous and then carnivorous animals. Where they differ is that the terrestrial habitat is more or less static, while the marine habitat is in constant vertical and horizontal movement. The nutrients that fertilize much of the plant growth (phytoplankton production) in the ocean may have been brought up to the surface by water that has upwelled from some depth, and the herbivores and carnivores may mature in areas far away, both vertically and horizontally, from where they were spawned (e.g., Hardy, 1956; Pickard, 1964). Such mobility suggests that a marine habitat will be less stable than a terrestrial one—that the water types would be ephemeral and continually intermingling and that the species assemblies associated with them would be mere temporary groupings rather than true communities. In fact, this is not so. Water types remain surprisingly stable because the ocean circulation patterns that ultimately give rise to them either remain more or less constant or vary only in predictable ways. The highly productive upwelling systems off Peru and southwest Africa are present almost all year round; those off northwest Africa and southeast Arabia occur only during certain seasons, but these seasons vary little from year to year (Cushing, 1971). There are, of course, occasional perturbations, the best known of which is the El Niño phenomenon off Peru. Every few years an intrusion of warm tropical water extends southward along the Peruvian coast and, in ways not fully understood, suppresses the upwelling system there. Anchovetas *Engraulis ringens* become scarce, and there is a catastrophic decline in the populations of seabirds—especially of Guanay Cormorants *Phalacrocorax bougainvillii*, Piquero Boobies *Sula variegata*, and Peruvian Pelicans *Pelecanus thagus*—whose principal food is this fish (Valdivia, 1978; Walsh, 1978). Despite this, the Peruvian upwelling system is stable enough to support a large and distinctive community of marine animals, including at least 12 endemic or near-endemic species of seabirds (Murphy, 1936).

The occurrence of these endemics is an indication that seabirds, despite their great mobility, can be associated with a specific water type. Seabird distributions in the Chilean fjords showed this in more detail. The Blue-eyed Shag *Phalacrocorax atriceps* was dominant in the north, and its sibling species the King Shag *P. albiventer* was dominant in the south and east. The boundary coincided with that between the cold, relatively saline water of the Southern Inlets in the Magellan Strait area, probably of Atlantic origin, and the warmer or less saline Pacific water types to the north (Brown *et al.*, 1975*a*; Pickard, 1971; see also Fig. 7). This would not have been apparent if the species' occurrence had been related either to temperature or salinity separately. In fact, an interpretation in terms of water types made sense of what at first seemed a geographically puzzling distribution not only of these siblings, but also of the Magellan Penguin *Spheniscus magellanicus*, Black-browed Albatross *Diomedea melanophris*, Rock Shag *Phalacrocorax magellanicus*, and Dominican Gull *Larus dominicanus*. All were commonest in the Northern and Southern Inlets, where the water was most saline and food most abundant, but scarce in the central fjords, where the water was almost fresh, or contained much silt.

Recently Pocklington (1979) has compared pelagic seabird distributions with water types over the Indian Ocean as a whole and has found some interesting differences between the marine habitat preferences of closely related species. Bulwer's and Jouanin's Petrels *Bulweria bulweria* and *B. fallax* seemed to have identical preferences for very warm waters, but a temperature–salinity plot showed that, in fact, there was very little overlap between the two; they were virtually confined to waters of salinity under and over about 35‰, respectively. The preferences of the Red-billed and Red-tailed Tropicbirds, *Phaethon aethereus* and *P. rubricauda*, and of the two subspecies of the White-tailed Tropicbirds, *P. lepturus lepturus* and *P. l. fulvus*, were just as discrete. *Phaethon l. fulvus* occurred in cooler, less saline water types than did *l. lepturus; aethereus* was the tropicbird of the highly saline water types. The preferences of *rubricauda* were similar to those of *l. fulvus*, but the two were geographically isolated; they occurred in comparable water types in the western and eastern Indian Ocean, respectively. The biological bases for these preferences were not clear, but the fact that the distributions of the various species of tunas and euphausiids also tended to be associated with different water types made it likely that each species was adapted to feed on a different community of marine animals.

To some extent the seabirds' great mobility allows them to make long migrations and yet maintain a paradoxical constancy in their habitat preferences throughout the year. In the Atlantic the transequatorially migrating Greater and Sooty Shearwaters travel between the cool–temperate waters of the northern and southern hemispheres, and Cory's Shearwaters move between the comparable subtropical zones (Brown, 1977; Brown *et al.*, 1975a,b). In terms of food the Greater Shearwaters, which prey extensively on *Illex* squid (Barker 1976; Rees, 1961), the migration would exchange *I. argentinus* of the Patagonian Shelf for the very similar *I. illecebrosus* and *I. coindetti* of the North Atlantic (Clarke, 1966). Some of the Arctic Terns *Sterna paradisaea* that migrate between polar seas may exchange ice-edge feeding habitats in the Arctic for those in the Weddell Sea (Cline *et al.*, 1969; Hartley and Fisher, 1936). However, by no means do all transequatorial migrants do this. Wilson's Storm Petrel *Oceanites oceanicus* breeds close to rich Antarctic and sub-Antarctic waters, and one might expect them to migrate up to the Arctic; instead they travel to productive waters off southeast Arabia and New England (Bailey, 1966; Palmer, 1962; Roberts, 1940). Arctic waters are far less productive than are the Antarctic waters (Dunbar, 1968), and so, in this case, it may be quantitative rather than qualitative similarities between summer and winter habitats that are significant.

It is difficult at present to interpret seabird distributions over a broad geographic scale in more precise terms than these. To demonstrate coincidences between the occurrence of seabirds and of physical and biological oceanographic phenomena is hardly a satisfactory basis for an understanding of the birds' pelagic ecology. We are still a long way from being able to trace all the connections from the bird to its food, and then on down the chain to the fundamental physical oceanographic processes, although this is also true of most other marine biological studies of animals near the summits of their food webs. The difficulty of collecting seabirds at sea for stomach analyses restricts our knowledge of their pelagic feeding habits, as does the fact that it is easier to define a water type in terms of its physical parameters than to collect the plankton and other biota associated with it. One refinement might be to adapt the statistical definitions that have been used to determine the species composition of marine communities (e.g., Fager and Longhurst, 1968; Fager and McGowan, 1963). Most seabirds feed in the surface layers, and it is technically fairly simple to collect the smaller plankton that occur at the surface (termed "neuston"). A statistically significant association between certain neuston species collected and the seabirds seen at the time of collection would imply some kind of link between birds and plankton through the food web, even if at one or two removes.

This section has discussed possible links between seabird distributions and marine habitats, with the implication that the latter differ not just in temperature and salinity, but in the types of potential prey they contain as well. But there are also quantitative differences in prey abundance. On the small scale of the Chilean fjords, Baker Inlet was the most barren of the water types and, as one would expect, few birds were seen in it (Brown *et al.*, 1975a; see also Fig. 7). On a broader scale, tropical waters are more • barren than are polar and subpolar seas or systems such as the Peruvian upwelling. These differences influence the population sizes and breeding ecology of seabirds in ways beyond the scope of this chapter, but that have been ably reviewed by N. P. Ashmole (1971) and Nelson (1970, 1978). The point to note here is that they also influence not just what a seabird will have available to it as food in a given habitat, but the ways in which it can catch its prey there. Seabirds in tropical waters must travel long distances to find food. They are able to do so economically because of their *large, broad* wings relative to their body size and their structural adaptations toward energy-efficient gliding, as opposed to energy-expensive flapping flight (Kuroda, 1954; Warham, 1977). Such birds feed either at the surface or by plunging briefly below it (Ainley, 1977). They cannot pursue their prey below the surface because seabirds swim underwater with beats of their partly folded wings (e.g., Brown *et al.*, 1978), and for maximum efficiency that requires a *short, narrow* wing (the flipper wings of the flightless Great Auk and penguins are extreme examples of this). Underwater pursuers are more efficient foragers in the sense that they are not limited to feeding only at the surface. However, this technique is energetically expensive in itself (e.g., Watson, 1968), and the narrowness of the wings commits the species to flapping flight if not flightlessness, which adds to the energetic costs. It is not surprising that the more specialized diving seabirds—penguins, alcids, diving petrels *Pelecanoides* spp., and almost all the more specialized diving shearwaters—are confined to polar and subpolar waters and upwelling systems, where the abundance of prey makes the technique energetically economical.

IV. THE IMPORTANCE OF LOCAL CONCENTRATIONS OF FOOD

Even in their preferred habitat, seabirds can exploit only a fraction of the potential prey. The prey, in theory at least, can be anywhere in the water column, yet most seabirds can feed only at or near the surface. The specialized divers are obvious exceptions; Emperor Penguins *Aptenodytes forsteri* can reach a depth of 260 m. However, most penguins feed at shallower depths (Kooyman, 1975). In general, foraging seabirds must rely heavily on situations in which, for a variety of physical and biological reasons, their prey is concentrated close to the surface. Such local concentrations will be important not just for their accessibility; the density of prey present will reduce the energetic costs of foraging. Brodie *et al.* (1978) estimated that a Finback Whale *Balaenoptera physalus* feeding on euphausiids could do so economically only where that prey was locally concentrated at more than 175 times the average density for Nova Scotian waters as a whole. Filter-feeding seabirds such as Dovekies probably have comparable requirements. Zelickman and Golovkin (1972) describe local aggregations of copepods and other zooplankton off Novaya Zemlya in which the biomass was up to 10 times greater than in adjacent areas; 76% of the Dovekies seen feeding were doing so at these aggregations. Seabirds may travel considerable distances to exploit such favorable foraging zones. Black-legged Kittiwakes in Spitsbergen traveled 50 km to feed at a glacier face where euphausiids were concentrated (Hartley and Fisher, 1936), and Greenland Dovekies may forage as far as 100 km from their colonies (Brown, 1976).

This section will examine four sets of circumstances in which food is likely to be concentrated close to the surface, as well as the extent to which seabirds take advantage of them.

A. Surface Shoals

Many marine animals form shoals at the surface. Seabird exploitation of shoals of pelagic fish is well known—Atlantic Mackerel *Scomber scom-*

brus and Atlantic Herring *Clupea harengus* by Northern Gannets *Sula bassana* (Nelson, 1978); Pilchards *Sardina pilchardus* by Manx Shearwaters (Lockley, 1953); sticklebacks Gasterosteidae by Common *Sterna hirundo* and Arctic Terns (Lemmetyinen, 1973); and Sandlance *Ammodytes* spp., Capelin *Mallotus villosus*, and Arctic Cod *Boreogadus saida* by the seabird communities of the North Sea, Newfoundland, and the eastern Canadian Arctic, respectively (Bradstreet, 1976; Pearson, 1968; L. M. Tuck, 1961). But invertebrate shoals, especially of euphausiids and other large crustaceans, are also very important. Lobster-krill *Munida gregaria* are preyed upon by Sooty Shearwaters (Bourne, 1975); *Euphausia pacifica* by Black-tailed Gulls *Larus crassirostris* and Rhinoceros Auklets *Cerorhinca monocerata* (Komaki, 1967); *Meganyctiphanes norvegica* by Greater Shearwaters and other Bay of Fundy seabirds (Barker, 1976); and euphausiids and squid (along with Sandlance) are the staple foods of seabirds along much of the Pacific coast between California and British Columbia (Baltz and Morejohn, 1977; Sealy, 1973; Wiens and Scott, 1975). More generally, the huge surface swarms of krill *Euphausia superba* in Antarctic waters are central to the food webs of most Antarctic seabirds and marine mammals (e.g., Bierman and Voous, 1950; Marr, 1962), as are swarms of euphausiids and hyperiid amphipods to the seabird and reef-fish communities of the Snares Islands (Fenwick, 1978).

These examples are all of shoals that form at the surface in the daytime. Much more is available at night. Many crustaceans spend the daytime at depths of several hundred meters and migrate up to the surface after dark; euphausiid densities at the surface in the North Pacific, for example, can be up to 1000 times greater at night (Brinton, 1967). For obvious practical reasons it is difficult to judge the extent to which seabirds exploit these nocturnal vertical migrations, but there are indications that many of them do. Wedge-tailed Shearwaters *Puffinus pacificus* and Sooty Terns *Sterna fuscata* have been seen feeding at night (Gould, 1967), and nocturnal feeding can be inferred from the circadian activity patterns of Dovekies in Iceland (Foster *et al.*, 1951) and Galápagos Swallow-tailed Gulls *Creagrus furcatus* (Snow and Snow, 1968) and from the occurrence of Northern Fulmars at Spitsbergen in the dark of the High Arctic winter night (Løvenskiold, 1964). It can also be inferred from the occurrence in seabird stomachs of prey that usually comes to the surface only at night, for example, squid (e.g., Baker, 1960), recorded from many gadfly petrels *Pterodroma* spp. (Imber, 1973); Wandering Albatross *Diomedea exulans* (Imber and Russ, 1975); Jouanin's Petrel and the Pale-footed Shearwater *Puffinus carneipes* (Bailey, 1966); myctophid fish in Leach's Storm Petrel *Oceanodroma leucorhoa* (Linton, 1978) and Craveri's Murrelet *Endomychura craveri* (de Weese and Anderson, 1976); and the copepod *Euchaeta antarctica* in the Common Diving Petrel *Pelecanoides urinatrix*

(Ealey, 1954). On the other hand, one would suppose that seabirds that rely heavily on visual cues (e.g., the sulids, which plunge onto their prey from a height) would be unable to feed at night.

How the prey is located is another matter. Gould's (1967) terns and shearwaters were feeding by moonlight. Light from the bioluminescent organs of myctophids and many crustaceans (e.g., Hardy, 1956) would be a likely cue. Location by scent is another possibility. Some shearwaters and storm petrels can locate food by smell, and the latter, at least, can do so by night as well as by day (Grubb, 1972). Because the scent of plankton-rich water can be apparent even to the human nose (E. L. Mills, personal communication), it seems quite possible that night-feeding seabirds could use it to locate concentrations of prey.

B. Marine Predators

N. P. Ashmole and Ashmole (1967) drew attention to the association between seabirds and tuna schools in relatively barren tropical waters, and suggested that birds that feed well offshore rely heavily on them in order to chase prey up to the surface. Flying-fish Exocoetidae were important prey for their birds, and these are most accessible when they "fly" in order to avoid predatory fish. Jehl (1974a) described similar associations off the Pacific coast of Central America, and Gould (1974) noted that central Pacific seabirds will also follow baleen whales and dolphins. In the subtropical Atlantic the breeding of Cory's Shearwater is closely linked to the arrival of tuna in summer (de Naurois, 1969; Zino, 1971).

However, associations between seabirds and marine predators are not confined to relatively barren waters. Bailey (1966) described seabird concentrations over schools of predatory fish, dolphins, and large whales in the productive southeast Arabian upwelling; in the Peruvian upwelling large flocks of Piquero Boobies formed over schools of dolphins and plunged in just ahead of them to take the fish they were chasing (R. G. B. Brown, unpublished data); seabirds in the North Pacific and in the Bering and Tasman Seas regularly follow whales and seals (Barton, 1979; Gudkov, 1962; Harrison, 1979; Ryder, 1957); in the North Atlantic, Greater Shearwaters regularly follow pods of Pilot Whales *Globicephala melaena*, either hunting for the squid on which these whales feed or scavenging their feces (Wynne-Edwards, 1935).

In productive waters the principal advantage of these associations is probably not so much fact that the prey is brought to the surface as its concentration, so that the seabirds can forage economically. In the Gulf of St. Lawrence, Minke Whales *Balaenoptera acutorostrata* concentrate Capelin up against the surface for their own feeding by circling the shoals in a narrowing upward spiral; Black-legged Kittiwakes exploit this by taking the fish from above (P. C. Beamish, personal communication). Up to a point, seabirds can themselves herd prey in this way; similar underwater behavior has been reported for the Magellan Penguin (Boswall and McIver, 1975) and Brandt's Cormorant *Phalacrocorax penicillatus* (Palmer, 1962), and herding fish at the surface has been reported for flocks of Guanay Cormorants, Double-crested and Little Black Cormorants *Phalacrocorax auritus* and *P. sulcirostris*, and White Pelicans *Pelecanus erythrorhynchos* (Murphy, 1936; Palmer, 1962; Serventy *et al.*, 1971). Such behavior would be difficult for birds that feed at the surface or plunge in from the air, although the apparently coordinated plunging of Blue-footed Boobies *Sula nebouxii* (Nelson, 1968) might have a herding effect. Many seabirds use the feeding behavior of high-flying aerial plungers in order to *locate* concentrations of food (e.g., Gould, 1974), but that is a different matter.

C. "Fronts"

Seabird feeding areas can often be related to physical oceanographic mechanisms that concentrate prey, or the prey's prey, at the surface. "Fronts" are good examples. Cromwell and Reid (1956) define "fronts" as narrow bands of water where density changes abruptly, although they are more easily recognizable by the sharp changes in temperature and salinity which also occur (see Fig. 5). Convergence fronts occur at the boundary between two water bodies, where one sinks below the other; N. P. Ashmole and Ashmole (1967) regard them as being particularly important in concentrating food at the surface. Pingree *et al.* (1974) describe such a front in the English Channel—a feeding area for terns, shearwaters, and Atlantic Puffins *Fratercula arctica*. At the boundary between cooler offshore and warmer inshore water, surface temperatures changed by over 1°C across a horizontal distance of as little as 50 m. A convergence line, marked by lines of rippled and calm water and by accumulations of weed and other flotsam, formed where the cooler water sank. Zooplankton and young fish also accumulated at the surface along the front; crustaceans were on average 74.5 times denser there than in the surface waters on either side.

Fronts are relatively stable, predictable phenomena found in all oceans, and seabird concentrations occur at many of them. One of the best known ornithologically, if not oceanographically, is that on the Patagonian Shelf off Argentina, roughly at 40°S 55°W, at the boundary between the cold Falkland and warm Brazil Currents (Anon., 1971; Warnecke *et al.*, 1971). This has been known at least since 1768 (Beaglehole, 1963, pp. 207–208), and seabird concentrations have been reported there during most months of the year; Greater Shearwaters are dominant in summer, and such Antarctic migrants as the Black-browed Albatross and Cape Pigeon *Daption capense* are dominant in winter (Brown *et al.*, 1975a). It is not known what birds feed on there, but there are indications for fronts elsewhere. Figure 5 shows the front at the outer edge of the Labrador Current, the Subarctic equivalent of the Falkland Current. At its northern end at about 60°N 60°W, concentrations of Greater Shearwaters, Northern Fulmars, and Black-legged Kittiwakes occur in August (R. G. B. Brown, unpublished manuscript). Bottle-nosed Whales *Hyperoodon ampullatus* were formerly hunted extensively there (Benjaminsen, 1972), which implies that there is a local abundance of their principal food, the squid *Gonatus fabricii*, which in turn feeds on euphausiid, amphipod, and copepod crustaceans (Clarke, 1966). All three seabirds eat gonatid squid and euphausiids, and the fulmars and kittiwakes, at least, take amphipods and copepods as well (Barker, 1976; Bradstreet, 1976). It is reasonable to conclude that the birds are attracted either to the squid or to crustaceans concentrated at the front in the way described by Pingree *et al.* (1974).

Seabirds also make use of divergence fronts, where water upwells to the surface and may bring food with it. One example is at the rather complex current boundaries in the central Pacific, where the west-flowing North and South Equatorial Currents are separated by the east-flowing Equatorial Countercurrent. Here, both convergences and divergences occur. For reasons that are unclear, the highest numbers of seabirds occur not at the convergences within the Equatorial Currents, as one might expect from observations in other areas, but at the boundary between the Countercurrent and the North Equatorial, which is a divergence (J. E. King and Pyle, 1957; W. B. King, 1974). It is possible that N. P. Ashmole and Ashmole (1967) are correct in suggesting that there is an indirect relationship with prey abundance here. The birds may not be reacting to prey trapped at the surface in the convergences, but rather to that chased to the surface by tuna while feeding in the enriched area of the divergence.

On a much smaller scale, Thick-billed Murres in Novaya Zemlya feed at the fronts that form between the freshwater entering the heads of narrow fjords from streams and the seawater outside (Uspenski, 1958), Blue-eyed Shags apparently do the same in Chilean fjords (Brown *et al.*, 1975a), and Sandwich Terns *Sterna sandvicensis* feed at the freshwater/saltwater boundary at the estuary of a stream in Yucatan (J. Burger and M. Gochfeld, personal communication). Birds may also exploit plankton concentrated at the surface in Langmuir circulation cells. These cells are formed by the action of the wind on the surface of the water, which sets up a series of shallow upwellings alternating with convergences, the latter apparent as parallel, often flotsam-filled lines of calm water aligned along the axis of the wind direction (Owen, 1966; Pollard, 1977; Stavn, 1971). Zooplankton also accumulates in them, much as in a conventional convergence; off Peru, phalaropes, terns, and small gulls often feed in the convergences, but never in the waters in between them (R. G. B. Brown, unpublished data). Comparable small-scale effects associated with turbulence over underwater canyons and downstream of islands apparently also concentrate food at the surface and are exploited by seabirds (N. P. Ashmole and Ashmole, 1967; Baltz and Morejohn, 1977; Ingham and Mahnken, 1966; Jehl, 1974*a*).

D. Ice

In both the Arctic and Antarctic, seabirds are often associated with pack ice. The underside of the pack supports a locally rich concentration of phytoplankton, zooplankton, and such fish as Arctic Cod (Bradstreet, 1976; Bradstreet and Finley, 1977; Horner, 1976; Johnson *et al.*, 1976; McRoy and Goering, 1974). Pack ice is available to the birds only when the ice is broken up into leads, or at the edge of the consolidated pack, because even diving seabirds are limited by their need for open-water areas where they can surface to breathe. In the Weddell Sea most species were commonest in areas of light ice, as opposed to open water or consolidated pack, although Adelie Penguins *Pygoscelis adeliae* apparently preferred the edges of near-solid pack (Cline *et al.*, 1969). In the Arctic, in Lancaster Sound, Thick-billed Murres, which feed on Arctic Cod, and Northern Fulmars, Dovekies, and Black Guillemots *Cepphus grylle*, which feed on the ice-associated amphipod *Apherusa glacialis*, were often locally concentrated at the edge of the pack (Bradstreet, 1976; Bradstreet and Finley, 1977; Nettleship and Gaston, 1978), as were Ivory Gulls *Pagophila eburnea* and Ross's Gulls *Rhodostethia rosea*, feeders on Arctic Cod and *Apherusa*, respectively, in the Chukchi Sea (Divoky, 1976). Nonbreeding Dovekies were associated with the pack ice in southern Baffin Bay (Fig. 3), as were pre- and post-breeding birds in northern Baffin Bay (R. G. B. Brown, unpublished data; Johnson *et al.*, 1976).

For different reasons food is also concentrated at glacier faces. In the long term, upwelling at these faces will bring nutrients up to the surface to fertilize a phytoplankton and, ultimately, a zooplankton "bloom" (Apollonio, 1973; Dunbar, 1968; Jacobs *et al.*, 1979). This upwelling occurs when freshwater, mainly melted from the glacier, but probably also flowing up from beneath it, moves away from the glacier face, and subsurface water rises to replace it (Dunbar, 1951). Zooplankton is also brought up and concentrated in a "brown zone." Northern Fulmars, Black-legged Kittiwakes, Thick-billed Murres, Atlantic Puffins, and Arctic Terns fed on euphausiids and amphipods at two such faces in Spitsbergen (Hartley and Fisher, 1936; Stott, 1936), as did Blue-eyed Shags and South American

Terns *Sterna hirundinacea* at a Chilean glacier (Brown *et al.*, 1975*a*). The glaciers of southeast Devon Island, Lancaster Sound, are also important feeding areas for seabirds (Nettleship, 1977*a*). The melt of freshwater ice at the faces of the very large Antarctic icebergs induces upwelling that might have comparable effects (Neshyba, 1977), and Petersen (1977) claims that even the smaller Arctic icebergs may induce upwelling from as deep as 200 m. Petersen recorded kittiwakes and terns feeding beside icebergs; in Baffin Bay during late summer it is common to see Northern Fulmars and, to a lesser extent, Red Phalaropes and Black Guillemots, feeding right at the bases of bergs (R. G. B. Brown, unpublished data).

V. RED PHALAROPES: A CASE STUDY

The last two sections have shown that seabird distributions at sea can be related to marine phenomena, both on the larger scale of water types and the smaller scale of fronts and other physical and biological mechanisms that make prey both accessible and concentrated at the surface. However, the examples quoted have been fragmentary and the links often speculative. It is worth describing an actual case study to show how our understanding of the pelagic ecology of seabirds may develop as ornithology and oceanography become more integrated.

The easiest seabirds to work with are those that feed at or near the surface on prey small enough or inactive enough to be collected without the risk of bias due to "net avoidance" (e.g., Brinton, 1967). The slow-moving copepods, amphipods, and pteropods are to be preferred to the faster-moving, hard-to-sample euphausiids and small fish and squid. The most suitable seabird groups are therefore the phalaropes, prions *Pachyptila* spp., and perhaps the smaller, filter-feeding alcids and diving petrels. The Canadian Wildlife Service and the University of Guelph have been investigating the very large concentrations of Red Phalaropes and other seabirds that form in late summer off Brier Island, Nova Scotia, in the Bay of Fundy (ca. 44°15′N, 66°23′W), and we have tried to relate these to the physical and biological oceanography of the area (Barker, 1976; R. G. B. Brown, unpublished manuscript).

The tidal streams in Fundy are notoriously strong. Just north of Brier Island they run against a group of steep-sided underwater ledges that reach up almost to the surface. Cool subsurface water is forced to the surface over these ledges in a miniature upwelling, drifts downstream, and sinks in a series of convergences; both upwelling and convergences are marked by areas of calm water termed "streaks" (see Fig. 8). The system is probably very similar to the "tidally generated wave packets" generated by the

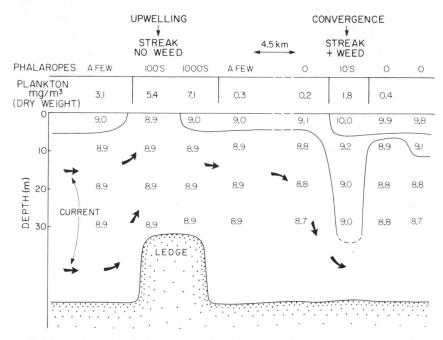

Fig. 8. Distributions of Red Phalaropes and zooplankton at the surface over an area of tidally induced upwelling off Brier Island, Nova Scotia, August 20, 1975. Estimates of phalarope numbers are qualitative. Plankton dry weights refer to copepod-size zooplankton collected in a net of 0.5-m entrance diameter and 571-μm mesh, handheld in the water and towed at ~ 2 kt. Temperatures were measured with a Beckman RS-5 salinometer. Note the well-mixed water upwelling over the submarine ledge, and sinking below the warmer, stratified water of the Bay of Fundy ~ 4.5 km downstream in a convergence front. Note also that the phalaropes were virtually confined to the front and the area of upwelling, where plankton was most abundant at the surface. Bottom contours and current arrows are schematic. For further details, consult text. (R. G. B. Brown, unpublished manuscript.)

passage of tidal streams over Stellwagen Bank, Massachusetts Bay (Haury *et al.*, 1979). Copepods are brought up along with the subsurface water and are concentrated in both the upwelling and convergence streaks. The phalaropes feed exclusively in the streaks, which they probably use as indications of high food concentration. Meanwhile, the euphausiid *Meganyctiphanes norvegica* also swarms at the surface, either brought passively up by the upwelling or actively following the copepods up to feed on them. These, in turn, are preyed on by Mackerel, Herring, *Illex* squid, Humpback *Megaptera novae-angliae* and Finback Whales, Herring and Great Black-backed Gulls *Larus argentatus* and *L. marinus*, and Greater and Sooty Shearwaters; the birds also feed on the fish and squid. In short, this tidal upwelling is the basis of a small web of marine feeding relationships in which seabirds figure as predators at three levels: (a) phalaropes on copepods, (2) shearwaters and gulls on euphausiids, and (3) shearwaters and gulls on fish and squid. A very similar upwelling–copepod–euphausiid system occurs at the same time of year across the Bay of Fundy off Deer Island, New Brunswick, but there, for reasons that are not yet clear, Northern Phalaropes *Phalaropus lobatus* replace Reds, and Bonaparte's Gulls *Larus philadelphia* replace the shearwaters.

The attractions of such a system to phalaropes are obvious. These birds feed exclusively at the surface, by picking at individual food items. The combination of upwelling and convergence brings prey up to the surface and presumably concentrates it sufficiently for the birds to forage economically by individual picking. The fact that prey is made available at the surface is particularly important for the birds in a marine environment. During the breeding season phalaropes feed in shallow tundra pools on readily accessible insect larvae. The birds' food during the three-quarters of the year they spend at sea is inaccessible except at fronts and similar concentrating systems, and the Red Phalarope's pelagic distribution is very largely determined by the presence of such areas. Figures 4 and 5 show the occurrence of the birds at the outer edge of the Labrador Current. It may well be that the birds are found along the entire 1000-km length of this front. Migrating or wintering flocks have also been reported at fronts off British Columbia (Martin and Myres, 1969), at the edges of the Gulf Stream off New England (Lamb, 1964) and of the Guinea Current off West Africa (Tåning, 1933), at tidally induced fronts in Hudson Strait (Gross, 1938), and at the fronts at the outer edges of the upwellings off northwest and southwest Africa and, probably, Peru (Brown, 1979; Lambert, 1971; Murphy, 1936; Stanford, 1953). On a smaller scale, their use of Langmuir circulation cell convergences and upwelling areas beside icebergs have already been described. In other words, by linking phalaropes with fronts, it has become possible, in a very preliminary way, to understand the basis of their pelagic ecology. Because these fronts are clearly shown on satellite photographs, it should be possible to predict phalarope distributions at sea. When that happens, pelagic ornithology will have come all the way from Columbus to the Space Age.

VI. ENERGY-FLOW MODELS

It is clear that the Bay of Fundy tidal upwellings are used by postbreeding phalaropes and Bonaparte's Gulls and prebreeding shearwaters as important staging areas, much as mudflats at the head of the Bay are used by postbreeding Semipalmated Sandpipers *Calidris pusilla* (Morrison, 1977). But although the marine communities in the upwellings are important to the birds, it does not necessarily follow that seabirds have a significant effect on the communities, measured in terms of the proportion of the annual production they remove as food.

To estimate this, one needs to construct quantitative models of the flow of energy through the community. Until recently these have been hampered by the lack of adequate information on seabird population sizes and feeding habits and the difficulty of making absolute estimates of bird numbers at sea is still a problem (see Section II.B). However, in two preliminary quantitative assessments, Evans (1971) estimated that in the North Sea the seabirds and the other large predators took between them roughly 10% of the annual production of the smaller fish, while Sanger (1972) suggested 8% for the North Pacific seabird community. More detailed estimates give higher figures. Wiens and Scott (1975) presented a detailed model for energy flow in the breeding and nonbreeding seabird populations along the Oregon coast, where the principal species were Leach's Storm Petrel, Brandt's Cormorant, Sooty Shearwater, and Common Murre, the last two being dominant. Wiens and Scott estimated that the birds took about 22% of the annual local production of small pelagic fish in the inshore zone; 11% were taken by the murres alone. In absolute terms, an estimated summer population of 4.4 million seabirds took 39,700 tons of prey from the sea. An estimated 23,000 tons of this was made up of Northern Anchovies *Engraulis mordax*, preyed on by Sooty Shearwaters—four times the weight of the annual commercial anchovy catch off that coast. More recently Furness (1978) has modeled the energy requirements of the seabird community breeding in the Shetland Islands, and has estimated that they consumed 29% of the fish production. Clearly, seabirds consume a significant fraction of their local marine production.

This is not a completely one-sided process because, according to Wiens and Scott's estimates, 30% of the energy content of the food is voided as feces, much of it straight back into the sea. The nutrient content of these would contribute to the fertilization of phytoplankton production, especially close to colonies where deposition would be most concentrated, as Zelickman and Golovkin (1972) have described for seabird colonies in Novaya Zemlya. This is probably not very important in a generally rich area like the California upwelling off Oregon, but it might significantly enhance production in the vicinity of seabird colonies in relatively barren tropical waters (Hutchinson, 1950).

VII. MAN AND SEABIRDS AT SEA

The impact of man on seabirds through oil and chemical pollution of the sea has received considerable publicity during the last decade. There have been many reviews—most recently by Bourne (1976b), Brown (in press), and Holmes and Cronshaw (1977)—and it would be superfluous to repeat them here. Instead this chapter will conclude by summarizing the ways in which man and seabirds act as partners and competitors through their common interest in removing food from the sea.

At the simplest level, man and seabirds have literally been partners, as in the traditional oriental fishery employing Japanese Cormorants *Phalacrocorax capillatus* (e.g., Alexander, 1928), or in the attempts of modern marine biologists to use seabirds as samplers of fish and squid (M. J. Ashmole and Ashmole, 1968; Imber, 1975) or as indicators of productive areas of the oceans (e.g., Gudkov, 1962; Sutcliffe *et al.*, 1976). Man has also enhanced the food supply of many seabirds, in both the short and long term. A ship in tropical waters, for example, puts up flying fish, which are hunted by Masked and Brown Boobies *Sula dactylatra* and *S. leucogaster* and by Pomarine Jaegers *Stercorarius pomarinus* (R. G. B. Brown, 1979; unpublished data; Palmer, 1962). Ships breaking ice in the Arctic are followed by flocks of kittiwakes, terns, and Ivory Gulls, which swoop down on the Arctic Cod exposed by the upturned floes (e.g., Divoky, 1976). Masked Boobies exploit the reef-fish communities that form around offshore oil rigs (Ortego. 1978). Antarctic penguin populations have recently expanded, apparently exploiting an increase in krill that has followed the overhunting of their competing predators, the baleen whales (Conroy, 1975). But the principal food supply has been provided by man's own wastes; at sea these are usually in the form of offal from the whaling and fishing industries (Fisher, 1952).

The potential size of this food supply is enormous. Boswall (1960) and Rees (1963) estimated that processing techniques in use in the North Atlantic during the 1950s resulted in up to 30% of the weight of the catch going to waste. Not all of this food supply would be scavenged; livers are highly preferred, but heads and anything attached to them quickly sink out of reach (R. G. B. Brown, unpublished data). However, Rees conservatively estimated that offal from the fishery off northeast Newfoundland in October–November 1958 could have supported 120,000 seabirds. Modern factory trawlers convert much of this material into fish meal nowadays but, even so, spillages of various kinds will probably always provide a fairly substantial food supply. Recent observations (Garcia Rodriguez, 1972; Watson, 1978) have shown some interesting differences in the behavior of the various species that scavenge behind fishing boats. Substantial quantities of fish offal have only been available since the late nineteenth century, yet these opportunistic scavengers seem to be partitioning their new food resource in ways analogous to those of highly evolved obligatory scavenger communities, such as the vultures of Africa (Kruuk, 1967).

The most detailed discussion of the long-term effects of this increased food supply has centered on the Northern Fulmar. Fisher (1952, 1966) has suggested that the massive expansion in range and population of this species in the northeast Atlantic is based on the availability of offal from the coincidentally expanding fishing industry. However, Salomonsen (1965), noting that the expansion is apparently confined to the warmer parts of the fulmar's range, has suggested instead that the primary cause of the expansion is the birds' exploitation of a hitherto vacant marine habitat, and that scavenging is secondary to this. Fulmar distributions off eastern Canada support Salomonsen's view; all the southern parts of the continental shelf are heavily fished, yet fulmars are abundant only in waters most strongly influenced by the Labrador Current (Brown, 1970).

Unfortunately, the harmful effects man has had as a fisherman on the seabird communities probably outweigh the benefits. It was once common practice to take shearwaters at sea for bait in the hand-line fishery off Newfoundland (Collins, 1884). The principal fishery-related mortality today is of alcids—mainly murres—drowned in gill nets in the North Atlantic and Pacific salmon fisheries (Ogi and Tsujita, 1973; Tull et al., 1972). Tull and coworkers estimated an annual kill of 500,000 Thick-billed Murres off west Greenland in the early 1970s—a very high figure for a species that, like most seabirds, has a low reproductive rate (N. P. Ashmole, 1971). Birds from the colony at Cape Hay, Lancaster Sound, winter off west Greenland (L. M. Tuck, 1961). The size of this colony has declined by 20–40% over the last 20 years (D. N. Nettleship, personal communication), although excessive hunting off Greenland may have contributed to this decrease (Evans and Waterston, 1976).

Over the longer term, man and seabirds compete for a common fish resource. Fishermen have always advocated the killing of supposedly destructive seabirds, such as cormorants, on the grounds that they destroy commercial fish stocks. [After investigating one such claim, Steven (1933) was able to show that although Great Cormorants *Phalacrocorax carbo* in England did, in fact, prey extensively on commercial species, the equally persecuted Green Shag *P. aristotelis* did not.] On a much wider scale it has been seriously suggested that a cull of the abundant guano-producing seabirds of the Peruvian upwelling would leave more Anchovetas for the fishermen to catch (Paulik, 1971; Schaefer, 1970). In stark economic terms this is undeniable—if the fishermen could harvest even one-half the Anchovetas eaten by the birds, the value of this catch would be three times that of the lost guano (Gulland et al., 1970, quoted by Paulik, 1971). The authors do not recommend exercising such a trade-off, but this situation illustrates the kind of conflict that can arise in this and other fisheries.

However, as Nettleship (1977b) has pointed out, the conflicts are more likely to be indirect, with man and seabirds competing for a restricted food supply. For example, the Northern Anchovy fishery off Oregon is a small one, but if the catch were to expand significantly, the Sooty Shearwater population would inevitably suffer (Wiens and Scott, 1975; see Section VI). Similarly, Furness (1978) has pointed out that 50% of pelagic fish production in the North Sea is taken commercially. Since 29% is taken by the seabirds, only 21% is left for other predators. Man and the two groups of animal predators are in competition, and there is little doubt as to who will win. The current trend in the fishing industry is to take fish for fish meal, as opposed to taking them for direct human consumption—a strategy forced on the industry by the gross overfishing of traditional commercial stocks. This means that there will be increasing pressure on such fish as Capelin, Arctic Cod, Sand-lance, and the various anchovy species, as well as *Illex* and other squids and Antarctic krill, which are central to the food webs of seabirds and much else in the oceans.

In theory, fisheries such as these are regulated so that the size of the harvest does not endanger either the species in question or other species that depend on it for food. For example, El-Sayed and McWhinnie (1979) estimate that 400 million tonnes of krill a year are eaten by predators (about 10% of this by seabirds). Theoretically, we could take an additional 150 million tonnes without endangering the species or the other predators, as this constitutes a surplus portion of the stock, which would have been eaten by baleen whales before man so drastically reduced the whale population. But so little is known in practice about krill biology or even about overall population size that it would be unsafe to rely too much on such an assumption. Unfortunately, these uncertainties have never prevented a fishery from being exploited if it was profitable in the short term, as the disastrous histories of the many overexploited fish stocks have proved. Even if the population statistics were accurate, and the proportion to be harvested set at an acceptable level, the competing predators would still be at a disadvantage. Gross estimates of the size of an exploitable fish population take no account of the local variations in density whose importance to the other predators was stressed in Section IV. As El-Sayed and McWhinnie point out, these are the very concentrations that will be the fishery's logical target. Man, like the other predators, must forage economically.

Competition with fisheries, along with the effects of pollution and other human activities, has already led to significant declines in the populations of many seabirds [e.g., alcids in the British Isles and Atlantic Canada (Cramp *et al.*, 1974; Nettleship, 1977*b*), Jackass Penguins *Spheniscus demersus* in South Africa (Frost *et al.*, 1976)]. It seems inevitable that these declines will not merely continue, but that they will accelerate.

ACKNOWLEDGMENTS

I would like to thank A. J. Erskine, D. N. Nettleship, R. Pocklington, and the editors for their comments on this manuscript. I am also grateful to

N. P. Ashmole, W. R. P. Bourne, and my colleagues in both the Bedford Institute of Oceanography and the Canadian Wildlife Service for much help and advice over the years. This chapter is associated with the Canadian Wildlife Service's program "Studies on Northern Seabirds" (Report No. 76).

REFERENCES

Ainley, D. G., 1976, The occurrence of seabirds in the coastal region of California, *West. Birds* 7:33–68.

Ainley, D. G., 1977, Feeding methods in seabirds: A comparison of polar and tropical nesting communities in the eastern pacific Ocean, in: *Adaptations within Antarctic Ecosystems* (G. A. Llano, ed.), pp. 669–685, Smithsonian Institution, Washington, D.C.

Alexander, W. B., 1928, *Birds of the Ocean*, G. P. Putnam's Sons, New York.

Anonymous, 1958, *Oceanographic Atlas of the Polar Seas. Part II. Arctic*, U.S. Navy Hydrographic Office, Washington, D.C.

Anonymous, 1971, *Monatskarten für den Südatlantischen Ozean* (*Dritte Auflage*), Deutsches Hydrographisches Institut, Hamburg.

Apollonio, S., 1971, Glaciers and nutrients in arctic seas, *Science* **180**:491–493.

Ashmole, N. P., 1971, Seabird ecology and the marine environment, in: *Avian Biology* (D. S. Farner and J. R. King, eds.), Vol. 1, pp. 223–286, Academic Press, New York.

Ashmole, N. P., and Ashmole, M. J., 1967, Comparative feeding ecology of seabirds of a tropical oceanic island, *Peabody Mus. Nat. Hist. Yale Univ. Bull.* **24**:1–131.

Ashmole, M. J., and Ashmole, N. P., 1968, The use of food samples from sea birds in the study of seasonal variation in the surface fauna of tropical oceanic areas, *Pac. Sci.* **22**:1–10.

Bagenal, T. B., 1951, Birds of the North Atlantic and Newfoundland Banks in July and August 1950, *Br. Birds* **44**:187–195.

Bailey, R. S., 1966, The sea-birds of the southeast coast of Arabia, *Ibis* **108**:224–264.

Bailey, R. S., 1968, The pelagic distribution of sea-birds in the western Indian Ocean, *Ibis* **110**:493–519.

Bailey, R. S., 1971, Sea-bird observations off Somalia, *Ibis* **113**:29–41.

Bailey, R. S., and Bourne, W. R. P., 1972, Counting birds at sea, *Ardea* **60**:124–126.

Baker, A. de C., 1960, Observations of squid at the surface in the NE Atlantic, *Deep-Sea Res.* **6**:206–210.

Baker, R. H., 1947, Observations on the birds of the North Atlantic, *Auk* **64**:245–259.

Baltz, D. M., and Morejohn, G. V., 1977, Food habits and niche overlap of seabirds wintering on Monterey Bay, California, *Auk* **94**:526–543.

Barker, S. P., 1976, Comparative feeding ecology of *Puffinus* (order Procellariiformes) in the Bay of Fundy, M.Sc. thesis, University of Guelph.

Barton, D., 1979, Albatrosses in the western Tasman Sea, *Emu* **79**:31–35.

Bartonek, J. C., and Gibson, D. D., 1972, Summer distribution of pelagic birds in Bristol Bay, Alaska, *Condor* **74**:416–422.

Bartonek, J. C., and Nettleship, D. N. (eds.), 1979, Conservation of marine birds of northern Northern North America, *U.S. Fish Wildl. Serv., Wildl. Res. Rep.* **11**, Washington, D.C.

Bary, B. McK., 1963, Distributions of Atlantic pelagic organisms in relation to surface water bodies, *R. Soc. Can. Spec. Publ.* **5**:51–67.

Beaglehole, J. C. (ed.), 1963, *The Endeavour Journal of Joseph Banks. 1768–1771*, Vol. 1, 2nd ed., Angus and Robertson, Sydney, New South Wales

Bédard, J., 1969, Feeding of the Least, Crested and Parakeet Auklets around St. Lawrence Island, Alaska, *Can. J. Zool.* **47**:1025–1050.

Belopol'skii, L. O., 1961, *Ecology of Sea Colony Birds of the Barents Sea*, (transl.), Israel Program for Scientific Translation, Jerusalem.

*Benjaminsen, T., 1972, On the biology of the Bottlenose Whale, *Norw. J. Zool.* **20**:233–241.

Bierman, W. H., and Voous, K. H., 1950, Birds observed and collected during the whaling expeditions of the 'Willem Barendsz' in the Antarctic, 1946–1947 and 1947–1948, *Ardea* Suppl.:1–124.

Boswall, J., 1960, Observations on the use by sea-birds of human fishing activities, *Br. Birds* **63**:212–215.

Boswall, J., and MacIver, D., 1975, The Magellanic Penguin *Spheniscus magellanicus*, in: *The Biology of Penguins* (B. Stonehouse, ed.), pp. 271–306, Macmillan Press, London and Basingstoke.

Bourne, W. R. P., 1963, A review of oceanic studies of the biology of seabirds, *Proc. XIII Int. Ornithol. Congr.* **1962**:831–854.

Bourne, W. R. P., 1975, Birds feeding on lobster-krill off the Falkland Islands, *Sea Swallow* **24**:22–23.

Bourne, W. R. P., 1976*a*, Birds of the North Atlantic Ocean, *Proc. XVI Int. Ornithol. Congr.* **1974**:705–715.

Bourne, W. R. P., 1976*b*, Seabirds and pollution, in: *Marine Pollution*, (R. Johnston, ed.), pp. 403–502, Academic Press, London.

Bourne, W. R. P., and Warham, J., 1966, Geographical variation in the Giant Petrels of the genus *Macronectes*, *Ardea* **54**:45–67.

Bradstreet, M. S. W., 1976, *Summer Feeding Ecology of Seabirds in Eastern Lancaster Sound, 1976*, Report for Norlands Petroleum, Calgary, LGL Ltd., Environmental Research Associates, Toronto.

Bradstreet, M. S. W., and Finley, K. J., 1977, *Distribution of Birds and Marine Animals along Fast-ice Edges in Barrow Strait, N.W.T.*, Report for Polar Gas Project Environmental Program, Ottawa, LGL Ltd, Environmental Research Associates, Toronto.

Brinton, E., 1967, Vertical migration and avoidance capability of euphausiids in the California Current, *Limnol. Oceanogr.* **12**:451–483.

Brodie, P. F., Sameoto, D. D., and Sheldon, R. W., 1978, Population densities of euphausiids off Nova Scotia as indicated by net samples, whale stomach contents, and sonar, *Limnol. Oceanogr.* **23**:1264–1267.

Brown, R. G. B., 1970, Fulmar distribution: A Canadian perspective, *Ibis* **112**:44–51.

Brown, R. G. B., 1976, The foraging range of breeding Dovekies *Alle alle*, *Can. Field-Nat.* **90**:166–168.

Brown, R. G. B., 1977, *Atlas of Eastern Canadian Seabirds*. Suppl. 1: *Halifax—Bermuda Transects*, Canadian Wildlife Service, Ottawa.

Brown, R. G. B., 1979, Seabirds of the Senegal upwelling and adjacent waters, *Ibis* **121**:283–292.

Brown, R. G. B., in press, Birds, oil and the Canadian environment, in: *The Fate and Effects of Oil Pollution in the Canadian Marine Environment* (J. Sprague, ed.), Environmental Protection Service, Ottawa.

Brown, R. G. B., Cooke, F., Kinnear, P. K., and Mills, E. L., 1975*a*, Summer seabird distributions in Drake Passage, the Chilean fjords and off southern South America, *Ibis* **117**:339–356.

Brown, R. G. B., Nettleship, D. N., Germain, P., Tull, C. E., and Davis, T., 1975*b*, *Atlas of Eastern Canadian Seabirds*, Canadian Wildlife Service, Ottawa.

Brown, R. G. B., Bourne, W. R. P., and Wahl, T. R., 1978, Diving by shearwaters, *Condor* **80**:123–125.

Butcher, W. S., Anthony, R. P., and Butcher, J. B., 1968, *Distribution Charts of Oceanic Birds in the North Atlantic*, Woods Hole Oceanography Institute Technical Report No. **68—69**.

Chapman, S. E., 1969, The Pacific winter quarters of Sabine's Gull, *Ibis* **111**:615–617.

Clarke, M. R., 1966, A review of the systematics and ecology of oceanic squids, *Adv. Mar. Biol.* **4**:91–300.

Cline, D. R., Siniff, D. B., and Erickson, A. W., 1969, Summer birds of the pack ice in the Weddell Sea, Antarctica, *Auk* **86**:701–716.

Collins, J. W., 1884, Notes on the habits and methods of capture of various species of sea birds that occur on the fishing banks off the eastern coast of North America, and which are used for catching codfish by New England fishermen, *Rep. U.S. Fish. Comm.*, *1882*, pp. 311–335.

Conroy, J. W. H., 1975, Recent increases in penguin populations in Antarctica and the Subantarctic, in: *The Biology of Penguins* (B. Stonehouse, ed.), pp. 321–336, Macmillan Press, London and Basingstoke.

Cooper, J., 1978, Results of beach patrols conducted in 1977, *Cormorant* **4**:4–9.

Cramp, S., Bourne, W. R. P., and Saunders, D., (eds.), 1974, *The Seabirds of Britain and Ireland*, Collins, London.

Cromwell, T., and Reid, J. L., 1956, A study of oceanic fronts, *Tellus* **8**:94–101.

Crossin, R. S., 1974, The storm petrels (Hydrobatidae), *Smithson. Contrib. Zool.* **158**:154–205.

Cushing, D. H., 1971, Upwelling and the production of fish, *Adv. Mar. Biol.* **9**:255–334.

de Naurois, R., 1969, Notes brèves sur l'avifaune de l'archipel du Cap-Vert. Faunistique, endémisme, écologie, *Bull. Inst. Fond. Afr. Noire, Sér. A*, **31**:143–218.

Devillers, P., and Terschuren, J. A., 1978, Midsummer seabird distribution in the Chilean fjords, *Le Gerfaut* **68**:577–588.

de Weese, L. R., and Anderson, D. W., 1976, Distribution and breeding biology of Craveri's Murrelet, *San Diego Soc. Nat. Hist., Trans.* **18**(9):155–168.

Divoky, G. J., 1976, The pelagic feeding habits of Ivory and Ross' Gulls, *Condor* **78**:85–90.

Dorval, M., 1969, Observations ornithologiques en Atlantique Nord durant les années 1964, 1966, 1967 et 1968, *Ar Vran (Brest)* **2**:133–155.

Dunbar, M. J., 1951, Eastern arctic waters, *Fish. Res. Board Can. Bull.* **8**:1–131.

Dunbar, M. J., 1968, *Ecological Development in Polar Regions*, Prentice-Hall, Inc., Englewood Cliffs, New Jersey.

Ealey, E. H. M., 1954, Analysis of stomach contents of some Heard Island birds, *Emu* **54**:204–210.

El-Sayed, S. Z., and McWhinnie, M. A., 1979, Antarctic krill: Protein of the last frontier, *Oceanus* **22**:13–20.

Evans, P., and Waterston, G., 1976, The decline of the Thick-billed Murre in Greenland, *Polar Rec.* **18**:283–293.

Evans, P. R., 1971, Avian resources of the North Sea, in: *North Sea Science* (E. D. Goldberg, ed.), pp. 400–412, Massachusetts Institute of Technology, Cambridge, Massachusetts.

Fager, E. W., and McGowan, J. A., 1963, Zooplankton species groups in the North Pacific, *Science* **140**:453–460.

Fager, E. W., and Longhurst, A. R., 1968, Recurrent group analysis of species assemblages of demersal fish in the Gulf of Guinea, *J. Fish. Res. Board Can.* **25**:1405–1421.

Fenwick, G. D., 1978, Plankton swarms and their predators at the Snares islands, *N.Z. J. Mar. Freshwater Res.* **12**:223–224.

Finch, D. W., Russell, W. C., and Thompson, E. V., 1978, Pelagic birds in the Gulf of Maine. *Am. Birds* **32**:140–155, 281–294.

Fisher, J., 1952, *The Fulmar*, Collins, London.

Fisher, J., 1966, The Fulmar population of Britain and Ireland, 1959, *Bird Study* **13**:5–76.

Fisher, J., and Lockley, R. M., 1954, *Sea-Birds*, Collins, London.

Foster, R. J., Baxter, R. L., and Ball, P. A. J., 1951, A visit to Grimsey (Iceland), July–August 1949, *Ibis* **93**:53–59.

Frost, P. G. H., Siegfried, W. R., and Cooper, J., 1976, Conservation of the Jackass Penguin (*Spheniscus demersus* (L.)), *Biol. Conserv.* **9**:79–99.

Furness, R. W., 1978, Energy requirements of seabird communities: A bioenergetics model, *J. Anim. Ecol.* **47**:39–53.

Garcia Rodriguez, L., 1972, Observaciones sobre aves marinas en las pesquerias del atlantico sudafricano, *Ardeola* **16**:159–192.

Gould, P. J., 1967, Nocturnal feeding of *Sterna fuscata* and *Puffinus pacificus*, *Condor* **69**:529.

Gould, P. J., 1974, Sooty Tern (*Sterna fuscata*), *Smithson. Contrib. Zool.* **158**:6–52.

Gross, A. O., 1938, Birds of the Bowdoin-MacMillan Arctic Expedition, *Auk* **54**:12–42.

Grubb, T. C., 1972, Smell and foraging in shearwaters and petrels, *Nature (London)* **237**:404–405.

Gudkov, V. M., 1962, Relationships between the distribution of zooplankton, sea birds and baleen whales, *Tr. Inst. Okeanol., Akad. Nauk SSSR* **58**:298–313 (U.S. Naval Oceanogr. Office transl. NOO T-7, 1974).

Gulland, J. A., Holmsen, A., Laing, A., Paulik, G. J., Popper, F. E., and Watzinger, H., 1970, Report of the panel on economic effects of alternative regulatory measures in the Peruvian anchoveta industry, FAO, Dept. Fisheries, Rome (unpub. mimeo, 34 pp.).

Hardy, A. C., 1956, *The Open Sea. The World of Plankton*, Collins, London.

Harris, M. P., and Hansen, L., 1974, Sea-bird transects between Europe and Rio Plate, South America, in autumn 1973, *Dan. Ornithol. Foren. Tiddskr.* **68**:117–137.

Harrison, C. S., 1979, The association of marine birds and feeding gray whales, *Condor* **81**:93–95.

Hartley, C. H., and Fisher, J., 1936, The marine foods of birds in an inland fjord region in west Spitsbergen. Part 2. Birds, *J. Anim. Ecol.* **5**:370–389.

Haury, L. R., Briscoe, M. G., and Orr, M. H., 1979, Tidally generated internal wave packets in Massachusetts Bay. *Nature (London)* **278**:312–317.

Hentschel, E., 1933, Allgemeine Biologie des südatlantischen Ozeans, Part 1: Das Pelagial der obersten Wasserschicht, *Wiss. Ergebn. D. Atlant. Exped. 'Meteor'* **11**:1–168.

Holmes, W. N., and Cronshaw, J., 1977, Biological effects of petroleum on marine birds, in: *Effects of Petroleum on Arctic and Subarctic Marine Environments*, (D. C. Malins, ed.), pp. 359–398, Academic Press, New York.

Horner, R. A., 1976, Sea ice organisms, *Oceanogr. Mar. Biol. Annu. Rev.* **14**:167–182.

Hutchinson, G. E., 1950, The biogeochemistry of vertebrate excretion, *Bull. Am. Mus. Nat. Hist.* **96**:1–554.

Imber, M. J., 1973, The food of Grey-faced Petrels (*Pterodroma macroptera gouldi* (Hutton)), with special reference to diurnal vertical migration of their prey, *J. Anim. Ecol.* **42**:645–662.

Imber, M. J., 1975, Lycoteuthid squids as prey of petrels in New Zealand seas, *N.Z. J. Mar. Freshwater Res.* **9**:483–492.

Imber, M. J., and Russ, R., 1975, Some foods of the Wandering Albatross, *Notornis* **22**:27–36.

Ingham, M. C., and Mahnken, C. V. W., 1966, Turbulence and productivity near St. Vincent Island, B. W. I. A preliminary report, *Caribb. J. Sci.* **6**:83–87.

Jacobs, S. S., Gordon, A. L., and Amos, A. F., 1979, Effects of glacial ice melting on the Antarctic Surface Water, *Nature (London)* **277**:469–471.

Jane, C. (ed.), 1960, *The Journal of Christopher Columbus*, Anthony Blond and Orion Press, London.

Jehl, J. R., 1973, The distribution of marine birds in Chilean waters in winter, *Auk* **90**:114–135.

Jehl, J. R., 1974a, The distribution and ecology of marine birds over the continental shelf of Argentina in winter, *San Diego Soc. Nat. Hist. Trans.* **17**(16):217–234.

Jehl, J. R., 1974b, The near-shore avifauna of the Middle American west coast, *Auk* **91**:681–699.

Jespersen, P., 1924, The frequency of birds over the high Atlantic Ocean *Nature (London)* **114**:281–283.

Jespersen, P., 1929, On the frequency of birds over the high Atlantic Ocean, *Verh, VI Int. Ornithol. Kongr.* **1926**:163–172.

Jespersen, P., 1930, Ornithological observations in the North Atlantic Ocean, *Oceanogr. Rep. "Dana" Exped. 1920–22* **7**:1–36.

Johnson, S. R., Renaud, W. E., Richardson, W. J., Davis, R. A., Holdsworth, C., and Hollingdale, P. D., 1976, *Aerial Surveys of Birds in Eastern Lancaster Sound, 1976.* Appendix A: Distribution maps. Report of Norlands Petroleum, Calgary, LGL Ltd., Environmental Research Associates, Toronto.

Johnstone, G. W., and Kerry, K. R., 1976, Ornithological observations in the Australian sector of the Southern Ocean, *Proc. XVI Int. Ornithol. Congr.* **1974**:725–738.

Joiris, C., 1976, Seabirds seen during a return voyage from Belgium to Greenland in July, *Le Gerfaut* **66**:63–87.

King, J. E., and Pyle, R. L., 1957, Observations on sea birds in the tropical Pacific, *Condor* **59**:27–39.

King, W. B., 1967, *Seabirds of the Tropical Pacific Ocean. Preliminary Smithsonian Identification Manual*, Smithsonian Institution, Washington, D.C.

King, W. B., 1970, The trade winds oceanography pilot study. Part VII: Observations of sea birds March 1964 to June 1965, *U.S. Fish Wildl. Serv. Spec. Sci. Rep.—Fish.* **586**:1–136.

King, W. B., 1974, Wedge-tailed Shearwater (*Puffinus pacificus*), *Smithson. Contrib. Zool.* **158**:53–95.

King, W. B., Watson, G. E., and Gould, P. J., 1967, An application of automatic data processing to the study of seabirds. I. Numerical coding, *Proc. U.S. Natl. Mus. 123*, **3609**:1–29.

Komaki, Y., 1967, On the surface swarming of euphausiid crustaceans, *Pac. Sci.* **21**:433–448.

Kooyman, G. L., 1975, Behaviour and physiology of diving, in: *The Biology of Penguins* (B. Stonehouse, ed.), pp. 115–138, Macmillan Press, London and Basingstoke.

Kruuk, H., 1967, Competition for food between vultures in East Africa, *Ardea* **55**:171–193.

Kuroda, N., 1954, *On the Classification and Phylogeny of the Order Tubinares, Particularly the Shearwaters (Puffinus), with Special Considerations on Their Osteology and Habit Differentiation* (published by the author, Tokyo).

Kuroda, N., 1955, Observations on pelagic birds of the northwest Pacific, *Condor* **57**:290–300.

Kuroda, N., 1960, Analysis of seabird distribution in the northwest Pacific Ocean, *Pac. Sci.* **14**:55–67.

Lamb, K. D. A., 1964, Sea birds at the confluence of the Gulf Stream and Labrador Current east of New York, *Sea Swallow* **16**:65.

Lambert, K., 1971, Seevogelbeobachtungen auf zwei Reisen im ostlichen Atlantik mit besonderer Berucksichtigung des Seegebietes vor Sudwestafrika, *Beitr. Vogelkd. Leipzig* **17**:1–32.

Lemmetyinen, R., 1973, Feeding ecology of *Sterna paradisaea* Pontopp. and *S. hirundo* L. in the archipelago of southwestern Finland, *Ann. Zool. Fenn.* **10**:507–525.

Linton, A., 1978, The food and feeding habits of Leach's Storm-Petrel (*Oceanodroma leucorhoa*) at Pearl Island, Nova Scotia, and Middle Lawn Island, Newfoundland, M.Sc. thesis, Dalhousie University.

Lockley, R. M., 1953, On the movements of the Manx Shearwater at sea during the breeding season, *Br. Birds* **46**(suppl.):1–48.

Lockley, R. M., and Marchant, S., 1951, A midsummer visit to Rockall, *Br. Birds* **44**:373–383.

Løvenskiold, H. L., 1964, *Avifauna Svalbardensis*, Norsk Polarinstitutt skr. 129, Oslo.

Lysaght, A. (ed.), 1971, *Joseph Banks in Newfoundland and Labrador, 1766: His Diary, Manuscript and Collections*, University of California Press, Berkeley.

Marr, J., 1962, The natural history and geography of the Antarctic Krill (*Euphausia superba* Dana), *Discovery Rep.* **32**:33–364.

Martin, P. W., and Myres, M. T., 1969, Observations on the distribution and migration of some seabirds off the outer coasts of British Columbia and Washington State, 1946–1949, *Syesis* **2**:241–256.

McRoy, C. P., and Goering, J. J., 1974, The influence of ice on the primary productivity of the Bering Sea, in: *Oceanography of the Bering Sea* (D. W. Wood and E. J. Kelley, eds.), pp. 403–521, University of Alaska, Fairbanks.

Moore, H. B., 1941, Notes on the distribution of oceanic birds in the North Atlantic, *Proc. Linn. Soc. N.Y., 1937–1941* **52—53**:53–62.

Moore, H. B., 1951, Seasonal distribution of oceanic birds in the western North Atlantic, *Bull. Mar. Sci. Gulf Caribb.* **1**:1–14.

Morrison, R. I. G., 1977, Use of the Bay of Fundy by shorebirds, in: *Fundy Tidal Power and the Environment* (G. R. Daborn, ed.), Acadia University Institute, Wolfville, N. S.

Mörzer Bruijns, W. F. J., 1965, Birds seen during an east to west trans-Pacific crossing along the Equatorial Countercurrent around latitude 7°N in the autumn of 1960, *Sea Swallow* **17**:57–66.

Murphy, R. C., 1914, Observations of birds of the South Atlantic, *Auk* **31**:439–457.

Murphy, R. C., 1924, The marine ornithology of the Cape Verde Islands, with a list of all the birds of the archipelago, *Bull. Am. Nat. Hist.* **50**:211–278.

Murphy, R. C., 1936, *Oceanic Birds of South America*, American Museum of Natural History, New York.

Murphy, R. C., 1964, Systematics and distribution of Antarctic petrels, in: *Biologie Antarctique. Proceedings of the 1st SCAR Symposium on Antarctic Biology* (R. Carrick, M. Holdgate, and J. Prévost, eds.), Hermann, Paris.

Nelson, J. B., 1968, *Galapagos. Islands of Birds*, Longmans, London.

Nelson, J. B., 1970, The relationship between behaviour and ecology in the Sulidae with reference to other sea birds, *Oceanogr. Mar. Biol. Annu. Rev.* **8**:501–574.

Nelson, J. B., 1978, *The Gannet*, Buteo Books, Vermillion, S. D.

Neshyba, S., 1977, Upwelling by icebergs, *Nature (London)* **267**:507–508.

Nettleship, D. N., 1972, Breeding success of the Common Puffin (*Fratercula arctica* L.) on different habitats at Great Island, Newfoundland, *Ecol. Monogr.* **42**:239–268.

Nettleship, D. N., 1977a, Seabird colonies and distributions around Devon Island and vicinity, *Arctic* **27**:95–103.

Nettleship, D. N., 1977b, Seabird resources of eastern Canada: status, problems and prospects, in: *Canada's Threatened Species and Habitats* (T. Mosquin and C. Suchal, eds.), Canadian Nature Federation, Ottawa.

Nettleship, D. N., and Gaston, A. J., 1978, Patterns of pelagic distribution of seabirds in western Lancaster Sound and Barrow Strait, Northwest Territories, in August and September 1976, Canadian Wildlife Service, Occasional paper 39, Ottawa.

Ogi, H., and Tsujita, T., 1973, Preliminary examination of stomach contents of murres (*Uria* spp.) from the eastern Bering Sea and Bristol Bay, June–August, 1970 and 1971, *Jpn. J. Ecol.* **23**:201–209.

Ortego, B., 1978. Blue-faced Boobies at an oil production platform, *Auk* **95**:762–763.

Owen, R. W., 1966, Small-scale, horizontal vortices in the surface layer of the sea, *J. Mar. Res.* **24**:56–66.

Ozawa, K., 1967, Distribution of sea birds in austral summer season in the Southern Ocean, *Antarct. Res. Ser.* **29**:1–36.

Paessler, R., 1909, Beiträge zur Verbreitung der Seevögel. *Orn. Monatsb.* **17**:99–103.

Palmer, R. S. (ed.), 1962, *Handbook of North American Birds*, Vol. 1, Yale University Press, New Haven, Conn.

Paulik, G. J., 1971, Anchovies, birds and fishermen in the Peru Current, in: *Environment: Resources, Pollution and Society* (W. W. Murdoch, ed.), pp. 156–185, Sinauer Assocs. Inc., Sunderland, Massachusetts.

Pearson, T. H., 1968, The feeding biology of sea-bird species breeding on the Farne Islands, Northumberland, *J. Anim. Ecol.* **37**:521–552.

Petersen, G. H., 1977, Biological effects of sea-ice and icebergs in Greenland, in: *Polar Oceans* (M. J. Dunbar, ed.), Arctic Institute of North America, Calgary.

Phillips, J. H., 1963, The pelagic distribution of the Sooty Shearwater *Procellaria grisea*, *Ibis* **105**:340–353.

Pickard, G. L., 1964, *Descriptive Physical Oceanography*, Pergamon Press, Oxford.

Pickard, G. L., 1971, Some physical oceanographic features of inlets in Chile, *J. Fish. Res. Board Can.* **28**:1077–1106.

Pingree, R. D., Forster, G. R., and Morrison, G. K., 1974, Turbulent convergent tidal fronts, *J. Mar. Biol. Ass. U.K.* **54**:469–479.

Pocklington, R., 1979, An oceanographic interpretation of seabird distributions in the Indian Ocean, *Mar. Biol.* **51**:9–21.

Pollard, R. T., 1977, Observations and theories of Langmuir circulations and their role in near surface mixing, in: *A Voyage of Discovery* (M. Angel, ed.), pp. 235–251, Pergamon Press, Oxford.

Rankin, M. N., and Duffey, E. A. G., 1948, A study of the bird life of the North Atlantic, *Br. Birds* **41(suppl.)**:1–42.

Redfield, A. C., 1941, The effect of the circulation of the water on the distribution of the Calanoid community in the Gulf of Maine, *Biol. Bull.* **80**:86–110.

Rees, E. I. S., 1961, Notes on the food of the Greater Shearwater, *Sea Swallow* **14**:54–55.

Rees, E. I. S., 1963, Marine birds in the Gulf of St. Lawrence and Strait of Belle Isle during November, *Can. Field Nat.* **77**:98–107.

Richdale, L. E., 1963, Biology of the Sooty Shearwater *Puffinus griseus*, *Proc. Zool. Soc. London* **141**:1–117.

Roberts, B. B., 1940, The life cycle of Wilson's Petrel *Oceanites oceanicus* (Kuhl), *Br. Graham Land Exped. 1934-37, Sci. Rep.* **1**:141–194.

Rumboll, M. A. E., and Jehl, J. R., 1977, Observations on pelagic birds in the South Atlantic Ocean in the austral spring, *San Diego Soc. Nat. Hist., Trans.* **19(1)**:1–16.

Ryder, R. A., 1957, Avian-pinniped feeding associations, *Condor* **59**:68–69.

Salomonsen, F., 1955, The food production in the sea and the annual cycle of Faeroese marine birds, *Oikos* **6**:92–100.

Salomonsen, F., 1965, The geographical variation of the Fulmar (*Fulmarus glacialis*) and the zones of marine environment in the North Atlantic, *Auk* **82**:327–355.

Sanger, G. A., 1970, The seasonal distribution of some seabirds off Washington and Oregon, with notes on their ecology and behavior, *Condor* **72**:339–357.

Sanger, G. A., 1972, Preliminary standing stock and biomass estimates of seabirds in the Subarctic Pacific Region, in: *Biological Oceanography of the Northern North Pacific Ocean* (A. Y. Takenouti, ed.), pp. 589–611, Idemitsu Shoten, Tokyo.

Schaefer, M. B., 1970, Men, birds and anchovies in the Peru Current—Dynamic interactions, *Trans. Am. Fish. Soc.* **99**:461–467.

Sealy, S. G., 1973, Interspecific feeding assemblages of marine birds off British Columbia, *Auk* **90**:796–802.

Serventy, D. L., Serventy, V., and Warham, J., 1971, *The Handbook of Australian Sea-Birds*, A. H. Reed and A. W. Reed, Sydney, Melbourne, Wellington, Auckland.

Shuntov, V. P., 1974, *Sea Birds and the Biological Structure of the Ocean*, U.S. Department of the Interior transl. TT 74-55032, Washington, D.C.

Snow, B. K., and Snow, D. W., 1968, Behaviour of the Swallow-tailed Gull of Galápagos, *Condor* **70**:252–264.

Stanford, W. P., 1953, Winter distribution of the Grey Phalarope *Phalaropus fulicarius, Ibis* **95**:483–492.

Stavn, R. H., 1971, The horizontal–vertical distribution hypothesis: Langmuir circulations and *Daphnia* distributions, *Limnol. Oceanogr.* **16**:453–466.

Steven, G. A., 1933, The food consumed by shags and cormorants around the shores of Cornwall (England), *J. Mar. Biol. Ass. U.K.* **18**:277–292.

Stott, F. C., 1936, The marine foods of birds in an inland fjord region in west Spitsbergen. Part 1. Plankton and shore benthos, *J. Anim. Ecol.* **5**:356–369.

Stresemann, E., and Stresemann, V., 1970, Über Mauser and Zug von *Puffinus gravis, J. Ornithol.* **111**:378–393.

Summerhayes, C. P., 1969, Seabirds seen in the northern Tasman Sea in winter, *N.Z. J. Mar. Freshw. Res.* **3**:560–570.

Summerhayes, C. P., Hofmayr, P. K., and Rioux R. H., 1974, Seabirds off the southwestern coast of Africa, *Ostrich* **45**:83–109.

Sutcliffe, W. H., Loucks, R. H., and Drinkwater, K. F., 1976, Coastal circulation and physical oceanography of the Scotian Shelf and Gulf of Maine, *J. Fish. Res. Board Can.* **33**:98–115.

Swartz, L. G., 1967, Distribution and movements of birds in the Bering and Chukchi Seas, *Pac. Sci.* **21**:332–347.

Swinburne, S., 1886, Notes on birds observed on various voyages between England and the Cape of Good Hope, *Proc. R. Phys. Soc. Edinburgh* **9**:193–201.

Szijj, L. J., 1967, Notes on the winter distributions of birds in the western Antarctic and adjacent Pacific waters, *Auk* **84**:366–378.

Tåning, A. V., 1933, The winter quarters of the phalaropes, *Ibis* **3**(13):132–133.

Tickell, W. L. N., and Woods, R. W., 1972, Ornithological observations at sea in the South Atlantic Ocean, 1954–64, *Br. Antarct. Surv. Bull.* **31**:63–84.

Tuck, G. S., 1978, *A Field Guide to the Seabirds of the World*, Collins, London.

Tuck, L. M., 1961, *The Murres*, Canadian Wildlife Service, Monograph Series No. 1, Ottawa.

Tull, C. E., Germain, P., and May, A. W., 1972, Mortality of Thick-billed Murres in the west Greenland salmon fishery, *Nature (London)* **237**:42–44.

Uspenski, S. M., 1958, The bird bazaars of Novaya Zemlya, *Canad. Wildl. Serv. Transl. Russ. Game Rep.* **4**:1–159.

Valdivia G., J. E., 1978, The anchoveta and El Niño, *Rapp. P.-v. Reun. Cons. Int. Explor. Mer* **173**:196–202.

van Oordt, G. J., and Kruijt, J. P., 1953, On the pelagic distribution of some Procellariiformes in the Atlantic and Southern Oceans, *Ibis* **95**:615–637.

van Oordt, G. J., and Kruijt, J. P., 1954, Birds observed on a voyage in the South Atlantic and Southern Oceans in 1951–52, *Ardea* **42**:245–280.

Veitch, C. R., 1978, Seabirds found dead in New Zealand in 1976, *Notornis* **25**:141–149.

Voous, K. H., and Wattel, J., 1963, Distribution and migration of the Greater Shearwater, *Ardea* **51**:143–157.

Wahl, T. R., 1975, Seabirds in Washington's offshore zone, *West. Birds* **6**:117–134.

Walsh, J. J., 1978, The biological consequences of interaction of the climatic, El Niño, and event scales of variability in the eastern tropical Pacific, *Rapp. P.-v. Reun. Cons. Int. Explor. Mer* **173**:182–192.

Warham, J., 1977, Wing loadings, wing shapes, and flight capabilities of Procellariiformes, *N.Z. J. Zool.* **4**:73–83.

Warnecke, G., Allison, L. J., McMillin, L. M., and Szekielda, K.-H., 1971, Remote sensing of ocean currents and sea surface temperature changes derived from the Nimbus II satellite, *J. Phys. Oceanogr.* **1**:45–59.

Watson, G. E., 1963, *Preliminary Field Guide to the Birds of the Indian Ocean*, Smithsonian Institution, Washington, D.C.

Watson, G. E., 1966, *Seabirds of the Tropical Atlantic Ocean*, Smithsonian Institute, Washington, D.C.

Watson, G. E., 1968, Synchronous wing and tail molt in diving petrels, *Condor* **70**:182–183.

Watson, G. E., 1975, *Birds of the Antarctic and Sub-Antarctic*, American Geophysical Union, Washington, D.C.

Watson, G. E., Angle, J. P., Harper, P. C., Bridge, M. A., Schlatter, R. P., Tickell, W. L. N., Boyd, J. C., and Boyd, M. M., 1971, *Birds of the Antarctic and Subantarctic*, American Geographical Society Antarctic Map Folio 14, New York.

Watson, G. E., and Divoky, G. J., 1972, Pelagic bird and mammal observations in the eastern Chukchi Sea, early fall 1970, *U.S. Coastgd. Oceanogr. Rep.* **50**:111–172.

Watson, P. S., 1978, Seabirds at commercial trawlers in the west Irish Sea, *Ibis* **120**:107–108.

Webb, B. F., 1973, Report on the investigations of the 'Lloret Lopez II" 8 January to 2 April 1970. Section 7. Bird distribution and relative abundance, *N.Z. Min. Ag. Fish., Fish. Tech. Rep.* **104**:1–24.

Wiens, J. A., and Scott, J. M., 1975, Model estimation of energy flow in Oregon coastal seabird populations, *Condor* **77**:439–452.

Wynne-Edwards, V. C., 1935, On the habits and distribution of birds on the North Atlantic, *Proc. Boston Soc. Nat. Hist.* **40**:233–346.

Zelickman, E. A., and Golovkin, A. N., 1972, Composition, structure and productivity of neritic plankton communities near the bird colonies of the northern shores of Novaya Zemlya, *Mar. Biol.* **17**:265–274.

Zino, P. A., 1971, The breeding of Cory's Shearwater *Calonectris diomedea* on the Salvage Islands, *Ibis* **113**:212–217.

Chapter 2

CHEMORECEPTION IN SEABIRDS

Bernice M. Wenzel

Department of Physiology and
  Brain Research Institute
UCLA School of Medicine
Los Angeles, California 90024

I. INTRODUCTION

In a world rich in odors, only the birds as a major group have been generally regarded as insensitive or oblivious to such stimuli—even, in Wood Jones's (1937*a*) phrase, as emancipated from olfactory dominance. Yet olfaction is one of the three great modalities for distance communication, functioning as effectively in an aerial habitat as in terrestrial or aquatic habitats. Perhaps our own diminished sensitivity to external chemical messages and our inability to sustain an aerial existence have influenced our opinions about olfactory functioning in birds. Notably, however, we feel no such hesitation in the case of flying insects, which are readily accepted as being critically guided by odor signals. In contrast, taste seems to present less of a problem. Birds, like most other creatures, must ingest food through the mouth, and nothing seems more reasonable than that this food would be selected partly on the basis of taste.

Except for the kiwi, which can be heard to sniff, the overt behavior of birds does not suggest olfactory awareness to the casual observer, especially to one who is himself microsmatic. With nostrils at the base of the beak, again except for the kiwi, olfactory scanning seems inefficient and unlikely. Seabirds are especially easy to overlook in this respect because of their less frequent contacts with human observers. Their behavior patterns typically unfold at remote points far from human habitats, and reports on which to base conjecture have been sparse in number and erratic in detail. With biologists' increased access to their breeding grounds, more systematic descriptions of behavior have become available in recent years. As these are added to the lore gradually accumulated from various sources, a provocative possibility of odor usage emerges, especially in the case of one order, Procellariiformes.

While suggestive behavioral evidence was beginning to mount, some attention was also being paid to anatomical and physiological characteristics of chemoreceptive systems in birds. Some of these findings, particularly regarding olfactory structures, proved so impressive that interest was heightened in collecting more directly relevant behavioral data. In summarizing all the information discussed in this chapter, the anatomical material is reviewed first, in light of the obvious dependence of function on structure, followed by the physiological and behavioral evidence. The meager data on taste in each category precede those on smell.

II. ANATOMY

A. Taste

Among the tiny handful of studies on taste receptors in birds, none deals with seabirds. Although taste buds are far less plentiful than in most mammals, they have been shown to exist in Anna's Hummingbird (*Selasphorus sasin*) (Weymouth, Lasiewski, and Berger, 1964), pigeon (*Columba livia*), chicken (*Gallus gallus*), European Bullfinch (*Pyrrhula pyrrhula*), and Japanese Quail (*Coturnix coturnix*) in sufficient numbers to support taste perception. Furthermore, their general structure is comparable to that seen in all mammals examined (Wenzel, 1973). It is a safe assumption that seabirds, like those birds already studied, possess taste buds probably distributed in the rear of the tongue and possibly in the pharynx as well.

In contrast to mammals, the nerve fibers involved with taste are restricted in birds to cranial nerve IX and do not occur in nerve VII. The specific branch of the glossopharyngeal nerve involved with taste is the laryngolingual, which innervates the posterior buccal and pharyngeal areas. Cutaneous information is also carried by these fibers. They terminate in the medulla oblongata, possibly in association with large numbers of fibers from the viscera. The details of their distribution at the level of the first relay and at all higher points in the brain are as yet undetermined.

B. Smell

Considerably more attention has been given to the peripheral olfactory structures of a wide range of birds (Bang, 1971; Bumm, 1883), and especially to procellariiforms (Bang, 1966; Burne, 1908; Klinckowström, 1891; Wood Jones, 1937*a,b*). Seabirds, like all the rest of the 151 species of birds that have been examined, possess olfactory mucosa typical of vertebrates, olfactory nerves, and more or less prominent olfactory bulbs. The variation among avian species in the relative amounts of these tissues is striking. Generally speaking, seabirds are among the better-endowed forms. A useful measure for expressing relative olfactory mass is the ratio of olfactory bulb diameter to the largest diameter of the cerebral hemisphere (Cobb, 1960). If the species studied are arranged in order of this index, from largest to smallest, all but two of the first 12 are procellariiforms. The exceptions are the kiwi, in second place, and the Turkey Vulture, in tenth. Of the procellariiforms measured, only the Diving Petrel (*Pelecanoides georgicus*), in rank 54, falls outside this superior range. Other pelagic forms are scattered widely through the remaining ranks, as shown in Table I.

Table I.   Percent of Largest Diameter of Cerebral
Hemisphere Represented by Diameter of Olfactory Bulb
in Various Seabirds[a]

Species	Percent	Rank[b]
Pagodroma nivea	37.0	1
Oceanodroma leucorhoa	33.0	3.5
Oceanites oceanicus	33.0	3.5
Puffinus pacificus	30.0	4.5
Puffinus gravis	30.0	4.5
Pachyptila desolata	29.5	6
Diomedea nigripes	29.0	7.5
Puffinus opisthomelas	29.0	7.5
Daption capense	27.5	9
Fulmarus glacialis	27.0	10.5
Phaethon aethereus	20.0	39
Pelecanoides georgicus	18.0	55
Pygoscelis adelie	17.0	60.5
Larus argentatus	16.3	64
Phalacrocorax niger	15.8	67
Fregata magnificens	15.0	72.5
Uria lomvia	15.0	72.5
Phalacrocorax carbo	14.5	78
Fratercula arctica	13.9	86
Phalacrocorax auritus	10.0	97.5
Pelecanus occidentalis	9.6	103
Phalacrocorax urile	9.5	104
Phalacrocorax pelagicus	8.0	109

[a] Data from Bang (1971).
[b] Ranked among 124 avian species for which such values have
been determined.

Earlier writers have remarked on the impressive amount of olfactory mucosa, requiring extensive turbinal development, and on the massive olfactory bulbs in shearwaters and albatrosses. To quote Wood Jones (1937*b*), "It requires . . . only one glance at the brain of *D. [iomedea] cauta* to realize the strange, almost mammalian appearance imparted to it by the enormous olfactory bulbs." The brain that inspired that comment is pictured in Fig. 1. Internal nasal architecture is also variable with respect to the extent and placement of sinuses, the existence and opening of nasal glands, the number and size of conchae, and whether the septum is perforated or not. Of special interest is the extent of the olfactory epithelium containing the olfactory receptor cells, the axons of which constitute the fibers of the olfactory nerves. Figure 2 shows frontal sections through the heads of four procellariiform species, all drawn to the same scale. The olfactory epithelium can be seen to differ markedly, even among those forms in which the olfactory bulbs are relatively large. It seems logical to expect to find functional variation when structural differences are as notable as these. A view of the olfactory area of the fulmar is shown in Fig. 3, in which the relatively massive olfactory bulb can be clearly seen.

The first synapse in the olfactory system occurs in the olfactory bulb, where olfactory nerve fibers connect with mitral cells. The size of the bulb,

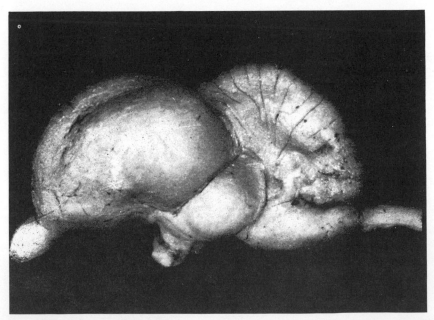

Fig. 1. Brain of White-capped Albatross (*Diomedea cauta* Gould) showing very large olfactory bulbs anterior to the cerebral hemispheres. [From Wood Jones (1937*b*) by permission of the Royal Australasian Ornithologists Union.]

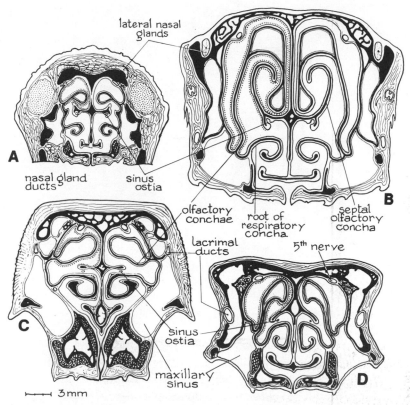

Fig. 2. Diagrammatic frontal sections through the olfactory region of the nasal cavities of (A) Diving Petrel (*Pelecanoides georgicus*), (B) Snow Petrel (*Pagodroma nivea*), (C) Black-footed Albatross (*Diomedea nigripes*), and (D) Wedge-tailed Shearwater (*Puffinus pacificus*). Drawn to natural scale. Note similarity in head size at this point between the albatross and Snow Petrel although the latter is a much smaller bird overall. Cross-hatching indicates muscle; black areas are bone and cartilage. [From Bang (1966) by permission of the author and S. Karger AG, Basel.]

therefore, reflects the number of incoming olfactory nerve fibers that must be accommodated, and the number of fibers, in turn, reflects the number of receptor cells in the olfactory mucosa. From the olfactory bulb, secondary neurons course into the anterior portion of the brain. Their precise destinations have been identified experimentally only in the pigeon (Hutchison and Wenzel, 1978; Hutchison *et al.*, 1977; Macadar *et al.*, 1980; Rieke and Wenzel, 1978), and even here the story is not yet complete. Because the relative size of the pigeon's olfactory bulb is only in the middle of the range, it is possible that birds with vastly larger amounts might have even more extensive central connections, but this remains to be determined. In the case

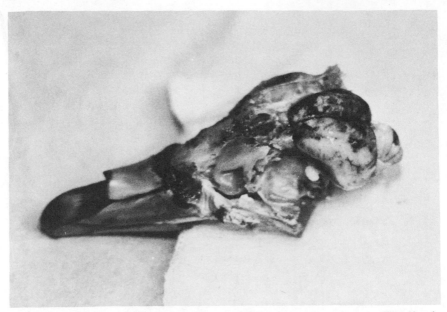

Fig. 3. Lateral view of the brain of Northern Fulmar (*Fulmarus glacialis*) *in situ*. The olfactory bulb and extensive olfactory mucosa can be clearly seen.

of the pigeon, the principal direct projection sites from the olfactory bulb are the prepiriform cortex, the ventral hyperstriatum, the parolfactory lobe ipsilaterally, and the parolfactory lobe and the paleostriatal complex contralaterally. Further connections are made to other telencephalic and diencephalic nuclei, and centrifugal fibers reach the bulb from various cellular groups. Many areas of the avian brain are not yet identified as to function, nor are the analogies to mammalian structures entirely clear. On the basis of current information, it is reasonable to say that the olfactory system is connected with structures that are presumably involved in life-sustaining and emotional functions, as is generally true in mammals.

The nasal cavity is innervated by the trigeminal nerve also, which is stimulated by the thermal and cutaneous qualities of the molecules contained in inspired air. The trigeminal fibers enter the brain stem and connect synaptically with secondary neurons that course toward higher areas in the diencephalon and telencephalon. Since odorous molecules probably stimulate trigeminal fibers as well as olfactory cells, both pathways would usually be involved when odorous air is inspired. The trigeminal system is not considered in this discussion. Not only is the nature of its interaction, if any, with the olfactory system unknown even in general, but also the literature is silent about any of its characteristics in seabirds.

A closer relative of the olfactory system in other vertebrates—the vomeronasal system—is absent in all birds. Whether its function, which is not well understood, is also absent or is carried out by the olfactory system is a completely unanswered question.

III. PHYSIOLOGY

A. Smell

The determination of sensory function is a difficult problem in non-verbal animals, which cannot respond to questioning about their perceptual experiences. One type of useful indirect data is electrophysiological. A small amount has been collected from seabirds, but only for the olfactory system. By showing that olfactory nerve fibers and olfactory bulb cells respond differentially to the presentation of odors, critical evidence is provided for olfactory function. Results of this sort neither tell us how these stimuli smell to the bird nor what, if anything, they signify. They do tell us that the neural mechanism exists for perception of discrete odors.

Extracellular recordings have been made in the olfactory mucosa of the shearwater (*P. leucomelas*) using odor stimulation (Shibuya and Tonosaki, 1972). The amplitude and stability were comparable to responses recorded from amphibians, reptiles, and mammals. Among 12 avian forms in which recordings have been made from small bundles of olfactory nerve fibers while either pure or odorized air was being inhaled, only one marine genus was represented. In the Ring-billed Gull (*Larus delawarensis*) and the Black-tailed Gull (*L. crassirostris*), the olfactory nerve twigs responded to such chemical stimuli as amyl acetate in essentially the same way as did those of all other reptiles, birds, and mammals studied in this type of experiment (Shibuya *et al.*, 1970; Tucker, 1965). Similarly, recordings of electrical activity from the olfactory bulb of the Black-footed Albatross (*Diomedea nigripes*) and Black-vented Shearwater (*Puffinus puffinus opisthomelas*) show no essential differences from those obtained from a variety of other vertebrate forms (Wenzel and Sieck, 1972). Some characteristic records from the latter species are shown in Fig. 4. It is obvious that the pattern changes when an odor is introduced, and the new pattern varies with the odor. It is also clear that the bulb receives input from other parts of the brain, as shown by the abrupt increase in frequency when a light suddenly came on in the dark testing enclosure with no change being made in the incoming air.

Another type of recording that can be made from the olfactory bulb or any other area of the brain is the potential evoked by stimulation at a site directly connected with the recording point. Thus if the olfactory nerve is stimulated, an evoked potential can be recorded in the olfactory bulb, with its specific outline varying according to the exact location of the recording electrode in the bulb. Such a potential, recorded from a Northern Fulmar (*Fulmarus glacialis*), is presented in Fig. 5. In addition to recording bulbar evoked potentials, responses have been recorded from individual cells in various telencephalic areas of the fulmar brain following stimulation of the olfactory nerve. Although the olfactory pathway is only in the process of being identified in this species, it is already evident that it involves a number of areas and is at least as extensive as that of the pigeon, which has only a medium-size olfactory bulb.

These different electrophysiological data, although extremely limited, imply that the avian olfactory system is functional and that it shows no striking differences in overall responsiveness from mammalian forms that have been studied more extensively. Aside from the different patterns of bulbar activity associated with different odors presented to albatrosses and shearwaters (Fig. 4), we have no physiological evidence about the ability of

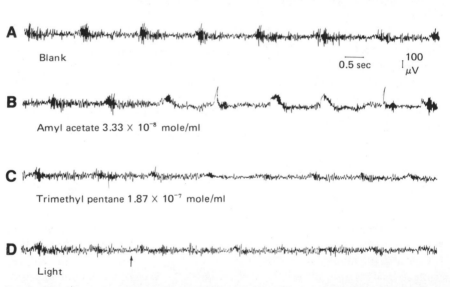

Fig. 4. Electrical activity recorded from olfactory bulb of Black-vented Shearwater (*Puffinus puffinus opisthomelas*). (A) Intrinsic activity. (B,C) Effects of adding, respectively, amyl acetate and trimethyl pentane to air stream. (D) Effect of sudden onset of light at arrow mark. Bursts in (A) are characteristic of recordings from olfactory bulbs of vertebrates and are synchronized with inspiration. Their frequency and amplitude change in specific ways related to inhalation of odors. [From Wenzel and Sieck (1972) by permission of Brain Research Publications.]

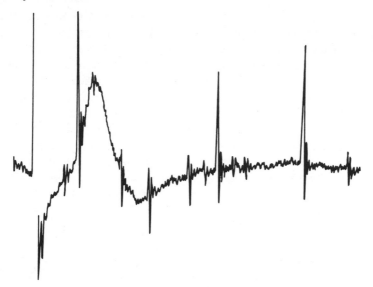

Fig. 5. Mitral cell responses superimposed on an evoked potential recorded from olfactory bulb of Northern Fulmar after electrical stimulation of ipsilateral olfactory nerve. Onset latency of the evoked potential is ~ 20 msec, and amplitude is 600 μV.

any seabird's olfactory system to differentiate among odors or about possible special sensitivity to odors that might be significantly involved in the daily lives of these birds.

IV. BEHAVIOR

A. Taste

1. Laboratory Studies

Direct behavioral evidence of perception and of its significance to the bird is much more difficult to obtain than are physiological recordings. A few attempts have been made to measure gustatory thresholds in seabirds using laboratory techniques. As early as 1925, Rensch and Neunzig tested several species of birds, mainly passerines, in the Berlin Zoo by offering salt and sour solutions of different concentrations after 12–24 hr of thirst and noting which concentration, if any, was rejected. For bitter and sweet tastes, they presented only a 30% solution of each, which was scored as accepted, rejected, or not tasted at all. The only bird in their group with any amount of marine habitat was a Common Tern (*Sterna hirundo*). In comparison with the rest of the group, its salt threshold was relatively high, the sour threshold was moderately high, and the bird gave no reaction to the bitter or sweet solutions.

A somewhat similar method was used to compare starlings, grackles, and gulls with respect to preference and tolerance for salt solutions at different concentrations (Harriman, 1967; Harriman and Kare, 1966a,b). After a period of water deprivation, each bird was given access to drinking bottles in its cage; one bottle contained distilled water and the other a solution of sodium chloride with a range of 0.05–4.0 M. The Herring Gulls (L. argentatus smithsonianus) chose water over the 0.20 M salt solution, as compared to 0.15 M for the starlings and 0.50 M for the grackles. Below these concentrations, the birds were as likely to drink the salt solution as the water. These results were replicated with the Laughing Gull (L. atricilla), which they considered to be a more pelagic species. Both adults and chicks showed a strong aversion to the lowest hypertonic solution used, 0.20 M, although their survival when given only seawater or salt solution to drink was far superior to that of the Herring Gulls. It appears that birds are able to detect the taste of salt and that its acceptability does not differ between birds that normally ingest seawater and those that normally drink less salty fluids.

2. Field Studies

Accounts of gustatory behavior in the field by experienced observers of seabirds are very rare. In describing the first experience of a shearwater fledgling in the sea, Lockley (1961) characterized the bird as "delighted with [the] sharp taste" of seawater. He felt that young marine birds, which are almost totally dependent on seawater, would be expected to relish it. Miller (1940) has provided descriptive comments, based on many observations at sea, about flavor preferences of the Black-footed Albatross, which he describes as a "feathered pig . . . [but] not without a certain discriminative judgment." While foraging among the kitchen scraps thrown overboard, the Black-footed Albatross showed great interest in anything associated with fat. A small amount of fat converted bread from a rejected to a preferred item. Miller says, "Masses of congealed bacon fat . . . caused the greatest excitement among the birds as soon as the flavor was tested. . . . [Food was taken in the] extreme tip of the beak where very sensitive taste buds must be situated, so quickly was the recognition of desirability of the object tested" (pp. 235–236). The "most palatable scraps" from halibut boats have been taken by Sooty Shearwaters (P. griseus) in the northern Pacific, although they caught most of their food themselves (Martin, 1942). Miller reported that albatrosses were seen to drink seawater on several occasions, but makes no comment about any apparent hedonic reaction. Penguins are said to drink fresh water eagerly after having been at sea for extended periods (Warham, 1971).

B. Smell

1. Laboratory Studies

Demonstrating that the presentation of an odor leads to certain changes in electrical activity of the primary olfactory system, as discussed above, indicates that the sensory channel into the central nervous system has functioned, but it says nothing about any effect of such activity on motor output. One very simple way of obtaining the latter information is to record such involuntary responses as heart rate and respiration while a bird is at rest in a quiet, darkened enclosure where only an occasional odor is made to intrude. Measurements of this sort have been collected from a number of birds that represent different habitats and all points on the scale of olfactory bulb–cerebral hemisphere ratios, and nearly all show some clear sign of altered heart or respiratory rate when odors are presented.

The only seabirds in this series are a shearwater (*P. puffinus opisthomelas*) (Wenzel, 1967; Wenzel and Sieck, 1972) and a penguin (*Spheniscus humboldti*).* Penguins are difficult to accommodate in this type of test. The responses obtained were suggestive of olfactory perception but were not conclusive. Their possible evolutionary link with procellariiforms (Baker and Manwell, 1975; Jacob, 1976) gives special interest to their olfactory physiology, as well as to the fact that their olfactory mucosa is variously shielded from direct access in different forms. Unlike the pigeon, which has been studied most extensively in this type of experiment, the shearwater's responsiveness appears more clearly in its respiratory rate than in its heart rate. When the smell of pyridine was presented, for example, respiratory rate increased by almost 10 cycles/min. This change was of the same general magnitude as that shown when a light was turned on in the dark chamber. Thus the presentation of the odorous stimulus elicited as large a response as the visual one. Perhaps paradoxically, presenting the odor of fish pulp to shearwaters led to only a modest increase in respiratory rate in this laboratory situation. As is discussed below, shearwaters have shown clear behavioral responsiveness to fish odors released under natural conditions at sea.

* This work was done in collaboration with Dr. Bernard Stonehouse at the San Diego Zoo and University of British Columbia. In the latter case, penguins were made available by the Stanley Park Zoo in Vancouver. Special thanks are due to all these institutions for providing birds and facilities.

2. Field Studies

References to olfactory behavior among field observations are somewhat more common than are those to taste. The great distances and inaccessible sites that characterize the habitats of many seabirds have hindered the collection of large amounts of either casual or systematic data. Ship followers, of course, have been known and described by ocean travelers for centuries. The albatross's fame 200 years ago made it a readily understood symbol for Samuel Coleridge in his epic poem, "The Rime of the Ancient Mariner." The development of a large whaling industry expanded such contacts in that the ships remained in one area for relatively long periods of time, and their discards served as richly attractive bait for a greater number and variety of birds than were typically seen traveling with moving boats. In this century, the spread of scientific observation posts and military installations into Antarctic regions and remote oceanic islands has contributed a great deal to our knowledge of seabird biology and behavior.

Except for a very few studies, discussed below, that were deliberately designed to test hypotheses about the use of olfaction by specific procellariiform birds, the information in the field literature is largely circumstantial. Some of the feeding, nesting, and social habits of seabirds indicate the possibility of olfactory involvement, but to test any of the suggested hypotheses is a very challenging task. It is well known that these birds feed on odorous material such as fish, crustaceans, other marine organisms, and even garbage, when available. They are usually colonial nesters, and in many cases burrowers; some are crepuscular or nocturnal, and some interact with mates and young in displays that include mutual preening and other close contact involving their beaks. In addition, many tube-nosed forms secrete a stomach oil that can be ejected for some distance and is very odorous.

Because of the notable olfactory endowment of so many representatives of procellariiforms among the birds available for measurement, there is special interest in the habits of these species. Petrels, shearwaters, and storm petrels are almost uniform both in their choice of burrows or crevices rather than open sites for nesting and in their primarily nocturnal visits to land. Some are said to approach land only in darkness and to show less activity in moonlight than under cloudy skies. The Little Shearwater (*P. assimilis*), for instance, lands at night very close to its own burrow; occasionally it crashes into bushes and then finds its way to its nest. After courtship and egg laying, homing is done silently, even in "abominable weather." Both the birds and their burrows were found to have a smell characteristic of their species, an observation made more suggestive by their homing performance in the complete absence of calling (Warham, 1955). The Short-tailed and the Pale-footed Shearwaters (*P. tenuirostris* and *P. carneipes*) approach their nesting islands always from the downwind direction and only after twilight is waning or past (Warham, 1958, 1960). Detailed analysis of such behavior in a petrel is discussed below. Tickell (1962) reports that of all the birds nesting on the South Orkney Islands, the Dove Prion (*Pachyptila desolata*) and the Black-billed Storm Petrel (*Fregetta tropica*) are the only ones that are completely nocturnal. Wilson's Storm Petrel (*Oceanites oceanicus*), although predominantly so, sometimes visits its nest during daylight; such occurrences may be promoted by the very limited amount of darkness during the peak of summer at this latitude. Roberts (1940) tells of Wilson's Storm Petrels flying in from sea each evening to their nests on the Argentine Islands, and remarks especially on their ability to dig at precisely the right spot after a snowfall "when the whole appearance of the cliff face is changed." Such adeptness was confirmed by observations at other breeding grounds on the South Orkney Islands (Beck and Brown, 1971, 1972). Nocturnal homing without calling, an unusually strong and distinctive odor, and very little aerial display behavior have been presented as evidence for a close relationship between Bulwer's Petrel (*Bulweria bulwerii*) and the Tahiti Petrel (*Pterodroma rostrata*) (Thibault and Holyoak, 1978). These characteristics have been interpreted as a common development of olfactory-based behavior in these two species of different genera that are otherwise less conclusively related.

Nocturnal, or even crepuscular, behavior could be a defense against predation as well as an adaptation to permit feeding on species that occupy surface waters only during darkness. It requires appropriate sensory mechanisms for guidance to the birds' own burrows, mates, and chicks. All these species have characteristic calls and are noisy in their burrows, although not necessarily on arrival. In some cases there might be more calling on dark nights. A great deal of vocal exchange occurs between Dove Prion mates, so that the colony sounds like "distant traffic" all night (Tickell, 1962). Wilson's Storm Petrels call so constantly between the incoming nesting birds that the calls can be heard for at least a mile (Ardley, 1936; Roberts, 1940).

Beck and Brown (1972) report that Wilson's Storm Petrels were often found visiting in nests other than their own, and Tickell refers to both male and female Dove Prions as "keeping company" with other birds in either's nest. In speculating about how ownership of the burrow is reestablished and how pair bonds are maintained, Beck and Brown suggest the possibility that individual recognition based on scent plays some role in this process. Whenever they removed one of these birds from its nest, it grasped at the investigator's fingers and often ejected oil. They wondered if this aggressive reaction was triggered by odor from other petrels they had handled shortly before.

Lockley (1961) also invoked smells as helpful signals in the life of the

Manx Shearwater (*Puffinus puffinus*). He proposed that the bird's small eye is adapted primarily for sensitivity in brilliant ocean light and that the "complete darkness" in the burrow demands nonvisual senses including smell. The chick pecks at its newly arrived parent's bill, which Lockley believes it could not see, perhaps guided by the strong smell of recently ingested fish. The parent birds fly home to the island and then walk to their individual burrows, just as well in fog as in moonlight or starlight, again possibly aided by distinctive odors. A similar statement, even more definite, occurs in Swales's (1965) account of seabirds on Gough Island. "More than any other species of night-bird studied, the Broad-billed Prion [(*P. vittata vittata*)] behaved as though it possessed a sense of smell. . . . When on the ground at night, the bird's behaviour was reminiscent of a young mammal, such as a dog, smelling its way back to the nest" (p. 41). The possibility that the well-known musky odor associated with petrels might play a role in recognition at night has been mentioned for the Band-rumped Storm Petrel (*O. castro*) as well (Allen, 1962). The depositing of excrement just outside the burrow entrance, a habit of the Wedge-tailed (*Puffinus pacifious*) and Greater (*P. gravis*) Shearwaters, for example (Rowan, 1952; Shallenberger, 1973), would leave an odorous label of possible value in a convenient location. Any large amounts of ejected stomach oil adhering to nests could function in a similar way. Lockley felt that voice was probably the critical factor in pair recognition. Vocal differences among the birds in a given colony are discernible even by human ears.

Under normal conditions, Warham (1964) believes that most petrels home successfully by use of a visual map that was established during visits to the area in the two or more prebreeding years after the fledglings first leave the colony. He admitted that the ability of a bird to find its own burrow under unfavorable visual conditions or when snow covers the landscape demands special explanation, and that the role of smell has not been studied. Field experiments described below have made a small start on this problem. Warham also points out the obvious need for information about visual sensitivity in procellariiforms. Nothing is known of their acuity under different levels of illumination or of their ability to discriminate visual patterns. That some are blinded by light at night has been documented (Bishop, 1949; Hagen, 1952; Swales, 1965; Travers, 1872). No obvious difference was found between the retinas of the diurnal fulmar and the Manx Shearwater, which function under both photopic and scotopic conditions (Lockie, 1952).

Because of the exceptionally large amount of olfactory tissue in the Snow Petrel (*Pagodroma nivea*) (Bang, 1965), Murphy's comment is especially interesting. He described the Snow Petrel as a short-day bird, moving "reluctantly from antarctic darkness for about three months of the year . . . feeding much at night. . . ." (Murphy, 1964, p. 357). Brown, on the other hand, felt ". . . that there is no evidence to indicate that Snow Petrels feed in darkness" (Brown, 1966, p. 59). Murphy reports the contents of 17 stomachs he examined to have been 95% fish, an analysis that is in agreement with an earlier study (Bierman and Voous, 1950). Both analyses conflict, however, with Murphy's original (1936) statement that euphausids filled their stomachs like those of their tube-nosed neighbors and with Beck's (1970) comment that Snow Petrels rely mainly on krill but feed on fish as well. The differences may reflect seasonal variations, with fish being more important during the summer (Maher, 1962). According to Ainley's (1977) analysis, reliance on fish as the principal food item would be very unlikely in any Antarctic breeder, and especially in one that never leaves the region. This bird seems to represent still another exception in being a strong flier and probably a plunging feeder. Some of its nesting sites have been found as far as 70 mi from the sea (Murphy, 1936), and it is very constant in maintaining a high elevation up to 80 ft while flying over water (Routh, 1949). It has been seen flying silently around a moving ship even in dense fog or on dark nights (Bierman and Voous, 1950). It nests in natural holes and crevices, which it clears of snow by digging in a manner analogous to that of more northerly burrow nesters (Brown, 1966). Chicks defend the nest by ejecting oil up to a meter away. According to Lovenskiold (1960), this oil is not offensive, but Murphy (1936) describes it as having the "usual petrel odor," which is generally characterized as nauseating.

Inspired by such striking olfactory bulbs, Jouventin (1977) conducted semicontrolled tests with two birds collected at Pointe Géologie in Antarctica. Both scored very well in finding the pieces of herring concealed in some of many paper napkins or behind opaque barriers on the floor of a testing room. Performance was equally good in an outside wire enclosure when the fish were buried in the snow. By contrast, a South Polar Skua (*Catharacta maccormicki*) never found any of the buried fish.

Considering the extremely low temperatures in which the Snow Petrel spends its entire life, might its oversized olfactory system be designed to provide sensitivity under conditions that permit very little volatility of odorous molecules? The olfactory systems of the two other procellariiforms that remain south of the Antarctic convergence (Carrick and Ingham, 1970; van Oordt and Kruijt, 1954; G. E. Watson *et al.*, 1971)—the Antarctic Fulmar (*F. glacialoides*) and Antarctic Petrel (*Thalassoica antarctica*)— have not been studied; it is important to know how the size of the bulbs compares with that of *Pagodroma*.

Fisher (1952) reminded us that procellariiforms are the "only group of birds which defends itself by what may be styled organised vomiting" (p. xi). The ejected material is principally stomach oil, but it can also include recently ingested food. With its typically offensive smell, it is assumed to be a useful defense against potential mammalian predators. A smell so revolting to microsmatic humans, however, might repulse such other microsmats as predatory gulls and skuas. In addition, some of the larger procellariiforms, such as the Giant Petrel (*Macronectes giganteus*) prey on smaller species of the same order (Warham, 1962).

Fisher (1952) argued that this waxy oil is also used for preening, in addition to the secretion of the uropygial gland. It flows from the nostrils down the grooves of the upper mandible so that it is spread evenly on feathers as the beak is run against them. He considered tubular nostrils to be an adaptation for such preening, noting that they are found only on birds that produce stomach oil. After an extensive study of Laysan (*D. immutabilis*) and Black-footed Albatrosses, Rice and Kenyon (1962) concluded that their stomach oil serves only as food for chicks. Whatever oil is used for preening, it could conceivably provide a strong odorous cue for individual recognition, a possibility raised by both Warham (1964) and Stager (1967). Murphy (1936) emphasizes the strong musky smell that clings to procellariiform skins for decades even after being washed in water and benzene. Breeding grounds are so permeated by the odor that they can be smelled by sailors many kilometers downwind. Even the eggs are scented with it (Potts, 1872). It is detectible in burrows that have been unoccupied for at least a year. Present information suggests no basis for constant individual differences in the chemical makeup of either stomach oil or preen gland oil. The former appears to be derived from ingested food (Warham, 1977) and hence varies unpredictably. The latter differs in composition among species (Jacob, 1976); possible differences among individual conspecifics have not been examined. A number of procellariiforms engage in courtship and recognition displays that involve different combinations of billing, mutual preening, head nibbling, and similar types of close proximity (Murphy, 1936; Palmer, 1962). In some instances, vocal signals are also exchanged as well as stereotyped motor patterns executed without physical contact between the birds. The fact that sight, sound, and touch may be stimulated to varying degrees at different times in the sequence does not preclude the occurrence of olfactory stimulation as well.

Perhaps because of their size and consequent greater freedom from predation, albatrosses and fulmars are generally diurnal in courtship and breeding behavior, and they nest in open areas. Their courtship displays are predominantly visual and auditory but include such contact features as nibbling the feathers on the mate's head, throat, and neck, and, in the case of the Wandering Albatross (*D. exulans*), touching the tip of the other's bill after inserting the bird's own bill into its own breast feathers (Matthews, 1929). In the case of Laysan and Black-footed Albatrosses, at least, there is evidence that billing and mutual preening are elicited by visual and tactile stimuli (Rice and Kenyon, 1962), which does not preclude the exchange of olfactory information. These species, however, fail to recognize their chicks or their nests if they are moved only 1 or 2 m away; the adults return to the original site and might even begin constructing a new nest. Reliance on olfaction by albatrosses is especially emphasized in feeding. On the basis of many very close observations of the Black-footed Albatross, Miller commented, "I was repeatedly impressed by their seeming acuity of olfactory perception" (1942, p. 8). He based his opinion on such observations as their instant ability to distinguish between an oil film on the ocean surface and a film of bacon fat without touching either one, their investigation of non-phosphorescent items thrown overboard in darkness, and their arrival at a deposit of bacon fat from 20 mi away. The fondness of the albatross for fatty foods has been noted by many observers. Miller (1942) described the effect of bacon grease as creating "almost . . . a frenzy," and Matthews (1929) wrote of the special treat for them in capturing the round plug of blubber cut from a dead whale's tail to make a hole for the tow rope. High fat content is usually associated with fairly strong odor, especially if the fat is warmed.

No amount of correlative data between habits and olfactory tissue can be conclusive without experimental verification of a relationship and unless analogous habits are less common in birds with fewer olfactory cells. Although all seabird orders have not been studied as intensively as procellariiforms, one distinction seems to emerge very clearly from the available evidence. Other birds are predominantly diurnal, whereas most procellariiforms are nocturnal. Furthermore, the habit of nesting in burrows—some of them very deep and long—is largely restricted to petrels, shearwaters, and storm petrels. Other habits of courtship, breeding, feeding, and general social interaction appear to be less distinctive. Many forms feed principally on fish and squid, and some degree of billing with mutual feather stroking and nibbling is engaged in by various species. The experimenter's task is to sort out the relative priorities among the several sensory channels that would appropriately be involved in regulating the many behavior patterns of interest. It should be remembered that every bird seems to possess a functional olfactory system, which can be assumed to contribute something to its experience. The role played by each type of sensory signal is the key question.

Some interesting beginnings have been made in this type of research, all concerned with the behavior of procellariiforms. The returning maneuvers of burrow nesters have been described in varying detail (e.g., Richdale, 1963; Rowan, 1952; Warham, 1955, 1958, 1960), stressing the accuracy and speed of entry, but with no analysis of wind direction at ground level. In a direct study, Leach's Storm Petrels (O. leucorhoa) were found to approach their breeding island in the Bay of Fundy entirely after dark and while flying upwind (Grubb, 1979). This evidence, coupled with their regular arrival in fog and overcast as well as on clear nights, supports olfaction's contribution to guidance over the last portion of the homing flight. The fact that procellariiforms must land into the wind also encourages some use of odors.

The landing behavior of the inhabitants of this small colony was also carefully observed (Grubb, 1974). The typical pattern was for each bird to circle one or more times above the canopy of trees before "crashing through" to land on the ground in the general vicinity of its home burrow. It then shuffled upwind, often swinging its head from side to side with the tip of the bill near the ground. During the observation periods, 29 petrels arrived and entered breeding burrows, 49 entered nonbreeding burrows, and 32 wandered off apparently without entering burrows. Analysis of the birds' directional movements showed that those who entered burrows moved significantly into the wind, whereas those that did not go to burrows moved at random with respect to the wind. There was a strong tendency, however, for all birds to continue moving in one direction once they set out. Many birds, while walking, bumped into trees or into the bank in which burrows were dug. After these collisions, they moved slowly around the tree while seeming to maintain contact with it or, in the case of the bank, swung their heads from side to side and then went directly toward home. There was heavy vocalization among the returning birds above the tree canopy, but only one ever called after landing on the ground, and no calls were made from the burrows that were eventually entered. All these observations suggested very strongly that significant olfactory guidance was involved in at least the final stage of homing at this site.

Grubb then plugged the nostrils of 12 birds recovered from burrows and completely sectioned the olfactory nerve in 23 more. Equal numbers of control birds were also prepared, one group in which the nostrils were probed with the same instrument used to insert the plugs, and one in which the birds were anesthetized and treated surgically in the same way as the experimental birds, except that no nerves were actually cut. Three of the nares-plugged birds returned but without their plugs, and none of the nerve-sectioned birds returned. Only one of the controls for nostril plugging and only two of the sham-operated controls failed to return.

The ability of petrels in this colony to discriminate between their own nest material and litter from the floor of the forest was then tested in a Y-maze. With one of the test materials at each end of the Y-shaped enclosure, the birds showed a significant preference for the nest material. Further tests in which stomach oil and preen gland oil were tested separately with the opposite arm of the maze empty showed no consistent choices. Thus the bases for discriminating the nest material remained unidentified.

An attempt to replicate these results with the Wedge-tailed Shearwater (*P. pacificus chlororhynchus*) on Manana Island off Oahu was less successful in certain respects (Shallenberger, 1973, 1975). Aside from the difference in species, there were fundamental differences between the two environmental situations. On Manana Island, there was no brush or tree cover above the burrows, and the whole area was illuminated by light from the shore installations nearby on Oahu. The birds showed no tendency to move into the wind in walking to their burrows. Applying a masking odor to selected nests had no effect on their behavior. Shallenberger also released some birds after either olfactory nerve section or sham operation. Three of the 14 experimental birds returned, 7 of 10 sham-operated, and 18 of 20 handled but unoperated birds. This outcome was close to that of Grubb's experiment. It should be noted that a similar experiment had been done many years ago with Noddy Terns (*Anous stolidus*) on Dry Tortugas Island (J. B. Watson, 1910). The nostrils of three birds were completely plugged; all returned promptly from both distant and adjacent release points. This result hints that plugged nostrils do not in themselves prevent normal flight.

Shallenberger's data from laboratory experiments were provocative in suggesting an ability to differentiate familiar and significant odors. Using the technique of recording heart and respiratory rates while presenting odors to individual birds, he found that four out of nine shearwaters showed very different changes in respiration to the odor of their own mates compared to that of another bird. It is possible that these four birds differed from the others in either their reliance on olfactory cues or their adaptability to the confining test situation. Incidentally, Shallenberger's results confirmed the earlier observation by Wenzel (1967) that respiratory changes to stimuli are more reliable and systematic than are heart rate changes in the shearwater.

Another way of testing olfactory responsiveness is to release odors at sea and record the appearance of birds in the vicinity. The greater attractiveness of heated fats and oils for Cape Petrels (*Daption capense*), albatrosses, and "Giant Fulmars" was discovered by Beck (Murphy, 1936). This technique was used semisystematically by Miller (1942) with Black-footed Albatrosses, and Kritzler (1948) regularly attracted fulmars by spreading a slick of hot grease. Slicks of vegetable oil have been found to attract storm petrels (*O. tethys*), which approached overwhelmingly from downwind and reached much higher densities than under normal conditions (Crossin, 1969). Grubb (1972) applied the method with a more natural stimulus. Working from a small boat in the Bay of Fundy, he attached a sponge, soaked either in cod liver oil or seawater, to the top of a pole held on a floating base. He reported that the oil sponge was approached from downwind more often than the water-soaked sponge by Wilson's Storm Petrels, Leach's Storm Petrels, and Greater Shearwaters. Other seabirds (*Morus bassanus, L. argentatus, Sterna paradisaea, Fratercula arctica*) in the vicinity showed no systematic responses to either sponge, but only one of each was seen. Sooty Shearwaters failed to show a preference for the cod liver oil odor, but their numbers also were small.

Much more extensive experimentation of this type has been reported by Hutchison and Wenzel (1980) using a variety of odors. These experiments took place over a period of 16 months in the Pacific Ocean approximately 6 mi offshore at 120°–122°W and 35°–36°N. The stimulus materials consisted of the various compounds listed in Table II. They were presented in three different ways, viz., as a surface slick, as a slick contained in a clear plastic enclosure floating on the ocean, and by a saturated wick in air attached to a vertical floating pole. In the case of the surface slicks and the enclosed slicks, puffed cereal was added in some cases to provide a discrete visual stimulus. The enclosed slicks avoided the possibility that any attraction of birds might be attributable to the presence of small marine

Table II. Percent of All Procellariiform and Nonprocellariiform Birds Sighted Approaching from Downwind and Upwind Directions[a]

Stimulus	Sightings P[b]	Sightings NonP[b]	<10 m[c] P	<10 m[c] NonP	Landed/Fed[d] P	Landed/Fed[d] NonP
		From Downwind				
Controls						
Natural ocean	51	48	12	17	8	12
Feeding P	60	50	90	86	83	86
Puffed rice	55	50	49	100	32	100
Surface slicks						
Tuna oil	74	52	88	73	84	12
Bacon fat/ puffed rice	54	50	94	90	94	90
Vegetable oil	61	45	88	59	77	3
Enclosed slicks						
Same time						
Ocean water	68	48	9	21	2	4
Tuna oil			84	21	26	3
Separate						
Ocean water	53	48	10	35	0	8
Tuna oil	75	49	77	39	34	6
Floating wicks						
Ocean water	56	51	12	14	0	0
Motor oil	53	52	12	19	0	2
Mineral oil	51	50	22	30	0	2
Hexane	51	49	0	6	0	0
Tuna oil	66	50	87	12	22	3
Tuna oil fraction	70	50	82	7	24	1
Squid homogenate	60	45	87	24	17	0
Vegetable oil	66	50	79	13	15	0

Table II. (*Continued*)

Stimulus	Sightings		$<10\ m^c$		Landed/Fed[d]	
	P[b]	NonP[b]	P	NonP	P	NonP
			From Upwind			
Controls						
Natural ocean	49	52	13	24	5	16
Feeding P	40	50	87	93	87	93
Puffed rice	45	50	64	100	34	100
Surface slicks						
Tuna oil	26	48	81	68	71	10
Bacon fat/ puffed rice	46	50	100	90	100	90
Vegetable oil	39	55	82	46	71	9
Enclosed slicks						
Same time						
Ocean water	32	52	8	21	2	4
Tuna oil			53	24	8	3
Separate						
Ocean water	47	52	9	36	0	4
Tuna oil	25	51	75	42	5	2
Floating wicks						
Ocean water	44	49	12	14	0	0
Motor oil	47	48	8	12	0	0
Mineral oil	49	50	9	25	0	0
Hexane	49	51	0	3	0	0
Tuna oil	34	50	48	14	3	4
Tuna oil fraction	30	50	40	8	3	1
Squid homogenate	40	55	29	15	1	0
Vegetable oil	34	50	47	12	1	0

[a] Data from Hutchison and Wenzel, 1980.
[b] P, Procellariiform; NonP, nonprocellariiform.
[c] Percentage flew within 10 m of source.
[d] Percentage on surface slicks or adjacent to enclosed slicks or floating wicks.

organisms that had been drawn, in turn, to the surface material. The wick provided the minimum amount of general visual cues and none whatsoever that differed among the various odorous materials. Experimental and control materials were deployed in daylight under all sorts of climatic and ocean conditions, but always when no birds were visible without binoculars. Through the next 1–2 hr, all birds sighted were counted simply as procellariiforms in approaching from the downwind direction and in greater according to an upwind or downwind approach as well as by proximity of approach to the stimulus materials, i.e., whether they flew within 10 m of it or not. In some tests, an experimental and a control material were present simultaneously but separated by 75–100 m; in other tests, they were presented in succession, with order reversed on a subsequent day.

As shown in Table II, procellariiform species far outnumbered non-procellariiforms in approaching from the downwind direction and in greater numbers when food-related odors (e.g., tuna oil, squid homogenate, vegetable oil, and bacon fat) prevailed. Another important difference relates to the effect of weather conditions. Procellariiforms were much more frequent during and immediately following storms, during reduced visibility resulting from low clouds, fog, or overcast, when winds were blowing from the west and northwest with at least force 5, and when the ocean surface presented ground swells of 5 ft or more. It was characteristic of these species to fly in foraging patterns around the food-related stimuli, and in the case of open slicks they sometimes settled on the surface in the immediate vicinity. In contrast, nonprocellariiforms (gulls, terns, pelicans) approached from all directions, under various weather conditions, and showed no particular orientation toward any of the chemical stimuli unless puffed cereal had been added. In such cases, these species arrived in large numbers from all directions, alighted on the water, and ate the cereal. They appeared to be attracted only by the sight of the particles, for they were as likely to approach the cereal alone as when it was associated with a slick. Procellariiforms, by contrast, paid no attention in such cases.

V. GENERAL CONSIDERATIONS

Among birds in general, and certainly among seabirds in particular, wide variation in olfactory equipment is the rule. From the Snow Petrel to the Pelagic Cormorant (*Phalacrocorax pelagicus*), the gamut of structure is so impressive as to imply a comparable range in functional significance. Such diversity in structure is usually taken to indicate adaptive radiations associated with different ecological niches. The special endowment of one order, Procellariiformes, emphasizes the importance of studying the differences in all aspects of biology between its members and those of other marine groups. Although a number of studies have been made of tube-nosed birds, few of the investigators were aware of their almost unique olfactory potential. The observations, therefore, were not always as oriented toward the function of this system as would be desirable. From the summary of many of the reports discussed earlier, it is clear that no definitive evidence yet exists about the nature of adaptive advantages of olfaction for procellariiforms.

For at least a century, scientific writers have been alerting us to the possible importance of smells for tube-nosed forms. Descriptions, sometimes illustrated, of their nasal cavities and olfactory bulbs have appeared sporadically since 1883 (Bang, 1966; Bumm, 1883; Burne, 1908; Klinckowström, 1891; Wood Jones, 1937a). In introducing Godman's *A*

Monograph of the Petrels in 1907, Pycraft wrote, "The great size of the olfactory chamber in the Petrel, has been generally overlooked; but in this particular it is significant to note that they approach the *Cathartidae*" (p. xxi). After reviewing the anatomical literature on avian olfaction and conducting some simple laboratory experiments on olfactory discrimination in a few land birds, Strong (1911) commented on the large size of the fulmar's olfactory lobe and regretted that he had not had time to study its behavior.

Seemingly independently, field workers have occasionally made either definite comments about olfactory reliance or strong suggestions based on behavior difficult to explain in any other way. The Cape Petrel arrives from nowhere to feed on the crustaceans vomited by a dying whale. Vallentin (1924) described their appearance as seeming "to spring from space" around a whaling ship about 10 mi off the Falkland Islands, an appearance all the more surprising to him because he never saw these birds from the islands, where he spent many months. The Antarctic Fulmar resembles the Cape Petrel in its habits, and is often seen with it at sea (Ardley, 1936). It is also described as always among the first arrivals at food (Routh, 1949). Murphy (1936) allowed several procellariiforms a more or less keen sense of smell, including the Giant Petrel, the Cape Petrel, and the White-chinned Petrel (*Procellaria equinoctialis*). Roberts (1940) added Wilson's Storm Petrel, "like other Petrels," to that list. Further implied support comes from the statement that more than 300 gathered extremely rapidly to feed on oily slicks, although there had been very few in the vicinity just before (Beck and Brown, 1972). Miller's (1942) respect for the Black-footed Albatross's prowess has been mentioned above, as has the Broad-billed Prion's mammallike homing behavior (Swales, 1965). Kritzler (1948) said that a fulmar attracted to his 100-yd-wide, half-mile-long slick of warm pork fat coursed "up drift along the slick not unlike a hound on a fresh trail, seeking the source of the fat" (pp. 5–6).

When all the available evidence is collected, it points first to the value of odor sensitivity in locating food, and second to its help in returning home. Both are critical life-support functions and are therefore of great significance in evolutionary selection. In spite of the recurrent reminders, however, almost no attention has been paid to olfaction and adaptation in these avian forms. Bang's (1966) article stands alone in its concern with this problem and in its attempt to promote serious work on it. No one seems to have wondered why diving petrels and auks, well-known instances of convergent evolution, differ so sharply in respect to the olfactory system. If penguins and procellariiforms are as closely related as chemical studies suggest (Baker and Manwell, 1975; Jacob, 1976), why is olfaction so well developed only in the latter?

Many questions could be asked about these and other functions, such as the several behavioral stages of the reproductive cycle. Is it possible that one or more breeding steps could be influenced, if not controlled, by olfactory stimulation, even to the point of triggering gonadal activity, as happens in many mammals with large olfactory structures? Hagen (1952) observed nibbling of the preen gland area, apparently preferentially directed, during mutual preening by a pair of Greater Shearwaters. Tactile releasers are assumed to be critical for nocturnal and cavity-nesting birds and their derivatives (Armstrong, 1951); olfactory modulation or even preeminence has yet to be tested. The inhibition of gametogenesis for 2 yr in albatrosses that have produced a chick in a given season remains a mystery with respect to controlling mechanisms (Tickell, 1970). Territoriality is said to be notably absent among procellariiforms (Rowan, 1965), but it is also characterized as being vigorous in many reports, especially among species with oil-ejecting chicks (e.g., Armstrong, 1951; Pryor, 1968; Travers, 1872). In order to find the most productive questions for analysis and experiment, careful field observations must be made, always bearing in mind the possibility that olfactory perception could be occurring.

To imagine the sensory experiences of a distant species is a most difficult task, full of dangerous traps. Were we the object of close scrutiny by intelligent beings whose biology and language we could not share, our behavior would seldom suggest even awareness of odors. The occasional sniffing of a flower might be obvious, but what would spraying the armpits be taken to signify? Our morphology does not emphasize olfaction as a source of critical information, and yet its total loss is a dismaying experience. For creatures more generously endowed, it must surely speak, with even greater drama and clamorous urgency, of truly momentous events.

REFERENCES

Ainley, D. G., 1977, Feeding methods in seabirds: A comparison of polar and tropical nesting communities in the eastern Pacific Ocean, in: *Adaptations within Antarctic Ecosystems* (G. A. Llano, ed.), pp. 664–685, Smithsonian Institution, Washington, D.C.

Allan, R. G., 1962, The Madeiran Storm Petrel *Oceanodroma castro*, *Ibis* **103b**:274.

Ardley, R. A. B., 1936, The birds of the South Orkney Islands, *Discovery Rep.* **12**:349.

Armstrong, E. A., 1951, Discharge of oily fluid by young fulmars, *Ibis* **93**:245.

Baker, C. M. A., and Manwell, C., 1975, Penguin proteins: Biochemical contributions to classification and natural history, in: *The Biology of Penguins* (B. Stonehouse, ed.), pp. 43–56, Macmillan, London.

Bang, B. G., 1965, Anatomical adaptations for olfaction in the Snow Petrel, *Nature (London)* **205**:513.

Bang, B. G., 1966, The olfactory apparatus of tubenosed birds (Procellariiformes), *Acta Anat.* **65**:391.

Bang, B. G., 1971, Functional anatomy of the olfactory system in 23 orders of birds, *Acta Anat. Suppl.* **58**:1.

Beck, J. R., 1970, Breeding seasons and moult in some smaller Antarctic Petrels, in *Antarctic Ecology*, Vol. 1 (M. W. Holdgate, ed.), pp. 542–550, Academic Press, London.

Beck, J. R., and Brown, D. W., 1971, The breeding biology of the Black-bellied Storm-petrel *Fregetta tropica*, *Ibis* **113**:73.

Beck, J. R., and Brown, D. W., 1972, The biology of Wilson's Storm Petrel, *Oceanites oceanicus* (Kuhl), at Signy Island, South Orkney Islands, *British Antarctic Survey Scientific Reports*, No. 69.

Bierman, W. H., and Voous, K. H., 1950, Birds observed and collected during the whaling expeditions of the "Willem Barendsz" in the Antarctic, 1946–1947 and 1947–1948, *Ardea* **37(suppl)**:1.

Bishop, L. B., 1949, Catching petrels by flashlight, *Condor* **51**:272.

Brown, D. A., 1966, Breeding biology of the Snow Petrel *Pagodroma nivea* (Forster), *ANARE Scientific Reports, Ser. B*(1), No. 89.

Bumm, A., 1883, Das Grosshirn der Vögel, *Z. Wiss, Zool.* **38**:430.

Burne, R. H., 1908, Exhibition of, and remarks upon, preparation of the Olfactory Organs of a Sea-Lamprey (*Petromyzon marinus*), Sea-Bream (*Pagellus centrodontus*), and an Albatross (*Diomedea exulans*), *Proc. Zool. Soc. London* **1**:65.

Carrick, R., and Ingham, S. E., 1970, Ecology and population dynamics of Antarctic sea birds, in *Antarctic Ecology*, Vol. 1 (M. W. Holdgate, ed.), pp. 505–525, Academic Press, London.

Cobb, S., 1960, Observations on the comparative anatomy of the avian brain, *Perspect. Biol. Med.* **3**:383.

Crossin, R. S., 1969, Apparent sense of smell in the family Hydrobatidae, Unpublished data from Pacific Ocean Biological Survey Program, Smithsonian Institution, Washington, D.C.

Fisher, J., 1952, *The Fulmar*, Collins, London.

Grubb, T. C., Jr., 1972, Smell and foraging in shearwaters and petrels, *Nature (London)* **237**:404.

Grubb, T. C., Jr., 1974, Olfactory navigation to the nesting burrow in Leach's Petrel (*Oceanodroma leucorrhoa*), *Anim. Behav.* **22**:192.

Grubb, T. C., Jr., 1979, Olfactory guidance of Leach's Storm Petrel to the breeding island, *Wilson Bull.* **91**:141.

Hagen, Y., 1952, Birds of Tristan de Cunha, in *Results of the Norwegian Scientific Expedition to Tristan da Cunha 1937–1938*, Vol. 3, No. 20 (E. Christophersen, ed.), 248 pp., Norske Videnskaps-Akademi, Oslo.

Harriman, G. E., 1967, Laughing Gull offered saline in preference and survival tests, *Physiol. Zool.* **40**:273.

Harriman, G. E., and Kare, M. H., 1966a, Tolerance for hypertonic saline solutions in Herring Gulls, Starlings, and Purple Grackles, *Physiol. Zool.* **39**:117.

Harriman, G. E., and Kare, M. H., 1966b, Aversion to saline solutions in Starlings, Purple Grackles, and Herring Gulls, *Physiol. Zool.* **39**:123.

Hutchison, L. V., Rausch, L. J., and Wenzel, B. M., 1977, Single unit activity in the olfactory pathway in the pigeon, *Soc. Neurosci. Abstr.* **3**:80.

Hutchison, L. V., and Wenzel, B. M., 1978, Unit activity in the ipsilateral and contralateral olfactory pathway in the pigeon, *Soc. Neurosci. Abstr.* **4**:87.

Hutchison, L. V., and Wenzel, B. M., 1980, Attraction of procellariiforms to food-related odors at sea, *Condor* (in press).

Jacob, J., 1976, Chemotaxonomical relationships between penguins and tubenoses, *Biochem. Syst.* **4**:215.

Jouventin, P., 1977, Olfaction in Snow Petrels, *Condor* **79**:498.

Klinckowström, A., 1891, Les lobes ofactoires du *Fulmarus glacialis, Biol. Foren. Stockholm* **3**:10.

Kritzler, H., 1948, Observations on behavior in captive fulmars, *Condor* **50**:5.

Lockie, J. D., 1952, A comparison of some aspects of the retinae of the Manx Shearwater, Fulmar Petrel, and House Sparrow, *Q. J. Microsc. Sci.* **93**:347.

Lockley, R. M., 1961, *Shearwaters*, Doubleday, Garden City, N.Y.

Lovenskiold, H. L., 1960, The Snow Petrel *Pagodroma nivea* nesting in Dronning Maud Land, *Ibis*, **102**:132.

Macadar, A. W., Rausch, L. J., Wenzel, B. M., and Hutchison, L. V., 1980, Electrophysiology of the olfactory pathway in the pigeon, *J. Comp. Physiol. A*. (in press).

Maher, W. J., 1962, Breeding biology of the Snow Petrel near Cape Hallett, Antarctica, *Condor* **64**:488.

Martin, P. W., 1942, Notes on some pelagic birds of the coast of British Columbia, *Condor* **44**:27.

Matthews, L. H., 1929, The birds of South Georgia, *Discovery Rep.* **1**:561.

Miller, L., 1940, Observations on the Black-footed Albatross, *Condor* **42**:229.

Miller, L., 1942, Some tagging experiments with Black-footed Albatrosses, *Condor* **44**:3.

Murphy, R. C., 1936, *Oceanic Birds of South America*, American Museum of Natural History, New York.

Murphy, R. C., 1964, Systematics and distribution of Antarctic Petrels, in *Biologie Antarctique* (R. Carrick, M. W. Holdgate, and J. Prévost, eds.), pp. 349–358, Hermann, Paris.

Palmer, R. S., (ed.), 1962, *Handbook of North American Birds*, Vol. 1, Yale University Press, New Haven.

Potts, T. H., 1872, On the birds of New Zealand, *Trans. Proc. N.Z. Inst.* **5**:171.

Pryor, M. E., 1968, The avifauna of Haswell Island, Antarctica, in *Antarctic Bird Studies* (O. L. Austin, ed.), *Antarctic Research Series*, Vol. 12, pp. 57–82, American Geophysical Union, Washington, D.C.

Pycraft, W. P., 1907, On the systematic position of the petrels, in: *A Monograph of the Petrels* (F. du C. Godman, author), pp. xv–xxi, Witherby, London.

Rensch, B., and Neunzig, R., 1925, Experimentelle Untersuchungen über den Geschmackssinn der Vögel II, *J. für Ornithol.* **73**:633.

Rice, D. W., and Kenyon, K. W., 1962, Breeding cycles and behavior of Laysan and Black-footed Albatrosses, *Auk* **79**:517.

Richdale, L. E., 1963, Biology of the Sooty Shearwater *Puffinus griseus, Proc. Zool. Soc. London* **141**:1.

Rieke, G. K., and Wenzel, B. M., 1978, Forebrain projections of the pigeon olfactory bulb, *J. Morphol.* **158**:41.

Roberts, B., 1940, The life cycle of Wilson's Petrel *Oceanites oceanicus* (Kuhl), *Br. Graham Land Exped. 1934–37* **1**:141.

Routh, M., 1949, Ornithological observations in the Antarctic seas 1946–47, *Ibis* **91**:577.

Rowan, M. K., 1952, The Greater Shearwater *Puffinus gravis* at its breeding grounds, *Ibis* **94**:97.

Rowan, M. K., 1965, Regulation of sea-bird numbers, *Ibis* **107**:54.

Shallenberger, R. J., 1973, Breeding biology, homing behavior, and communication patterns of the Wedge-tailed Shearwater, *Puffinus pacificus*, Ph.D. dissertation, University of California, Los Angeles.

Shallenberger, R. J., 1975, Olfactory use in the Wedge-tailed Shearwater (*Puffinus pacificus*) on Manana Island, Hawaii, in: *Olfaction and Taste 5* (D. A. Denton and J. P. Coghlan, eds.), pp. 355–360, Academic Press, New York.

Shibuya, T., and Tonosaki, K., 1972, Electrophysiological activities of single olfactory receptor cells in some vertebrates, in: *Olfaction and Taste 4* (D. L. Schneider, ed.), pp. 102–108, Wissenschaftliche Verlagsgesellschaft, Stuttgart.

Shibuya, T., Iijima, M., and Tonosaki, K., 1970, [Responses of the olfactory nerve in the seagull, *Larus crassirotris*], *Zool. Mag. Tokyo* **79**:237.

Stager, K. E., 1967, Avian olfaction, *Am. Zool.* **7**:415.

Strong, R. M., 1911, The sense of smell in birds, *J. Morphol.* **22**:619.

Swales, M. K., 1965, The sea-birds of Gough Island, *Ibis* **107**:215.

Thibault, J.-C., and Holyoak, D. T., 1978, Vocal and olfactory displays in the petrel genera *Bulweria* and *Pterodroma*, *Ardea* **66**:53.

Tickell, W. L. N., 1962, The Dove Prion, *Pachyptila desolata* Gmelin, *Falkland Islands Dependencies Survey Scientific Reports*, No. 33, 55 pp.

Tickell, W. L. N., 1970, Biennial breeding in albatrosses, in: *Antarctic Ecology* Vol. 1 (M. W. Holdgate, ed.), pp. 551–557, Academic Press, London.

Travers, H. H., 1872, On the birds of the Chatham Islands, *Trans. Proc. N.Z. Inst.* **5**:212.

Tucker, D., 1965, Electrophysiological evidence for olfactory function in birds, *Nature (London)* **207**:34.

Vallentin, R., 1924, Zoology, in: *The Falkland Islands* (V. F. Boyson, ed.), pp. 283–382, Clarendon Press, Oxford.

van Oordt, G. J., and Kruijt, J. P., 1954, Birds observed on a voyage in the South Atlantic and Southern Oceans in 1951/1952, *Ardea* **42**:245.

Warham, J., 1955, Observations on the Little Shearwater at the nest, *West. Aust. Nat.* **5**:31.

Warham, J., 1956, The breeding of the Great-winged Petrel *Pterodroma macroptera*, *Ibis* **98**:171.

Warham, J., 1958, The nesting of the Shearwater *Puffinus carneipes*, *Auk* **75**:1.

Warham, J., 1960, Some aspects of breeding behaviour in the Short-tailed Shearwater, *Emu* **60**:75.

Warham, J., 1962, The biology of the Giant Petrel *Macronectes giganteus*, *Auk* **79**:139.

Warham, J., 1964, Breeding behaviour in Procellariiformes, in: *Biologie Antarctique* (R. Carrick, M. W. Holdgate, and J. Prévost, eds.), pp. 389–394, Hermann, Paris.

Warham, J., 1971, Aspects of breeding behaviour in the Royal Penguin *Eudyptes chrysolophus schlegeli*, *Notornis* **18**:91.

Warham, J., 1977, The incidence, functions and ecological significance of petrel stomach oils, *Proc. N.Z. Ecol. Soc.* **24**:84.

Watson, G. E., Angle, J. P., Harper, P. C., Bridge, M. A., Schlatter, R. P., Tickell, W. L. N., Boyd, J. C., and Boyd, M. M., 1971, Birds of the Antarctic and Subantarctic, in: *Antarctic Map Folio Series* (V. C. Bushnell, ed.), Folio 14, pp. 1–8 with 15 plates, American Geophysical Union, Washington, D.C.

Watson, J. B., 1910, Further data on the homing sense of noddy and sooty terns, *Science* **32**:470.

Wenzel, B. M., 1967, Olfactory perception in birds, in: *Olfaction and Taste 2* (T. Hayashi, ed.), pp. 203–217, Pergamon Press, Oxford.

Wenzel, B. M., 1973, Chemoreception, in: *Avian Biology*, Vol. 3 (D. S. Farner, and J. R. King, eds.), pp. 389–415, Academic Press, New York.

Wenzel, B. M., and Sieck, M., 1972, Olfactory perception and bulbar electrical activity in several avian species, *Physiol. Behav.* **9**:287.

Weymouth, R. D., Lasiewski, R. C., and Berger, A. J., 1964, The tongue apparatus in hummingbirds, *Acta Anat.* **58**:252.

Wood Jones, F., 1937a, The olfactory organ of the Tubinares. I., *Emu* **36**:281.

Wood Jones, F., 1937b, The olfactory organ of the Tubinares. III., *Emu* **37**:128

Chapter 3

HABITAT SELECTION AND MARINE BIRDS

Francine G. Buckley and Paul A. Buckley

North Atlantic Regional Office
National Park Service
Boston, Massachusetts 02109

I. INTRODUCTION

Marine birds are those species that depend on marine environments for all or part of their food (Fisher and Lockley, 1954; Lack, 1967). Lack (1968) numbers over 260 species of seabirds, belonging to four taxonomic orders (Sphenisciformes, Procellariiformes, Pelecaniformes, Charadriiformes) embracing 13 families, and including several species that breed inland, or near freshwater and that may also feed on terrestrial organisms for part of the year. Additional but primarily freshwater orders and families include grebes, loons, and ducks, that feed on marine organisms, but they will not be considered here.

Seabirds may be conveniently placed into two categories on the basis of their feeding habitats: inshore or neritic species—such as most gulls, terns, and penguins, which normally feed in waters within sight of land; and offshore or pelagic species—such as albatrosses, tropicbirds, and shearwaters, which normally feed in waters out of sight of land. Some workers (Ashmole, 1971) have separated offshore or pelagic species into two separate categories—offshore species—those which feed out sight of land as far as the continental shelf, although spending night on land; and pelagic—those species that generally feed over deep waters, returning to land only to breed. Factors that influence feeding habitats and ranges are discussed in Chapter 1.

Despite their wide ranges of feeding habitats, all marine birds—even the most highly pelagic species such as kittiwakes, Sooty Terns, Dovekies, albatrosses, and shearwaters—are dependent on terrestrial habitats for their breeding sites, and must return to land even if only to isolated oceanic islands. It is when seabirds have their greatest ties to terrestrial habitats that they may experience their greatest exposure to natural selective factors such as predation, disease, and inter- and intraspecific competition for available food and nest sites.

Selection of a breeding site that provides optimal conditions for the successful production and survival of young is of prime importance to individual and species survival. Although there is no disputing that natural selection encourages mechanisms whereby breeding habitats appropriate to the requirements of a species are selected by individuals, identification of those mechanisms and their operations is another matter. Much work remains to be done along these lines, especially in seabirds, about which even in comparatively well-studied species little is known of the processes used in habitat selection. Their food preferences, colonial mode of nesting, and comparatively long maturation periods make experimental investigation of habitat selection difficult.

Thus, while this chapter deals with the ecological and evolutionary significance of breeding habitat selection by seabirds, it is of necessity almost exclusively an inferential synthesis of descriptive rather than experimental studies.

II. CONCEPTS AND LIMITS

A. The Concept of Habitat Selection

Habitat selection has been defined as "the choice of a type of place in which to live" (Partridge, 1978) and as "the perception of adequate environment" (Miller, 1942). Habitat is used here in a broad sense to embrace those physical and biotic factors that combine to form the characteristics of the nesting sites chosen by seabirds. Thus microclimate, microtopography, substrate, nest-building materials, food resources, and the presence or absence of conspecifics or other species (social factors) are important habitat components.

Most seabirds nest colonially on lands bordering or lying very near the oceans (see Fig. 1) or on islands surrounded by marine waters, but some species (certain gulls and terns, skuas and jaegers) nest inland. Some species may nest either solitarily or colonially (e.g., Black Guillemot), while others are just solitary nesters (e.g., Abbott's Booby).

Despite the fact that numerous seabirds are cosmopolitan and range over broad geographic areas, seemingly limited only by geophysical barriers and their own biological limits to dispersal, they can generally be associated

Fig. 1. Seabird nesting cliffs at St. Paul, Pribilof Islands, Alaska.

with specific types of breeding habitat within their ranges. Examples of this would be the nesting of kittiwakes on cliffs or Cahows in burrows. The diversity of habitats and social possibilities available to seabirds is not selected randomly by individual birds, but is chosen in species-consistent patterns.

The evolutionary mechanisms of habitat selection are believed to be influenced by two types of causative factors—proximate and ultimate (Lack, 1954; Hildén, 1965; Orians, 1971). Ultimate factors, as defined by Immelman (1972) are "environmental factors which in the course of evolution have led, through natural selection, to the relevant restriction" and would include food availability, morphology, protection offered by the habitat from predation and adverse environmental conditions. In sum, those factors dealing with the survival and reproductive success of a species in differing habitats are considered ultimate factors.

Proximate factors, according to Immelman (1972) are "those external stimuli which initiate or maintain biological processes under most favourable ecological conditions." Proximate factors are thus concerned with adaptations of physiology and behavior (Lack, 1954) or the (still virtually unknown) characteristic stimuli of species-specific habitats (Hildén, 1965). Orians (1971) considers proximate factors to be those environmental stimuli that motivate settling behavior. This category includes psychological factors described by Lack (1937) and involves the mental reaction of the bird to its environment. Klopfer and Hailman (1965) recognized two problems in identifying proximate factors—the determination of (1) the portions of the habitat relevant to the animal and involving its sensory modalities and perceptual world (Umwelt), and (2) the mechanisms that establish the preferences or constraints that direct behavior to the relevant portion of the preferred habitat. Hildén (1965) enumerates several proximate factors, including stimuli of landscape; terrain; sites for nest, song, lookout, feeding, and drinking; other animals; internal motivation (contributing to the release of the selection response); and, in some species, food.

The identification of proximate factors used in habitat selection is most easily made by studying correlations between species abundance or population density, and features of the habitats where the greatest numbers are found (Orians, 1971; Partridge, 1978). It is generally accepted that the optimal environment of a species will coincide with the most productive types of its habitat range (Hildén, 1965), and this is taken to mean that the greatest number of individuals within a given species will select that habitat providing the optimum stimuli or conditions (nest sites, food, protection from predators, shelter, social stimulation) for survival and reproduction. Analysis of habitat features at several breeding sites allows identification of those features likely selected for by individuals of a given species.

Buckley and Buckley (1972) examined colony-site characteristics of eight Royal Tern colonies in Virginia and North Carolina in order to determine the topographic features common to all despite variability among them (see Table I). Four topographic features of Royal Tern colony sites emerged that also seemed typical of colony sites elsewhere: (1) complete absence of quadruped predators; (2) general inaccessibility and excellent visibility of surroundings; (3) extensive areas of adjacent shallows; and (4) location at or very near inlets between bay and ocean. Despite the presence in the study area of additional sites having some but not all of the above features, Royal Terns used none of them during the five years of the study, even though colony-site changes are typical of Royal Terns. In some instances certain unused sites had been used previously and were again used in ensuing years, but with periods of nonuse in intervening years. Recent additional support for these features as requisites for Royal Tern colonies came when the first known Royal Tern colony in Puerto Rico was described by Vicente (1979) as also having the same characteristics.

Social stimulation may play a role in site selection when more than one suitable site is available (Klopfer and Hailman, 1965). Laughing Gulls in Virginia and North Carolina were found to favor one salt marsh site over another despite their seemingly identical nature and proximity and previous or subsequent use. It was postulated that the particular choice of one site over another in such situations was attributable to the more or less fortuitous selection of the birds arriving earliest at the colony sites, with later arrivals choosing the site because conspecifics were present. It is, of course, possible that certain features of the habitat perceptible to Laughing Gulls, but not to the investigator, may have been modified from year to year.

Burger and Lesser (1978a,b) examined Common Tern colonies in Ocean County, New Jersey, comparing the 34 occupied colony sites with the 225 islands available. These investigators determined that all islands selected by the terns faced a minimum of 2 mi of open water in at least one direction, and that other important selective factors included vegetation, island size, distance to the nearest island, distance to the nearest shore, and exposure to open water. Predation and tidal flooding were also factors in nest-site selection. According to these criteria, only three suitable islands were unoccupied in 1976. Further investigation in subsequent breeding seasons (1978 and 1979) provided additional support that these factors were being used because the terns still were nesting only on previously used islands and on the three remaining "usable" islands. Only one so-called unsuitable island was used and it was just a few square meters smaller than those considered "suitable" (Burger, personal communication).

B. Variability in Habitat Selection

In certain instances, optimal habitats are unavailable and habitats normally considered suboptimal for a given species then become the preferred or "optimal habitat." Lack (1937) noted that recognition features used in selection of habitat did not necessarily have to be those essential to species existence, and cited examples of atypical nesting habitats that indicated the use of so-called psychological factors, rather than environmental factors, as the limiting restrictions in habitat selection. Miller (1942) stated that mobile higher vertebrates are forced to disperse and pioneer because of increasing population pressures "shouldering them out." He attributed the adoption of new areas (habitats) to selection through instinctive and habitual preferences of the animal. The latter are comparable to the psychological factors of Lack (1937).

The occurrence of numerous seabirds nesting successfully in several different types of habitat provides ample proof that species are able to exploit different habitats, although some may be superior to others in terms of reproductive success. For example, Roseate Terns use three habitats with varying success on Long Island, New York, where they have shown alarming decreases in population between 1974 and 1978 (Buckley and Buckley, 1980). The largest and easternmost colony was on Great Gull Island, where they nested in taller grasses and rock crevices on the island's slopes (Cooper et al., 1970). The island, a sanctuary, also supports a large Common Ternery,

Table I. Characteristics of

Maximum No. pairs (est. or counted)	Size of area of colony location (ha)	Location	Approx. distance and direction to nearest deep water (>2 m) at low tide	Water proximity to colony edges
Fisherman's Island, Va.				
Location A 4500	0.1	Spoil bank	150 m W	Top: 5–6 m N, E, S, W Sides: 1 m E, S
Location B 2000	1	Sandspit	300 m N	6 m E
Location C 250	1	Sandspit	60 m E	6 m E
Location D 2000	2	Sandspit	60 m E	10 m W/10 m E
Wachapreague Inlet, Va. 300	0.3	Shoal/ sandbar	15–30 m N, S 150–300 m E, W	3 m S; 15 m N 30 m E; 30 m W
Oregon Inlet, N.C. 2000	0.4	Spoil bank	60 m N 150 m E, W, S	2 m N, S, W 60 m E
Hatteras Inlet, N.C. 2000	0.4	Spoil bank	90 m W	30 m N, E, W 150 m S
Ocracoke Inlet, N.C. 500	0.1	Spoil bank	15 m N 60 m S	Exact nest sites unknown

[a] From Buckley and Buckley (1972).
[b] Location A, 2nd (main) 1967 site; Location B, 1st 1967 site; Location C, C Colony (1967); Location D, 2nd 1968 site.
[c] Asterisk (*), but to a quadruped-free island.

and is protected from mammalian predators. Great Gull Island has consistently supported the greatest numbers of Roseate Terns on Long Island, although numbers have steadily decreased in recent years. Further west on Long Island, Roseate Terns nest in sandy, grassy areas on or near barrier beaches and islands, preferring denser and taller vegetation than the Common Terns also present at these sites. Roseate numbers have also diminished in this habitat. Their third habitat, the salt marshes at Shinne-

Royal Tern Colony Sites[a-c]

Mean height of colony above m.h.w.	Flooded by storm/ spring tides	Connected to larger land mass?	Substrate	Vegetation	Nesting associaes
3 m	No	No	White sand	Practically none	Oystercatcher Black Skimmer
1.3 m	Yes			Moderate to heavy	Common Tern Gull-billed Tern
±0.3 m	Yes	Yes*	White sand	*Ammophila* patch	Do.; (Laughing Gull)
±0.3 m	Yes	Yes*	White sand	Bare	Do.
±0.3 m	Yes	Yes*	White sand	Bare	Do.
±0.3 m	Yes	No	Gray sand	Bare	Do.
(0.3 m −) 2 m	Partially	No	White sand	Moderate to heavy	Oystercatcher Herring Gull (1 pair) Sandwich Tern (2 pair)
1–5 m	No	No	White sand	Almost none	Gull-billed Tern Sandwich Tern (15–20 pair) Common Tern Black Skimmer Brown Pelican (4 pairs)
±0.3 m	Yes	No	Oyster shells (over sand)	Heavy	Laughing Gull (many) Common Tern (5 pairs)

cock and Moriches Bays where their nests were placed atop *Spartina alterniflora* wrack or tunneled in tall *Spartina* grasses were the least productive because of springtide flooding and rat predation. Many of the Roseate Tern nests were located near rat mounds, where eggs and carcasses were found in greater numbers than were those of the Common Terns also nesting there. It is likely that the Roseate Terns' more obscured nest sites in taller grasses made them more vulnerable to nocturnal rats moving through runways in the vegetation, preventing the incubating birds' successful escape. By 1978 no Roseate Terns were found nesting in the largest marsh colony, Lane's Island, where as many as 561 pairs had nested in 1974.

Common Terns and Black Skimmers also nest both in salt marshes and on open, lightly grassed, sandy beaches in New York (Buckley and Buckley, 1980) and New Jersey (Buckley, 1979; Burger and Lesser, 1978a). Although neither species seems to show any obvious adaptations to salt marsh nesting such as are found in Black and Forster's Terns, Common Terns have nested in salt marshes in New Jersey at least since the early 1800s, when recorded by Alexander Wilson. Flooding from storm and high tides in 1977 caused great losses of chicks and eggs in New Jersey and, in other years, on Long Island, New York. Nonetheless, an average of about 20% of the 12,000–14,000 pairs breeding on Long Island between 1974 and 1978 were marsh nesters, indicating if not some success, at least great persistence (Buckley and Buckley, 1980). Their ability to re-lay may be one adaptation to nesting in low-lying areas, providing at least some adaptation to salt marsh nesting.

III. ECOLOGICAL CONSIDERATIONS

A. Competition

Competition, both intra- and interspecific, plays and important role in the habitat range of a species (Svardson, 1949; Hildén, 1965; Cody, 1974; Partridge, 1978). Intraspecific competition often leads to the occupation of a wide range of habitats. In populations with low density there is decreased competition for available optimum habitat and there is little competitive pressure to exploit less than optimal habitat, assuming its local availability. For this reason, birds are generally *stenotopic* (restricted to a single habitat) at the edges of their geographic distribution (Hildén, 1965). Conversely, in areas of high population density there is greater competition for available optimal habitat, resulting in the use of less preferred, or suboptimal, habitat types. Species that exploit a wide range of habitats are called *eurytopic*; birds are generally eurytopic in the centers of their breeding ranges (Hildén, 1965).

Examples of the ability of seabirds to nest in diverse habitats are most readily found in species that have shown rapid population increases and expansion and expansion into new nesting habitats (Kadlec and Drury, 1968; Nisbet, 1978; Erwin, 1979b). Bent (1921) describes the Herring Gull as breeding as far south as Maine on the U.S. Atlantic Coast, nesting on cliffs, on the ground, in grassy fields, sand, gravel or rocks, and in trees. By 1946 (Buckley, in press) they were nesting as far south as New Jersey, and today they nest south to South Carolina. Drury (1973) suggests that the New England populations are growing more slowly in recent years compared to the exponential growth rates of the 1930s to 1950. Formerly nesting on coastal islands, they are now commonly found in other habitats, especially in New York, where they nest on grassy median strips along highways, atop abandoned buildings and on the ground of rocky islands in lower New York Bay and the East River, on sandy dredge spoil from Jamaica Bay to Montauk, on sand cliffs on Gardiner's and Plum Islands, and in salt marshes along almost the entire south shore of Long Island (Buckley and Buckley, 1980). In New Jersey, they also occupy a wide range of nesting habitats (Buckley, 1978; Burger, 1977; Kane and Farrar, 1977), nesting on Osprey platforms, on diked dredge spoil islands, on sandy beaches with little vegetation, on grassy islands in tall grasses to the bases of dense bushes and on *Spartina* mats in salt marshes, where they were competing with Laughing Gulls (Burger and Shisler, 1978*b*) and with Common Terns (Burger and Lesser, 1978*b*).

Svardson (1949) discussed the effects of high population densities on Black-headed Gulls in Finland where their normal habitat was in *Carex* vegetation on eutrophic lakes. During the 1920s when its population expanded, the species was also found nesting on open bare rocky islets in the marine archipelago, but when the population decreased, these colonies were the first to be abandoned.

Normally cliff-breeding Black-legged Kittiwakes in Britain, Ireland, and Scandinavia are currently increasing in numbers as well, and are now occupying new habitats (Coulson, in Cramp et al., 1974). Colonies have been found at lower cliff elevations (under 30 ft) than previously, and some birds are now nesting on building ledges. In two areas where they nest on buildings (Newcastle and Gateshead), the birds are 11 mi from the sea, and Coulson (op. cit) reports their occasional feeding on freshwater fish, an extraordinary change. The occurrence of a breeding colony on sand dunes in Denmark and sand dune breeding attempts in Britain, are also extraordinary for a species that is highly adapted to cliff nesting (Cullen, 1957).

Despite an exploding population of Northern Fulmars in Britain and Ireland (an estimated 7% increase per year and a 280% increase between 1949 and 1970), the fulmar seems to be retaining its preferred cliff habitat. However, they are nesting on the ground in more isolated areas such as the Outer Hebrides, and they have been noted around artificial cliffs, building ruins and wall tops elsewhere (Cramp et al., 1974; Nørrevang, 1960). It is possible that the already increasing fulmar populations, in addition to intraspecific competition, have prevented the more recently increasing kittiwake populations from nesting in the latter's preferred habitat of high cliffs. The increases in kittiwake populations are attributed to greater protection and reduced human predation, while the fulmar population increases have been tied to increases in food resulting from human fishing procedures (Cramp *et al.*, 1974).

Intraspecific competition for limited nest sites in a large population of Red-billed Tropicbirds on Daphne Island in the Galapagos (Murphy, 1936; Snow, 1965) has been held responsible for a 12-month breeding season and only 32% breeding success. Snow (1965) contrasted this situation with that of South Plaza Island (16 mi away) where, with less competition for nest sites, breeding success was 55% and there was a discrete annual breeding season. Stonehouse (1962*a,b*) also noted intraspecific nest-site competition in both Red-billed and White-tailed Tropicbirds on Boatswain Bird Island off Ascension Island in the South Atlantic Ocean. Here, they nested on the slopes and cliffs of the island in shallow cavities, rock crevices, under boulders, and, in the case of the Red-billed Tropicbirds, also in burrows. Both species showed a continuous breeding season with intra- and interspecific competition affecting nesting success.

Despite the ability to re-lay after loss of an egg or a chick, most unsuccessful breeders were displaced from their nest sites (Stonehouse, 1962*b*). Competition for nest and resting sites was also cited by Stonehouse as being the probable basis for the continuous breeding season, rather than seasonal influences on the breeding cycle or seasonal fluctuations in food supply. He observed that in most other localities within the breeding ranges of both species, breeding is restricted to discrete 3–6–mo spring or summer seasons.

Further work on Red-billed Tropicbirds by Harris (1969) on Tower Island in the Galapagos showed that food supplies were important in the timing and synchrony of breeding and production of young. He noted that between 1965 and 1967 the Plaza Island young were healthy when other Galapagos populations seemed to be experiencing a food shortage. The Tower Island birds laid eggs during most months of the year, but there was a peak between June and August 1966, when food supplies were more readily available. Harris did not discuss intraspecific competition on Tower Island, and he attributed breeding success and timing to food availability. However, on South Plaza Island, which had a lower population density than that of Daphne Island, Harris did note egg loss caused by intraspecific fighting in addition to an overall nesting success rate of 44%, which was lower than Snow (1965) had found on South Plaza. Additional work on tropicbirds (Schreiber and Ashmole, 1970; Diamond, 1975) has shown that the three species (Red-billed, White-tailed, and Red-tailed) nest both annually (or seasonally) and continuously (or nonseasonally) in different locations. Diamond (1975), working with the latter two species that breed year round on Aldabra Island off the East Coast of Africa, found that "where nest sites are scarce and competition for them is intense, breeding may be continuous with individuals returning to breed not annually but at intervals corresponding to the length of the individual sexual cycle, i.e., they breed as often as possible." This also suggests that despite population pressures, the birds are unable to move into different habitats, either because of nonadaptation or nonavailability, and that they were utilizing the available suitable habitat to the fullest extent possible.

The competitive exclusion principle, or Gause's Rule (Kendeigh, 1961), states that "an ecological niche cannot be simultaneously and completely occupied by stabilized populations of more than one species." Seabirds nesting in mixed-species colonies of either closely related or unrelated species may often find themselves competing for available nesting habitat or sites, given the finite number of suitable breeding locations available in many seabird breeding ranges and the overlapping of these ranges. In areas where species with similar habitat requirements overlap, each will show greater stenotopy toward the species-specific optimal conditions, given a limited availability of the habitat in question. If there is a superabundance of the habitat, competition will be reduced allowing greater generalization, or eurytopy. Thus intense interspecific competition tends to promote specialization or adaptation to a narrower habitat range where one species is better suited to survive and reproduce than others, tending toward the eventual exclusion of the less suitably adapted species from that habitat. Lack (1966) attributes avoidance of such competition between species to the evolution of habitat selection.

Generally, colonially nesting seabirds obtain food some distance away from their nest sites and have feeding habitat specializations exhibiting temporal, spatial, size, and species differences (Fisher and Lockley, 1954; Bourne, 1963; Ashmole, 1963, 1968; Ashmole and Ashmole, 1967; Lack, 1968; Cody, 1973), which act to reduce competition among them for food resources on their breeding grounds. Similar habitat partitioning occurs among seabirds in mixed-species colonies in the selection of nest sites.

The partitioning of the seabird cliffs or bazaars according to species-specific habitat preferences in northern seas presents a dramatic example of the possible results of habitat selection in response to interspecific competition. Fisher and Lockley (1954) present an excellent description of the vertical zonation of nest sites on the seabird cliffs of the eastern North Atlantic:

> Whether the rocks be volcanic or intrusive or extrusive or sedimentary, we are sure to find the *Larus* gulls breeding on the more level ground a little way back from the tops of the cliffs—fulmars on the steeply sloping turf and among the broken rocks at the cliff edge, puffins with their burrows honeycombing the soil wherever this is exposed at the edge of a cliff or a cliff buttress, Manx shearwaters or Leach's petrels in long burrows, storm-petrels in short burrows and rock crevices, razorbills in cracks and crannies and on sheltered ledges, shags in shadowy pockets and small caves and hollowed-out ledges dotted about the cliff, kittiwakes on tiny steps or finger-holds improved and enlarged by the mud construction of their nests, tysties or black guillemots in talus and boulders at the foot of the cliff.

The above description is a generalized account. Of course, not all species are present at all sites, and there is some variation in individual species' nest sites depending on local conditions. Figure 2 from Sowls *et al.* (1978) presents a generalized distribution of species on seabird cliffs in Alaska similar to that described by Fisher and Lockley (1954) and Lack (1968) on Bear Island, although species composition differs in each area, comparable zonation occurs. Variation in species nest-site locations according to local conditions is exemplified by the findings of several workers in different locales, e.g., Tuck (1960), who noted that gannets and murres nested close together on Funk Island, Newfoundland, and that kittiwakes and Thick-billed Murres nested on the same ledges at Cape Hay and in colonies of the Barents Sea (Uspenski, 1958; Belopol'skii, 1961).

Investigations of the Novaya Zemlya coast on the Barents Sea by

Fig. 2. Seabird cliff habitat partitioning by colonial seabirds in Alaska. [From Sowls *et al.* (1978).]

Fig. 3. Seabird cliff bare-ledge nesting sites of Thick-billed Murres at St. Paul, Pribilof Islands, Alaska.

Uspenski (1968) led to his conclusion that a shortage of suitable nest sites was preventing the expansion of the bird bazaars despite seemingly abundant and underexploited food resources, especially for murres. Interspecific competition for nest sites does exist in this region, among the species described below. Thick-billed Murres are the most numerous birds of the Novaya Zemlya bazaars. They spend most of their life in the water, except to nest. They nest on bare rock ledges that vary in width from 10 cm to 10 +m, with most nest ledges unevenly surfaced and sloping (see Fig.3). No nests are built, although placement of small stones or pebbles beneath an egg does occur. Closely related Common Murres occur in much smaller numbers in this area. They nest on the wider, longer ledges in small, more densely packed colonies in the bazaars dominated by Thick-billed Murres. Kittiwakes selected the same kinds of ledges on the steep coastal cliffs in this area as the murres did (Uspenski 1958). However, they also nested in niches, small caves, and crevices in the rocks, sites not generally chosen by the murres. In some instances, the kittiwakes nested on narrow projections on vertical cliffs, considered precarious by Uspenski (1958) because nests at these sites were blown down by wind or were washed away by rains.

Common Murres have more specialized feeding habits than do Thick-billed Murres, and Novaya Zemlya is at the northern end of their range.

Uspenski attributed the low numbers here of the former to their "ecological complexities" limiting expansion. He did not consider Common Murres a serious competitor of Thick-billed Murres in that region at that time. However, in the Murmansk region, further south, Common and Thick-billed Murres did compete for nest sites, and Common Murres were more numerous there than were Thick-billed Murres. Belopol'skii (1961) states that the central, broad, comfortable ledges were occupied by Common Murres, while Thick-billed Murres were pushed to the edges and upper ledges. Common Murres are larger and stronger than the Thick-billed Murres, which gives them a competitive edge, as evidenced by the position occupied by their nest sites on the Novaya Zemlya cliffs, reported by Belopol'skii (1961) as the most suitable sites interspersed among Thick-billed Murres, which also occupied the large, wide, open ledges there. It is interesting to note that where the Common Murres occurred in greater numbers (Murmansk region), the Thick-billed Murres occupied narrower cliff terraces, sometimes peat covered, or edges of large, open or semiopen ledges (see Fig. 4). Uspenski (1958) suggested that the peat-covered ledges were unfavorable for the incubation of the porous-shelled Thick-billed Murre eggs, and that there was probably greater egg loss from nest sites closer to the cliff edges. However, similar situations described by Tuck (1960) in New-

Fig. 4. Thick-billed Murre nesting sites on vegetated cliff face at St. Paul, Pribilof Islands, Alaska.

foundland, where both species also occurred together, indicated that Thick-billed Murres were able to nest successfully under these conditions and were also able to nest successfully in a wider range of habitats than was true of Common Murres. Tuck also suggested that the Newfoundland populations on Green and Funk Islands showed a differential selection of nest-site habitat, thereby reducing competition between them.

Belopol'skii (1961) believed that competition between the two species of murres and kittiwakes for ledge-nesting sites as a result of increasing populations of murres at the Seven Islands Sanctuary, in the Barents Sea, resulted in (1) a decrease in the kittiwake population nesting in the bird bazaars, and (2) the establishment of new kittiwake colonies at some locations. Occupation of the kittiwake nests by both species of murres and the subsequent crushing of kittiwake eggs because of nest-site competition was described in detail by Belopol'skii: the kittiwakes were driven off and the murres laid their own eggs in the kittiwake nests, which were not maintained by the murres, and so eventually disintegrated, leaving the murre chicks on their usual bare ledge. Uspenski (1958) stated that only Thick-billed Murres occupied kittiwake nest sites, whereas Belopol'skii observed both species of murres displacing kittiwakes.

In periods of increasing population, Belopol'skii also noted the displacement of Razor-billed Auks and Atlantic Puffins from their nest sites, and the occupation of these sites by murres, which normally did not nest under the rocks and in tunnels used by the auks and puffins. In 1950, after a substantial increase in the Common Murre population, Thick-billed Murres, forced to the periphery of the murre colonies, had pushed the Razor-billed Auks out from their nest sites under rock ledges. The Razor-billed Auks, in turn, laid their eggs at puffin burrows that had been destroyed by Common Murres, digging up the soil. The puffins, crowded out by the Razorbills and the destruction of their nest holes by the murres, dug nest holes at new sites, higher than usual and closer to the cliff tops and potential predation.

Northern Gannets, the largest seabird nesting on northern seacliffs apparently are able to overcome any competition for suitable nest sites. Nelson (1978) noted that they are successful in competition with kittiwakes, fulmars, Herring and Lesser Black-backed Gulls, guillemots (murres), shags, and cormorants. Fulmar is the only species apparently able to oppose Northern Gannet on the ledges, but the fulmars' more frequent absences from the nest site and the gannets' greater persistence eventually results in the occupation of the site by the gannets.

Further examples of the tendency toward specialization in habitat preference in response to direct competition, while retaining the ability to adapt to differing local conditions through the use of a wider range of habitats, is illustrated by the three tropical sympatric sulids—Masked,

Brown, and Red-footed Boobies. Nelson (1978) attributed the Red-footed Booby's nesting in arboreal habitats, at varying heights in low shrubs and tall trees, to the evolution of its body proportions (short legs and small, light body) associated with its specialized fishing techniques of catching its prey above or at the water surface. Use of arboreal habitat allows this species to avoid nest-site competition with the other two species. Masked Booby and Brown Booby both nest in a wide range of ground habitats, and although both species nest on cliffs, slopes, flat ground, and on and among vegetation, Brown Booby shows a greater preference for steep sites and has greater ease of movement through and under vegetation. Nelson (1978) noted that generally there is not an absolute shortage of nest sites for either species and that their habitat preferences become important only where there is a scarcity, such as on Boatswain Bird Island, where each shows strong species-specific habitat preferences.

Further adaptations to interspecific competition for nesting sites might involve responses other than the narrowing of a habitat preference. Lockley (1974) discusses the use of the same breeding burrows at different seasons by related shearwater species. In the Salvage Islands, north of the Canary Islands, three small tube-noses—Little Shearwater, White-faced Storm Petrel, and Hartcourt's Storm Petrel—occupy the same burrows (cf. Bannerman, 1963), but each species lays its eggs at different times so that each occupies the burrow for 4 months.

Despite numerous authors attributing such adaptations as consecutive nest-site use by several species and the lack of defined nesting seasons as found in tropicbirds to interspecific competition for nest sites, other investigators (Bourne, 1955; Ashmole, 1963; Lack, 1968; and Harris, 1969) ascribe such adaptations to competition for food during the breeding season. Because food availability is an important aspect of the breeding habitat, it probably does play a more important role in areas where food resources are not as abundant as in the Barents Sea and North Atlantic.

B. Food as a Factor in Breeding Habitat Selection

Seabirds, dependent for the most part on mobile and, in some cases, far-ranging food sources, may face special problems in obtaining appropriate nest sites within range of suitable food supplies. Orians (1971) states that "for any given level of food availability there will be a point at which the flying time and energy expended in the flight will exactly offset the energy realizable by foraging at that distance from the nest. . . . If food levels are so low that foraging must extend beyond that radius in order for the young to be nourished, [successful] breeding will not be possible at that locality." Therefore, seabirds must find nest sites that provide suitable conditions for their successful breeding close enough to food sources abundant enough to feed themselves and their young without exceeding their own physical and biological capabilities. Such sites are scarce (Nettleship, 1977), and this results in a nonrandom species distribution with breeding seabird concentrations at suitable locations in large single- or mixed-species colonies.

Nettleship and Gaston (1978), for example, found the largest concentrations of breeding seabirds near inlets and the mouths of large bays on the coastline of the Barrow Straits, Lancaster Sound region of Canada. They attributed this high concentration to the more abundant food resources of these areas, particularly the large numbers of Arctic Cod that occur in these waters. Cod constitute a major portion of the diet of most of the breeding seabirds (fulmars, kittiwakes, and Thick-billed Murres) in this area and are found in the low saline, shallow waters of certain inlets and bays during the summer.

Five years of helicopter censusing of Long Island colonial waterbirds by the authors revealed that the largest Common/Roseate Tern colonies on the ocean or south shore were always located at four major undeveloped inlets between bays and ocean, and that year-to-year variation in colony sizes at the inlets was not great. Assuming that the inlets facilitated concentration of the terns' major food (*Menidia*, *Ammodytes*, and other small fishes) and that the tidal prism at each inlet (the amount of water passing through the inlet throat each tidal cycle) was a good measure of an inlet's supply of prey fishes, we used regression analysis to see if colony size could be related to tidal prisms. Although the resulting r^2 for the combined 5-yr data was only 0.60, that value was very highly significant (Fig. 5). Further evidence that the relationship was a real one was provided by separate regression lines run for each of the 5 years. Three of these lines yielded r^2 values varying between 0.86 and 0.98, and although the r^2 values for the two remaining years were not significant, certain events probably account for the breakdown in relationship those years [rat predation and human disturbance (Buckley and Buckley, 1980)]. Therefore, it would seem reasonable to conclude that the inlets were providing a major source of the terns' food and that colony size each year at each inlet was related to tidal prism. The unexplained variance about the 5-yr line and the failure to demonstrate a relationship between colony sizes and tidal prisms in two of the individual years could be explained by annual fish-population fluctuations in the inlets, as well as by predation, disturbance, and weather factors. Efforts to quantify fish populations at each of the inlets were unsuccessful, as there were no commercial fisheries tied to the inlets, and local sport fisheries generally kept unreliable or no data. Moreover, dichotomization of all 1977 terneries into inlet and noninlet locations revealed that inlet

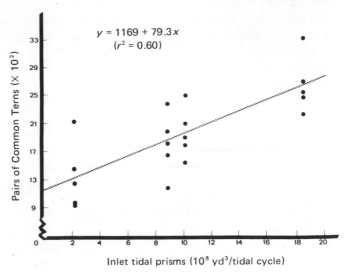

$$y = 1169 + 79.3x$$
$$(r^2 = 0.60)$$

Fig. 5. Regression analysis of tern population size on inlet tidal prisms. Tern numbers from 1974 to 1978 at the four major undeveloped inlets on the south (ocean) shore of Long Island, N.Y., which are, reading left to right on the scattergram: Shinnecock, Moriches, Jones, and Fire Island. [From Buckley and Buckley (1980).]

colonies were significantly larger than were noninlet colonies, reinforcing the importance of these inlets to terns in the selection of their major colony sites.

Differential nest-site selection in six species of alcids, such as that found by Lack in Iceland (1934), and on Bear Island (1968), usually attributed to interspecific competition for nest sites, has been reinterpreted by Cody (1973, 1974) following his work on alcid communities off the coast of Washington, as more properly influenced by interspecific competition for feeding areas at sea and predation on both adult and chick. Cody cited the inshore feeders' open, exposed nest sites, with good sea access, which requires the presence of a parent to guard chicks from gull predation, as being economically feasible because of the proximity to the food supply. Cody also believed that the inshore fishing species must be the least selective as to nest site in order to utilize all available inshore waters, with no increase in flight distances. The larger size of these species was also linked to their proximity to food sources and successful deterrence of predatory gulls. Offshore feeders, according to Cody (1974), could not economically afford to have one parent guard the chick; therefore, chicks must be hidden in deep crevices or burrows so that both parents could forage at the same time. Thus, their colony-site selection would be more discriminating because of their more rigorous nest-site requirements and because it is not as essential for them to be as close to their food source offshore, since they are able to forage longer distances. Bédard (1976) has criticized Cody's data and interpretation, citing Cody's small data base and the possibly inappropriate extension of observations on only a small portion of the large North Pacific seabird population and Cody's comparison of his sample with the North Atlantic seabird population. Bédard (1976) gives a complete discussion of these criticisms, which are beyond the purview of this chapter.

Most authors do agree that, in general, food availability plays a major role in habitat selection. However, while food availability may be more important than nest-site competition in less rich tropical waters (Ashmole, 1963; Lack, 1968), the relative importance of competition for food and nest sites in temperate and polar regions, where food is more abundant, remains to be unraveled. Most workers to date attribute at least a major role to optimal nesting habitat availability (Fisher and Lockley, 1954; Salomonsen, 1955; Robertson, 1964). All factors must be weighed in determining the selective pressures involved in habitat selection for each species, as local conditions may vary from year to year, so that different selective factors will assume greater significance under changing circumstances.

C. Predation

One of the ultimate factors involved in habitat selection is the protection that habitat offers from predators. Lack (1968) divided seabirds into two groups on the basis of a combination of characteristics concerned with degree of development of the young at hatching, feeding zone, and behavioral and ecological adaptations to predation. Members of his first group typically breed in large colonies at relatively safe sites, generally do not defend their nests against predators, and feed offshore. Included are birds in the orders Sphenisciformes, Procellariiformes, Pelecaniformes, and Charadriiformes (in the family Alcidae). Members of his second group nest in more accessible sites, defend their nest with relatively greater aggression, generally have cryptically colored eggs and young, and generally feed inshore. Included are members of the Laridae, Rynchopidae, and Stercorariidae in the Charadriiformes.

There is some overlap in the characterizations of the two groups, and the second group includes some major predators of the first group. Many seabirds will not nest at sites with quadruped mammalian predators, and it is assumed that there has been selective pressure from both aerial and terrestrial predation, resulting in ecological and behavioral adaptations such as nesting in inaccessible sites (trees, steep cliffs) and colonial nesting.

Smith (1966) called attention to the role that accessibility to Arctic Fox predation played in the nest-site distribution of three species of gulls on Baffin Island. He determined that the lower, smaller cliff ledges were suboptimal nest sites because Arctic Foxes were able to climb the talus slopes and destroy eggs and nests located there. Normally, the larger, higher, and thereby safer, ledges were occupied by Glaucous Gulls, with either Kumlien's or Thayer's Gulls below them on lower ledges. When the Glaucous males defending upper ledges were removed early in the season, Kumlien's or Thayer's males from the lower ledges usually moved up and occupied the preferred upper ledges.

Extensive adaptations to nesting on tiny cliff ledges by Black-legged Kittiwakes, as a response to ground and aerial predation, have been well documented by Cullen (1957), who credits kittiwake nesting on small perches on steep cliffs to the comparative safety of these sites from large gull predation.

Ground-nesting species may be somewhat more vulnerable to predation, and it is among these species that spacing of nests, the presence or absence of vegetation to conceal eggs, chicks, or both, and suitable antipredator behavior are best developed. Black-headed Gulls in a British colony that was preyed upon by foxes, hedgehogs, Carrion Crows, Herring Gulls, Lesser Black-backed Gulls, and other Black-headed Gulls showed a marked preference for nest sites in areas already densely occupied. Birds nesting in the denser parts of the colony were found to have higher success rates than did those outside of the colony or on the periphery (Patterson, 1965). Clearly, the nest sites in the denser, central portions of the colony were to be preferred as being presumably safer from predation because of both the increased aggressive encounters a predator would have compared to the outlying, less densely settled areas, and the comparative ease of reaching a nest on the edge rather than the middle of the colony (cf. Tinbergen, 1967).

Ground-nesting Royal Terns were found to be nesting as densely as possible (see Fig. 6), their nests being hexagonally packed in portions of their colonies in Virginia and North Carolina (Buckley and Buckley, 1977a). Their high nesting densities were believed to result from (1) their habit of selecting colony sites on quadruped-free, small bare sand islands with limited areas suitable for nest sites above mean high water, and (2) extreme egg predation by Laughing Gulls on the periphery of their colonies (Buckley and Buckley, 1972). Royal Terns also show the same colony-site desertion proclivities that Cullen (1960) described for Sandwich Terns and Caspian Terns disturbed early in the nesting season, which is also believed to be an antipredator response.

Further evidence of the role that predators may play in the selection of habitat exists in the Ross's Gulls recently found nesting on bare gravel on low stony reefs near polynias in the Queen Elizabeth Islands in N.W.T. Canada (MacDonald, 1979). In Siberia, Ross's Gull is reported to usually nest on tundra in grass or on grass tussocks, building nests of grasses, moss,

Fig. 6. Densely nesting Royal Terns at Cape Charles, Virginia.

and leaves (Dement'ev *et al.*, 1969). In Canada Ross's Gulls nested on bare gravel with little or no vegetation for chick concealment. Polar Bears preyed upon the chicks in Canada in 1978 (MacDonald, 1979) causing 100% chick mortality, so the return of the colonizing gulls would seem doubtful.

Southern and Southern (1979) have noted that most so-called anti-predator adaptations in gulls are, in fact, anti-*diurnal*-predator responses, and that several species of gulls evidence no response to nocturnal predators beyond flight or colony abandonment. They suggest that gulls' only anti-*nocturnal*-predator adaptation may be selection of breeding sites inaccessible to such predators, a factor already identified by Buckley and Buckley (1972) as critical in Royal Terns' choice of nesting colony islands (see Table I). This may be widespread among colonially nesting marine birds and deserves further investigation.

Differences in nesting habitat can play a role in providing shelter for predators, therby rendering aggressive protection of their nests by marine birds effective in one habitat, ineffective in another. Lemmetyinen (1971), contrasting the flat, treeless, grassy/rocky/sandy coastal North Sea and southern Swedish Arctic and Common Tern colonies with those on islets in the Finnish Archipelago, found that the presence of trees and shrubs on the islets provided shelter to predators, resulting in high mortality rates on tern broods. He pointed out that mink, unable to withstand tern attacks in open terrain, were effective predators protected from aerial attack where they were able to move under cover of bushes. Crow predation in Finland was similarly effective, because they could seek shelter from attack in trees and shrubs. Lemmetyinen also noted that Finnish terneries were much smaller than North Sea Coast colonies, thereby losing the advantage of nesting in dense, large colonies where predation loss is greater at the edges than at the center (Kruuk, 1964).

Burrow-nesting seabirds are especially vulnerable to predation once their burrows have been located by predators. The Bermuda Petrel or Cahow, thought to be extinct after the 1600s, was rediscovered in 1951 nesting in small numbers on offshore rocky islets in Bermuda. Formerly nesting in burrows in sand dunes, they were found nesting in burrows in soil or decomposed rock (Greenway, 1958; Wingate, 1977). Unfortunately, rats were able to reach their new locations and caused serious population losses. Elimination of rats from these islets by use of anticoagulants (warfarin) resolved the rat-predation problem, but a scarcity of suitable nesting habitat and direct nest-site competition from more agrressive White-tailed Tropicbirds has caused other problems for the Cahows (cf. Wingate, 1977).

Thus, seabird predators clearly can influence the selection of nest sites and even of colony sites by their presence or absence at a particular location. Their actions may cause desertion and subsequent disuse of a successful nesting location. Many seabirds have few or no defenses against a predator on land, depending instead on inaccessible or hidden nest sites. Others have specialized nesting habits that reduce predation, some show great aggressiveness toward predators, while still others just abandon their nests and move to safer locations.

D. Kleptoparasitism

Kleptoparasitism is a parasitic form of predation—the stealing of food from other birds—and is prevalent in several families of seabirds (Fregatidae, Stercorariidae, Laridae) whose members are generally excellent fliers who often have structural and behavioral adaptations for their mode of feeding. Many are also true predators of eggs and chicks, including conspecifics and even smaller adult birds. Both pirate and victim may develop habitat preferences resulting from their interrelationship.

Nettleship (1972) has shown that Common Puffins nesting at Great Island, Newfoundland, on both the steep slopes and level ground at the cliff tops, preferred the steep-slope nest sites; he attributed this preference to their relative breeding success in each habitat in the presence of heavy Herring and Great Black-backed Gull kleptoparasitism of fish, which the puffins were bringing to their young. Nearby islands with low gull populations did not exhibit the differential reproductive success between habitats found on Great Island. Birds nesting on level ground had to travel further to the cliff edge in order to become airborne and, on returning, took longer to reach their burrows than did those on the slopes, thereby increasing the opportunity for gull harassment and kleptoparasitism. Puffins nesting on the slopes were able to return to their burrows more rapidly and directly, landing at them immediately on their approach from the sea. Direct predation by gulls upon eggs and chicks was also greater on puffins in level habitats where nests were unguarded for longer periods of time, due also in part to the puffins' increased nervousness and more frequent panics (the mass departure of birds from their nests) that Nettleship observed there. Examination of chick growth rates revealed that those on level ground had lower prefledging body weights, probably the result of a lower feeding rate because of greater kleptoparasitism and interference with fish-laden returning adults. Higher nest density on the slope habitat was attributed to preference for that habitat following successful breeding there and to the higher numbers of slope-reared young returning to breed in their natal habitat. The fact that both habitats were occupied is indicative of a very localized philopatry as well as overflow from slope-reared birds, unsuccessful in obtaining a preferred slope nest site. At colonies where gull predation, kleptoparasitism, and interference were not responsible for different reproductive success rates, preferential selection of one habitat over the other would not be as critical, and other factors might assume greater importance in site-selection processes.

Skuas and jaegers are generalized feeders able to obtain their own prey, but they are serious aerial predators of many seabirds, and are kleptoparasites as well. Their great agility in the air often gives them maximal advantage in harassing other seabirds until they drop or regurgitate their prey, which is then caught by the skuas. Arctic Skua–Common Puffin interactions along these lines have been described by Grant (1971) and by Arnason and Grant (1978) in great detail. In a predator–prey evolutionary contest of classic dimensions, measure and countermeasure have developed in predator and prey, respectively, in the one case to practice, and in the other to avoid, kleptoparasitism. The reader is referred there for detailed discussions of those mechanisms. However, selection of their nesting habitat is influenced by proximity to food and, in the case of the skuas, proximity to their piracy victim is important. Some Arctic Skuas that prey on puffins also seem dependent on successfully breeding Arctic Terns for their own breeding success. Arnason and Grant (1978) attribute low breeding success for Arctic Skuas in 1973 at Vik, Iceland, in part, to the failure of Arctic Terns to breed that year, as they and Black-legged Kittiwakes constitute an apparently essential additional food source for the skuas. At Foula, in the Shetlands, Furness (1977) found that Arctic Skua chases were most successful with Arctic Terns, which also made more feeding trips per day than did the kittiwakes or Razorbills breeding there (Pearson 1964), which were also kleptoparasitized. Furness suggested that the large Arctic Tern and kittiwake breeding populations may induce Arctic Skuas to nest in the area. Also on Foula, Great Skuas have been increasing, forcing Arctic Skuas from their preferred nesting habitats in areas free from human disturbance, with high vegetation providing protective cover; to suboptimal sites in areas nearer human disturbance with shorter vegetation. Furness suggested that suboptimal nesting conditions and abnormally high density were compensated by the proximity of a nearby large Arctic Ternery providing a large food supply, which minimized the amount of time adult skuas had to forage away from their nest territory and maximized their time to guard their nests from intraspecific egg and chick predation. The locality-specific nature of many of these associations is emphasized by Nørrevang's (1960) report that Arctic Terns in the Faeroes got most of their food by kleptoparasitizing puffins, while skuas robbed the kittiwakes and puffins.

IV. BEHAVIORAL INTERACTIONS

A. Social Factors

Expansion into new areas often occurs when a few individuals pioneer in new habitats or nesting locations. They, in turn, attract others and a new colony is thus formed. The powerful mutual attraction of conspecifics is evidenced in the clumped or clustered species distributions found in mixed colonies where despite seemingly species-specific habitat over wide areas, members of most species will occur in proximity to one another rather than be dispersed over the site. An example of this is found in Virginia and North Carolina, where Sandwich Tern nests in clusters among the more numerous Royal Terns (Buckley and Buckley, 1972; see Fig. 7). Interspecific social attractions were also evident in these two species, as Sandwich Terns did not nest in these locales except in association with Royal Terns and it is believed that they selected large, densely packed, Royal Tern colonies for protection against predators as well as for social stimulation. On the Gulf Coast, Sandwich Terns are more numerous than Royal Terns, but both species are still found frequently nesting together (Portnoy, 1977). In Europe, Cullen (1960) and others have commented on the almost obligate association between Sandwich Terns and the smaller but more aggressive Common Terns and Black-headed Gulls. Similar nesting associations are known for Arctic and Common Terns (Cullen, 1960; Lemmetyinen, 1971), where Arctic Tern is the more aggressive species. In each instance (Royal–Sandwich; Arctic–Common), both species seemed to exhibit the same general habitat preferences except in the cases of the

Fig. 7. (A, B) Cluster of Sandwich Terns in Royal Tern colony on the outer banks of North Carolina. Arrow in A indicates Sandwich Tern cluster.

Sandwich and Common Terns, where the presence of Royal and Arctic Terns, respectively, represented additional selective factors in choice of habitat.

Gochfeld (1978) recorded the nesting association of Black Skimmers and Common Terns at almost all colony sites along the Atlantic Coast. Erwin (1979a) attributed their association to protection afforded the skimmers by the more aggressive Common Terns. In New Jersey in 1977 when a few pairs of Black Skimmers nested in association with Common Terns at salt marsh colonies (Buckley, 1979), skimmer eggs were laid in small depressions on wet eelgrass mats (*Zostera*) near tern nests. As skimmers normally nest on open sandy beaches, this was almost certainly a suboptimal habitat to which they had been attracted by the presence of Common Terns, most of the latter nesting in salt marshes in New Jersey (Buckley, 1978). On Long Island, Buckley and Buckley (1980) and Gochfeld (1978) also found a few pairs of skimmers nesting in salt marshes, but also only in association with Common Terns.

Koskimies (1957) explains the attraction of tern and gull colonies to other species (in his case, ducks, grebes, and shorebirds in Scandinavia) as due to a combination of two principles. The first emphasized the biological significance of the phenomenon of protection from predators, gained through close association with a more aggressive species; the second is based on a purely social attraction for one species by another. Koskimies hypothesized that as there is survival value is nesting with more aggressive species, natural selection would favor any tendencies promoting communal nesting. The larids themselves would constitute the environmental cues used in attracting nonlarids to their colonies, and a preference for them would be acquired by imprinting of young raised in their colonies. Advantages accruing to the gulls in such cases, as well as other benefits of colonial nesting, are beyond the scope of this chapter; the reader is therefore referred to Koskimies (1957), Cullen (1960), Belopol'skii (1961), Wynne-Edwards (1962), Lack (1968), and Erwin (1978) for extended discussions.

B. Imprinting

Imprinting is a specialized learning process occurring only during specific sensitive periods (critical period), resulting in a stable and apparently irreversible preference for the stimulus or releasing object imprinted on during the critical period. Current theory holds that the ontogeny of habitat selection is effected by a combination of inherited factors and imprinted habitat preferences which work together to produce the selective features of habitats preferred by birds (Orians, 1971). Therefore, a habitat preference acquired through imprinting upon habitat parameters of the nest site selected by parents would be passed on to the chick, which in turn would also select that nesting habitat. Sudden preference for, and rapid occupation of, new habitats were considered by Hildén (1965) to be proof that imprinting does influence habitat selection, because natural selection could not affect habitat selection so quickly. He credited the initial occupation of new habitat types to such factors as population increases, limited supply of usual nest sites, destruction or alteration of habitat, and sudden availability of abundant food supply. Birds breeding successfully in new habitats because of these or other factors would give rise to young imprinted upon the new habitat, then preferred over others. The repeated nesting of gulls on man-made structures would be one such example.

Chick-rearing experiments along these lines produced mixed results. While some Herring Gulls raised in zoological gardens inland from their coastal hatching sites returned as nesters to other inland sites resembling their rearing places, some also returned to nest on the coast and still others returned to the zoos where they were raised [cf. R. Drost (1955, 1958) cited in Hildén (1965)]. Experiments by Noseworthy et al. (1973) on habitat preferences of Herring Gulls at Little Bell Island, Newfoundland, showed a tendency for chicks to prefer vegetation types in which their nests were located. Chicks captured from three vegetation types (grass, shrubs, and herbs), in which their nests were located, were released in other vegetation types and their movements followed. Significant numbers of chicks returned to the vegetation types of their capture. The younger chicks (1–3 weeks) tended to return to their capture territories (their nests) while older chicks did not. However, the selection of distinct vegetation types not necessarily in their nesting territories by released older chicks was interpreted by Noseworthy et al. to be a preference for a particular habitat based on vegetation type. These findings suggest that these preferences are based on imprinting of nest-site habitat.

The question of whether imprinting is upon species-specific habitat features, or general site locations, or both, is one needing further study, as experience is known to modify and influence habitat preferences. (cf. Klopfer and Hailman, 1965).

C. Site Tenacity and Group Adherence

Site tenacity, the tendency of birds to return to their hatching site for their first breeding and once breeding there, to return to the same site in subsequent breeding seasons (Austin, 1949), is also known as Ortstreue (Klopfer and Hailman, 1965); it is akin to, or possibly synonymous with, philopatry, which Mayr (1963) defined loosely as an urge to stay in the native locality. Others (Hildén, 1965; McNicholl, 1975; Southern, 1977) used "site tenacity" to include return to nesting sites where previous breeding occurred, not necessarily confining it to the natal site.

Austin (1949) found that Common Terns on Cape Cod had a tendency to return to their natal colonies for their first nesting, and once establishing a breeding site in that colony, tended to return to it in subsequent years. Attachment to the breeding site increased with the bird's age and experience. Old and young birds at the smaller sites exhibited the strongest site tenacity despite repeated nesting failures, while the others joined larger colonies where conditions were more advantageous. Austin also noted that certain birds tended to nest near one another, and when birds shifted to other locations because of changing local conditions, very often the same individuals were found nesting near one another at the new sites. Austin (1949) termed this group adherence. Group adherence may play a role in site selection at colonies other than the natal colony when birds are attracted to the presence of other birds rather than their former nesting site.

Southern (1977) found similar site tenacity in stable Ring-billed Gull colonies on the Great Lakes, while at unstable sites high levels of group adherence was typical of older birds, but not of younger birds, contra Austin. Southern also found that return to the natal colony at stable sites was not well marked, but that return to the site after an initial successful breeding was well developed in the Ring-billed Gulls. He hypothesized that the group adherence shown by experienced birds nesting at unstable sites offered some colony stability by providing a nucleus of experienced birds able to move to new sites, thereby reestablishing an efficient colony. The situations described by Southern (1977) and Austin (1949) both support McNicholl's (1975) hypothesis that nest site tenacity should be well developed in stable habitats and reduced in unstable habitats, where group adherence would be more advantageous, and that in colonies of intermediate stability both phenomena would be important. He cited the slow growth of existing colonies and slow, gradual development of new colonies in such stable habitats as cliffs (alcids), and rocks in tundra lakes (gulls) as examples of site tenacity in stable habitats. Contrariwise, the rapid colonization of unstable sites by such species as Royal Terns (sand bars), Black-billed Gulls (mud banks: Beer, 1966) and Least Terns (fresh sand) in large numbers are suggested here as examples of group adherence encouraging selection and colonization of new sites, and perhaps new habitats.

Site tenacity and group adherence each have advantages and disadvantages for habitat selection. If site-tenacity tendencies are too strong, attachment to one particular site can be detrimental if that site becomes less suitable or overcrowded. Rowan (1952, 1965) found that Greater Shearwaters at Nightingale Island in the Tristan de Cunhas would lay their eggs on the bare ground of the colony surface when they were unable to obtain a nest burrow in the densely populated colony, rather than move to another location with suitable habitat, which was available. Native islanders were able to collect large numbers of these eggs for consumption. Rowan (1965) estimated the annual potential loss at 200,000–300,000 eggs, or 10% of the breeding pairs.

Austin (1949) cited Common Terns' continued occupation of sites so rat infested and subject to flooding that there was no successful reproduction. Buckley and Buckley (1980) documented the similar decline of a Common–Roseate Tern marsh colony on Long Island, which had 1175 pairs of Common and 561 pairs of Roseate Terns in 1974, but only 675 pairs of Common, and no Roseates by 1978. The continued use of sites no longer suitable for successful nesting seems to be a major disadvantage of site tenacity but Austin and others have found that most individuals seem to eventually shift to more suitable sites, doubtless aided by group adherence. The advantages of site tenacity are similar to those described by Tinbergen (1957) for territories—familiarity with places to feed, to flee from predators, and to seek or receive social contacts—and would seem to be powerful incentives for birds to at least attempt to return to the same nesting area each season. Another major effect of site tenacity on habitat selection would be a broadening of a species' habitat range when changing conditions at a nesting site are tolerated by older birds because of their site tenacity, thus allowing imprinting of new or different habitat conditions upon birds hatched under them.

V. MISCELLANEOUS FACTORS

A. Morphological Determinants

The evolution of specialized structural characters in some seabirds (such as murres' pyriform egg shape, which decreases the eggs' chances of rolling off narrow ledges) may reinforce the habitat preferences of certain species through increased adaptation, but some structural characters may also limit species' habitat choices. The body size of burrow and crevice nesters is one obvious example.

Adequate air currents and updrafts necessary for certain birds (gannets, albatrosses, frigatebirds, alcids) to take off and land easily because of their wing structure and mode of flight, are probably an important factor in the selection of cliffs or high nesting sites adjacent to the sea (Murphy, 1925; Nelson, 1978; Vermeer et al., 1979). Wing structure and flight aerodynamics of some, such as shearwaters and albatrosses, require areas where they are able to run along the ground or water for take-offs and landings; therefore the availability of such suitable areas would have to be an important feature of the selected nesting habitat. Hailman (in Klopfer and Hailman, 1965) observed in certain locales that Laughing Gulls required take-off and landing sites wide enough to spread their wings and within a suitable distance from their nests. He noted that birds nesting in a marsh never landed or took off from their nests, instead using rafts near their nests, and dune-nesting birds were observed to take off and land on paths adjacent to their nests. The susceptibility of the local grasses to being trampled into tunnels or paths seemed to determine the distance of the nest from the landing sites.

Coulson and Brazendale (1968) credited Great Cormorant colony distribution in the British Isles and Ireland to a reluctance of the cormorants to cross large expanses of open sea (attributed to the "poor waterproofing" of cormorant body feathers and "their need to roost on land"), except during certain seasons and in certain directions. They found that cormorants usually returned to breed in their natal colony and that there was little movement between colonies or into new areas where suitable habitat was available if flight over extensive expanses of open water was involved.

Following comparison of Common and Thick-billed Murre skeletal and muscular systems, Spring (1971) was able to correlate certain ecological differences with anatomical differences between the two species. He attributed the ability of Common Murres to out-compete or exclude Thick-billed Murres from broader nesting ledges to Commons' upright incubation stance and greater walking dexterity, when compared to Thick-billeds' more prone incubation posture and greater difficulty in walking. Further anatomical differences concerning flying, diving and swimming locomotor abilities were related to real and potential differential exploitation of available food resources which could also account for some habitat differences between the two. Comparisons by Bédard (1969) of Crested and Least Auklet nesting habitat on St. Lawrence Island, Alaska, revealed that Least Auklet's smaller body size allowed it to occupy a broader range of nest sites than larger Crested Auklets. Rock size on the talus slopes where they nested was determined to be the prime factor in nest site selection: Leasts nested mostly in areas of small rocks, Cresteds nested most densely in areas with large rocks (see Fig. 8).

B. Weather

Apart from the obvious limitations on a species' range intrinsic to its physiology, local weather conditions can play a role in nesting habitat selection. In the bird bazaars of the Bering Sea, murres are able to occupy nest sites when their nesting ledges are still covered by ice (Belopol'skii 1961): they lay their eggs on the ice, the incubating bird and egg sinking down to

Fig. 8. Rock crevice nest sites of Least Auklets in Pribilof Islands, Alaska. (Note puffin rock crevice nest and pathway.)

the bare ledges as the ice melts. However, those species nesting in tunnels and rock crevices of the cliff faces cannot gain their entrances until the ice has melted from them. If weather conditions are unusually severe, these species will have to nest elsewhere, or nest later in the season, or perhaps not nest at all in a given year.

We observed this phenomenon during the first week of June 1979 on the cliffs of Sevuokok Mountain near the village of Gambell, St. Lawrence Id., Alaska. On May 31 the sea-facing north cliffs were comparatively free of snow and ice, and Pigeon Guillemots, Horned and Tufted Puffins, Least, Crested and Parakeet Auklets, Common and Thick-billed Murres, and Black-legged Kittiwakes were flying in and out of nest sites, while the landward-facing, west cliff faces were still covered with snow and ice, with few if any bare surfaces showing. By June 6, after an extraordinary seven days of sunny weather and temperatures well above freezing, hundreds of puffins and auklets were sitting in crevices and on ledges up to 1.5 miles ($<$ 2 km) along the landward (west) side of the mountain, only now free of snow and ice. None had been observed in these locations prior to June 6, so it is likely that these birds were unable to obtain nest sites on the already crowded sea-facing north cliffs, and had been forced to wait until local weather conditions finally permitted them access to additional nesting habitat. Bédard (1969) noted that July was maximal cliff-occupation time on Sevuokok, and he did not record the colony extending so far south, so it would seem that conditions were significantly changed by favorable weather in 1979. On the Yukon–Kuskokwin Delta, waterfowl and shorebirds were nesting about 3 weeks earlier than usual, following an unprecedented warm spring in 1979 (C. Lensink, personal communication). Drury (1960) also noted that Arctic Terns did not arrive or nest in, certain locales until snow and ice conditions were suitable.

Rigorous local weather conditions can also determine the suitability of nesting sites used previously under more salubrious conditions. In 1977, Black Skimmers were unable to nest at a N.J. roadside landfill used for several years previously, because a particularly wet June combined with settling fill had flooded the site rendering it unsuitable for use by skimmers (Buckley, 1978). In 1967, a Royal Tern colony on the low-lying natural, ocean-front beach on Fisherman's Island, Cape Charles, Virginia was flooded out by a combination of spring high tides and storm tides. Instead of renesting in the same locale as they had done previously, the birds moved to a high, safe spoil island on the bay side, where they nested successfully for several years (Buckley and Buckley, personal observation).

The impact of El Niño on Peruvian coastal seabird nesting in proximity to the abundant food resources of the rich ocean upwellings there has long been known (Murphy, 1925, 1936). The effect occurs when the intensity of the southwest tradewinds drops, coastal upwellings weaken, and surface water temperatures rise. As an influx of warm water (El Niño) moves near the coast, natural upwellings and organisms dependent on them are severely reduced. The resultant loss of food resources causes serious nesting failures and mortality from starvation. Recent work by Boersma (1978) on Galapagos Penguins indicates that these impacts may be more widespread than was previously believed. In the 1972 El Niño, surface water temperatures surrounding the Galapagos Island of Punta Espinosa, normally about 22°C during the breeding season, rose to above 25°C, and there was total breeding failure (no fledged chicks), weight loss and nest desertions among Galapagos Penguins, which Boersma attributed to the rise in water temperature and associated reduced food supply. Brown Pelican experienced similar nesting failures at Punta Espinosa in 1972 that Boersma also attributed to increased water temperatures and decreased food supplies. The role of local weather conditions in habitat selection deserves increased attention from investigators.

C. Ectoparasites

Little is known about the effects that avian ectoparasites may have upon seabird habitat selection. However, two recent papers draw attention to the role that ectoparasites might play in choice of habitat; the topic is included here to call attention to it.

In 1973, C. J. Feare (1976) noted a mass desertion of eggs and newly hatched chicks in part of a Sooty Tern colony on Bird Island in the Seychelles. Numerous ticks, not observed in previous years, were found at the colony site. In 1974 the same area of the colony was not occupied and it was found to have a higher tick density than was found in surrounding occupied areas. Feare suggested that the desertion resulted from the high tick infestation and associated irritation from tick bites or transmittal of viral infection to the terns by the ticks.

Houston (1979) speculated that tree nesting by Fairy Terns without building any nests may be an antiparasite response. The lack of nest material would preclude the harboring of nest parasites and the location of nesting sites in trees would reduce the spread of parasites by increasing inter-nest distances. His comparison of bare-branch-reared Fairy Tern chicks with tree-nest-reared Black Noddy chicks, at Cousin Island in the Seychelles in 1977, although not conclusive, did indicate a lack of ectoparasites on Fairy Terns and the presence of some (but not many) on the noddies.

D. Human Factors

Direct human predation on seabirds; disturbance and alteration of nesting habitat; introduction of predators and/or domestic animals to islands and coastal areas; pollution of waterways and salt marshes; and development and filling of coastal areas and wetlands have all affected seabirds and their habitats. Seabirds have responded in different ways: some have become extinct (Great Auk), others may face extinction (Cahows), unless human management efforts are successful. Some others have adapted too well, such as the Herring Gulls in the northeastern United States, where they have greatly expanded their breeding ranges and are now showing greater catholocity in their selection of nesting habitat, successfully using a wide range of disparate habitats (Buckley, 1978). Some have taken to using buildings as artificial cliffs (Herring Gulls, see Fig. 9; Black-backed Gulls; Black-legged Kittiwakes; fulmars), while Least Terns and Black Skimmers are using flat roof tops as "beaches" (Fisk, 1978) following increasing human disturbance of more natural sites in Florida. In Virginia, North Carolina, New Jersey, and New York, terns, gulls, and skimmers have been forced off barrier island beaches by recreational activities and beach development, sometimes into habitats that they are less suited to (Buckley and Buckley, 1977b). In some instances, dredging activities have unintentionally provided suitable alternate nesting habitats in the form of manmade islands behind barrier beaches similar to those previously available at

Fig. 9. Roof-top nest sites of Herring Gulls at North Brother Island, East River, Bronx Co., New York City.

abandoned natural locations (Buckley and Buckley, 1975, 1977b; Buckley, 1979). Although beyond the scope of this chapter (and badly in need of an updated, comprehensive review), ditching of salt marshes for mosquito control has led to profound changes in marsh physiognomy, vegetation, general ecology and use by waterbirds (cf. Buckley and Buckley, 1977b; Burger et al., 1979 for recent brief discussions). The spoil produced by several methods of salt marsh ditching has proven beneficial to Herring and Laughing Gulls by providing, at least at some sites in New Jersey, high areas where nests were safe from devastating inundation by spring and storm tides (Burger and Shisler, 1978a, 1979, 1980). However, it should be pointed out that Herring Gulls' marsh breeding success has been in some instances at the expense of Common Terns (Burger and Lesser, 1978b) and Laughing Gulls (Burger and Shisler, 1978b), and most ecologists have grave reservations about ecosystem manipulation on the scale practiced by both mosquito-control officials and navigational dredging authorities—even though both practices may on occasion provide habitat benefits to certain marine and coastal birds.

   Economic exploitation of breeding seabirds for food, feathers, guano, and oil still occurs (Uspenski, 1958; Belopol'skii, 1961; Manuwal, 1978; Buckley and Buckley, personal observation). Continuing disruption of their nesting grounds have forced many species from accessible sites and habitat into inaccessible and more restricted sites (Murphy, 1925) and caused severe population losses (Manuwal, 1978).

Despite recent protective measures and management efforts (e.g., Nettleship, 1976; Buckley and Buckley, 1976) to preserve and conserve seabirds, the terrestrial and marine habitats necessary to their survival are still in danger from human activities. Chemical pollution and coastal development are currently major threats to their breeding habitats, although introduced quadrupeds may be as important. Man has affected seabird choice of habitat through disruption and destruction of favored sites and by providing new sites (buildings, landfills, roofs, dredge spoil) as alternatives. Detailed studies of seabird populations are necessary to determine just how much latitude each species has in its choice of habitats, but unless management efforts include massive habitat protection, other measures will be meaningless, and additional habitat selection research will be too late.

VI. SUMMARY

Habitat selection in marine birds has been examined directly in only a few species, so that concepts and terms have largely been taken from studies on other birds. Information about marine birds' habitat selection is thus drawn from descriptive rather than experimental studies, as seabird biology renders experimental studies, especially long-term ones, intrinsically difficult.

Habitat selection is influenced by so-called ultimate and proximate factors. The former are evolutionary adaptations ensuring reproductive success and species survival, whereas the latter are the environmental cues and triggers controlling these adaptations. Ultimate factors (e.g., cliff-nesting adaptations in Black-legged Kittiwakes) are reasonably well-known, so attention has focused on identification of proximate factors; examples from several species are discussed. Variability in habitat requirements has been noted in many species, and is frequently related to inter- and intraspecific competition for limited nest or colony sites (most seabirds being colonial). Several examples relate increased variability in habitats used to dramatic population increases (e.g., Herring Gull, Black-legged Kittiwake), although the exact relationship between cause and effect is not always clear. One principle seems unarguable, viz. that high population densities are most frequently associated with great variability in habitats used (eurytopy): when populations shrink, habitat variability declines as well (stenotopy). Generally speaking, populations at the geographic limits of their range tend to be stenotopic, becoming more eurytopic as the center of their range is approached.

Some dispute has arisen over the relative importance of competition for nest sites versus food, as factors limiting seabird populations or causing abnormally short breeding cycles (certain tropicbird populations, for example). Nonetheless, it is clear that many mixed-species colonies have managed to finely partition breeding habitats, and although microhabitat differences seem consistent, data from other colonies with different species compositions, habitat availability, or food resources soon destroy blanket generalities. Alcids present a striking example of this effect, and data from tropical sulids indicate how several closely related species are able to simultaneously exploit breeding sites of seemingly limited variability.

Securing adequate sources of food has played a major role in seabird habitat selection, whether on temperate coasts or tropical oceanic islands, although surprisingly few studies have attempted more than superficial associative analyses.

Data are presented tying the sizes of several large tern colonies in New York to tidal prisms at major oceanic inlets, confirming similar observations involving Canadian Arctic species. Another study has examined the significance of inshore versus offshore feeding to nesting habitat of coastal Washington alcids, drawing unexpected conclusions. At present it seems safe to state that most authors agree that food availability plays a major role in nest habitat selection and that it is more important in nutrient-poor tropical than in nutrient-rich temperate waters. However, it is in temperate waters that the relative importance of nest habitat availability begins to play a greater role, one still largely unquantified, if not essentially unknown.

Predation and kleptoparasitism on seabirds may be unexpectedly important in breeding habitat selection, and clearly deserve greater attention. Several studies involving Northern Hemisphere gulls, terns, alcids, and skuas demonstrate the effect elegantly and the considerable amount of geographic variation probably occurring is suggested in several papers.

Behavioral factors, including social simulation, mutual benefits from coloniality, habitat imprinting, site tenacity, and group adherence are seen to be of critical importance in the establishment of new colonies, and the maintenance of old ones in the face of changing environmental conditions. Several studies, some of them experimental, and involving, gulls, terns, skimmers, and shearwaters, suggest that behavioral factors may be as important as competition for food and nest sites in seabird habitat selection, although many more species need to be looked at in detail.

Several "miscellaneous factors" are also shown to have unanticipated impacts on habitat selection. Certainly bird and egg size and shape are well known, but other more subtle determinants include alcid "clumsiness" on cliffs, cormorant waterproofing, and the need for strong air currents for albatross take-offs and landings. The effects of weather on habitat selection range from obvious (changes in the timing of the El Niño current resulting in wholesale nesting failures of Peruvian seabirds) to subtle (murres have a competitive edge on seabird cliffs by being able to lay their eggs on ice), and may be more pervasive than heretofore appreciated. Even ectoparasites have been implicated in the habitat selection of Indian Ocean Sooty and Fairy Terns. Finally, the effects of human interaction with seabirds are touched upon briefly, although this too is now one of the more important determinants of seabird habitat selection—by both creation and destruction of habitat, by introduction of exotic animals and plants, and by management (or mismangement) of seabird populations.

ACKNOWLEDGMENTS

The authors' studies discussed in this chapter have been funded by several National Science Foundation grants, a contract from the U.S. Army Corps of Engineers, and financial and logistic support have been provided by the following units of the National Park Service: Office of the Chief Scientist; North Atlantic Regional Office; Gateway National Recreation Area; Fire Island National Seashore; Cape Hatteras National Seashore and Cape Lookout National Seashore. Helpful editorial comments were received from M. Gochfeld, J. Burger, and B. Olla, and the library resources of the Museum of Comparative Zoology, Harvard University, and the Massachusetts Audubon Society were invaluable. B. Kachadoorian provided aid in manuscript preparation that is greatly appreciated. Collins Publishers and Academic Press gave permission for copyrighted material to be reproduced herein. We wish to thank all these persons and institutions for their assistance.

APPENDIX: SCIENTIFIC AND COMMON NAMES OF SEABIRDS MENTIONED IN TEXT

SCIENTIFIC NAME	COMMON NAME
Sphenisciformes	
Spheniscidae	Penguins
Spheniscus mendiculus	Galapagos Penguin
Procellariiformes	
Diomedeidae	Albatrosses
Procellariidae	Fulmars, shearwaters, petrels
Fulmarus glacialis	Northern Fulmar

SCIENTIFIC NAME	COMMON NAME
Puffinus assimilis	Little Shearwater
Puffinus gravis	Greater Shearwater
Puffinus puffinus	Manx Shearwater
Pterodroma cahow	Bermuda Petrel, Cahow
Hydrobatidae	Storm petrels
Pelagodroma marina	White-faced Storm Petrel
Hydrobates pelagicus	(British) Storm Petrel
Oceanodroma castro	Harcourt's Storm Petrel
Oceanodroma leucorhoa	Leach's Storm Petrel
Pelecaniformes	
Phaethontidae	Tropicbirds
Phaethon athereus	Red-billed Tropicbird
Phaethon lepturus	White-tailed Tropicbird
Phaethon rubricauda	Red-tailed Tropicbird
Pelecanidae	Pelicans
Pelecanus occidentalis	Brown Pelican
Sulidae	Boobies, gannets
Sula abbotti	Abbott's Booby
Sula bassana	Northern Gannet
Sula dactylatra	Masked Booby
Sula leucogaster	Brown Booby
Sula sula	Red-footed Booby
Phalacrocoracidae	Cormorants
Phalacrocorax aristotelis	Shag
Phalacrocorax carbo	Great Cormorant
Fregatidae	Frigatebirds
Charadriiformes	
Stercorariidae	Skuas, jaegers
Stercorarius parasiticus	Arctic Skua, Parasitic Jaeger
Catharacta skua	Great Skua
Laridae	Gulls, terns
Larus argentatus	Herring Gull
Larus atricilla	Laughing Gull
Larus bulleri	Black-billed Gull
Larus delawarensis	Ring-billed Gull
Larus fuscus	Lesser Black-backed Gull
Larus glaucoides kumlieni	Iceland (Kumlien's) Gull
Larus hyperboreus	Glaucous Gull
Larus marinus	Great Black-backed Gull
Larus ridibundus	Black-headed Gull
Larus roseus	Ross's Gull

SCIENTIFIC NAME	COMMON NAME
Larus thayeri	Thayer's Gull
Rissa tridactyla	Black-legged Kittiwake
Sterna albifrons	Least Tern
Sterna caspia	Caspian Tern
Sterna dougallii	Roseate Tern
Sterna forsteri	Forster's Tern
Sterna fuscata	Sooty Tern
Sterna hirundo	Common Tern
Sterna maxima	Royal Tern
Sterna paradisaea	Arctic Tern
Sterna sandvicensis	Sandwich Tern
Chilidonias niger	Black Tern
Anous tenuirostris	Black Noddy
Gygis alba	Fairy (White) Tern
Rynchopidae	Skimmers
Rynchops niger	Black Skimmer
Alcidae	Auks, auklets, murres
Alle alle	Dovekie
Alca impennis[a]	Great Auk[a]
Alca torda	Razorbill, Razor-billed Auk
Uria aalge	Common Murre
Uria lomvia	Thick-billed Murre
Cepphus columba	Pigeon Guillemot
Cepphus grylle	Black Guillemot
Cyclorrhynchus psittacula	Parakeet Auklet
Aethia cristatella	Crested Auklet
Aethia pusilla	Least Auklet
Fratercula arctica	Atlantic Puffin, Common Puffin
Fratercula corniculata	Horned Puffin
Lunda cirrhata	Tufted Puffin

[a] Extinct.

REFERENCES

Arnason, E., and Grant, P. R., 1978, The significance of kleptoparasitism during the breeding season in a colony of Arctic Skuas *Stercorarius parasiticus* in Iceland, *Ibis* **120**:38–54.

Ashmole, N. P., 1963, The regulation of numbers of tropical oceanic birds, *Ibis* **103b**:458–473.

Ashmole, N. P., 1968, Body size, prey size, and ecological segregation in five sympatric tropical terns (Aves: Laridae), *Syst. Zool.* **17**:292–304.

Ashmole, N. P., 1971, Seabird ecology and the marine environment, in: *Avian Biology* (D. S. Farner and J. R. King, eds.), Vol. I, pp. 223–282. Academic Press, New York.

Ashmole, N. P., and Ashmole, M. J., 1967, Comparative feeding ecology of seabirds of a tropical oceanic island, *Peabody Mus. Nat. Hist. Yale Univ.* Bull. **24**:1–131.

Austin, O. L., 1949, Site Tenacity, A behaviour trait of the Common Tern (*Sterna hirundo* Linn.), *Bird-Banding* **20**:1–39.

Bannerman, D. A., 1963, *Birds of the Atlantic Islands*, Vol. I, Oliver and Boyd, Edinburgh, 358 pp.

Bédard, J., 1969, The nesting of the Crested, Least, and Parakeet Auklets on Saint Lawrence Island, Alaska, *Condor* **71**:386–398.

Bédard, J., 1976, Coexistence, coevolution and convergent evolution in seabird communities: A comment, *Ecology* **57**:177–184.

Beer, C., 1966, Adaptations to nesting habitat in the reproductive behaviour of the Black-billed Gull, *Larus bulleri*, *Ibis* **108**:394–410.

Belopol'skii, L. O., 1961, *Ecology of Sea Colony Birds of the Barents Sea*, English Translation by Israel Program for Scientific Translations, Jerusalem, 346 pp.

Bent, A. C., 1921, Life histories of North American Gulls and Terns, *U.S. Nat. Mus. Bull.* **113**:337 pp.

Boersma, P. D., 1978, Breeding patterns of Galapagos Penguins as an indicator of oceanographic conditions, *Science* **200**:1481–1483.

Bourne, W. R. P., 1955, The birds of the Cape Verde Islands, *Ibis* **97**:508–556.

Bourne, W. R. P., 1963, A review of oceanic studies of the biology of seabirds, *Proc. XIII Int. Ornithol. Congr.*, Oxford, pp. 831–854.

Buckley, F. G., 1978, The use of dredged material islands by colonial seabirds and wading birds in New Jersey, Unpublished, Final Technical Report, United States Army Engineers, Waterways Experiment Station, Vicksburg, Mississippi, 300 pp.

Buckley, F. G., 1979, Colony site selection by colonial waterbirds in coastal New Jersey, *Proc. Colonial Waterbird Group*, October 1978, New York, N.Y., pp. 17–26.

Buckley, F. G., In press. *A History of Colonial Waterbirds in Coastal New Jersey*, N.J. Audubon Society, Bernardsville, N.J.

Buckley, F. G., and Buckley, P. A., 1972, The breeding ecology of Royal Terns *Sterna* (*Thalasseus*) *maxima maxima*, *Ibis* **114**:344–359.

Buckley, P. A., and Buckley, F. G., 1975, The significance of dredge spoil islands to colonially nesting waterbirds in certain national parks, in: *Proceedings of a Conference on the Management of Dredge Islands in North Carolina Estuaries* (J. Parnell and R. Soots, eds.), pp. 35–45, University of North Carolina Sea Grant Publication, UNC-SG-75-01, 142 pp.

Buckley, P. A., and Buckley, F. G., 1976, *Guidelines for the Protection and Management of Colonially Nesting Waterbirds*, Boston, United States National Park Service, 54 pp.

Buckley, P. A., and Buckley, F. G., 1977a, Hexagonal packing of Royal Tern nests, *Auk* **94**:36–43.

Buckley, P. A., and Buckley, F. G., 1977b, Human encroachment on barrier beaches of the northeastern United States and its impact on coastal birds, in: *Coastal Recreation Resources in an Urbanizing Environment: A Monograph* (J. H. Noyes and E. H. Zube, eds.), pp. 68–76, University of Massachusetts, Amherst.

Buckley, P. A., and Buckley, F. G., 1980, Population and colony site trends in Long Island, New York's colonial waterbirds: A five year study in the mid 1970's, *Trans. Linn. Soc.*, *N.Y.* **9** (in press).

Burger, J., 1977, Nesting behavior of Herring Gulls: invasion into *Spartina* salt marsh areas of New Jersey, *Condor* **79**:162–169.

Burger, J., and Lesser, F., 1978a, Selection of colony sites by Common Terns *Sterna hirundo* in Ocean County, New Jersey, *Ibis* **120**:433–449.

Burger, J., and Lesser, F., 1978*b*, Determinants of colony site selection in Common Terns (*Sterna hirundo*), *Proc. Colonial Waterbird Group*, October 1977, De Kalb, Illinois, pp. 118–127.

Burger, J., and Shisler, J., 1978*a*, The effects of ditching a salt marsh on colony and nest site selection by Herring Gulls (*Larus argentatus*), *Am. Midl. Nat.* **100**:54–63.

Burger, J., and Shisler, J., 1978*b*, Nest site selection and competitive interactions of Herring and Laughing Gulls in New Jersey, *Auk* **95**:252–266.

Burger, J., and Shisler, J. 1979, The immediate effects of ditching a saltmarsh on nesting Herring Gulls *Larus argentatus*, *Biol. Conserv.* **15**:85–103.

Burger, J., and Shisler, J., 1980, Colony and nest site selection in Laughing Gulls *Larus atricilla* in response to tidal flooding, *Condor* **82** (in press).

Burger, J., Shisler, J., and Lesser, F., 1979, The effects of ditching salt marshes on nesting birds of New Jersey, *Proc. Colonial Waterbird Group*, October 1978, New York, N.Y., pp. 27–37.

Cody, M. L., 1973, Coexistence, coevolution and convergent evolution in seabird communities, *Ecology* **54**:31–44.

Cody, M. L., 1974, *Competition and the Structure of Bird Communities*, Monographs in population biology, No. 7, Princeton University Press, Princeton, 318 pp.

Cooper, D., Hays, H., and Pessino, C., 1970, Breeding of the Common and Roseate Terns on Great Gull Island, *Proc. Linn. Soc. N.Y.* **71**:83–104.

Coulson, J. C., and Brazendale, M. G., 1968, Movements of cormorants ringed in the British Isles and evidence of colony-specific dispersal, *Br. Birds* **61**:1–21.

Cramp, S., Bourne, W. R. P., and Saunders, D., 1974, *The Seabirds of Britain and Ireland*, Taplinger, New York, 287 pp.

Cullen, E., 1957, Adaptations in the kittiwake to cliff nesting, *Ibis* **99**:275–302.

Cullen, J. M., 1960, Some adaptations in the nesting behaviour of terns, *Proc. 12th Int. Ornith. Congr.*, Helsinki, pp. 153–157.

Dement'ev, G. P., Gladkov, N. A., and Spangenberg, E. P., 1969, *Birds of the Soviet Union*, Vol. III, Israel Program for Scientific Translations, Ltd., Jerusalem, 756 pp.

Diamond, A. W., 1975, The biology of tropicbirds of Aldabra Atoll, Indian Ocean, *Auk* **92**:1–39.

Drury, W. H., Jr., 1960, Breeding activities of Long-tailed Jaeger, Herring Gull, and Arctic Tern on Bylot Island, Northwest Territories, Canada, *Bird-Banding* **31**:63–79.

Drury, W. H., 1973, Population changes in New England seabirds, *Bird-Banding* **44**:267–313.

Erwin, R. M., 1978, Coloniality in terns: The role of social feeding, *Condor* **80**:211–215.

Erwin, R. M., 1979*a*, Species interactions in a mixed colony of Common Terns (*Sterna hirundo*) and Black Skimmers (*Rynchops niger*), *Animal Behav.* **27**:1054–1062.

Erwin, R. M. 1979*b*, *Coastal Waterbird Colonies: Cape Elizabeth, Maine to Virginia*, Fish and Wildlife Service, Office of Biological Services, United States Department of Interior, FWS/OBS-79/10, 212 pp.

Feare, C. J., 1976, Desertion and abnormal development in a colony of Sooty Terns *Sterna fuscata* infested by virus-infected ticks, *Ibis* **118**:112–115.

Fisher, J., and Lockley, R. M., 1954, *Seabirds*, Collins, London, 320 pp.

Fiske, E. J., 1978, Roof-nesting terns, skimmers, and plovers in Florida, *Fl. Field Nat.* **6**:1–8.

Furness, R., 1977, Effects of Great Skuas on Arctic Skuas in Shetland, *Br. Birds* **70**:96–107.

Gochfeld, M., 1978, Colony and nest site selection by Black Skimmers, *Proc. Colonial Waterbird Group*, October 1977, De Kalb, Illinois, pp. 78–90.

Grant, P. R., 1971, Interactive behaviour of puffins (*Fratercula arctica* L.) and skuas (*Stercorarius parasiticus* L.) *Behaviour* **XL**:263–281.

Greenway, J. C., Jr., 1958, *Extinct and Vanishing Birds of the World*, American Committee for International Wild Life Protection, Special Publ. Number 13, New York, 518 pp.

Harris, M. P., 1969, Factors Influencing the breeding cycle of the Red-billed Tropicbird in the Galapagos Islands, *Ardea* 57:149–157.

Hildén, O., 1965, Habitat selection in birds, *Ann. Zool Fenn.* 2:53–75.

Houston, D. C., 1979, Why do Fairy Terns *Gygis alba* not build nests? *Ibis* 121:102–104.

Immelmann, K., 1972, Erörterungen zur Definition and Anwedbarkeit der Begriffe, Ultimate Factor, "Proximat Factor" und "Zeitgeber," *Oecologica (Berlin)* 9:259–264.

Kadlec, J. A., and Drury, W. H., 1968, Structure of the New England Herring Gull population, *Ecology* 49:644–676.

Kane, R., and Farrar, R. B., 1977, 1977 Coastal colonial bird survey of New Jersey, *Occas. Pap. #131*, New Jersey Audubon Society, Bernardsville, N.J.

Kendeigh, S. C., 1961, *Animal Ecology*, Prentice-Hall, Englewood Cliffs, N.J., 468 pp.

Klopfer, P. H., and Hailman, J. P., 1965, Habitat selection in birds, in: *Advances in the Study of Behavior* (D. S. Lehrman, R. A. Hinde, and E. Shaw, eds.), Vol. I, pp. 279–303, Academic Press, New York.

Koskimies, J., 1957, Terns and gulls as features of habitat recognition for birds nesting in their territories, *Ornis. Fenn.* 34:1–6.

Kruuk, H., 1964, Predators and anti-predator behaviour of the Black-headed Gull, *Larus ridibundus* L., *Behav. Suppl.* 11:1–129.

Lack, D., 1934, Habitat distribution in certain Icelandic birds, *J. Anim. Ecol.* 3:81–90.

Lack, D., 1937, The psychological factor in bird distribution, *Br. Birds* 31:130–136.

Lack, D., 1954, *The Natural Regulation of Animal Numbers*, Clarendon Press, Oxford, 343 pp.

Lack, D., 1966, *Population Studies of Birds*, Clarendon Press, Oxford, 341 pp.

Lack, D., 1967, Interrelationships in breeding adaptations as shown by marine birds, *Proc. 14th Int. Ornithol. Congr.*, Oxford, pp. 3–41.

Lack, D., 1968, *Ecological Adaptations for Breeding in Birds*, Methuen, London, 409 pp.

Lemmetyinen, R., 1971, Nest defense behaviour of Common and Arctic Terns and its effects on the success achieved by predators, *Ornis Fenn.* 48:13–22.

Lockley, R. M., 1974, *Ocean Wanderers*, David and Charles, Ltd., Newton, Abbot Devon, 168 pp.

MacDonald, S., 1979, First breeding record of Ross's Gull in Canada, *Proc. Colonial Waterbird Group*, October 1978, New York, N.Y., Abstr., p. 16.

Manuwal, D., 1978, Effect of man on marine birds: A review, *John S. Wright Forestry Conference Proceedings 1978*, Wildlife and People, (C. M. Kirkpatrick, ed.), pp. 140–160, Department of Forestry and Natural Resources and Cooperative Extension Services, Purdue University.

Mayr, E., 1963, *Animal Species and Evolution*, Belknap Press, Cambridge, 795 pp.

McNicholl, M. K., 1975, Larid site tenacity and group adherence in relation to habitat, *Auk* 92:98–104.

Miller, A. H., 1942, Habitat selection among higher vertebrates and its relation to intraspecific variation, *Am. Nat.* 76:25–35.

Murphy, R. C., 1925, *Bird Islands of Peru*, G. P. Putnam's Sons, New York, 362 pp.

Murphy, R. C., 1936, *Oceanic Birds of South America*, Vols. I, II, American Museum of Natural History, New York.

Nelson, J. B., 1978, *The Sulidae—Gannets and Boobies*, Oxford University Press, Oxford, 1012 pp.

Nettleship, D. N., 1972, Breeding success of the Common Puffin (*Fratercula arctica* L.) on different habitats at Great Island, Newfoundland, *Ecol. Monogr.* 42:239–268.

Nettleship, D., 1976, Census techniques for seabirds of Arctic and Eastern Canada, Canadian Wildlife Service, Ottawa, *Occas. Pap. #25*, 33 pp.

Nettleship, D., 1977, Seabird resources of Eastern Canada: Status, problems and prospects,

Proceedings Symposium on Canada's Threatened Species and Habitats, pp. 96–108, Canadian Nature Federation, Ottawa, Canada.

Nettleship, D. N., and Gaston, A. J., 1978, Patterns of pelagic distribution of seabirds in western Lancaster Sound and Barrow Strait, NWT, Canadian Wildlife Service, Ottawa, *Occas. Pap.* #39, 40 pp.

Nisbet, I. C. T., 1978, Recent changes in gull population in the western North Atlantic, *Ibis* **120**:129–130.

Nørrevang, A., 1960, Søfuglenes udvaegellse of ynglebiotop pa Mykines, Faeroerne, (Habitat selection of seabirds in Mykines Faeroes), (English summary) Dansk. *Ornithol. Foren. Tidsskr.* **54**:9–35.

Noseworthy, C., Stoker, S., and Lien, J., 1973, Habitat preferences in Herring Gull chicks, *Auk* **90**:193–194.

Orians, G., 1971, Ecological aspects of behaviour in: *Avian Biology* (D. S. Farner and J. R. King, eds.), Vol. I, pp. 513–546, Academic Press, New York.

Partridge, L., 1978, Habitat selection, in: *Behavioural Ecology an Evolutionary Approach* (J. R. Krebs and N. B. Davies, eds.), pp. 351–376, Sinauer Associates, Inc., Sunderland, Mass.

Patterson, I. J., 1965, Timing and spacing of broods in the Black-headed Gull *Larus ridibundus*, *Ibis* **107**:433–459.

Pearson, T. H., 1968, The feeding biology of seabird species breeding on the Farne Islands, Northumberland, *J. Anim. Ecol.* **37**:521–557.

Portnoy, J. W., 1977, Nesting colonies of seabirds and wading birds—Coastal Louisiana, Mississippi, and Alabama, Fish and Wildlife Service, United States Department of the Interior, FWS/OBS-77/07, 126 pp.

Robertson, W. B., Jr., 1964, The terns of the Dry Tortugas, *Bull. Fla. State Mus.* **8**:1–94.

Rowan, M. K., 1952, The Greater Shearwater *Puffinus gravis* at its breeding grounds, *Ibis* **94**:97–121.

Rowan, M. K., 1965, Regulation of seabird numbers, *Ibis* **107**:54–59.

Salomonsen, F., 1955, The food production in the sea and the annual cycle of Faeroese marine birds, *Oikos* **6**:92–100.

Schreiber, R., and Ashmole, N. P., 1970, Sea-bird breeding seasons on Christmas Island, Pacific Ocean, *Ibis* **112**:363–394.

Smith, N. G., 1966, Evolution of some arctic gulls (*Larus*): An experimental study of isolating mechanisms, *Ornithological Monographs* Number 4, American Ornithologists' Union, 99 pp.

Snow, D. W., 1965, The breeding of the Red-billed Tropicbird in the Galapagos Islands, *Condor* **67**:210–214.

Southern, W. E., 1977, Colony selection and site tenacity in Ring-billed Gulls at a stable colony, *Auk* **94**:469–478.

Southern, L. K., and Southern, W. E., 1979, Absence of nocturnal predator defense mechanisms in breeding gulls, *Proc. Colonial Waterbird Group*, October 1978, New York, N.Y., pp. 157–162.

Sowls, A. L., Hatch, S. A., and Lensink, C. J., 1978, *Catalog of Alaskan Seabird Colonies*, United States Department of the Interior, Fish and Wildlife Service, FWS/OBS-78/78.

Spring, L., 1971, A comparison of functional and morphological adaptations in the Common Murre (*Uria aalge*) and Thick-billed Murre (*Uria lomvia*), *Condor* **73**:1–27.

Stonehouse, B., 1962a, Ascension Island and the British Ornithologists' Union centenary expedition 1957–59, *Ibis* **103b**:107–123.

Stonehouse, B., 1962b, The tropicbirds of Ascension Island, *Ibis* **103b**:124–161.

Svardson, G., 1949, Competition and habitat selection in birds, *Oikos* **1**:157–174.

Tinbergen, N., 1957, The functions of territory, *Bird Study* **4**:14–27.

Tinbergen, N., 1967, Adaptive features of the Black-headed Gull *Larus ridibundus* L., *Proc. XIV Int. Ornithol. Congr.*, Oxford, pp. 43–59.

Tuck, L. M., 1960, *The Murres*, Queen's Printer, Ottawa, 260 pp.

Uspenski, V. S., 1958, *The Bird Bazaars of Novaya Zemlya* (English translation by Canadian Wildlife Service, translations of Russian game reports), Vol. 4, Queen's Printer, Ottawa, 159 pp.

Vermeer, K., Vermeer, R., Summers, K. R., and Billings, R. R., 1979, Numbers and habitat selection of Cassin's Auklet breeding on Triangle Island, British Columbia, *Auk* **96**:143–151.

Vincente, V., 1979, The occurrence of a nesting colony of the Royal Tern, *Sterna* (*Thalasseus*) *maxima* on the south coast of Puerto Rico, *Am. Birds* **33**:147.

Wingate, D. W. 1977, Excluding competitors from Bermuda Petrel nesting burrows, in: *Endangered Birds* (S. A. Temple, ed.), pp. 93–102, University of Wisconsin Press, Madison.

Wynne-Edwards, V. C., 1962, *Animal Dispersion in Relation to Social Behaviour*, Hafner, New York, 653 pp.

Chapter 4

MATE SELECTION AND MATING SYSTEMS IN SEABIRDS

George L. Hunt, Jr.

Department of Ecology and Evolutionary Biology
University of California
Irvine, California 92717

I. INTRODUCTION

This chapter examines the evolutionary and ecological factors that have influenced courtship and mating patterns of marine birds. The literature on the reproductive behavior of seabirds is vast, and it is not my purpose to review it here. Rather, I focus attention on several questions and seek patterns that may help us understand aspects of variation in reproductive biology. In particular, I will be concerned with the assessment of mate quality and how these qualities affect reproductive success and mating systems.

Two basic assumptions underlie modern evolutionary approaches to the study of reproductive biology. One is that the reproductive biology of a species has evolved towards maximization of individual or inclusive fitness; the second is that time and energy are limited and are partitioned among the various components of the reproductive effort.

While it is reasonable to assume there is a limited resource base available for reproduction, there are virtually no critical studies showing a trade-off between allocation of time or energy to courtship and to other facets of reproduction. (For the purposes of this presentation, I include as "courtship" all activities prior to egg-laying that are primarily evolved to promote the acquisition, selection and retention of a mate.) The assumption of trade-offs among different schemes of allocation is basic to much of optimization theory in ecology. This assumption has been traced by Dr. John Schultz (personal communication) to Cody (1966), who references an unpublished manuscript of R. Levins and R. MacArthur, but it seems otherwise largely undocumented or tested. Williams (1966) also recognized this trade-off in time or energy allocation when he discussed how the male, with less commitment to gamete production, should devote the major portion of his reproductive effort to securing additional females. The energetic cost of mate attraction, assessment and defense, and its impact on the energy or time available for gamete production and parental care, has yet to be adequately studied.

In contrast to the paucity of information on the trade-offs of resource allocation between courtship and other aspects of reproduction, there has been considerable interest in how organisms maximize their fitness. Maximizing reproductive output must be seen in terms of a life-history strategy in which not only fecundity, but also age of first breeding and future potential reproductive value must be considered (Fisher, 1930; Lack, 1954, 1966, 1968; Williams, 1966; Gadgil and Bossert, 1970; Ashmole, 1971; Charnov and Krebs, 1974; Goodman, 1974; Wiley, 1974; Wittenberger, 1980). A critical distinction must be made between fecundity in any one year and life-long reproductive effort. Evidence is accumulating that life expectancy may be inversely proportional to annual reproductive output (Stonehouse, 1960; Perrins, 1965; Williams, 1966; Cody, 1971; but see Harris, 1970; Nelson, 1978, p. 366, for a different view). The possibility that delayed breeding or reduced seasonal effort may result in a larger overall lifetime output is particularly important in the case of long-lived marine birds, which may have life spans exceeding 30 years (Ashmole, 1971; Goodman, 1974). The demonstration in any one season that pairs are capable of raising more offspring than they attempt to raise (Vermeer, 1963; Nelson, 1964, 1978; Harris and Plumb, 1965; Harris, 1970; Ward, 1973) does not invalidate this hypothesis. It remains to be shown that any seabird limits its lifetime reproductive output below that which it could produce.

Thus for the purpose of this chapter I will assume that the seabird systems thus far studied reflect a long-term maximization of reproductive output and the allocation of time and energy to the components of reproduction is the result of optimization through natural selection. The challenge, then, is to identify the manner by which the variations in behavior and ecology are interrelated to yield maximum fitness.

The study of apparently "maladaptive" behavior may provide insights or provoke questions about more conventional behavior. During my ongoing studies of female–female pairing in gulls, I have become increasingly curious about how seabirds assess the quality of prospective mates. Assessment of mate quality requires the investment of time and energy. Too large an investment in this phase of the reproductive effort may diminish time or energy resources needed for other aspects of reproduction; too small an investment may result in taking an inappropriate or poor-quality partner.

Some characteristics of mate quality, such as age or experience, have been correlated with reproductive performance. But the mechanism by which the age or experience of the members of a pair actually translate into more or healthier young is poorly understood. There are some clues, discussed below, but the behavioral mechanisms are virtually unstudied.

II. COURTSHIP

Courtship behavior has been the object of a wide variety of studies of differing approaches. Ethologists have described courtship displays and focused on the analysis of their components and their derivation (Huxley, 1923; Lorenz, 1941; Tinbergen, 1952; Blest, 1961) and their motivation (Moynihan, 1955a,b, 1958; Bastock, 1967). Briefly, these authors and others have shown that displays have been ritualized through evolution for communication, and have generally been derived from activities that previously served a function unrelated to courtship. Analyses of "motivation" suggest that many of the displays associated with courtship result from conflicting tendencies of fear, aggression and mating (Bastock, 1967).

An alternative approach to the study of courtship is to ask questions about the function or selective value of courtship. Many species of seabirds have elaborate and lengthy courtship ceremonies and long delays between the initiation of courtship and the laying of eggs. These courtship activities therefore cannot be without cost (Nelson, 1977; Daly, 1978). It is thus instructive to investigate the balance between the costs and benefits of courtship activity. Courtship occurs in two contexts, that of finding a mate for the first time and that of reestablishing old pair bonds, but the basic problem of assessing the potential value of a possible future mate remains similar. The major difference in the two contexts is the information base on which the decision is made.

A. Cost of Courtship

The ultimate measure of cost in an evolutionary context is the reduction in the number of offspring contributed to future generations below the number the organism might have contributed had the cost not been incurred. This cost may manifest itself by a reduction in the number of years in which an organism reproduces, a reduction in the reproductive output in any one season, or both. For instance, in lizards of the genus *Cnemidophorus*, clutch size is larger in parthenogenetic species than in bisexual species of the same size (Congdon *et al.*, 1978). Congdon *et al.* (1978) believe this difference is attributable to an increase in energy allocated to competition in the bisexual species, but another hypothesis would be that the unisexual species does not require a large investment in courtship.

In species with larger investments in eggs or parental care, one would expect there to be a more careful assessment of mate quality than in those with smaller investments, all other things being equal. Thus it would be extremely interesting to estimate the costs associated with the protracted courtship periods of seabirds. In practice, it will be extraordinarily difficult to identify and segregate the energetic costs associated with courtship as a whole, let alone to assign costs to particular aspects of courtship. In contrast, it may be more feasible to measure the cost of time investment, because the cost of delay can be measured if we have an estimate of reproductive success as a function of laying date.

1. Cost of Reproduction

Four studies provide an indication of the cost of reproduction in terms of changes in mortality rates of adults. The effort of raising young increases adult mortality, and in the King Penguin (*Aptenodytes patagonica*) the mortality rates of adults raising chicks is greater than that of adults failing to raise chicks (Stonehouse, 1960). Similarly, Perrins (1965) found that the adult life span of titmice was shortened as the number of young raised per clutch increased.

Attempts by birds to breed at unusually young ages have also been associated with increased mortality. Mortality, as judged by return to the colony the year following breeding, of "young" breeding Adelie Penguins (*Pygoscelis adeliae*) was greater than that of "established" pairs (LeResche and Sladen, 1970), and Northern Fulmars (*Fulmar glacialis*) breeding for the first time at age six years had higher mortality rates than those commencing to breed at nine or twelve years (Ollason and Dunnet, 1978).

Loss of weight during the breeding season may also serve as a measure of the cost of reproduction (Nelson, 1977). Ingolfsson (1967) has shown for three species of gulls in Iceland that adult weights change seasonally. There was a steady decline in weight of both male and female Great Black-backed (*Larus marinus*), Iceland (*L. hyperboreus*) and Lesser Black-backed Gulls (*L. fuscus*) throughout the breeding season. He interpreted these changes to result from food supplies being limited as the biomass of gulls increased during the breeding season. However, changes in population biomass cannot account for loss of weight early in the spring before eggs were laid. This early weight loss may reflect the cost of preparation for breeding, although alternative hypotheses cannot be rejected without additional data comparing weight changes in breeding and nonbreeding birds.

Least Auklets (*Aethia pusilla*) on St. Lawrence Island also showed a decrease in adult weight as the breeding season progressed (Bédard, 1969*a*). Working on St. Paul Island in the Pribilofs, I too found a decrease in the weights of adult Least Auklets during the breeding season. These decreases were measured for the population as a whole and for a few banded individuals that were caught two or more times during the same season (G. L. Hunt, unpublished data). Weight loss during the reproductive season has been shown for numerous species of penguins including Adelie Penguins (Johnson and West, 1973), Yellow-eyed Penguins, *Megadyptes antipodes*, (Richdale, 1957), and Galapagos Penguins, *Spheniscus mediculus* (Boersma, 1977). Ricklefs (1974) summarizes recent studies on the energetic costs of reproduction, but the studies available provide more information on the cost of egg production and parental care than they do on the cost of courtship.

2. Estimating the Cost of Courtship

Several studies provide a means of estimating the cost of reproductive effort prior to egg laying. This estimate provides some indication of the cost of courtship, although, in the case of females, the cost of gamete production may be included. In the Black-legged Kittiwake (*Rissa tridactyla*), Coulson and Wooller (1976) have been able to show increased mortality in males during January–March, when kittiwakes are reestablishing nest site territories. In this season males have the highest mortality of any season, and there is the greatest disparity between male and female mortality rates (male mortality being 10 times female mortality). Although it is unclear whether such territorial behavior should be considered courtship, the holding of a nest site is a prerequisite to attracting a mate (Cullen, 1957), and the males suffer a cost for this aspect of mate attraction.

One group of marine birds, the penguins, seems particularly suited for estimating the cost of courtship. Many, if not most, species of penguins fast from the time they first reoccupy a colony until the clutch is complete. This period of territory establishment, courtship and pair formation can be interpreted as an investment in mate selection and future reproductive potential, and the time and energy (weight loss) associated with these activities can be separated from time and energy committed to reproductive effort after clutch initiation. Benefits of mate selection need not offset costs for the behavior to evolve. Costs need only be offset by overall reproductive output, but if benefits do not offset costs at each step, that step should become greatly reduced (J. Wittenberger, personal communication).

Table I assembles data available to me on investment in courtship by

Table I. Investments of Penguins in Reproduction

Species	Adult wt (g), arrival at colony		Time elapsed between arrival and egg laying (days)		Weight of clutch (g)	Length of incubation (days)	Length of chick care (days)	Percent wt loss in courtship		Source
	Male	Female	Male	Female				Male	Female	
Aptendodytes fosteri	34,000	26,000	30		—	62–64	150			Stonehouse (1960)
A. patagonica	16,000	14,000	19.1	13.9	410	53–55	300	20	22	Stonehouse (1960)
Pygoscelis adeliae	4,600	4,450	25—Signy 13—Hope Bay 28.2	10.2	223	33–37	56–63		26	Sladen (1957) Ainley (1975a)
Spheniscus magellanicus	3,800	3,200	21–28		225	43–44	85			Boswell and MacIver (1975)
S. mendiculus	2,183	1,870	1–14		—	38–42	30–50	8.7	8.1	Boersma (1977)
Eudyptes chrysocome	2,700	2,500	18	11	—	33	70			Warham (1963, 1971)
Eudyptes chrysolophus schlegeli			28	20	248.5	35	73			Warham (1971)

penguins. For this purpose, the period between arrival at the colony and egg laying is considered the investment in courtship. The amount of information is very small, but there is a modest trend toward increased investment in courtship as reproductive commitment increases, when all species excluding *Aptenodytes fosteri* are considered. In *Aptenodytes* there is an inverse relationship, but *A. fosteri* breeds at much higher latitudes and may have less time available because of a shorter summer season. In contrast, *Spheniscus magellanicus*, with a longer time commitment to parental care, invests more in courtship than does *S. mendiculus*. Additional information is needed, particularly from species breeding sympatrically.

A comparison of two species of albatross nesting at South Georgia Island suggests that in this group investment in the pre-egg stage may increase with increasing commitment to parental care. *Diomedea melanophris* breeds annually, spends 184 days in parental care and has a pre-egg period of 15.6 days (Tickell and Pinder, 1975). *D. chrysostoma*, in contrast, breeds every other year, if successful, and has a period of 213 days of parental care and 26.1 days elapse between return to the colony and egg-laying ($p < 0.001$). During the pre-egg stage, on the average *D. melanophris* males are present on the territory 12 days and females 2 days, while for *D. chrysostoma* the periods are 15 and 3 days for males and females, respectively (Tickell and Pinder, 1975). Thus in this pair of species breeding on the same island, the species with the greater investment in parental care may devote more time to the pre-egg stage.

A second point of equal interest is the observation that the investments of females in mate selection are not obviously greater than those of males. The female King Penguin apparently invests more energy than does the male, losing 230 g/day versus 170 g/day for the male from the time they return to the colony until the egg is laid. However, the males on the average invest about 5 days more during this period than do the females, so the weight loss of females is 22% in contrast to 20% in males [calculated from data in Stonehouse (1960)]. When polynya, nearshore openings in the ice, are absent, weight loss may approach 33% (Stonehouse, 1967). Weight loss in male and female Galapagos Penguins is almost equal during the period on the colony prior to egg laying, with males losing slightly more than females [calculated from data in Boersma (1977)].

Theorists have emphasized the unequal investment of the two sexes in gametes, and thus the need for females to exercise greater discrimination in mate selection (Bateman, 1948; Verner, 1964; Orians, 1969a; Trivers, 1972; Altmann *et al.*, 1977). In monogamous marine birds with large, nearly equal investments in a long period of parental care (incubation and chick raising), the overall investment of the two sexes may be nearly equal in spite of the unequal investment in gametes. Under these circumstances, it would not be surprising to find males and females investing equally in assessment of mate quality (see Section II.B.2).

In summary, costs associated with courtship may include increased mortality of one or both sexes due to stress or predation, and a decrease in the time or energy available for other aspects of reproduction. To date we have little information on the time or energy devoted to courtship, let alone data on the extent to which these expenditures affect the number of offspring produced over a lifetime. Data on these costs and how they are divided between the two sexes will help provide us with a much clearer understanding of the relative significance of courtship for males and females.

B. Benefits of Courtship

Courtship is an important part of the reproductive process in sexually reproducing species. There is a finite point below which reduced investment in courtship may result in reproductive failure. I now wish to examine some of these adaptive functions of courtship behavior as they pertain to seabirds.

1. Advertisement

It is standard textbook fare that courtship and the associated defense of a breeding territory serve to advertise the species, sexual identity, and status of an individual seeking a mate. Delay in acquiring a mate may result in late nesting, the possible loss of the best mates, and a concomitant reduction in reproductive success (Hunt and Hunt, 1976) or in failure to find a good nest site in a given season if nesting space is limited. Thus there is a premium for an unmated individual to acquire contact with appropriate prospective mates quickly.

Most initial pair formation in seabirds apparently takes place at the colony or nest site. In numerous species either males return to the colonies before females, or advertising displays have been described for birds on the colony. These observations suggest pair formation takes place in the colony and not at sea. The existence of advertisement displays at the breeding site does not, however, preclude the possibility of preliminary pair formation before return to the breeding area; pair formation in ducks is known to occur before they reach the nesting grounds (Dane et al., 1959; Johnsgard, 1965). The nature of social behavior patterns at sea away from the colony is an area in which our knowledge of seabird behavior is sorely lacking.

At large, crowded colonies, birds seeking mates have two options: they may establish a nesting territory and seek to attract a mate, or they may gather together to seek mates. If a central location is used, then defense of a territory is compromised and the individual, once mated, may find it difficult to obtain a nest site. Although use of an area segregated from the future nesting territory is apparently uncommon in seabirds, pair formation in "clubs" (as well as on nesting territories) is known for gulls [Herring Gull, *L. argentatus* (Tinbergen, 1953); Western Gull, *L. occidentalis* (A. Newman, unpublished observation); but not Lesser Black-backed Gulls (*L. fuscus*) (Brown, 1967)], on the water for some alcids [murres, *Uria* (Tuck, 1961); Pigeon Guillemots, *Cepphus columba* (Drent, 1965); Common Puffins, *Fratercula arctica* (Nettleship, 1972)], or on snow banks before nest sites are exposed [Parakeet Auklet, *Cyclorrhynchus psittacula* (Sealy and Bédard, 1973)]. *Sterna sandvicensis* pairs before taking up a territory, and then settles in synchronized subcolonies (Langham, 1974). The same is true for *S. dougallii*; such behavior is not maladaptive because nesting space is not limiting in these species (I. Nisbet, personal communication). King Penguins pair while part of a large group before pairs move off to establish nest sites (Stonehouse, 1960). Emperor Penguins (*A. forsteri*) also pair while part of a large group (Stonehouse, 1953).

In contrast, the males of most sphenisciform, procellariform, and pelicaniform species establish a nesting territory before attracting a female. The males are thereby guaranteed a nest site if they obtain a mate. The female, in turn has an opportunity to assess the quality of her prospective mate based on the quality of the territory or nest site that he has been able to secure. We may then ask under what circumstances advertisement for a mate should be done in a group situation rather than on individual territories.

A shortage of available nest sites or material with which to build nests should select for pairing behavior at the nest site or territory. Breeding opportunities of many burrow-nesting species may be limited by competition for burrows (Swales, 1965; Bédard, 1969b). Likewise, some Pelicaniformes are nortorious stealers of nest material, and unattended nests may be quickly dismantled [cormorants (Hunt, personal observation); gannets (Nelson, 1978)]. Nest-stone stealing is also common in Adelie Penguins for which stone nest platforms are essential for breeding (Yeates, 1975; Spurr, 1975; Ainley, 1975b). Thus it is advantageous to both the male and the female to have a place to nest, or material with which to nest, before investing heavily in pair formation and courtship. In addition as mentioned above, the female can use information about the quality of the territory to assess the quality of the male.

If pair formation on the nesting site is advantageous to both males and females, we may ask why pair formation should ever take place in a communal setting. In the case of the Emperor Penguin it is no surprise, as this species does not build a nest and conserves energy by staying in groups (Stonehouse, 1953). With ground-nesting gulls, pairing can take place either in clubs or on the nesting territory, which is not likely to be usurped when birds are absent for brief periods (Hunt, personal observation). Why gull territories are less vulnerable to invasion, or more easily reclaimed by returning owner than may be true of nest sites of other species, is not known.

The observations of group pairing activities in alcids are harder to rationalize, as many of the species involved at least occasionally compete for an apparently limited number of nest sites (personal observation; Bédard, 1969b). One explanation, based on group selection, would be that these are epideictic displays which allow birds to assess their numbers relative to food supplies (Wynne-Edwards, 1962). An alternative explanation is that gatherings of birds seeking mates facilitates rapid pairing. Nelson (1978), working with gannets, points out that in a large, congested colony, females seeking a mate may have difficulty in locating unmated males. Locating unmated murres in their vast loomeries surely would be no less difficult, and possibly an alternative solution to mate location has been to evolve communal displays.

2. Assessment of Mate Suitability

Species Identification. Choosing a mate requires not only identification of the species and sex of a prospective partner, but also assessment of the potential for enhancement of fitness of offspring. The importance of specific reproductive isolation and the potential for reduced production of offspring or the production of maladapted or infertile offspring by hybrids has long been recognized by evolutionists. Smith (1966), Patton and Weisbrod (1974), and Hoffman et al. (1978) discuss species isolating mechanisms in gulls and the problem of hybridization. Of interest is the lack of indications of reduced fertility in the hybrid gulls studied by Ingolfsson (1970) and Hoffman et al. (1978). Hybridization appears to be more frequent in gulls than in other groups of seabirds, and in gulls it may have been facilitated because range expansion, due to the availability of refuse, has allowed previously isolated populations to come in contact (Ingolfsson, 1970).

Sex Identification. Until recently, little attention has been paid to the problem of identifying the sex of a potential mate. However, with the finding of female–female pairs in three species of gulls [Western Gull (Hunt and Hunt, 1977); Ring-billed Gull, *L. delewarensis* (Ryder and Somppi, 1979); Ring-billed and California Gull, *L. californicus* (Conover et al., 1979)], attention needs to be focused on the mechanism for identifying sex. Plumages of the two sexes in Sphenisciformes, most Procellariiformes, most Pelicaniformes, Stercoradiae, Laridae, Sternidae, and Alcidae are virtually identical. Exceptions include two boobies [*Sula leucogaster* and *S. nebouxii* (Nelson, 1970, 1978)], frigatebirds (*Frigata*), tropicbirds [*Phaethon lepturus* (Stonehouse, 1962)], and the Wandering Albatross [*Diomedea exulans* (Murphy, 1936, p. 558)]. In many groups, males are larger or heavier than females (Harris and Hope Jones, 1969; Warham, 1963, 1975), but in other species there is no apparent means of recognizing sex on the basis of external characteristics.

Behavioral differences between males and females provide a mechanism by which partners of the opposite sex may be chosen. Prior to the return of females, male occupancy of a territory within which courtship will take place should, in itself, provide sexual identity. Sexual identification will be reinforced in these cases if males are noticeably larger than females.

Brown (1967) provides an excellent account of pair formation in the migratory Lesser Black-backed Gull, in which males take up territories upon arrival in the colony. Territorial males initially tend to drive off any gull, of either sex, landing in their territory. However, females landing in a male's territory use distance-reducing, nonaggressive displays, whereas intruding males usually give aggressive or assertive displays. As the season progresses, a persistent female will come to be recognized and accepted by the male, and a pair will form. These behavioral sources of information might not be available to species of gulls in which pair formation occurred in clubs. The mechanisms used by gulls pairing in clubs are not known.

Pairing mechanisms occasionally fail to prevent homosexual matings. Despite the low incidence of female–female pairings in Western, Ring-billed, and California Gulls, it can be assumed that females of these and other species have a means of distinguishing appropriate partners that normally operates to produce heterosexual pairs. Given the existence of such a mechanism, it is then curious as to how female–female pairs can form.

One way individuals may choose their partners is to identify sex *per se*, and then to choose among individuals of different quality within the appropriate sex class. If this hypothesis were true, homosexual pairings would never occur. From an evolutionary perspective, pairings with a member of the opposite sex is so fundamental for the production of offspring that there would be a very strong mechanism to ensure heterosexual matings. The presence of distinctively different male and female plumages or qualitative differences in behavior between males and females in many species is consistent with this hypothesis of sex recognition.

An alternative hypothesis may apply to sexually monomorphic species. The presence of female–female pairing may indicate the lack of a mechanism whereby the sex of a potential partner is recognized. Rather, potential partners may be judged on a variety of characteristics that vary on an essentially quantitative basis between males and females.

For monomorphic species there may be no qualitative character that absolutely and unequivocably distinguishes male from female during pair formation. The lack of unequivocable sex-identification signals suggests that, in the evolution of pairing behavior of some species, sexual identification *per se* was not required. Heterosexual pairing resulted from the selection of individuals characterized by a suite of desirable characteristics. As long as appropriate partners of the opposite sex were available, all pairings would be heterosexual; when the "best" available individuals were of the same sex, homosexual matings would ensue.

Our recent investigations of Western Gulls suggest that such an hypothesis is not unreasonable. These birds pair not only on the male's territory, but also in clubs. Newman *et al.* (unpublished manuscript) have examined size differences and behavioral differences of males and females in heterosexual and homosexual pairs. They have measured numerous Western Gulls for which sex has been ascertained by laparotomy and have found males to be significantly larger than females ($p < 0.001$), with ratios of male-to-female weight of 1.24; bill length, 1.10; bill depth, 1.11; and head length plus bill length, 1.09. However, they have also found a considerable degree of overlap between unusually large females and unusually small males.

If size is one of the characteristics on which female gulls base their choice of mate (Harris and Hope Jones, 1969), then it is not surprising that females always mate with a male larger than themselves. If males are not available, then females would choose to mate with the largest females available—however, in female–female pairs, because both females will be seeking large mates, the largest mate that either could get might be a bird of its own size. In this case, members of female–female pairs would be of similar size. This might explain why it has proved difficult to tell members of female–female pairs apart when they are not marked.

We have also obtained both qualitative and quantitative descriptions of male and female behavior in Western Gulls. Although our sample of observations of the initial stages of pair formation is very small, we have found no behaviors unique to either sex in the pre-egg stage. Likewise, we found no significant differences in the rates of "long-calling," "mew calling," or "choking" by the two sexes [see Tinbergen (1953, 1959) for description of these behaviors]. Females do, however, head-toss more often, and males more commonly display courtship feeding and mounting (Newman *et al.*, unpublished manuscript). These results are consistent with the hypothesis that in Western Gulls there are no qualitative differences in size or behavior by which males and females can be unequivocably identified. Likewise in terns, there are behavioral characteristics that are 90% associated with sex, but some males will perform "female" behaviors for a substantial period in order to obtain short-term advantages (I. Nisbet, personal communication).

Individual Quality. The quality of a partner may influence not only the number and quality of offspring raised, but also the ease with which they are brought to fledging. Evidence that reproductive success improves with age or experience has been obtained for a number of species (see Section

III). At least a portion of this improvement results from improved reproductive effort related to birds remaining paired in successive breeding seasons.

The competence of a bird's partner may result in reduced stress for a given level of reproductive success. In the Glaucous-winged Gull, *L. glaucescens*, begging by the chick influences the effort of parents in the provision of food (Henderson, 1972). If the partner is more efficient at foraging [older birds are more efficient (Orians, 1969b; Recher and Recher, 1969; Dunn, 1972; Ainley and Schlatter, 1972; Morrison *et al.*, 1978; Searcy, 1978)], the chicks can be raised with reduced stress on both parents, and adult postbreeding survival may be greater. It would be valuable to compare adult provisioning effort and parental postbreeding mortality.

Given the obvious advantages of mating with a high-quality partner, it would be interesting to know the criteria by which birds select mates. Ainley (1975a) suggests that female Adelie Penguins may be able to recognize immaturity in a male's actions, and thus avoid pairing with less experienced males. Experienced birds select partners of similar levels of experience more often than would occur by chance (Coulson, 1966; Brooke, 1978), and experienced females may mate with males of greater experience than their own (Richdale, 1957; Ollason and Dunnet, 1978). On the basis of the assumption that females have a greater reproductive investment owing to the production of eggs, one might have expected this result (Orians, 1969a). However, in Black-legged Kittiwakes (Coulson, 1966), Red-billed Gulls, *L. novaehollandiae* (Mills, 1973), and Parasitic Jaegers (*Stercorarius parasiticus*: Davis, 1976) females generally mate with less experienced males. This finding is at odds with the theoretical prediction above, although in the kittiwake it may result from the slight shortage of males caused by higher male mortality rates (Coulson and Wooller, 1976; J. C. Coulson, personal communication).

3. Courtship Feeding

Occurrence. Among marine birds, passing of food from males to females prior to clutch completion (courtship feeding) is a significant feature of the courtship of skuas (Stonehouse, 1956; Burton, 1968; Anderson, 1971), gulls (Lack, 1940; Tinbergen, 1953; Brown, 1967; G. L. Hunt, personal observation) and terns (Cullen and Ashmole, 1963; Brown, 1973; Nisbet, 1977). I have been unable to find reference to courtship feeding in the Sphenisciformes, Procellariiformes, or Alcidae. Murphy (1936, p. 395) does describe male Adelie Penguins providing their mates with snow, which is apparently accepted for its water content. Likewise Waite (1909) (also cited in Murphy, 1936, p. 553), in describing courtship in the Wandering Albatross, noted that drops of oil ooze from the beak (sex of bird not stated) during "billing" between mates, but Murphy does not state that there is a transfer of oil to the partner. In Dovekies (*Plotus alle*) "presenting stones" may serve in courtship and mate assessment (Ferdinand, 1970).

There exist reports of "courtship feeding" in Thick-billed Murres, *Uria lomvia* (Pennycuick, 1956) and White-tailed Tropicbirds, *Phaethon lepturus* (Gross, 1912; also cited in Van Tets, 1965). The former report apparently referred to passing fish from an arriving adult to a brooding adult, as all of Pennycuick's observations were made during the chick phase. Gross' observation of "courtship feeding" in White-tailed Tropicbirds was of an adult feeding its incubating mate and is apparently the same observation as the one attributed to Murphy (1936) by Lack (1968). Provisioning of an incubating female by the male should be distinguished from courtship feeding, because it relates to partitioning of investment in parental care rather than in mate selection.

Van Tets (1965) categorically states that there is no courtship feeding in the Pelicaniformes, and Nelson (1978) reports no courtship feeding in the Sulidae. He describes "incipient courtship feeding" in the Brown Booby (*Sula leucogaster*) in which, after "bill touching," one bird may put its bill inside the other's mouth. Although "pumping" movements similar to those used in transferring food from parent to young are used, no food is transferred (Nelson, 1978, pp. 491–492). If courtship feeding occurs in any group of seabirds other than skuas, gulls, and terns, it is certainly not common.

Mate Selection. Lack (1940) was the first to synthesize the data available on courtship feeding. He emphasized its symbolic nature and its presumed function in maintaining the pair bond. In this early paper, Lack specifically denied a nutritive role for courtship feeding, although in his later writing, he was pursuaded that the passage of food from male to female could be of energetic importance to the female (Lack, 1968). I will return to the question of the energetic value of courtship feeding in the next section. First I wish to discuss its possible use in choice of partners.

Courtship feeding in gulls and terns provides a potential mechanism for the assessment of mate quality by the female (Nisbet, 1973). The female can obtain information on the size or quantity of the food the male is able to obtain and the frequency with which the male can provide it. This information may provide important clues as to a particular male's ability to provide for her or for her offspring (Nisbet, 1973, 1977). In addition, since terns and gulls may have preferences for particular food types or foraging methods, or hold foraging territories, information on the kinds of food brought may provide the female with data on the future access of offspring to food. Once young are fledged, they may gain entrance to parental territories and access to food at a critical time (Drury and Smith, 1968), although Davis (1975*b* did not find young Herring Gulls foraging with their parents (discussed by Burger, this volume).

There are three important elements to this early phase of courtship and the initial courtship feeding of the Common Tern (*Sterna hirundo*). First, the food items provided by the males are carried in their bills, and the size and quality of fish are readily identified by the females. Second, pairs frequently form and break up in this first phase; in the second phase ("honeymoon phase") pairs are more stable (Nisbet, 1977). Third, males are at first reluctant to give up fish to the females. This series of interactions clearly provides the female with the opportunity to exercise choice with respect to male foraging ability or to quality of nesting territory.

Males may also be exercising choice in this behavior. Whereas the emphasis has been on female investment (Orians, 1969a), male terns may provide a substantial amount of the energy for egg production (see below). Therefore, they will also have a strong vested interest in the quality of their mates. Their initial reluctance to surrender fish to females may be an indication of the exercise of this discrimination. I. Nisbet (personal communication) has found that male Common Terns may assess at least the "intentions" of a prospective mate by retaining their fish until the female has completed several courtship flights and has repeatedly landed in the male's territory and has joined him in scraping. Likewise, in his descriptions of courtship feeding in Lesser Black-backed Gulls, Brown (1967) comments on the reluctance of males to surrender food to females, particularly early in the season. In these species the carrying and presentation of food by males is very obviously a testing of the female's persistence and determination.

J. Wittenberger (personal communication) has suggested that male Bobolinks (*Dolichonyx oryzivorus*) chase newly arrived females from the nesting territories in order to test whether the females have a serious intention of mating there or are merely seeking food before selecting a mate elsewhere. The common pattern of male birds initially attacking and chasing females that enter nesting territories may serve the function of discouraging females with only a weak commitment to breeding with that male. This behavior would provide for a conservation of any resource the female might deplete and would prevent males from wasting courtship effort on females not receptive to mating (J. Wittenberger, personal communication). In the case of terns and gulls, the reluctance of the male to feed the female would save the male time lost from courtship while seeking a new supply of food. Male reluctance to feed females may also be related to sex recognition, as casually interested "females" may actually prove to be males masquerading as females in order to gain fish (I. Nisbet, personal communication). If the female were to take the food and leave, the male would lose his food offering. Only if there is a reasonable prospect of the female remaining with the male is it worth his while to invest in her. Thus courtship feeding provides both females and males with the opportunity to assess the qualities and intentions of a prospective mate. Recently,

Gladstone (1979) developed hypotheses concerning the importance of male investment and male choice in mate selection in monogamous colonial birds.

Provision of Energy to the Female. Assessment of mate quality should be important for females in virtually all species, and for males in monogamous species in most cases. If a prime function or selective value of courtship feeding were the assessment of the quality of prospective mates, one might expect courtship feeding to be a widespread phenomenon among seabirds. As shown above, it is not. We may then ask what sets apart those species in which courtship feeding occurs.

The distribution of courtship feeding among marine birds supports the hypothesis that its prime function is to provide energy to the female. For species in which the clutch represents a substantial proportion of the female's body weight, there should be special adaptations whereby the female can assemble the energy required for egg production. Two such adaptations are egg neglect and courtship feeding. Egg neglect, in which the nest is temporarily deserted by one or both parents after an egg is laid, is possible only in burrow- or cavity-nesting species. For surface-nesting species, in which temporary desertion of eggs would result in exposure to predation or overheating, egg neglect is not a viable alternative for gaining energy. Instead, in these species courtship feeding may allow the female to obtain the energy for the production of an energetically expensive clutch without having to leave the nest for long periods.

As discussed above, courtship feeding is common in skuas, gulls, and terns. Gulls and terns have exceptionally heavy clutches with respect to adult body weight, with only storm petrels (25% body weight: Lack, 1968) and murrelets (16–48% adult weight: Sealy, 1975b) having as heavy clutches in proportion to body weight (Table II). As gulls and terns may suffer egg mortality both to overheating and to predation, it is not surprising that females do not have an extended period of absence from the nest after laying the first egg of a multiple-egg clutch. Rather, courtship feeding, which is continued until completion of egg formation (Cullen and Ashmole, 1963; Brown, 1967; Nisbet, 1973), provides them with the necessary energy.

Burrow-nesting or cavity-nesting shearwaters, storm petrels, and murrelets do not rely on courtship feeding. In Procellariforms, which lay single-egg clutches, the females may be absent from the nest for several days to three weeks before egg laying, and they may leave again after the egg is laid (Lack, 1966; Boersma and Wheelwright, 1979). Even if the male leaves the egg unattended during a foraging bout, it will not be exposed to either overheating or most forms of predation. Similarly, Xantus' Murrelets (*Endomychura hypoleuca*) and Ancient Murrelets (*Synthliboramphus antiquus*) have a period of up to 7 days when the egg is left unattended until the female returns to lay the second egg (Sealy, 1975b; Murray et al.,

Table II. Relationship of Clutch Weight to Adult Body Weight

Family	Number of species on which calculation is based	Clutch weight as percent adult body weight		Source
		Mean	Extremes	
Sphenisciformes	15	5	2–10	Lack (1968)
Diomedeidae	5	7	5–11	Lack (1968)
Procellariidae	14	15	12–21	Lack (1968)
Hydrobatidae	6	25	20–29	Lack (1968)
Phaethontidae	3	11	9–13	Lack (1968)
Pelicanidae	1	9		Lack (1968)
Sulidae	9	6	3–8	Nelson (1978)
Phalacrocoracidae	5	6	6–9	Lack (1968)
Larinae	7	29	21–45	Lack (1968)
Sterninae	9	29	18–46	Lack (1968)
Alcidae	18	17	9–48	Sealy (1975)

unpublished manuscript). These small alcids produce the heaviest eggs in proportion to body weight of any of the alcids (Table II; see also Sealy, 1975b). With a clutch size of two eggs representing 40–50% of the female's weight, it is not surprising that egg neglect is practiced by the female. However, it is not clear why the males do not protect their investment in the nest site more closely by guarding the first-laid egg.

We may ask why these species, particularly the murrelets, rely on egg neglect rather than on courtship feeding. It may be that their food supplies are very patchy (Sealy, 1975a), and it would be inefficient for the male to attempt to bring food to the female. The fact that the precocial young of these species leave to forage with their parents at two days of age supports the hypothesis that it is more efficient to move individuals to the foraging grounds than it is to bring food to them (Sealy, 1972, 1973). Courtship feeding in the terns and gulls may be an energetically less efficient solution that is accepted because of the exposed nature of their nests. It would be interesting to know the frequency of courtship feeding in tropical terns, many of which have single-egg clutches and forage far from the colony. For example, Brown (1973) working on Manana Island, Hawaii found that Brown Noddy Terns (*Anous stolidus*) had courtship feeding, whereas he never recorded courtship feeding for Sooty Terns (*Sterna fuscata*). His data on incubation shifts suggest that Sooty Terns forage further offshore than do the Brown Noddies, although absences for foraging during the chick phase were equal in the two species. Dinsmore (1971) also reported courtship feeding as rare in Sooty Terns in the Dry Tortugas.

Royama (1966) was the first to stress the energetic importance of courtship feeding. Royama and later Brown (1967), Schreiber and Ashmole (1970), and Krebs (1970) suggested that the male provides the female with a significant amount of food. Male Great Tits (*Parus major*) and Blue Tits (*P. caeruleus*) feed the female off the nest early in the season, before eggs are laid, and also on the nest after incubation begins (Royama, 1966). Royama demonstrated that during this period the female did not have sufficient time in which to gather all the food she needed, and that the provision of food to the female by the male was essential for egg production. To this end, courtship feeding reached its peak frequency during egg-laying, not during early courtship or just before coition as would be expected if the food-giving had a primarily symbolic role. A similar peak of courtship feeding occurs just prior to egg-laying in the Black Noddy, *Anous tenuirostris* (Cullen and Ashmole, 1963) and Lesser Black-backed Gulls (Brown, 1967). These authors reached the conclusion that courtship feeding serves a primarily nutritional function. Krebs (1970), working with the Blue Tit, added the important finding that the female was able to gain more food per unit time by soliciting courtship feedings from her mate than she could have obtained by devoting her entire efforts to foraging for herself.

Courtship feeding makes a significant contribution not only to egg size, but also ultimately to reproductive success (Nisbet, 1973, 1977). Nisbet found that male Common Terns provided females with a substantial amount of food, particularly just before egg laying, when the female may be unable to forage as effectively as usual due to the added weight of developing eggs. Egg size and clutch size were both positively correlated with the amount of food provided by males.

Courtship feeding also influences egg size in Western Gulls. Members of female–female pairs presumably receive no courtship feeding by males and have no net input of food via courtship feeding, even when a female may rarely show behavior that can be construed as courtship feeding (Hunt and Hunt, 1977; Hunt *et al.*, unpublished manuscript; Newman *et al.*, unpublished manuscript). Table III and Fig. 1 show that eggs produced by female–female pairs are significantly smaller than those produced by heterosexual pairs. Although most of these eggs are infertile (Hunt and Hunt, 1973, 1977), even if they were fertile, they would be expected to have a lower hatching success and posthatching survival because of their smaller size (Parsons, 1970; Nisbet, 1978).

Parsons (1970) and Nisbet (1978) have shown by egg-exchange experiments with Herring Gulls and Common and Roseate Terns (*S. dougallii*), respectively, that early chick survival is a function of egg size, regardless of adult quality. In addition, Parsons's (1970) work shows that large eggs have greater hatching success. Thus courtship feeding in gulls and terns may influence reproductive success not only through changes in clutch size, but also through affecting hatching and posthatching survival of young.

A second indication that food provision by the male may be important

Table III. Volumes of Eggs Laid by Western Gulls in
Heterosexual and Female–Female Pairs[a]

Order of laying	Pairs	
	Heterosexual[b]	Female–female[c]
1	81.0 ± 5.0	80.4 ± 3.9
2	82.8 ± 4.5	75.8 ± 4.8
3	75.4 ± 2.8	77.3 ± 3.5
4	—	68.2 ± 8.1
5	—	69.6 ± 3.9
6	—	67.8

[a] Volumes (in milliliters) and standard deviations were calculated using the formula $v = kld^2$, where l = length, d = breadth, and k is a constant = 0.476 (Harris, 1964).
[b] Heterosexual pairs: 18 clutches (all c/3).
[c] Female–female pairs: 6 clutches (three c/4, two c/5, and one c/6).

to female Western Gulls is provided by a comparison of the amount of time females in heterosexual and female–female pairs spend on territory just prior to egg laying. During the two years in which we studied them, females in female–female pairs tended to spend less time on territory than did females in heterosexual pairs during the 10 days prior to clutch initiation. Heterosexually paired females showed a marked increase in territory attendance during this period while, at least in one year, homosexually mated females did not show a similar increase. Earlier in the breeding season there were no apparent differences in territory attendance between heterosexually paired females and females in female–female pairs (Newman et al., unpublished manuscript). In Red-billed Gulls, Mills (1979) has found a similar increase in female attendance on the territory during the last 10 days before egg laying to be associated with increased courtship feeding.

4. Synchronization of Breeding

Courtship provides other benefits in addition to mate assessment and transfer of energy to the female. In particular, courtship and other reproductive displays may be important for the synchronization of reproductive cycles and breakdown of distance barriers between members of a pair (Lehrman, 1955, 1959, 1964; Brown, 1967), and for the synchronization of the colony, or subsections of the colony (Darling, 1938; Collias et al., 1971; MacRoberts and MacRoberts, 1972; Southern, 1974; Nelson, 1978). This synchrony in breeding in a colony may have important implications for protection against attacks by predators (Kruuk, 1964; Patterson, 1965; Nisbet, 1975). These aspects of courtship are outside the scope of this chapter, but are discussed by M. Gochfeld in Chapter 7.

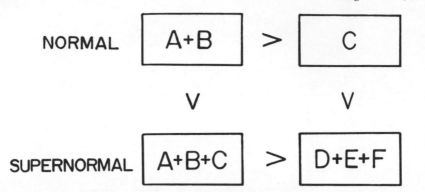

Fig. 1. The relationship of egg volume to order of laying within and between normal and supernormal clutches of Western Gulls. Differences between each box are statistically different ($p < 0.001$). Groups within each box were not statistically different ($p > 0.05$). Volumes of the C eggs in normal clutches were not significantly different from the A, B, and C eggs in supernormal clutches.

III. THE ROLE OF EXPERIENCE AND MATE FIDELITY

A particularly striking relationship between reproductive success and age or previous breeding experience has been demonstrated for a large number of seabird species. This aspect of mate quality has immense importance for long-lived marine birds, in which lifetime reproductive output can be significantly affected by the frequency of changing mates and the age or previous experience of mates. Although correlations between reproductive success and age or experience are available, we have a disappointingly poor understanding of the mechanisms whereby age or experience result in the improvement of reproductive output, or how the older or more experienced individual is identified during pair formation.

A. Age

Although Austin (1945) was the first to relate age to reproductive performance, most early work failed to differentiate the roles of age and experience (Richdale, 1957; Coulson and White, 1958, 1960; Ryder, 1975). Nelson (1966) and Coulson (1966) were the first to separate the effects of age and the experience of the partner. Coulson (1966) recognized the importance not only of the experience of the partner, but also the significance of continuing experience with a specific partner (see Section III.B.1).

1. Reproductive Success

Age, by itself, has a significant effect on reproductive success. Richdale's (1957) data show that Yellow-eyed Penguins breeding for the first time at age two did less well than those first breeding at age three. Ainley and Schlatter (1972) ascribed the increased reproductive success of six-year old Adelie Penguins, compared to three year olds, to improved foraging ability, regardless of previous breeding experience. Nelson (1966) discounted the role of additional foraging experience in gannets. Rather, Nelson emphasized that the poorer success of inexperienced birds was primarily the result of inadequate incubation and care of small young; he suggested that the behavior patterns associated with reproduction may mature slowly, as in the Ring Dove (Lehrman, 1955).

2. Egg Size

Variation in egg size has been correlated with the age both of females and their mates in the Red-billed Gull (Mills, 1979). In this species, egg size was not significantly affected by change or resumption of the pair bond with the previous year's mate. Black-legged Kittiwakes also show an increase in egg size with age (Coulson, 1963). In contrast, Davis (1975a), working with Herring Gulls, suggested that egg size declines with age after an initial increase during the first 7 or 8 years of age. This pattern is similar to that found in the Yellow-eyed Penguin (Richdale, 1957). Mills (1979) showed that timing of breeding in the Red-billed Gull has a greater influence on egg size than the age of the parents. Apparently, the age of the male or female affects egg size differently in different species; it would be of interest to know why these differences occur.

Increased egg size results in improved chick survival, at least during the first few days after hatching (Parsons, 1970, 1975; Nisbet, 1973, 1978; Davis, 1975a). In a series of manipulative experiments, Parsons demonstrated that hatching success, regardless of order of laying, was correlated with egg size. Furthermore, the egg-switching experiments of Parsons (1970, 1975) and Nisbet (1978) showed that the larger, more vigorous chicks associated with large eggs have higher survival rates in the first few days posthatching, regardless of whether their foster parents had laid large or small eggs. Nisbet's studies were controlled for the effects of timing of breeding within the season.

Although Davis (1975a) found a correlation between Herring Gull egg size and chick survival in one season, there was no correlation between egg size and chick survival, above a minimum egg size, in a second season. Davis therefore argued that parental age and not egg size was responsible for the apparent correlation between egg size and chick survival in Herring Gulls. Regardless of Davis's results in the second season, the work of Lundberg and Vaisanen (1979) and the manipulative experiments of Parsons (1970) and Nisbet (1978) make a clear case for the importance of egg size. That fluctuations in environmental conditions or correlations with parental age may occasionally mask the contribution of egg size to reproductive success does not prove that egg size is unimportant.

It is puzzling that egg size should decrease in older birds. Davis (1975a) speculates that the greater experience of the older birds compensates for the reduced survival of chicks hatching at lower weights. Davis suggests that the smaller eggs may reduce strain on the adult, thereby enhancing adult survival. This appears to be a reasonable hypothesis given the large investment in eggs relative to body weight in gulls, but it does not seem valid for penguins. Data on the long-term effects on fitness of trade-offs between adult and chick survival are needed.

These studies and others suggest that the age of a prospective mate is an important quality to assess in mate selection. With the possible exception of those species in which egg size (and possibly reproductive success) may decrease in older birds (Richdale, 1957; Davis, 1975a), birds should select older partners when they have to form a new pair. These older birds are likely to lay or promote laying earlier in the season (Nelson, 1966; Coulson, 1966; Davis, 1975a, 1976; Hays, 1978; Brooke, 1978; Blues and Keahey, 1978; Mills, 1979; except in Yellow-eyed Penguins, Richdale, 1957), have a larger clutch size (Richdale, 1957; Coulson, 1966; Coulson and Horobin, 1976; Blues and Keahey, 1978; Hays, 1978), lay larger eggs (Coulson, 1963; Nelson, 1966; Coulson and Horobin, 1976; Brooke, 1978; Nisbet, 1978; Mills, 1979), and enjoy greater overall reproductive success (Austin, 1945; Richdale, 1957; Coulson and White, 1958; Ainley and Schlatter, 1972; Mills, 1973; Coulson and Horobin, 1976; Davis, 1976; Brooke, 1978; Blues and Keahey, 1978; Dunnet and Ollason, 1978; Ollason and Dunnet, 1978). Because the effects of both age and timing of breeding appear to contribute to improved reproductive success, well-designed experimental studies are required to identify the contributions of these two factors.

B. Experience

1. Reproductive Success

Experience, particularly in terms of continued pairing with a given mate, has been recognized apart from age to have important influence on reproductive success. In the Black-legged Kittiwake, the retention of the same mate allows females to breed earlier than females of comparable age that take a new mate, to produce larger clutches, and to successfully hatch and raise more chicks (Coulson, 1966, 1972). Likewise, in Atlantic Gannets (*Sula basanus*, Nelson, 1966, 1972), Parasitic Jaegers (Davis, 1976), Manx Shearwaters, *Puffinus puffinus* (Brooke, 1978), and Northern Fulmars (Ollason and Dunnet, 1978), breeding success has generally been found to improve with the duration of the pair bond across breeding seasons. An exception to this rule is the decrease in reproductive success immediately before fulmars "divorce" or change nest site (Ollason and Dunnet, 1978). Kittiwakes are also more likely to change mates after reproductive failure (Coulson, 1966, 1972). For pairs breeding for the first time, pairing with an experienced partner may improve reproductive success [Gannet (Nelson, 1966), Parasitic Jaeger (Davis, 1976), Northern Fulmar (Ollason and Dunnet, 1978)]. In contrast, the Galapagos Flightless Cormorant (*Nannopterum harrisi*) usually changes mates, nest sites, and often nesting area after each breeding attempt (Harris, 1979). The reason for this behavior is unclear, but these changes do not appear to result in reduced reproductive success.

The experience of the male is as important as that of the female in contributing to increased reproductive success. Mating with an experienced male enables female gannets to breed at an earlier age and to lay earlier in the season than would be predicted by female age alone (Nelson, 1966). In Parasitic Jaegers and Manx Shearwaters, the sex of the experienced bird was not important in determining the extent to which experience enhanced reproductive success or hatching date. These results suggest that the contribution of the male, although qualitatively different from that of the female, is roughly of equal magnitude. In contrast to the other species discussed, the relative experience of male and female Northern Fulmars affects reproductive success differently (Ollason Dunnet, 1978). Breeding success improves with male experience up to at least 21 breeding years, but in females it begins to decline after 17 years. Why fulmars are different is unknown.

2. Mode of Action

Coordinating and performing incubation and early chick care appear to be affected most by experience (Coulson and White, 1958; Nelson, 1966, 1972; Coulson, 1972; Brooke, 1978), although timing of clutch initiation and clutch and egg size may also be involved (Coulson, 1966; Nelson, 1966; Mills, 1979). In gannets, Nelson (1972) suggested that the "special relationship with their partner"—not age—was crucial in determining the form of the greeting ceremony, a behavior by which pair bonds are maintained and which presumably would affect timing of clutch initiation. Egg size may be influenced by age or previous breeding experience of male Red-billed Gulls through improved courtship feeding (Mills, 1979), but this study provides no data on the linkage between male age or past breeding experience and the quality of courtship feeding. Older terns may have "better" nesting and feeding territories and may be more proficient in copulation and feeding young (I. Nisbet, personal communication).

Incubation by first time breeders is less effective in gannets because birds may hold the egg incorrectly (Nelson, 1966). In Black-legged Kittiwakes, Coulson and White (1958) have found that first-time breeders have problems coordinating incubation between members of a pair, completing incubation, and making the transition from incubation to brooding and feeding newly hatched chicks. Coulson (1972) used radioactive tags to obtain continuous records of incubation trade-offs in two pair of kittiwakes. One pair failed when the male developed an erratic incubation pattern and did not return for 5 days. Coulson did not state the ages or levels of experience of the birds involved. There is still much to be learned about the mechanisms whereby experience in general or experience with a particular partner is translated into reproductive success.

3. Mate Choice

Mates should be chosen on the basis of expected future reproductive value. If the experience of a mate has an important impact on reproductive success, or if reproductive output increases with age, then older partners should be chosen, whereas if reproductive output decreases after some point, younger mates should be then sought. Furthermore, because a considerable cost in terms of reduced reproductive success with a new partner is incurred when a mate is lost by death, the partner chosen should be young enough to have a reasonable life expectancy. The optimal age of a new partner will be a function of its influence on the chooser's reproductive potential.

In the two species in which female reproductive success declines at advanced age [Yellow-eyed Penguin (Richdale, 1957), and Northern Fulmar (Ollason and Dunnet, 1978)], males tend to pair with younger females. In contrast, where female reproductive output increases steadily or shows no decline with age, females tend to pair with younger males [Black-legged Kittiwake (Coulson, 1966), Red-billed Gull (Mills, 1973), Parasitic Jaeger (David, 1976)]. In at least the kittiwake, the female's choice of a younger male may be advantageous because of the higher mortality rates of males and the consequent need to remate sooner if an older male is chosen. Both sexes of kittiwakes suffer increasing mortality rates with age, but males have higher mortality rates than females in every age class (Coulson and Wooller, 1976). While mating with a less experienced male would result in a modest drop in reproductive output, the cost of remating after the death of a partner might result in an even greater drop in reproductive output.

It is not known how individuals recognize older or more experienced potential partners. If past breeding has been successful, individuals can repair or remain paired. If the past season's efforts resulted in failure or reduced success, changing mates may result in improved reproductive success (Coulson, 1966, 1972; Ollason and Dunnet, 1978). Although the choice to leave or remain with a previous year's partner can be based on past experience, that information is not available for selection of a new partner. Date of return to the colony may give a clue as to age or experience and may lead to mating of birds with similar experience or age [Black-legged Kittiwakes (Coulson, 1966), but not Manx Shearwaters (Brooke, 1978)].

IV. MATING SYSTEMS

Until recently, it was believed that marine birds were almost strictly monogamous, since the active participation of both parents appears required for successful incubation and chick rearing (Wittenberger, 1979; Lack, 1968; but see Nisbet *et al.*, 1978). However, polygyny, promiscuity, and female–female pairings can be found in gulls, and polygyny is known for skuas. It is not known if these mating patterns are recent phenomena, or if they were overlooked or misidentified in the past. To date, monogamy is the sole mating pattern identified in seabirds other than gulls and skuas, although promiscuous copulations are known in murres (Tuck, 1961) and in terns, especially Least Terns (*S. albifrons*: I. Nisbet, personal communication).

A. Polygyny

Polygyny may occur either as a response to an unbalanced sex ratio (Darwin, 1871; Mayr, 1939; Verner, 1964; Wittenberger, 1976; Altmann *et al.*, 1977) or because a female may choose to mate with an already mated male with the "expectation" that by doing so she will achieve higher fitness than if she mated monogamously (Verner, 1964; Verner and Willson, 1966; Orians, 1969a; Wittenberger, 1976; Altmann *et al.*, 1977; Emlen and Oring, 1977; Weatherhead and Robertson, 1979). Within the framework of the female choice model, two models exist; one in which there is cooperation between the females, and one in which there is competition (Altmann *et al.*, 1977). Polygyny of both the cooperative and competitive types exists in seabirds.

1. Occurrence of Polygyny

Polygyny in skuas is rare and results in reduced breeding success for both females. Polygyny, in which two females lay eggs in the same nest and share incubation duties with a single male partner, has been reported for Brown Skuas (*Catharacta skua lonnbergi*) from South Georgia Island (one case: Bonner, 1964) and from Signy Island, South Orkney Islands (two cases: Burton, 1968). The clutch size in this species is normally two eggs, and Burton found monogamous pairs at Signy frequently raised both chicks, unlike other populations of skuas in which the large chick always kills its smaller sibling (Young, 1963). High food abundance at Signy Island was apparently responsible for the survival of both young. In the two polygynously mated trios, both females laid eggs, and once in 8 years, a clutch of four eggs was found. However, in two of these seasons no eggs hatched and, because the number of eggs laid per female and hatching success in the supernormal clutches were reduced, polygynously mated females had lower success than monogamously mated birds. In view of this reduced fitness, it is particularly strange that one polygynous trio lasted for at least 8 years, during which period two of the original members were replaced (the sex of the replaced partners is not given: Burton, 1968).

Polygyny has also been recorded in Herring Gulls (Shugart and Southern, 1977) and in Ring-billed Gulls (Conover *et al.*, 1979). Shugart and Southern found 13 territories in two colonies in Michigan in which there were two nests with touching rims or almost-touching rims. At one territory the attending adults were captured and marked. Two females and a male shared incubation and chick-rearing duties of the two nests. In eastern Washington, Conover *et al.* (1979) trapped three females and a male from a nest containing six eggs. In this case, there was no evidence of a second nest.

2. Effects of Polygyny on Fitness

Given that there is frequently a sex ratio skewed in favor of excess females in gulls (Coulson, 1966, personal communication; Mills, 1973; Wingfield *et al.*, unpublished manuscript; Hunt *et al.*, 1980), one might have expected a higher incidence of polygyny. Unmated excess females and males obtaining a second mate would benefit by polygynous matings, and it is likely the trio would be able to raise four or five young (Ward, 1973; Hunt and Hunt, 1975). However, the primary female on a territory would suffer lowered fitness caused by reduced clutch size, resulting from premature cessation of egg laying in response to feedback from eggs in the shared nest (Paludan, 1951). (The existence of supernormal clutches in Western Gulls is in apparent contradiction to Paludan's results.) Reduced hatching success might also result from insufficient heat transfer in supernormal clutches

(Schreiber, 1970; Shugart and Southern, 1977). The constraints differ if the two females use separate nests (Shugart and Southern, 1977; J. Burger, personal communication). In this case, both females can lay full-size clutches, but the primary female will have to share the parental care provided by the male. Female Herring Gulls will drive off intruding birds, particularly other females (Tinbergen, 1953), and in Western Gulls, females are active in territory defense (Newman *et al.*, unpublished manuscript). The self-interest of the primary female expressed through territorial behavior may have prevented the spread of polygyny in gulls, even when the excess females and the males would both benefit by it.

B. Female–Female Pairing

An alternative means by which excess females may enter the breeding population is female–female pairing. Both females would have zero reproductive success if they were single, the findings of Nisbet *et al.* (1978) notwithstanding. They would increase their chance of reproducing if mated to another female and if they were the recipient of promiscuous mating (Hunt and Hunt, 1977; Conover *et al.*, 1979). In this system, both females and the male would enjoy increased reproductive success. Female–female pairing would therefore be more adaptive than polygynous matings. Preconditions necessary to promote female–female pairing include an excess of females, promiscuous mating by males, and the ability of females to establish nesting territories. Female–female pairing in the absence of a skewed sex ratio would result in lower fitness for both females.

1. Occurrence of Female–Female Pairing

Female–female pairing has now been documented for four populations of three species of gulls. Western Gulls (*L. o. wymani*) on the California Channel Islands have been found incubating supernormal clutches (Schreiber, 1970; Hunt and Hunt, 1973, 1977; Hunt *et al.*, 1979), and on Santa Barbara Island, these clutches have been shown to be the result of female–female pairing (Hunt and Hunt, 1977; Wingfield *et al.*, unpublished manuscript). Up to 14% of 1,200 nests on Santa Barbara Island contained more than the usual maximum of three eggs. Since we have found female–female pairs incubating clutches of three eggs, the figure of 14% is a minimum estimate of the extent of homosexual matings. Supernormal clutches have also been reported for the yellow-legged race of the Western Gull (*L. o. livens*) from islands in the Gulf of California (J. Hand, personal communication; T. Case, personal communication), but their association with female–female pairing has not been established.

Female–female pairing apparently has not occurred in the more northern populations of the Western Gull (*L. o. occidentalis*, J. Smail, personal communication). It has been present in the southern California Western Gull population since at least 1968 (Schreiber, 1970). There is no record of frequent supernormal clutches before that date (Hunt *et al.*, 1979).

Female–female pairing has been associated with supernormal clutches in two populations of Ring-billed Gulls (Ryder and Somppi, 1979; Conover *et al.*, 1979). At Granite Island in Lake Superior, 4–12% of clutches contained five to eight eggs (Ryder and Somppi, 1979). Four egg clutches were not considered supernormal (Vermeer, 1970; Ryder and Somppi, 1979). After allowing for "dump nesting," Ryder and Somppi estimated that 1.9% of clutches were the result of female–female pairing. Conover *et al.* (1979) found 3.3% of 1669 clutches of Ring-billed Gulls nesting in three colonies in eastern Washington to have more than three eggs. Only 1.3% of the clutches contained five or more eggs. Female–female pairs were captured on five and six egg clutches and on one three egg clutch.

Clutches of four eggs were found in 1.7% of 416 California Gull nests in these same three colonies in Washington. One of these four egg clutches was incubated by a heterosexual pair, another by a female–female pair. The frequency of female–female pairs, as opposed to supernormal clutches, was not established in the study by Conover *et al.* (1979).

In contrast to Western Gulls, the existence of supernormal clutches in Ring-billed Gull colonies has been known for over three decades (Moffitt, 1942; Johnston and Foster, 1954; Vermeer, 1970). If female–female pairing is responsible for these supernormal clutches in Ring-billed Gulls, then it is widespread and not of recent origin. It is also unlikely to have been caused by any single environmental change or perturbation.

2. Sex Ratios in Gulls

The existence of an excess of females in a population may facilitate, if not cause female–female pairing. In captive populations of birds with skewed sex ratios, homosexual matings are not infrequent (Collias and Jahn, 1959; Dilger, 1960; Sauer, 1972; P. J. B. Slater, cited in Jefferies, 1967; various aviculturalists, personal communications). Thus gull populations in which female–female pairing occurred should have an excess of females.

The only gull population in which female–female pairing occurs and for which sex ratios have been investigated is that of Western Gulls on Santa Barbara Island. Wingfield *et al.* (unpublished manuscript) and Hunt *et al.* (1980) have captured and sexed by laparotomy 1060 Western Gulls. The sex ratio of 239 chicks was 0.85 males per female. An estimated sex ratio for adults, based on the frequency of female–female pairs, the ratio of nonbreeding females to males in the clubs, and the total number of breeding and nonbreeding adults in the colony, was close to 0.67 males per female.

Skewed sex ratios have been reported for other seabird species in which only monogamous matings have been recorded. Mills (1973) found a surplus of females in the Red-billed Gull and noted that females have difficulty in obtaining mates. He found males mated at a younger age than females, a phenomenon we also have observed in Western Gulls (Wingfield *et al.*, unpublished manuscript; Hunt *et al.*, 1980). There may also be an excess of females in Black-legged Kittiwakes (Coulson and Wooller, 1976; J. C. Coulson, personal communication), and in this species males mate for the first time about 1 year earlier than do females (Wooller and Coulson, 1977; J. C. Coulson, personal communication). In contrast, the Yellow-eyed Penguin has an excess of males, and females breed at a younger age (Richdale, 1957); the same is also true with Adelie Penguins at Cape Crozier (Ainley, 1978).

Consequences of Unbalanced Sex Ratios. An unbalanced sex ratio leads to earlier mating for the less common sex (Wittenberger, 1980). As long as the population is expanding, there will be sufficient numbers of new recruits of the less common sex to provide partners for the excess of older birds of the more common sex. However, once population size reaches a steady state, an excess unmated population of adults of the more common sex may develop. Populations of both the Red-billed Gull (Mills, 1973) and the Black-legged Kittiwake (Wooller and Coulson, 1977) have been expanding, and, although the kittiwake colony has recently been fairly stable, presumably no excess of nonbreeding adult females has developed. In contrast, the Western Gull population on Santa Barbara has been stable or declining (Hunt *et al.*, 1979), and an excess of nonbreeding adult females exists.

Causes of Unbalanced Sex Ratios. Skewed sex ratios in the populations studied are primarily the result of differential mortality. Return of breeding adult Western Gulls to Santa Barbara Island has been followed for only one year. Of 111 breeding females in 1977, 90% were resighted in 1978; only 80% of 40 males breeding in 1977 were seen in 1978 (Hunt *et al.*, 1980). Although limited in number, these observations, in conjunction with our data on chick and adult sex ratios, strongly suggest a sexual bias in adult mortality. This finding is similar to that of Coulson and Wooller (1976) for the Black-legged Kittiwake. In a 20-yr study, they found males to have a lower survival rate than that of females (0.81 vs. 0.86).

A second and not mutually exclusive hypothesis is that sex ratios are skewed before termination of parental care. The sex ratio of Western Gull chicks at hatching, although not statistically significantly different from a sex ratio of 1, does show an excess of females (0.85). Fisher (1930), Williams (1975), and Trivers and Hare (1976) have demonstrated that parental investment in the two sexes should be equal at the termination of parental care. Trivers and Willard (1973) argue that females in poor condition would produce female offspring because male young in poor condition are likely to have poorer reproductive success than do females in poor condition—at least in polygynous species with male competition for mates. However, on Santa Barbara Island, with excess females one would expect female competition for mates and therefore only females in good condition would obtain mates. Our findings therefore do not support Trivers and Willard.

Myers (1978) disagreed with Trivers and Willard and suggested that poor environmental conditions should lead to production of the cheaper sex as a means of maximizing reproductive success. Since adult male Western Gulls are about 20% heavier than females (Newman *et al.*, unpublished manuscript), and a similar difference exists between male and female chicks at fledging (Hunt *et al.*, 1980), females are the cheaper sex to produce. Our evidence suggests that food has been limiting the Channel Islands Western Gull population (Hunt and Butler, 1980), consistent with Myers's (1978) hypothesis.

A third hypothesis that would predict skewing of the sex ratio prior to the termination of parental care is provided by Maynard-Smith (1978). Fisher's prediction of a sex ratio of 1 does not hold if mating is nonrandom (Hamilton, 1967). To the extent that inbreeding prevails, the sex ratio should be skewed toward females (Maynard-Smith, 1978). Whereas philopatry is well known in gulls, there are no data on the extent of inbreeding for any population, although Chabrzyk and Coulson's (1976) finding of considerable movement between colonies argues against extensive inbreeding.

3. Promiscuous Matings

Until the discovery of fertile eggs in nests attended by female–female pairs, it was assumed that colonial gulls were not promiscuous (Tinbergen, 1953; Lack, 1968). However, more than 14% of the eggs in supernormal clutches of Western Gulls on Santa Barbara Island (Hunt and Hunt, 1975) and 65–70% of the eggs in supernormal clutches in Washington and Great Lakes Ring-billed Gulls were fertile (Ryder and Somppi, 1979; Conover *et al.*, 1979). These results show that promiscuity is far more widespread in these presumptively monogamous species than was formerly believed.

Records of promiscuous matings in seabirds are few. Tuck (1961) reported promiscuous matings in both Common (*U. aalge*) and Thick-billed Murres and cites a description by Nørrevang (1958) of a female Common Murre being mated by three different males within a single 20-min period. We have seen promiscuous matings in Western Gulls in which the femal accepted the advances of the male (Hunt and Hunt, 1975). More recently, K. Winnett (unpublished data) has observed numerous promiscuous matings of Western Gulls at the clubs on Santa Barbara Island. We do not yet know if these matings are performed by a few or many males, if the females are already mated, or if their mates are male or female. As the proportion of individually marked, known-sex birds increases in this colony, answers to these questions should be forthcoming.

These reports of promiscuous matings are in marked contrast to descriptions of "rape" in gulls. MacRoberts (1973) examined cases of "extramarital" courtship in gulls and gives detailed descriptions of this behavior in lesser Black-backed and Herring Gulls. He found that male gulls attempted to rape females and that these advances were rejected by the females in all cases. No attempted copulations were successful. Vermeer (1963) described rape in Glaucous-winged Gulls in which copulations were successful, and Burger and Beer (1975) documented rape of incubating female Laughing Gulls (*L. atricilla*) by intruding males. In all instances of rape, males forced themselves on unwilling females. However, rape, like promiscuous mating, provides a potential explanation for the fertile clutches laid by female–female pairs. Gladstone (1979) provides arguments concerning the evolution of forced extrapair copulations and the limits to their occurrence in supposedly monogamous colonial birds.

V. SUMMARY

Mate selection is a subset of courtship and is the process of assessing mate quality. This view of mate selection provides a common denominator for many aspects of courtship, mate choice, mating systems, and the sensitivity of reproductive success to age and experience. I have presented a number of hypotheses and predictions about the adaptive functions of investment in mate selection, mate choice, the distribution of courtship feeding and the occurrence of unusual mating patterns. Most of these could be tested easily, but all too frequently the necessary data were either not available to me or have not yet been obtained. These data gaps provide a challenge and opportunity to marine ornithologists.

Some of the more important of these ideas are (1) investment in courtship should be proportional to the reproductive investment at risk; (2) courtship feeding is an alternative strategy to egg neglect; courtship feeding thus should be found in surface nesting species with multiple egg clutches that are large with respect to adult body weight; (3) sex identification in courtship may depend on a suite of characters defining mate quality, rather than on recognition of sex *per se*; and (4) polygyny alone or with female–female pairing may increase male and female fitness when sex ratios are skewed in favor of excess females.

Throughout this examination of the literature and of my own work, I have been impressed with the variations in reproductive behavior and ecology among populations of the same species as well as among different seasons for the same population. These differences are lost when we speak of "the Herring Gull" let alone, "the gull." The wealth of detailed variation in behavior and ecology of individual species provides fertile ground for examining the adaptive significance of behavior and suggests the need for genetic studies to separate genetic and phenotypic adaptations. Colonial seabirds, because of their numbers and ease of access, provide an unusually valuable resource for testing theories related to reproductive biology.

ACKNOWLEDGMENTS

I thank Peter Atsatt, Dee Boersma, Zoe Eppley, Audry Newman, Ian Nisbet, Gordon Orians, John Schultz, and James Wittenberger for stimulating discussions in which ideas presented in this chapter found their origin. Zoe Eppley, Ann Etgen, Donald Farner, Judith Myers, Ian Nisbet, John Wingfield, and James Wittenberger have provided valuable commentary on earlier drafts. Lucia Schnebelt typed the manuscript. Partial support for the research on Western Gulls was provided by a grant from the Chapman Fund of the American Museum of Natural History, and by National Science Foundation Grant PCM 77-05629 to D. S. Farner and G. L. Hunt.

REFERENCES

Ainley, D., 1975a, Development of reproductive maturity in Adelie Penguins, in: *The Biology of Penguins*, (B. Stonehouse, ed.), pp. 139–157, University Park Press, Baltimore, Md.

Ainley, D. G., 1975b, Displays of Adelie Penguins: A reinterpretation, in: *The Biology of Penguins* (B. Stonehouse, ed.), pp. 503–534, University Park Press, Baltimore, Md.

Ainley, D. G., 1978, Activity patterns and social behavior of non-breeding Adelie Penguins, *Condor* **80**:138.

Ainley, D. G., and Schlatter, R. P., 1972, Chick raising ability in Adelie Penguins, *Auk* **89**:559.

Altmann, S. A., Wagner, S. S., Lenington, S., 1977, Two models for the evolution of polygyny, *Behav. Ecol. Sociobiol.* **2**:397.

Andersson, M., 1971, Breeding behavior of the Long-tailed Skua *Stercorarius longicaudus* (Vieillot), *Ornis. Scand.* **2**:35.

Ashmole, N. P., 1971, Sea bird ecology and the marine environment, in: *Avian Biology* (D. S. Farner and J. R. King, eds.), Vol. I, pp. 223–286, Academic Press, New York.

Austin, O. L., 1945, The role of longevity in successful breeding by the Common Tern (*Sterna hirundo*), *Bird-Banding* **16**:21.

Bastock, M., 1967, *Courtship: An Ethological Study*, Aldine, Chicago, 220 pp.

Bateman, A. J., 1948, Intrasexual selection in Drosophila, *Heredity* **2**:349.

Bédard, J., 1969a, Feeding of the least, crested, and parakeet auklets around St. Lawrence Island, Alaska, *Can. J. Zool.* **47**:1025.

Bédard, J., 1969b, The nesting of the crested, least and parakeet auklets on St. Lawrence Island, Alaska, *Condor* **71**:386.

Blest, A. D., 1961, The concept of ritualization, in: *Current Problems in Animal Behavior* (W. H. Thorpe and O. L. Zangwill, eds.), pp. 102–124. Cambridge University Press, Cambridge, England.

Blues, L. J., and Keahey, J. A. 1978, Variation in reproductivity with age in the Brown Pelican, *Auk* **95**:128.

Boersma, P. D., 1977, An ecological and behavioral study of the Galapagos Penguin, *Living Bird* **15**:43.

Boersma, P. D., and Wheelwright, N. T., 1979, Egg neglect in the Procellariiformes: Reproductive adaptations in the Fork-tailed Storm-Petrel, *Condor* **81**:157.

Bonner, W. N., 1964, Polygyny and super-normal clutch size in the brown skua, *Catharacta skua lonnbergi* (Mathews), *Br. Antarct. Surv. Bull.* **3**:41.

Boswell, J., and MacIver, D., 1975, The Magellanic Penguin *Spheniscus magellanicus*, in: *The Biology of Penguins* (B. Stonehouse, ed.), pp. 397–409, University Park Press, Baltimore, Md.

Brooke, M. de L., 1978, Some factors affecting the laying date, incubation and breeding success of the Manx shearwater, *Puffinus puffinus*, *J. Anim. Ecol.* **47**:477.

Brown, R. G. B., 1967, Courtship behavior in the Lesser Black-backed Gull, *Larus fuscus*, *Behaviour* **29**:122.

Brown, W. Y., 1973, The breeding biology of Sooty Terns and Brown Noddies on Manana or Rabbit Island, Oahu, Hawaii, Ph.D. thesis, University of Hawaii.

Burger, J., and Beer, C. G., 1975, Territoriality in the Laughing Gull (*L. atricilla*), *Behaviour* **55**:301.

Burton, R. W., 1968, Breeding biology of the Brown Skua, *Catharacta skua lonnbergi* (Mathews) at Signy Island, S. Orkney Islands, *Br. Antarct. Surv. Bull.* **15**:9.

Chabrzyk, G., and Coulson, J. C., 1976, Survival and recruitment in the Herring Gull *Larus argentatus*, *J. Anim. Ecol.* **45**:187.

Charnov, E. L., and Krebs, J. R., 1974, On clutch-size and fitness, *Ibis* **116**:217.

Cody, M. L., 1966, A general theory of clutch size, *Evolution* **20**:174.

Cody, M. L., 1971, Ecological aspects of reproduction, in: *Avian Biology* (D. S. Farner and J. R. King, eds.), Vol. I, pp. 461–546, Academic Press, New York.

Collias, N. E., and Jahn, L. R., 1959, Social behavior and breeding success in Canada Geese (*Branta canadensis*) confined under semi-natural conditions, *Auk* **76**:478.

Collias, N. E., Victoria, J. K., and Shallenberger, R. J., 1971, Social facilitation in weaverbirds: Importance of colony size, *Ecology* **52**:823.

Congdon, J. D., Vitt, L. J., and Hadley, N. F., 1978, Parental investment: Comparative reproductive energetics in bisexual and unisexual lizards *Cnemidophorus*, *Am. Nat.* **112**:509.

Conover, M. R., Miller, D. E., and Hunt, G. L., 1979, Female–female pairs and other unusual reproductive associations in Ring-billed and California Gulls, *Auk* **96**:6.

Coulson, J. C., 1963, Egg size and shape in the Kittiwake (*Rissa tridactyla*) and their use in estimating age composition of populations, *Proc. Zool. Soc. London* **140**:211.

Coulson, J. C., 1966, The influence of pair-bond and age on the breeding biology of the Kittiwake Gull *Rissa tridactyla*, *J. Anim. Ecol.* **35**:269.

Coulson, J. C., 1972, The significance of the pair-bond in the Kittiwake, *Proc. XV Int. Ornithol. Congr.*, p. 424.

Coulson, J. C., and Horobin, J., 1976, The influence of age on the breeding biology and survival of the Arctic Tern *Sterna paradisaea, J. Zool.* **178**:247.

Coulson, J. C., and White, E., 1958, The effect of age on the breeding biology of the Kittiwake, *Rissa tridactyla, Ibis* **100**:40.

Coulson, J. C., and White, E., 1960, The effect of age and density of breeding birds on the time of breeding birds on the time of breeding of the Kittiwake *Rissa tridactyla, Ibis* **102**:71.

Coulson, J. C., and Wooller, R. D., 1976, Differential survival rates among breeding Kittiwake Gulls *Rissa tridactyla* (L.), *J. Anim. Ecol.* **45**:205.

Cullen, E., 1957, Adaptations in the kittiwake to cliff-nesting, *Ibis* **99**:275.

Cullen, J. M., and Ashmole, N. P., 1963, The Black Noddy *Anous tenuirostris* on Ascension Island. Part 2. Behavior, *Ibis* **103b**:423.

Daly, M., 1978, The cost of mating, *Am. Natur.* **112**:771.

Dane, B., Walcott, C., and Drury, W. H., 1959, The form and duration of the display actions of the Goldeneye (*Bucephala clangula*), *Behaviour* **14**:265.

Darling, F. F., 1938, *Bird Flocks and the Breeding Cycle*, Cambridge University Press, Cambridge, England.

Darwin, C. R., 1871, *The Descent of Man and Selection in Relation to Sex*, John Murray, London.

Davis, J. W. F., 1975a, Age, egg-size and breeding success in the Herring Gull *Larus argentatus, Ibis* **117**:460.

Davis, J. W. F., 1975b, Specialization in feeding location by Herring Gulls, *J. Anim. Ecol.* **44**:795.

Davis, J. W. F., 1976, Breeding success and experience in the arctic skua, *Stercorarius parasiticus* (L.), *J. Anim. Ecol.* **45**:531.

Dilger, W. C., 1960, The comparative ethology of the African parrot genus *Agapornis, Z. Tierpsychol.* **17**:649.

Dinsmore, J., 1971, Sooty tern behavior, *Bull. Fla. State Mus., Biol. Sci.* **16**(3):129.

Drent, R., 1965, Breeding biology of the Pigeon Guillemot, *Cepphus columba, Ardea* **53**:99.

Drury, W. H., and Smith, W. J., 1968, Defense of feeding areas by adult Herring Gulls and intrusion by young, *Evolution* **22**:193.

Dunn, E. K., 1972, Effect of age on fishing ability of Sandwich Terns, *Sterna sandvicensis, Ibis* **114**:360.

Dunnet, G. M., and Ollason, J. C., 1978, The estimation of survival rate in the fulmar, *Fulmarus glacialis, J. Anim. Ecol.* **47**:507.

Emlen, S. T., and Oring, L. W., 1977, Ecology, sexual selection, and the evolution of mating systems, *Science* **197**:215.

Ferdinand, L., 1970, Some observations on the behavior of the Little Auk (*Plotus alle*) on the breeding ground, with special reference to voice production, *Dansk. Ornithol. Foren. Copenh. Tidsskrift* **63**:19.

Fisher, R. A., 1930, *The Genetical Theory of Natural Selection*, Clarendon Press, Oxford, 2nd ed., Dover, New York, 1958.

Gadgil, M., and Bossert, W. H., 1970, Life historical consequences of natural selection, *Am. Nat.* **104**:1.

Gladstone, D. E., 1979, Promiscuity in monogamous colonial birds, *Am. Nat.* **114**:545.

Goodman, D., 1974, Natural selection and a cost ceiling on reproductive effort. *Am. Nat.* **108**:247.

Gross, A. O., 1912, Observations on the Yellow-billed Tropic-bird (*Phaethon americanus* Grant) at the Bermuda Islands, *Auk* **29**:49.

Hamilton, W. D., 1967, Extraordinary sex ratios, *Science* **156**:477.

Harris, M. P., 1964, Aspects of the breeding biology of the gulls *Larus argentatus*, *L. fuscus*, and *L. marinus*, *Ibis* **106**:432.

Harris, M. P., 1970, Breeding ecology of the Swallow-tailed Gull, *Creagrus furcatus*, *Auk* **87**:215.

Harris, M. P., 1979, Population dynamics of the Flightless Cormorant *Nannopterum harrisi*, *Ibis* **121**:135.

Harris, M. P., and Hope Jones, P., 1969, Sexual differences in measurements of Herring and Lesser Black-backed Gulls, *Br. Birds* **62**:129.

Harris, M. P., and Plumb, W. J., 1965, Experiments on the ability of Herring Gulls *Larus argentutus* and Lesser Black-backed Gulls *L. fuscus* to raise larger than normal broods, *Ibis* **107**:256.

Hays, H., 1978, Timing and breeding success in three- to seven-year-old Common Terns, *Ibis* **120**:127.

Henderson, B. A., 1972, The control and organization of parental feeding and its relationship to the food supply for the Glaucous-winged Gull, *Larus glaucescens*, unpublished M.Sc. thesis, University of British Columbia, Vancouver.

Hoffman, W., Wiens, J. A., and Scott, J. M., 1978, Hybridization between gulls (*Larus glaucescens* and *L. occidentalis*) in the Pacific Northwest, *Auk* **95**:441.

Hunt, G. L., and Butler, J. L., 1980, Reproductive ecology of Western Gulls and Xantus' Murrelets with respect to food resources in the Southern California Bight, *Cal. C.O.F.I.* **20** (in

Hunt, G. L., and Hunt, M. W., 1973, Clutch size, hatching success, and eggshell thinning in Western Gulls, *Condor* **75**:483.

Hunt, G. L., and Hunt, M. W., 1975, Reproductive ecology of the Western Gull: The importance of nest spacing, *Auk* **92**:270.

Hunt, G. L., and Hunt, M. W., 1976, Gull chick survival: The significance of growth rates, timing of breeding and territory size, *Ecology* **57**:62.

Hunt, G. L., and Hunt, M. W., 1977, Female–female pairing in Western Gulls (*Larus occidentalis*) in southern California, *Science* **196**:1466.

Hunt, G. L., Pitman, R. L., Naughton, M., Winnett, K., Newman, A., Kelley, P. R., and Briggs, K. T., 1979, Distribution, status, reproductive ecology and foraging habits of breeding seabirds, in: *Vol. III: Summary of Marine Mammal and Seabird Surveys of the Southern California Bight Area 1975–1978, Part III: Seabirds of the Southern California Bight*, Regents of the University of California, Santa Cruz and Irvine.

Hunt, G. L., Wingfield, J. C., Newman, A., and Farner, D. S., 1980, Sex ratio of Western Gulls on Santa Barbara Island, *Auk* **97**.

Huxley, J. S., 1923, Courtship activities in the Red-throated Diver (*Columbus stellatus* Pontopp); together with a discussion on the evolution of courtship in birds, *J. Linn. Soc. London* **25**:253.

Ingolfsson, A., 1967, The feeding ecology of five species of large gulls (*Larus*) in Iceland, Unpublished Ph.D. thesis, University of Michigan, Ann Arbor.

Ingolfsson, A., 1970, Hybridization of Glaucous Gulls *Larus hyperboreus* and Herring Gulls *L. argentatus* in Iceland, *Ibis* **112**:340.

Jefferies, D., 1967, The delay in ovulation produced by pp'-DDT and its possible significance in the field, *Ibis* **109**:266.

Johnsgard, P. A., 1965, *Handbook of Waterfowl Behavior*, Cornell University Press, Ithaca, 378 pp.

Johnson, S. R., and West, G. C., 1973, Fat content, fatty acid composition and estimates of energy metabolism of Adelie Penguins (*Pygoscelis adeliae*) during the early breeding season fast, *Comp. Biochem. Physiol.* **45**:709.

Johnston, W. D., and Foster, M. E., 1954, Interspecific relations of breeding gulls at Honey Lake, California, *Condor* **56**:38.

Krebs, J. R., 1970, The efficiency of courtship feeding in the Blue Tit *Parus caeruleus*, *Ibis* **112**:108.

Kruuk, H., 1964, Predators and anti-predator behavior of the Black-headed Gull (*Larus ridibundus* L.), *Behav. Suppl.* **11**:1–129.

Lack, D., 1940, Courtship feeding in birds, *Auk* **57**:169.

Lack, D., 1954, *The Natural Regulation of Animal Numbers*, Oxford University Press, Oxford, 343 pp.

Lack, D., 1966, *Population Studies of Birds*, Clarendon Press, Oxford, 341 pp.

Lack, D., 1968, *Ecological Adaptations for Breeding in Birds*, Chapman and Hall, London, 409 pp.

Langham, N. P. E., 1974, Comparative breeding biology of the Sandwich Tern, *Auk* **91**:255.

Lehrman, D. S., 1955, The physiological basis of parental feeding in the Ring Dove (*Streptopelia risoria*), *Behaviour* **7**:241.

Lehrman, D. S., 1959, Hormonal responses to external stimuli in birds, *Ibis* **101**:478.

Lehrman, D. S., 1964, The reproductive behavior of Ring Doves, *Sci. Am.* **211**:48.

LeResche, R. E., and Sladen, W., 1970, Establishment of pair and breeding site bonds by young known-age Adelie Penguins (*Pygoscelis adeliae*), *Anim. Behav.* **8**:517.

Lorenz, K., 1941, Vergleichende Bewegungsstudien an Anatinen, *J. Orn., Lpz.* **89** (English Vol. 3):194.

Lundberg, C. A., and Vaisanen, 1979, Selective correlation of egg size with chick mortality in the Black-headed Gull (*Larus Ridibundus*), *Condor* **81**:146.

MacRoberts, B. R., and MacRoberts, M. H., 1972, Social stimulation of reproduction in Herring and Lesser Black-backed Gulls, *Ibis* **114**:495.

MacRoberts, M. H., 1973, Extramarital courting in Lesser Black-backed and Herring Gulls, *Z. Tierpsychol.* **32**:62.

Maynard-Smith, J., 1978, *The Evolution of Sex*, Cambridge University Press, Cambridge, 222 pp.

Mayr, E., 1939, The sex-ratio in wild birds, *Am. Nat.* **73**:156.

Mills, J. A., 1973, The influence of age and pair bond on the breeding biology of the Red-billed Gull, *Larus novaehollandiae scopulinus, J. Anim. Ecol.* **42**:147.

Mills, J. A., 1979, Factors affecting the egg size of Red-billed Gulls *Larus novaehollandiae scopulinus*, *Ibis* **121**:53.

Moffitt, J., 1942, A nesting colony of Ring-billed Gulls in California, *Condor* **44**:105.

Morrison, M. L., Slack, R. D., and Skanley, E., 1978, Age and foraging ability relationships of Olivaceous Cormorants, *Wilson Bull.* **90**:414.

Moynihan, M., 1955*a*, Remarks on the original sources of displays, *Auk* **72**:240.

Moynihan, M., 1955*b*, Some aspects of reproductive behavior in the Black-headed Gull (*Larus ridibundus ridibundus* L.) and related species, *Behav. Suppl.* **4**, 201 pp.

Moynihan, M., 1958, Notes on the behavior of some North American gulls. III: pairing behavior, *Behaviour* **12**:112.

Murphy, R. C., 1936, *Oceanic Birds of South America*, American Museum of Natural History–MacMillan, New York, 1245 pp.

Myers, J. H., 1978, Sex ratio adjustment under food stress: maximization of quality or numbers of offspring? *Am. Nat.* **112**:381.

Nelson, J. B., 1964, Factors influencing clutch-size and chick growth in the North Atlantic Gannet *Sula bassana*, *Ibis* **106**:63.

Nelson, J. B., 1966, The breeding biology of the Gannet *Sula bassana* on the Bass Rock, Scotland, *Ibis* **108**:584.

Nelson, J. B., 1970, The relationship between behavior and ecology in the Sulidae with reference to other sea birds, *Oceanogr. Mar. Biol. Annu. Rev.* **8**:501.

Nelson, J. B., 1972, Evolution of the pair bond in the Sulidae, *Proc. XV Inter. Ornith. Congr.*, p. 371.

Nelson, J. B., 1977, Some relationships between food and breeding in the marine Pelecaniformes, in: *Evolutionary Ecology* (B. Stonehouse and C. Perrins, eds.), pp. 77–87, University Park Press, Baltimore, Md.

Nelson, J. B., 1978, *The Sulidae—Gannets and Boobies*, Aberdeen University Studies Series 154, Oxford, 1012 pp.

Nettleship, D. N., 1972, Breeding success of the Common Puffin (*Fratercula arctica* L.) on different habitats at Great Island, Newfoundland, *Ecol. Monogr.* **42**:239.

Nisbet, I. T. C., 1973, Courtship-feeding, egg-size and breeding success in Common Terns, *Nature (London)* **241**:141.

Nisbet, I. T. C., 1975, Selective effects of predation in a tern colony, *Condor* **77**:221.

Nisbet, I. T. C., 1977, Courtship-feeding and clutch size in Common Terns *Sterna hirundo*, in: *Evolutionary Ecology* (B. Stonehouse and C. Perrins, eds.), pp. 101–109, University Park Press, Baltimore, Md.

Nisbet, I. T. C., 1978, Dependence of fledging success on egg size, parental performance and egg-composition among Common and Roseate Terns, *Sterna hirundo* and *S. dougalii, Ibis* **120**:207.

Nisbet, I. T. C., Wilson, K. J., and Broad, W. A., 1978, Common Terns raise young after death of their mates, *Condor* **80**:106.

Nørrevang, A., 1958, On the breeding biology of the Guillemot [*Uria aalge* (Pont.)], *Dansk. Orn. Foren. Tidsskr.* **52**:48.

Ollason, J. C., and Dunnet, G. M., 1978, Age, experience and other factors affecting the breeding success of the Fulmar, *Fulmarus glacialis*, in Orkney, *J. Anim. Ecol.* **47**:961.

Orians, G. H., 1969a, On the evolution of mating systems in birds and mammals, *Am. Nat.* **103**:589.

Orians, G. H., 1969b, Age and hunting success in the Brown Pelican (*Pelecanus occidentalis*), *Anim. Behav.* **17**:316.

Paludan, K., 1951, Contributions to the breeding biology of *Larus argentatus* and *Larus fuscus*, *Vidensk. Medd. Dan. Naturhist. Foren. Kobenhaven* **114**:1.

Parsons, J., 1970, Relationship between egg-size and post-hatching chick mortality in the Herring Gull (*Larus argentatus*), *Nature (London)* **228**:1221.

Parsons, J., 1975, Asynchronous hatching and chick mortality in the Herring Gull *Larus argentatus, Ibis* **117**:517.

Patterson, I. J., 1965, Timing and spacing of broods in the Black-headed Gull (*Larus ridibundus* L.), *Ibis* **107**:433.

Patton, S., and Weisbrod, A. R., 1974, Sympatry and interbreeding of Herring and Glaucouswinged Gulls in southeastern Alaska, *Condor* **76**:343.

Pennycuick, C. J., 1956, Observations on a colony of Brünnich's Guillemot *Uria lomvia* in Spitsbergen, *Ibis* **98**:80.

Perrins, C. M., 1965, Population fluctuations and clutch-size in the Great Tit, *Parus major* L., *J. Anim. Ecol.* **34**:601.

Recher, H. F., and Recher, J. A., 1969, Comparative foraging efficiency of adult and immature little Blue Herons (*Florida caerulea*), *Anim. Behav.* **17**:320.

Richdale, L. E., 1957, *A Population Study of Penguins*, Clarendon Press, Oxford.

Ricklefs, R. E., 1974, Energetics of reproduction in birds, in: *Avian Energetics* (R. A. Paynter, Jr., ed.), pp. 152–292, *Nutt. Ornithol. Club Pub.* 15.

Royama, T., 1966, A re-interpretation of courtship feeding, *Bird Study* **13**:116.

Ryder, J. P., 1975, Egg-laying, egg size, and success in relation to immature–mature plumage in Ring-billed Gulls, *Wilson Bull.* **87**:534.

Ryder, J. P., 1978, Possible origins and adaptive value of female–female pairing in gulls, *Proc. Colonial Waterbird Group*, **1978**:138.

Ryder, J. P., and Somppi, P. L., 1979, Female-female pairing in Ring-billed Gulls, *Auk* **96**:1.

Sauer, E. G. F. 1972, Aberrant sexual behavior in the South African Ostrich, *Auk* **89**:717.

Schreiber, R. W., 1970, Breeding biology of Western Gulls (*Larus occidentalis*) on San Nicolas Island, California, 1968, *Condor* **72**:133.

Schreiber, R. W., and Ashmole, N. P., 1970, Sea-bird breeding seasons on Christmas Island, Pacific Ocean, *Ibis* **112**:363.

Sealy, S. G., 1972, Adaptative differences in breeding biology in the marine bird family Alcidae, Unpublished Ph.D. thesis, University of Michigan, Ann Arbor.

Sealy, S. G., 1973, Adaptative significance of post-hatching developmental patterns and growth rates in the alcidae, *Ornis Scand*, **4**:113.

Sealy, S. G., 1975a, Feeding ecology of the Ancient and Marbled Murrelets near Langara Island, British Columbia, *Can. J. Zool.* **53**:418.

Sealy, S. G., 1975b, Egg size of murrelets, *Condor* **77**:500.

Sealy, S. G., and Bédard, J., 1973, Breeding biology of the Parakeet Auklet (*Cyclorrhynchus psittacula*) on St. Lawrence Island, Alaska, *Astarte* **6**:59.

Searcy, W. A., 1978, Foraging success in three age classes of Glaucous-winged Gulls, *Auk* **95**:586.

Shugart, G. W., and Southern, W. E., 1977, Close nesting, a result of polygyny in Herring Gulls, *Bird-Banding* **48**:276.

Sladen, W. J. L., 1957, The Pygoscelid Penguins II. The Adelie Penguin *Pygoscelis adelie* (Hombron and Jacquinot) *Faulkland Is. Dept. Surface Sci. Rep.* **17**:23.

Smith, N. G., 1966, Evolution of some Arctic gulls (Larus): An experimental study of isolating mechanisms, *Ornithol. Mongr.* **4**:1–99.

Southern, W. E., 1974, Copulatory wing-flagging: A synchronizing stimulus for nesting Ring-billed Gulls, *Bird-Banding* **45**:210.

Spurr, E. B., 1975, Communication in the Adelie Penguin, in: *The Biology of Penguins* (B. Stonehouse, ed.), pp. 449–501, University Park Press, Baltimore, Md.

Stonehouse, B., 1953, The Emperor Penguin *Aptenodytes fosteri* Gray. I. Breeding behavior and development, *Falkland Is. Dept. Survey Sci. Rep.* **6**:1–35.

Stonehouse, B., 1956, The Brown Skua *Catharacta skua lonnbergi* (Mathews) of South Georgia, *Falkland Is. Dept. Survey Sci. Rep.* **14**:1–25.

Stonehouse, B., 1960. The King Penguin *Aptenodytes patagonica* of South Georgia. I. Breeding behavior and development, *Falkland Is. Dept. Survey Sci. Rep.* **23**:1–81.

Stonehouse, B., 1962, The tropic birds (genus *Phaethon*) of Ascension Island, *Ibis* **103B**:124.

Stonehouse, B., 1967, Penguins in high latitudes, *Tuatara* **15**:129.

Swales, M. K., 1965, The sea-birds of Gough Island, *Ibis* **107**:17.

Tickell, W. L. N., and Pinder, R., 1975, Breeding biology of the Black-browed Albatross *Diomedea melanophris* and Grey-headed Albatross *D. chrysostoma* at Bird Island, South Georgia, *Ibis* **117**:433.

Tinbergen, N., 1952, "Derived activies," their causation, biological significance, origin and emancipation during evolution, *Q. Rev. Biol.* **27**:1.

Tinbergen, N., 1953, *The Herring Gull's World*, Collins, London, 255 pp.

Tinbergen, N., 1959, Comparative studies of the behavior of gulls (*Laridae*): A progress report, *Behaviour* **XV**:1.

Trivers, R. L., 1972, Parental investment and sexual selection, in: *Sexual Selection and the Descent of Man 1871–1971* (B. Campbell, ed.), pp. 136–179, Aldine, Chicago.

Trivers, R. L., and Hare, H., 1976, Haplodiploidy and the evolution of the social insects, *Science* **191**:249.

Trivers, R. L., and Willard, D. E., 1973, Natural selection of parental ability to vary the sex ratio of offspring, *Science* **179**:90.

Tuck, L. M., 1961, *The Murres*, Vol. 1, pp. 1–260, Canadian Wildlife Service, Ottawa.

Van Tets, G. F., 1965, A comparative study of some social communication patterns in the Pelicaniformes, *Ornith. Monogr.* **2**:1–88.

Vermeer, K., 1963, The breeding ecology of the Glaucous-winged Gull (*Larus glaucescens*), on Mandarte Island, B. C., *Occas. Pap. B.C. Prov. Mus.* 13, 104 pp.

Vermeer, K., 1970, Breeding biology of California and Ring-billed Gulls, *Can. Wildl. Serv. Rep. Ser.* **12**:1–52.

Verner, J., 1964, Evolution of polygamy in the long-billed marsh wren, *Evolution* **18**:252.

Verner, J., and Willson, M. F., 1966, The influence of habitat on mating systems of North American Passerine birds, *Ecology* **47**:143.

Waite, E. R., 1909, Vertebrata of the subantarctic islands of New Zealand, in Chiltow's "Subantarctic Islands of New Zealand," **2**:542–600.

Ward, J. G., 1973, Reproductive success, food supply, and the evolution of clutch-size in the Glaucous-winged Gull, Ph.D. thesis, University of British Columbia, Vancouver.

Warham, J., 1963, The Rockhopper Penguin, *Eudyptes chrysocome*, at Macquarie Island, *Auk* **80**:229.

Warham, J., 1971, Aspects of breeding behavior in the Royal Penguin *Eudyptes chrysolophus schlegeli*, *Notornis* **18**:91.

Warham, J., 1975, The crested penguins, in: *The Biology of Penguins* (B. Stonehouse, ed.), pp. 189–269, University Park Press, Baltimore, Md.

Weatherhead, P. J., and Robertston, R. J., 1979, Offspring quality and the polygyny threshold: "The sexy son hypothesis," *Am. Nat.* **113**:201.

Wiley, R. H., 1974, Effects of delayed reproduction on survival, fecundity, and the rate of population increase, *Am. Nat.* **108**:705.

Williams, G. C., 1966, *Adaptation and Natural Selection*, Princeton University Press, Princeton, N.J., 307 pp.

Williams, G. C., 1975, *Sex and Evolution*, Vol. 8, *Princeton Monographs in Population Biology*, Princeton University Press, Princeton, N.J., 200 pp.

Wittenberger, J. F., 1976, The ecological factors selecting for polygyny in birds, *Am. Nat.* **97**:405.

Wittenberger, J. F., 1979, The evolution of mating systems in birds and mammals, in: *Handbook of Behavioral Neurobiology: Social Behavior and Communication*, Vol. 3 (P. Marler and J. Vandenbergh, eds.), pp. 271–349, Plenum Press, New York.

Wittenberger, J. F., 1980, A model of delayed reproduction in iteroparus animals, *Am. Natur.* **114**:439.

Wooller, R. D., and Coulson, J. C., 1977, Factors affecting the age of first breeding of the Kittiwake *Rissa tridactyla*, *Ibis* **119**:339.

Wynne-Edwards, V. C., 1962, *Animal Dispersion in Relation to Social Behavior*, Oliver and Boyd, Edinburgh, 653 pp.

Yeates, G. W., 1975, Microclimate, climate and breeding success in Antarctic penguins, in: *The Biology of Penguins* (B. Stonehouse, ed.), pp. 397–409, University Park Press, Baltimore, Md.

Young, E. C., 1963, The breeding behavior of the South Polar Skua *Catharacta maccormicki*, *Ibis* **105**:203.

Chapter 5

THE INFLUENCE OF AGE
ON THE BREEDING BIOLOGY OF
COLONIAL NESTING SEABIRDS

John P. Ryder

Department of Biology
Lakehead University
Thunder Bay, Ontario, Canada P7B 5E1

I. INTRODUCTION

In 1938 Fraser Darling enunciated his now famous hypothesis known as the Fraser Darling effect. He suggested, from investigations with Herring Gulls (*Larus argentatus*) and Lesser Black-backed Gulls (*L. fuscus*), that the main advantage to birds nesting in colonies was that the individuals received what he termed "social stimulation." The result of such stimulation was, "in almost all cases that the larger colonies not only start laying earlier, but the time taken by the whole colonies to lay their crop of eggs is shorter than in the colonies of lesser numbers" (Darling, 1938, p. 67). Darling stated ". . . the most obvious interpretation is to be found in the total value of visual auditory stimulation for each pair in the larger flocks, compared with the total amount of such stimulation in the smaller flocks." He concluded that breeding synchrony, realized by social stimulation, may be density related (Darling, 1938, p. 55).

Some authors subsequently found support for Darling's ideas (Klopfer and Hailman, 1965; Crook, 1968; Horn, 1970), whereas others could not (Davis, 1940; Armstrong, 1947; Weidmann, 1956; Coulson and White, 1960; Orians, 1961; Vermeer, 1963; Hailman, 1964; Tenaza, 1971; MacRoberts and MacRoberts, 1972; Langham, 1974). Richdale (1951) criticized the Darling hypothesis, emphasizing that it would be difficult to demonstrate without knowing the age and breeding status of individuals composing a colony.

That age may be an important factor in some of Darling's observations did not receive serious attention until the pioneering works of Austin (1945), Richdale (1949a,b), Fisher (1954), and Coulson and White (1956). These contributors were the first to discuss the significance of age composition of colonies and its relationship to the timing and duration of breeding in colonial nesting species, particularly seabirds. Following these communications, a number of researchers have investigated the influence of age on the breeding biology of birds. In most instances age was found to be an important variable affecting the breeding success.

This review considers the relevant literature on the effects of age on the breeding biology of seabirds. The groups considered include the Sphenisciformes (penguins), Procellariiformes (albatrosses, shearwaters, fulmars, storm petrels), Pelecaniformes (tropicbirds, pelicans, boobies, gannets, cormorants, darters, frigatebirds), and the Stercorariidae (jaegers and skuas), Laridae (gulls and terns), and Alcidae (auks, murres, and puffins) within the Charadriiformes. The chapter comprises four sections: timing of breeding, clutch size, incubation, and breeding success. The postfledging period is the subject of Chapter 10. I hope the review will serve to provide a basis for further study. I have not include figures and tables from publications of mine or others. I have omitted such detail because the data are available in the original articles documented in the text. However, the main ideas and summary data from the research papers are included. I have selected papers in which the age of birds was known from banding, and also those in which the authors stated they were working with young or old birds, based on plumage or behavior.

For the reader interested in a more general account of the distribution and biology of the seabirds, I recommend the following works: Fisher and Lockley (1954), Gordon (1955), Belopol'skii (1961), Hindwood *et al.* (1963), Swartz (1966), Ashmole and Ashmole (1967), Carrick and Ingham (1967), Lack (1967), Murphy (1967), Ashmole (1971), and Brown *et al.* (1975).

II. TIMING OF BREEDING

Coulson and White (1958) showed for the first time that, in individually color-banded kittiwakes (Black-legged Kittiwake, *Rissa tridactyla*) nesting at a warehouse colony at North Shields, Northumberland, England, birds breeding for the second time or more started their activities (i.e., nesting) an average of 7.5 days before conspecifics breeding for the first time. Moreover, birds with breeding experience returned to the colony 36 days earlier than did those returning to breed for the first time. This study on Black-legged Kittiwakes confirmed Coulson and White's (1956) earlier hypothesis that differences in timing of breeding are age related and are not strictly a result of social stimulation as advanced by Darling (1938). Coulson and White (1960) hypothesized from their two previous studies (Coulson and White, 1956, 1958) that colonies of Black-legged Kittiwakes with different age compositions should show differences in their time of breeding. This hypothesis was tested at a number of British colonies, 1952–1958. In addition to testing the effect of age on the timing of breeding, they compared nest densities and timing. This was done because in some colonies in which the age composition of the colonies was calculated, some of the older colonies that contained proportionately fewer young birds than did new colonies did not show a clear age–time of breeding relationship (Coulson and White, 1960, p. 73). Female Black-legged Kittiwakes breeding for at least the fifth time did *not* breed earlier than did females breeding for the fourth time. In 38 areas of seven colonies, densities ranged 1.5–9.6 nests within a radius of 5 ft (80 sq ft). The result was a strong correlation between density and time of breeding ($r = 0.81$; $p < 0.001$). Birds nesting at a density of nine nests per 80 sq ft, generally in the older colonies ($r = 0.49$; $p < 0.01$), bred about 18 days earlier than did those nesting at a density of two nests per 80 sq ft in the younger colonies, where there was a higher proportion of young birds. Although age appeared to be a contributing factor in timing, to some extent the Darling "social stimulation" hypothesis was upheld. However, this same investigation showed that the spread of breeding was greater in dense colonies of Black-legged Kittiwakes, which contradicted the Darling hypothesis.

The mechanism of how the age density and timing of breeding phenomenon operates is unclear from Coulson and White's (1960) study. They suggested that proximate conditions at the colony site were not the cause, but rather the result, of physiological differences operating before the birds first return to the breeding areas. The view that stimulation received in the previous breeding season(s) is cumulative and affects the start of breeding activities in subsequent years is difficult to confirm. Ashmole (1963a), Lack (1966), and Orians (1969) suggested instead that birds experienced at a particular colony would be familiar with the resources of nearby feeding grounds relative to inexperienced birds. This can shorten the amount of time the former individuals require to build up necessary food reserves for breeding relative to new young recruits. The idea seems credible and might explain the decreasing relationship between age and time of breeding with advancing age (Coulson and White, 1960), both of individuals and the colony as a whole. As individuals get older, their competitive ability levels off, to the extent that age no longer has a prime bearing on the time of breeding.

It is conceivable that most young individuals returning to their natal areas at about the same time as older birds, as in some terns (Austin, 1940;

Harrington; 1974), are driven away from suitable nesting sites by the older birds. In our studies of Ring-billed Gulls (*L. delawarensis*) on Granite Island in northwestern Ontario, Canada, we noted (Ryder, 1975) that 62% of pairs, in which at least one member had immature plumage, were relegated to peripheral areas of the colony and started their nesting activities on average 3 weeks later than did adult pairs located in the center of the colony. This occurred even though individuals in pairs of one or more immature members arrived on the colony at the same time each year as did pairs in which both members had mature plumage. Johnston (1956) reported that this happens to young California Gulls (*L. californicus*), and Coulson (1971) reported it among Shags (*Phalacrocorax aristotelis*). Such an incident was recently recorded by Chabrzyk and Coulson (1976) wherein 19% of 32 young Herring Gulls, although preferentially landing in dense parts of the colony, were unsuccessful in acquiring a territory there. Seventy-one percent of the young recruits were successful in establishing territories in an area that had recently been culled of territory holders, and thus were not subject to aggressive behavior of older birds.

Since the original detailed studies of Coulson and White, a number of authors have confirmed the relationship between advancing age and earlier nesting in a number of seabird species [Marshall and Serventy (1956): Slender-billed Shearwater (*Puffinus tenuirostris*); Richdale (1963): Sooty Shearwater (*Puffinus griseus*); Nelson (1966): Gannet (*Morus bassanus*); Coulson *et al.* (1969): Shag; Fisher (1969): Laysan Albatross (*Diomedea immutabilis*); Mills (1973): Red-billed Gull (*L. novaehollandiae scopulinus*); Harrington (1974): Sooty Tern (*S. fuscata*); Davis (1975) and Chabrzyk and Coulson (1976): Herring Gull; Blus and Keahey (1978): Brown Pelican (*Pelecanus occidentalis*); and Brooke (1978): Manx Shearwater (*Puffinus puffinus*)].

Two other factors have been shown to influence timing of breeding in seabirds. Coulson (1966) divided female Black-legged Kittiwakes breeding for the first time into two groups—those that arrived at the colony for the first time in the current breeding season and those that were present on the colony as nonbreeders during the previous year. Individuals of the former group commenced breeding on the average of 7 days later than did those in the latter group. Coulson (1966) noted that some of the young birds that were previously on the colony had formed nonbreeding pairs. Pair formation hardly explains the timing difference in the two groups of first-time breeders, as it is unusual for the same pairs to be reformed in the following year. The time delay of newly arriving recruits to the colony is not clear, but familiarity with food resources and nest sites the previous years would appear to be important (see Perrins, 1970).

The date of laying is also influenced by the character of the pair bond. When a female Black-legged Kittiwake retains her mate of the previous year, her laying advances by about 5 days (Coulson, 1966). If a female changes her mate, the date is similar to females that are less experienced (Coulson, 1966). A comparable situation occurs in Red-billed Gulls (Mills, 1973), in which females that retain mates from the previous breeding season start on average 10 days earlier than do those that change their mates. Brooke (1978) likewise reports newly formed pairs of Manx Shearwaters that breed together in the subsequent season advance their laying date by as much as 6 days. Pairs that have been established for more than 1 year advanced their laying date by an average of only 0.6 days. Conversely, when females moved from an established pair to form a new bond, the laying date was delayed by 1.7 days.

The factors that cause pair-bond breakage are unclear, but Coulson (1966, p. 278) suggests "that there may be a degree of incompatibility between individuals which results in unsuccessful breeding, and it is clearly an advantage for such pairs to split up in the hope that they will find a new partner who is more suitable."

When a female remates, the age of the new male plays a role in determining when breeding begins. Coulson (1966) found that females breeding for the first time with experienced males started their nesting activities more than 5 days earlier than the usual case of inexperienced pairs. Mills (1973) also reported a progressive advance of the average laying date with increasing age of the male in newly bonded pairs of Red-billed Gulls. Fisher (1969) reports results similar to those of Coulson (1966) and Mills (1973) with regard to age of the partners and the effects of mate change on breeding time by Laysan Albatrosses. The only apparent discrepancy is when females that have lost their mates and later bond with younger males lay progressively earlier. Fisher (1969) explains that this is to be expected because in the Laysan Albatross the eggs develop before the female returns to the nesting area, presumably physiologically ready for nesting immediately, regardless of the age of her mate.

The retardation or relative delay in breeding of young seabirds has been linked to a slower gonadal maturation relative to older, experienced individuals. Mills (1973) showed clearly that the testes of young male Red-billed Gulls matured later than did those of older birds, thus effectively delaying all behavior associated with breeding. It is conceivable that females that form bonds, either for the first time or with males younger than themselves, do so later than do females with older mates because of gonadal immaturity. Conversely, when young females are courted by experienced males, their response to male courtship behavior is less than that of older females (Lehrman and Wortis, 1960). Current knowledge concerning the timing of breeding in seabirds appears to show it is directly influenced by the age of both partners, characteristics of the pair bond, and colony density, either individually or in combination.

III. CLUTCH SIZE

That clutch size was affected by age was mentioned by Lack (1954) in reference to the work of Richdale (1949*b*) who elucidated this phenomenon first in seabirds. In Richdale's 12-year study (1936–1948) of Yellow-eyed Penguins (*Megadyptes antipodes*) nesting on the Otago Peninsula, near Dunedin, New Zealand, he was able to show that 2-year old penguins tended to lay fewer than the normal two eggs. In older members, age did not influence clutch size.

In addition to evincing a relationship between young birds and their clutch size, Richdale (1949*b*) clearly showed that the eggs of first-time breeders were narrower and significantly lighter than those laid by older birds. Coulson and White (1958) and Coulson (1972) showed that female Black-legged Kittiwakes nesting for the first time laid significantly fewer eggs than older birds. In their 1958 study, Coulson and White reported that the average clutch of females breeding for the first time was 0.5 ± 0.17 ($p < 0.01$) of an egg less than the clutch of those individuals breeding for at least the third time. Although most Black-legged Kittiwakes lay two eggs, the younger birds laid proportionately more one-egg clutches, whereas three-egg clutches were only laid by females that had bred at least twice before. In contrast, Davis (1975) found that the largest eggs were laid by 7- and 8-year-old Herring Gulls. Both older and younger gulls laid smaller eggs. This has also been shown in the Yellow-eyed Penguin (Richdale, 1957). The reasons for the decline with senility are currently unknown, but they may be correlated with a decreased competitive ability in acquiring sufficient food to form the normal number and size of eggs.

In his later study on Black-legged Kittiwakes, Coulson (1966) showed that the characteristics of the pair bond also play a part in the determination of clutch size. He reported that individuals that change mates show a marked reduction in clutch size relative to those that retain theirs. Females that changed mates through divorce reduced their clutches by 6.1%, as compared to 10.1% for females that had changed mates because of the death of the male. Mills (1973) found virtually the same trends with pair-bond characteristics affecting clutch size in Red-billed Gulls. Mills's results are rather more dramatic than those reported by Coulson in kittiwakes. Female Red-billed Gulls that changed mates laid 0.27 fewer eggs per clutch than did those that retained their mates. This represents a 13.4% drop in reproductive potential. It is therefore clear from Coulson and Mills's investigations that mate retention and advancing age serve to produce larger clutches.

In seabird species that lay more than one egg per season, there is a trend of increasing clutch volume with age. This has been demonstrated for the Yellow-eyed Penguin (Richdale, 1957), the Black-legged Kittiwake (Coulson, 1963), the Shag (Coulson *et al.*, 1969), and the Herring Gull (Davis, 1975). Interestingly, Coulson *et al.* (1969) stated that female shags of the same age that bred at different times of the season, laid different-size eggs; those laying later laid smaller eggs than did those laying earlier—an average difference of 4.5%. Thus clutch size appears to be influenced by a combination of the timing of breeding and the age of the female.

In seabirds that lay only one egg, young birds likewise tend to lay smaller eggs than do older birds. This has been established for the Gannet (Nelson, 1966) and the Razorbill (*Alca torda*) (Lloyd, 1976). The reason is likely to be tied closely to the general trend of nesting later during any one breeding season. A seasonal decline in clutch size of many seabird species tends to confirm this trend. The question of why younger birds lay smaller eggs is yet to be definitively investigated. For some seabirds there is no seasonal decline in food, although it may be irregular (Coulson *et al.*, 1969). Mills (personal communication) reported that early-nesting Red-billed Gulls generally lay two eggs, whereas those nesting during the peak of the season lay three. This apparent contradiction with the general trend has been correlated directly with the patchy food resource, which consists almost entirely of euphasid plankton. Before peak nesting, the euphasid population increases, providing prelaying females with sufficient food to form three-egg clutches. In line with these observations is the phenomenon of prebreeding courtship feeding. In the Red-billed Gull, the male starts courtship feeding the female up to 20 days before egg laying. If the male does not regurgitate food brought back to the territory, the female refuses cooperation in mounting. Similarly, in the McCormick Skua (*Catharacta maccormicki*) and the Chilean Skua (*C. chilensis*) (Spellerberg, 1971) courtship feeding is important to successful copulation. It is possible that females with young inexperienced mates suffer frequently from a lack of food before egg laying. This may serve to retard ovary development and breeding conditions with the resultant production of a smaller size and number of eggs (Mills, 1973; Nisbet, 1977). Brown (1967) reported that courtship feeding in the Lesser Black-backed Gull reaches a peak at the time of accelerated ovarian development. This suggests that if a male is inefficient at providing food at this time, the ovaries will not attain maximum development. The feeding efficiency of young birds and its resultant effects on the breeding biology of Herring Gulls and Greater Black-backed Gulls (*L. marinus*) is discussed in detail by Tolonen (1976). His conclusion was that "feeding efficiency differentials could be the primary cause of these [age-related] differences in reproductive performance."

IV. INCUBATION

Lack (1966) summarized the general trends of incubation periods in seabirds. The two most obvious relationships involve a direct association

between egg weight and fledging period with the length of incubation (but
see Feare, 1976). In addition, the incubation period is longer for offshore
feeders, such as members of the Diomedeidae (albatrosses, 70 days),
Procellariidae (shearwaters and petrels, 53 days), and Sulidae (gannets and
boobies, 45 days), than in the inshore feeding members of the Laridae (gulls
and terns, 28 days) and Pelicanidae (pelicans, 31 days). The rate of growth
of the embryo in larger eggs and the growth rate of hatched young has most
likely evolved with reference to the availability and abundance of food, both
to the parents and young of the various species. In his monographic treatment on the constancy of incubation, Skutch (1962) does not mention the
influence of age on the characteristics of the incubation period. Most in-
depth studies of incubation have stated only the approximate percentage of
time spent on the nest from laying to hatching of eggs. Gulls, for example,
sit about 90% of the time on their eggs [Baerends and Drent (1970): Herring
Gulls; Burger (1976): Laughing Gulls (*L. atricilla*); Somppi (1978): Ring-
billed Gull]. A few studies (i.e., Nelson, 1967; Serventy, 1967; Spellerberg,
1971) have considered the role of the sexes during incubation, specifically
the time spent by each partner on the nest incubating, without regard to
age. Richdale (1957) noticed that young Yellow-eyed Penguins characteristically have longer incubation periods (44.2 days) than do older birds (42.9
days). He intimated that the younger birds do not sit as tightly as do their
older colleagues, thus effectively retarding embryo growth and development
by repeated cooling of the egg. In contrast, Nelson (1966) found no difference in the length of incubation periods of different age classes of Gannets. However, he did note that some young female Gannets incubated
incorrectly. For example, one young female Gannet first incubated the egg
on top of her webs and later lost it. Le Resche and Sladen (1970) noted
several behavioral irregularities during incubation by young Adelie Penguins (*Pygoscelis adeliae*), including incubating eggs that were not their own
and rolling eggs from nest to nest.

   In the Herring Gull, larger eggs elicit more effective incubation, presumably from a relative increase in tactile stimulation between the eggs and
the brood patches, than do smaller eggs (Drent, 1973). In addition, smaller
clutches are subject to fewer sittings than are larger ones in the Black-billed
Gull (*L. bulleri*) (Beer, 1965). It is conceivable that because younger gulls
typically lay smaller and fewer eggs their incubation drive is less than that
of older birds because they lack sufficient stimulation from the eggs.

   Coulson and White (1958) give several instances in which Black-legged
Kittiwakes breeding for the first time failed to incubate efficiently and some
even deserted their clutches before hatching. They judged that the urge to
incubate was not necessarily fully developed in birds capable of laying eggs.
A possible advantage to young birds laying eggs and attempting to incubate,
even though abortive, may be to accelerate the maturation process provid-

ing the individuals involved with sufficient experience to breed successfully the following year (Brown and Baird, 1965).

V. BREEDING SUCCESS

In his review of timing of breeding in relation to age and season, Lack (1968, p. 298) states, "The available evidence therefore supports the view that younger individuals find it harder than older birds to raise young, and there is no evidence to the contrary." I have been able to find little evidence to refute Lack's statement from literature with data on known-age birds.

For purposes of clarification, I define breeding success as a combination of both hatching and fledging success. The former is defined as the proportion of eggs laid that hatch, and the latter as the proportion of eggs hatched that fledge. Breeding success is the proportion of eggs laid that produce fledged young.

One of the first and stronger indications that age rather than experience was a factor in breeding success was reported by Richdale (1957). He remarked that 2-year-old Yellow-eyed Penguins breeding for the first time were less efficient and successful than older birds in producing fledged young. However, individuals that bred first when 3 years old were as successful as experienced older birds. Hence, the value gained in experience by a young bird laying an abortive egg may have been overestimated (see Brown and Baird, 1965). Certainly a bird that lays such an egg does not gain any experience in raising young, and the overall increase in efficiency of breeding with age, at least in the Yellow-eyed Penguin, is likely strictly age related.

Cases of young birds hatching proportionately fewer eggs than older birds have been reported by Coulson and White (1958) and Coulson (1966) for the Black-legged Kittiwake; Nelson (1966) for the Gannet; Horobin (1969) for the Arctic Tern (*S. paradisaea*); Fisher (1975, 1976) for the Laysan Albatross; Chabrzyk and Coulson (1976) for the Herring Gull; and Blus and Keahey (1978) for the Brown Pelican.

Most problems associated with egg loss by young birds are behavior and habitat related. For example, Fisher (1975) recorded only 50% of the eggs of young Laysan Albatrosses hatched, as compared to 75% for older birds, because proportionately fewer of the young birds even completed incubation. Similarly, Coulson and White (1958) observed young Black-legged Kittiwakes often failing to incubate properly or deserting before incubation was completed. However, Blus and Keahey (1978) attribute the higher egg mortality of young Brown Pelicans to flooding of their nests, which were in low, suboptimal habitat, on the colony edge. A similar situation was reported by Dexheimer and Southern (1974) for young Ring-billed Gulls. Young birds successful in establishing territories and nest sites in the central parts of colonies (usually the preferred sites) generally hatch more eggs than do young individuals forced into the peripheral areas (usually suboptimal) (Coulson, 1971; Tenaza, 1971; Ryder, 1975, 1976; Wooller and Coulson, 1977). This phenomenon seems to be more related to the quality of individuals getting into various parts of the colony than is age *per se* (Wooller and Coulson, 1977). Acquiring a nest site in the center of a colony usually involves more effort on the part of the individual to overcome the higher densities and concomitant territorial aggression than is necessary in the more sparsely occupied periphery. The ability of some young individuals to succeed in nesting in the center of the colony may well be the result of superior vigor or size (Wooller and Coulson, 1977) or of the existence of familial groups wherein adults allow their young relatives to nest near them (Ryder, 1976).

The inability of young individuals to raise young has been attributed to a number of causes. Coulson *et al.* (1969) intimated that much of the nestling mortality in Black-legged Kittiwakes and Shags occurs during the first few days after hatching and part of this mortality may be related to the size and quality of the egg, mostly the yolk. As I have mentioned, younger birds generally lay smaller eggs than older birds and it is quite possible these eggs do not supply sufficient nutrients to ensure the survival of newly hatched young relative to a larger egg. Nisbet (1978) suggested a most important factor governing the success of Roseate Tern (*S. dougallii*) eggs is the amount of albumen in the eggs, the smaller the egg, the less albumen. He points out that this component includes about two-thirds of the protein available to the developing embryo and may be a critical agent involved in providing the chick with sufficient vigor to establish a competitive advantage within the brood for food acquisition. Variation in the amount of albumen within clutches has also been recorded in the Herring Gull (Parsons, 1976) and in the Laughing Gull (Ricklefs *et al.*, 1978). It appears that parental performance is most important only following the first few days after hatching (Nisbet, 1978).

Many of the pelagic seabirds forage far from their nesting area. Under such circumstances, foraging experience of the parents is correlated with chick growth and survival (Ainley and Schlatter, 1972). These authors calculated the percentage of two-chick broods of Adelie Penguins that survived through the creche stage increased from 0% for 3-year-old parents to 36% for 8-year-olds. The maximum chick rearing ability in this species is reached at 7–8 years of age. Ashmole (1963*b*) suggests that foraging experience is important in determining the reproductive success of tropical seabirds because of the general insufficiency of food to such species. For seabirds that forage closer to their nesting grounds (e.g., Laridae), chick survival appears to be influenced more by seasonal variation in food availability than by the foraging ability of the parents (Hunt and Hunt, 1976). In seasons or locations in which food for the young is rarely in short supply, it is possible that inadequate parental behavior contributes to chick mortality (Parsons, 1975). Coulson and White (1958) recorded that a few Black-legged Kittiwake chicks died because they were never fed by their parents, which were breeding for the first time. A similar situation was documented for the Laysan Albatross by Fisher and Fisher (1969) and the Adelie Penguin by Le Resche and Sladen (1970) wherein late-hatching young of first-time breeders were undernourished because of parental ineptitude.

The significance of mate change to breeding success has been discussed in detail by Coulson (1966), who found the success of female kittiwakes that retained their mates was higher than those that paired with a new mate. The selective advantage of pairing with a new male likely alleviates incompatability between members of a pair that fail to hatch their eggs. Such trends seem to hold for other species that have been investigated [Mills (1973): Red-billed Gull; Kepler (1969): Blue-faced Booby (*Sula dactylatra personata*)]. Of course, as I have discussed previously, pair-bond characteristics are tied closely to age, so that any marked differences in breeding success found in relation to pair bonds are very likely also related to the age of the partners.

VI. DISCUSSION

The main objective of this chapter is to review some of the relevant literature on the effects of age on the breeding biology of seabirds. It is clear in the literature that major trends include older established pairs usually arriving earlier at a breeding colony, laying larger clutches and eggs that, on average, produce more fledged young per season than do younger pairs or pairs nesting together for the first time. The psychological and physiological mechanisms underlying these trends are unclear and have not been tested rigorously in a wild population of seabirds. Such demonstrations will only be possible when the specific contributions of a number of factors known to control breeding success in social groups are quantified. The effects of weather (Ainley and Le Resche, 1973; Dunn, 1975; White *et al.*, 1976); predation (Nettleship, 1972; Manuwal, 1974; Hayward *et al.*, 1975; Patton and Southern, 1977); and more recently human disturbance (Ashmole, 1963*b*; Reid, 1968; Kadlec and Drury, 1968; Hunt, 1972; Robert and Ralph, 1975; Gillet *et al.*, 1975) all influence the annual breeding success of colonial bird species resulting in some nesting seasons being better than others in terms of a natural rate of increase. Of greater importance are the ultimate factors that control seabird numbers. According to major theses of Wynne-Edwards (1962), Ashmole (1963a), and Lack (1966), these populations are most likely regulated, in a density-dependent way, by the availability of food.

Dependence on a usually scarce commodity over the long term combined with low annual mortality seem to explain adequately the establishment of such typical seabird characteristics as deferred breeding and low clutch size.

The above attributes render it possible for man to search for and define short- and long-term conditions that govern the life of seabirds. The stimulating ideas and reports of Richdale, Ashmole, Coulson, and others cited in this chapter should be considered models for current and future attempts to elucidate new facts about this fascinating assemblage of species.

REFERENCES

Ainley, D. G., and Le Resche, R. E., 1973, The effects of weather and ice conditions on breeding in Adelie Penguins, *Condor* **75**:235.

Ainley, D. G., and Schlatter, R. B., 1972, Chick raising ability in Adelie Penguins, *Auk* **89**:559.

Armstrong, E. A., 1947, *Bird Display and Behaviour*, Dover, New York.

Ashmole, N. P., 1963a, The regulation of numbers of tropical oceanic birds, *Ibis* **103**:458.

Ashmole, N. P., 1963b, The biology of the Wideawake or Sooty Tern *Sterna fuscata* on Ascension Island, *Ibis* **103**:297.

Ashmole, N. P., 1971, Seabird ecology and the marine environment, in: *Avian Biology*, Vol. 1 (D. S. Farner and J. R. King, eds.), pp. 223–286, Academic Press, New York.

Ashmole, N. P., and Ashmole, M. J., 1967, Comparative feeding ecology of seabirds of a tropical oceanic island, *Yale Univ. Bull. Peabody Mus. Nat. Hist.* **24**:131.

Austin, O. L., 1940, Some aspects of individual distribution in the Cape Cod tern colonies, *Bird-Banding* **11**:155.

Austin, O. L., 1945, The role of longevity in successful breeding by the Common Tern (*Sterna hirundo*), *Bird-Banding* **16**:21.

Baerends, G. P., and Drent, R. H., 1970, The Herring Gull and its egg, *Behav. Suppl.* **17**:312.

Beer, C. G., 1965, Clutch size and incubation behavior in Black-billed Gulls (*Larus bulleri*), *Auk* **82**:1.

Belopol'skii, L. O., 1961, *Ecology of Sea Colony Birds of the Barents Sea*, (transl.), Israel Program for Scientific Translation, Jerusalem.

Blus, L. J., and Keahey, J. A., 1978, Variation in reproductivity with age in the Brown Pelican, *Auk* **95**:128.

Brooke, M. de L., 1978, Some factors affecting the laying date, incubation and breeding success of the Manx Shearwater, *Puffinus puffinus*, *J. Anim. Ecol.* **47**:477.

Brown, R. G. B., 1967, Courtship behaviour in the Lesser Black-backed Gull, *Larus fuscus*, *Behaviour* **24**:122.

Brown, R. G. B., and Baird, D. E., 1965, Social factors as possible regulators of *Puffinus gravis* numbers, *Ibis* **107**:249.

Brown, R. G. B., Nettleship, D. N., Germain, P., Tull, C. E., and Davis, T., 1975, *Atlas of Eastern Canadian Seabirds*, Canadian Wildlife Service, Ottawa.

Burger, J., 1976, Daily and seasonal activity patterns in breeding Laughing Gulls, *Auk* **93**:308.

Carrick, R., and Ingham, S. E., 1967, Antarctic seabirds as subjects for ecological research. *Jpn. Antarctic Res. Exp. Sci. Rep.* **1**:151.

Chabrzyk, G., and Coulson, J. C., 1976, Survival and recruitment in the Herring Gull *Larus argentatus*, *J. Anim. Ecol.* **45**:187.

Coulson, J. C., 1963, Egg size and shape in the Kittiwake (*Rissa tridactyla*) and their use in estimating age composition of populations, *Proc. Zool. Soc. London* **140**:211.

Coulson, J. C., 1966, The influence of the pair bond and age on the breeding biology of the Kittiwake Gull (*Rissa tridactyla*), *J. Anim. Ecol.* **35**:269.

Coulson, J. C., 1971, Competition for breeding sites causing segregation and reduced young production in colonial animals, *Proc. Adv. Study Inst. Dynamics and Popul.* **1971**:257.

Coulson, J. C., 1972, The significance of the pair bond in the Kittiwake, *Proc. 15th Intern. Ornithol. Congr.* **1972**:424.

Coulson, J. C., and White, E., 1956, A study of colonies of the Kittiwake, *Rissa tridactyla*, *Ibis* **98**:63.

Coulson, J. C., and White, E., 1958, The effect of age on the breeding biology of the Kittiwake *Rissa tridactyla*, *Ibis* **100**:40.

Coulson, J. C., and White, E., 1960, The effect of age and density of breeding birds on the time of breeding in the Kittiwake, *Rissa tridactyla*, *Ibis* **102**:71.

Coulson, J. C., Potts, G. R., and Horobin, J., 1969, Variation in the eggs of the Shag (*Phalacrocorax aristotelis*), *Auk* **86**:232.

Crook, J. H., 1968, The nature and function of territorial aggression, in: *Man and Aggression* (M. F. A. Montagu, ed.), pp. 141–178, Oxford University Press, New York and London.

Darling, F. F., 1938, *Birds Flocks and the Breeding Cycle. A Contribution to the Study of Avian Sociality*, Cambridge University Press, Cambridge, England.

Davis, D. E., 1940, Social nesting habits of the Smooth-billed Ani, *Auk* **57**:179.

Davis, J. W. F., 1975, Age, egg size and breeding success in the Herring Gull *Larus argentatus*, *Ibis* **117**:460.

Dexheimer, M., and Southern, W., 1974, Breeding success relative to nest location and density in Ring-billed Gull colonies, *Wilson Bull.* **86**:288.

Drent, R., 1973, The natural history of incubation, in: *Breeding Biology of Birds* (D. S. Farner, ed.), pp. 262–322, National Academy of Science, Washington, D.C.

Dunn, E. K., 1975, The role of environmental factors in the growth of tern chicks, *J. Anim. Ecol.* **44**:743.

Feare, C. J., 1976, The breeding of the Sooty Tern *Sterna fuscata* in the Seychelles and the effects of experimental removal of its eggs, *J. Zool. London* **179**:317.

Fisher, H. I., 1969, Eggs and egg-laying in the Laysan Albatross, *Diomedea immutabilis*, *Condor* **71**:102.

Fisher, H. I., 1975, The relationship between deferred breeding and mortality in the Laysan Albatross, *Auk* **92**:433.

Fisher, H. I., 1976, Some dynamics of a breeding colony of Laysan Albatrosses, *Wilson Bull.* **88**:121.

Fisher, H. I., and Fisher, M. L., 1969, The visits of the Laysan Albatrosses to the breeding colony, *Micronesica, J. Univ. Guam* **5**:173.

Fisher, J., 1954, Evolution and bird sociality, in: *Evolution as a Process* (J. Huxley, A. C. Hardy, and E. B. Ford, eds.), pp. 71–83, Allen and Unwin, London.

Fisher, J., and Lockley, R. M., 1954, *Sea-birds*, Houghton-Mifflin, Boston.

Gillett, W. H., Hayward, J. L., Jr., and Stout, J., 1975, Effects of human activity on egg and chick mortality in a Glaucous-winged Gull colony, *Condor* **77**:492.

Gordon, M. S., 1955, Summer ecology of oceanic birds off southern New England, *Auk* **72**:138.

Hailman, J. P., 1964, Breeding synchrony in the equatorial Swallow-tailed Gull, *Am. Nat.* **98**:79.

Harrington, B. A., 1974, Colony visitation behaviour and breeding ages of Sooty Terns (*Sterna fuscata*), *Bird-Banding* **45**:115.

Hayward, J. L., Jr., Amlaner, C. J., Jr., Gillett, W. H., and Stout, J. F., 1975, Predation on nesting gulls by a river otter in Washington State, *Murrelet* **56**:9.

Hindwood, K. A., Keith, K., and Serventy, D. L., 1963, Birds of the South-west Coral Sea, *CSIRO Div. Wildl. Res. Tech. Pap.* **13**:1.

Horn, H. S., 1970, Social behaviour of nesting Brewer's Blackbirds, *Condor* **72**:15.

Horobin, J. M., 1969, The breeding biology of an aged population of Arctic Terns, *Ibis* **111**:443.

Hunt, G. L., Jr., 1972, Influence of food distribution and human disturbance on the reproductive success of Herring Gulls, *Ecology* **53**:1051.

Hunt, G. L., Jr., and Hunt, M. W., 1976, Gull chick survival: The significance of growth rates, timing of breeding and territory size, *Ecology* **57**:62.

Johnston, D. W., 1956, The annual reproductive cycle of the California Gull. I. Criteria of age and the testis cycle, *Condor* **58**:134.

Kadlec, J. A., and Drury, W. H., 1968, Structure of the New England Herring Gull population, *Ecology* **49**:644.

Kepler, C. B., 1969, Breeding Biology of the Blue-faced Booby *Sula dactylatra personata* on Green Island, Kure Atoll, *Nuttall Ornithol. Club* **8**:1.

Klopfer, P. H., and Hailman, J. P., 1965, Habitat selection in birds, *Adv. Study Anim. Behav.* **1**:279.

Lack, D., 1954, *The Natural Regulation of Animal Numbers*, Clarendon Press, Oxford.

Lack, D., 1966, *Population Studies of Birds*, Clarendon Press, Oxford.

Lack, D., 1967, Interrelationships in breeding adaptations as shown by marine birds, *Proc. 14th Intern. Ornithol. Congr.* **1967**:3.

Lack, D., 1968, *Ecological Adaptations for Breeding in Birds*, Methuen, London.

Langham, N. P. E., 1974, Comparative breeding biology of the Sandwich Tern, *Auk* **91**:255.

Lehrman, D. S., and Wortis, R. P., 1960, Previous breeding experience and hormone-induced incubation behaviour in the Ring Dove, *Science* **132**:1667.

Le Resche, R. E., and Sladen, W. J. L., 1970, Establishment of pair and breeding site bonds by young known-age Adelie Penguins (*Pygoscelis adeliae*), *Anim. Behav.* **18**:517.

Lloyd, C. S., 1976, The breeding biology of the Razorbill, *Alca torda* L., Unpublished Ph.D. dissertation, Oxford University, London.

MacRoberts, B. R., and MacRoberts, M. H., 1972, Social stimulation of reproduction in Herring and Lesser Black-backed Gulls, *Ibis* **114**:495.

Manuwal, D. A., 1974, The natural history of Cassin's Auklet (*Ptychoramphus aleuticus*), *Condor* **76**:421.

Marshall, A. J., and Serventy, D. L., 1956, The breeding cycle of the Short-tailed Shearwater, *Puffinus tenuirostris* (Temminck), in relation to trans-equatorial migration and its environment, *Proc. Zool. Soc. London* **127**:489.

Mills, J. A., 1973, The influence of age and pair-bond on the breeding biology of the Red-billed Gull *Larus novaehollandiae scopulinus*, *J. Anim. Ecol.* **42**:147.

Murphy, R. C., 1967, *Distribution of North Atlantic Pelagic Birds. Serial Atlas of the Marine Environment*, Folio 14, American Geographical Society, New York.

Nelson, J. B., 1966, The breeding biology of the Gannet *Sula bassana* on the Bass Rock Scotland, *Ibis* **108**:584.

Nelson, J. B., 1967, The breeding behaviour of the White Booby *Sula dactylatra*, *Ibis* **109**:194.

Nettleship, D. N., 1972, Breeding success of the Common Puffin (*Fratercula arctica* L.) on different habitats at Great Island, Newfoundland, *Ecol. Monogr.* **42**:239.

Nisbet, I. C. T., 1977, Courtship feeding and clutch size in Common Terns *Sterna hirundo*, in: *Evolutionary Ecology* (B. Stonehouse and C. Perrins, eds.), pp. 101–109, University Park Press, Baltimore, Md.

Nisbet, I. C. T., 1978, Dependence of fledging success on egg-size, parental performance and egg-composition among Common and Roseate Terns, *Sterna hirundo* and *S. dougallii*, *Ibis* **120**:207.

Orians, G. H., 1961, Social stimulation within blackbird colonies, *Condor* **63**:330.

Orians, G. H., 1969, Age and hunting success in the Brown Pelican (*Pelecanus occidentalis*), *Anim. Behav.* **17**:316.

Parsons, J., 1975, Asynchronous hatching and chick mortality in the Herring Gull *Larus argentatus*, *Ibis* **117**:517.

Parsons, J., 1976, Factors determining the number and size of eggs laid by the Herring Gull, *Condor* **78**:48.

Patton, S. R., and Southern, W. E., 1977, The effect of nocturnal red fox predation on the nesting success of colonial gulls, *Proc. Col. Waterbird Group Conf.* **1977**:91.

Perrins, C. M., 1970, The timing of birds' breeding seasons, *Ibis* **112**:193.

Reid, B., 1968, An interpretation of the age structure and breeding status of an Adelie Penguin population, *Notornis* **15**:193.

Richdale, L. E., 1949a, A study of a group of penguins of known age, *Biol. Monogr., Dunedin, N.Z.* **1**:1.

Richdale, L. E., 1949b, The effect of age on laying dates, size of eggs and size of clutch in the Yellow-eyed Penguin, *Wilson Bull.* **61**:91.

Richdale, L. E., 1951, *Sexual Behaviour of Penguins*, University of Kansas Press, Lawrence.

Richdale, L. E., 1957, *A Population Study of Penguins*, Oxford University Press, Oxford.

Richdale, L. E., 1963, Biology of the Sooty Shearwater *Puffinus griseus*, *Proc. Zool. Soc. London* **141**:1.

Ricklefs, R. E., Hahn, D. C., and Montevecchi, W. A., 1978, The relationships between egg size and chick size in the Laughing Gull and Japanese Quail, *Auk* **95**:135.

Robert, H. C., and Ralph, C. J., 1975, Effects of human disturbance to the breeding success of gulls, *Condor* **77**:495.

Ryder, J. P., 1975, Egg-laying, egg size and success in relation to immature–mature plumage of Ring-billed Gulls, *Wilson Bull.* **87**:534.

Ryder, J. P., 1976, The occurrence of unused Ring-billed Gull nests, *Condor* **78**:415.

Serventy, D. L., 1967, Aspects of the population ecology of the Short-tailed Shearwater, *Puffinus tenuirostris*, *Proc. 14th Intern. Ornithol. Congr.* **1967**:165.

Skutch, A. F., 1962, The constancy of incubation, *Wilson Bull.* **74**:115.

Somppi, P. L., 1978, Reproductive performance of Ring-billed Gulls in relation to nest location, Unpublished M.Sc. thesis, Lakehead University, Thunder Bay.

Spellerberg, J. F., 1971, Breeding behaviour of the McCormick Skua *Catharacta maccormicki* in Antarctica, *Ardea* **59**:189.

Swartz, L. G., 1966, Sea-cliff birds, in: *Environment of the Cape Thompson Region, Alaska* (N. J. Wilimovsky, ed.), pp. 611–678, U.S. Atomic Energy Commission, Division of Technical Information, Washington, D.C.

Tenaza, R., 1971, Behaviour and nesting success relative to nest location in Adelie Penguins (*Pygoscelis adeliae*), *Condor* **73**:81.

Tolonen, K., 1976, Behavioral ecology of *Larus argentatus* and *Larus marinus:* Age-specific differential in feeding, Unpublished Ph.D. dissertation, Yale University, New Haven.

Vermeer, K., 1963, The breeding ecology of the Glaucous-winged Gull (*Larus glaucescens*) on Mandarte Island, B.C., *Occ. Pap. B.C. Prov. Mus.* **13**:1.

Weidmann, U., 1956, Observations and experiments on egg-laying in the Black-headed Gull (*Larus ridibundus* L.), *Br. J. Anim. Behav.* **4**:150.

White, S. W., Robertson, W. B., Jr., and Ricklefs, R. E., 1976, The effect of Hurricane Agnes on growth and survival of tern chicks in Florida, *Bird-Banding* **47**:54.

Wooller, R. D., and Coulson, J. C., 1977, Factors affecting the age of first-breeding of the Kittiwake *Rissa tridactyla*, *Ibis* **119**:339.

Wynne-Edwards, V. C., 1962, *Animal Dispersion in Relation to Social Behaviour*, Oliver and Boyd, Edinburgh.

Chapter 6

THE COMMUNICATION BEHAVIOR OF GULLS AND OTHER SEABIRDS

C. G. Beer

Institute of Animal Behavior
Rutgers University
Newark, New Jersey 07102

I. INTRODUCTION

A. Communication Behavior in General

Any behavior can be communicative. That is to say, whatever an animal might be doing can tell you something about it. A gull dropping clams on a beach is manifestly trying to get food; a gull preening on a post is just as manifestly not. A tern making a harsh clicking call above you as you walk near its nest signals that it is about to dive at and possibly strike you, and the same tern parading with head held high and bill pointed skyward in front of another conveys the information that it is ready for sexual attachment.

However, these two pairs of examples contrast in that in the case of the first pair communication is not the point of the behavior—what it is done for—but communication is the point in the case of the second pair. Investigation and discussion of animal communication have to do with behavior like that in the second pair of examples: behavior apparently "designed" to serve as signal or display—communication behavior.

Paradoxically, the information conveyed by communication behavior tends to be more difficult to arrive at than that conveyed by behavior not designed for communication, at least in the experience of ethologists who have worked on the subject. This is one reason why the study of animal communication, even though it has been a popular subject of ethological research from the beginning of ethology's history, has yet to achieve unity or coherence as a scientific discipline. There have been numerous approaches, lots of concepts and theories and interpretations, and some fat books compiled about the subject (e.g., Sebeok, 1968, 1977), but there is little agreement on anything. Even the descriptive terms and the criteria for defining units of communication behavior remain matters of dispute (cf. Beer, 1977).

Part of the problem is in the language that offers itself for use in the description and discussion of animal behavior. Many of the words in this language have meanings in talk about ourselves, which may or may not apply in talk about monkeys or cats or gulls. The bogy of anthropomorphism frightens many ethologists, as well other students of animal behavior who take pride in being tough-minded, into eschewing all mentalistic and humanistic connotations that attach to such words as intention, meaning, knowledge, ceremony, and ritual. I have even heard the use of the word communication criticized as implying intent. Yet these and other such words are frequently used in animal studies. We read of intention movements, message–meaning analysis, the innate knowledge that tells a newly-hatched gull chick to peck at red when it is hungry, greeting ceremonies, ritualization, and animal communication. Of course, we also recognize the implicit provisos in such usage. Nevertheless, the undercurrents of old connotations can undermine the basis that technical definition attempts to establish and cause cracks in whatever structure of concept and theory is supposed to be held in scientific common.

For example, ethologists sometimes give the impression of being equivocal with the word information. They invoke the essentially statistical concept of information theory when they want to appear rigorously operational or quantitative, but turn to the everyday meaning—facts or instructions about something or other—when that suits the sense to be conveyed. Such ambiguity tends to create uncertainty about what is meant and to tempt to invalid inference via the verbal bridge. Similarly, when terms such as semantics and syntax are applied to nonhuman cases, it is often difficult to gauge how much of their primary meanings in linguistics is carried over; and even when the terms are explicitly redefined, it is difficult to keep the primary meanings from coloring what is conveyed by their use in the secondary contexts.

However, it could be a mistake to demand too insistently that the students of animal communication purify the language of their tribe. Scientists, whether or not they recognize or admit the point, share with poets a motive for metaphor, especially in what Kuhn (1970) described as the "preparadigm" stage of the history of a science. As I have implied already, the study of animal communication has yet to find "its first universally received paradigm" (Kuhn, 1970, p. 13). Meantime several possibilities, derived largely by analogy from neighboring disciplines, have been tried. Each has led to discovery of facts, to formulation of concepts, and to proposal of hypotheses, but none has established itself as *the* way to go or arrived at a theory or synthesis sufficiently powerful or comprehensive to draw all together into a coherent relationship.

Part of what obstructs such unification is the variety in the questions raised by animal communication behavior. This is a problem that the study of communication shares with ethology in general and that has been equally hindering to aspiration to effect the modern synthesis. Tinbergen (1963) defined ethology in terms of four questions—proximate causation, ontogenetic development, survival value, and evolutionary descent—and said: "it is useful to distinguish between them and to insist that a comprehensive coherent science of ethology has to give equal attention to each of them and to their integration" (Tinbergen, 1963, p. 411). It is more than merely useful to make such distinctions; it is necessary if confusion is to be avoided, as the perennial muddle about nature and nurture testifies.

However, the distinctions as Tinbergen drew them may not fit the case of communication as comfortably as he supposed. Indeed, to get the topic into them it has to be trimmed to a conception of animal communication that they tacitly imply. To ask about proximate causation, for example, is to assume a causal conception, which directs research to look for what will ultimately come down to physiological mechanisms. Of course, physiological mechanisms are involved in animal communication. Even for human speech, knowledge of how our vocal apparatus and auditory system work is relevant to a full understanding of the matter. But there is more to conversation than the production and reception of sounds. There are also the rules of syntax and semantics, the lexical and logical proficiency in virtue of which these sounds are language. Linguistic principles, cannot, at least as yet, be expressed in, translated onto, or reduced to physiological terms. If there are analogous or parallel principles in animal communication— something corresponding to grammer and reference, for example—their discovery and comprehension might likewise require the use of concepts that lie, at least in the meantime, outside the universe of physiological discourse.

Suppose such principles to have been discovered to apply in some animal species. Their existence would affect formulation of the developmental questions, for these would now have to include the emergence of syntactic and semantic competence, the acquisition of proficiency in the use and apprehension of the signals in the repertoire, in addition to the ontogeny of the forms of the signal patterns, and of the sensorimotor mechanisms underlying their production and reception, to which the purely causal conception would restrict attention. Similarly, with the questions of survival value and phylogeny, the answers sought depend on the conception of what has to be explained.

The matter of survival value is complicated in another way as well, in that it is often taken to be equivalent to function. The problem here is that function has multiple meanings, at least two of which apply in the context of animal communication studies, and need to be distinguished. There is the sense in which function refers to the immediate consequences that performance of a signal is likely to have, as implied in describing such a signal as a threat display or as a distraction display, which is to say that threat or distraction is the function of the display in question. But knowing the function in this sense, which might be referred to as the immediate or proximate function, does not always tell us how use of the signal contributes to reproductive success, which is what survival value is more likely to be taken to mean, and which might be distinguished as the ultimate function. Thus the ultimate function of the threat display might be the contribution it makes to ensuring an adequate food supply via its role in the acquisition and maintenance of territory; and the ultimate function of the distraction display will be defense of the eggs or chicks against predation.

On the other hand, understanding of ultimate function is no more enlightening about proximate function than is what can be inferred in the reverse direction. For example, we now know a great deal about the ecological correlates of the various forms of territoriality—how such social systems answer to specific kinds of selection pressure and so, in the terms of sociobiology, maximize inclusive fitness. But from none of this can we deduce the nature of the signals used in territory establishment and defense, their forms, modes of presentation, patterns of effect, and so forth. Indeed, there are cases for which it would appear to be a mistake to ask for *the* survival value or ultimate function—signals that serve more than one such end, as, for example, a display that contributes to territorial possession *and* pair information *and* prevention of interspecies hybridization. There are signals that serve more than one proximate function also, because the effect they have can depend on the context in which they occur, including the other signals performed in concert or in sequence with them.

E. O. Wilson has argued that ethology is "destined to be cannibalized by neurobiology and sensory physiology from one end and sociobiology and behavioral ecology from the other" (Wilson, 1975, p. 6). According to the view I have been trying to express, at least that part of ethology that deals with the nature of animal communication and its proximate functioning is indigestible to both of these predatory sciences, now and for the foreseeable future. There appear to be principles at work in communication behavior that do not break down to those of either cellular biology or population biology. They are relevant even in the study of the evolutionary history of such behavior.

This last of Tinbergen's four questions has been cast in forms derived from two analogies—one to comparative anatomy, and the other to comparative philology. In comparative anatomy the key concept is that of homology—correspondence between elements in structures sharing the same design. The existence of such correspondence provided evidence for the theory of evolution, which, in turn, provided explanation for the correspondence. The encouragement to circular argument notwithstanding, modern textbooks consequently define homology in terms of descent from common evolutionary origins (e.g., Wilson, 1975, p. 586). Nevertheless, in practice, structural homologies are still often recognized on the basis of purely morphological criteria, such as position in a pattern of connected elements, usually, but not necessarily, with the aid of fossil evidence. Applied to behavior, however, as it has been especially in the study of communication behavior, the concept of homology is in want of morphological criteria equivalent to those that obtain in comparative anatomy, and there is virtually no fossil behavior to which appeal might be made. The comparative ethologist has to do what he can with the behavior of living species and has to rely largely on intuitive perception, or perceptual intuition, to judge what corresponds to what in the behavioral repertoires of related species. Such judgment is often predicated on prior belief or assumption about phylogenetic relatedness; but inference has sometimes gone in the other direction, arguing phylogenetic relatedness from independently based premises asserting homology. The independent grounds on which such an assertion has been founded vary from case to case and include similarity in some respect of form, correspondence of position in sequential patterns or in the set of behavioral patterns apparently shared by the several species in question, and the existence of intermediate forms linking divergent patterns in graded series.

In spite of the differences between structural homology and behavioral homology in the ways they are applied to cases, an argument can be made to the effect that essentially the same concept is intended in both contexts (Beer, 1977). At least some ethologists who take an interest in this matter would deny that we are dealing with an analogy. They would say that when they claim two patterns of behavior to be homologous they mean the same thing as when they claim that two structures are homologous, i.e., that they are dealing with a case of inheritance from a common phylogenetic origin. Such objection to being regarded as analogy would not be raised for the comparison with the history of language.

The linguistic analogy is less frequently and less literally invoked than that to anatomy, yet in some lights it is the closer and more pervasive of the two, especially for conceptions about the evolution of communication behavior. What the comparative linguist does in trying to identify cognate words in different living languages, and trace the changes in the sound and sense to their origins in a common ancestral language, gives perhaps the best available parallel to what the comparative ethologist does in trying to identify homologous displays in the repertoires of different living species, and trace the changes in form and function to their origins in common evolutionary antecedents. In neither case is there a consistent equivalent of the anatomical criterion of position in pattern—the "Principle of Connections" of the 18th century anatomist Geoffroy de St. Hilaire—as a formula for correspondence. Sequential patterning of displays can differ from species to species; grammatical rules for word order can differ from language to language. In both cases the historical interpretation variously draws on formal similarities, correspondences, or continuities of meaning or function; connection through intermediate links; and other less-than-conclusive lines of evidence.

Another feature common to animal communication and human language is the distinction between what can be loosely described as form and force, or structure and signification. In analyzing the communication behavior of a species, the ethologist usually begins by trying to sort out and describe the physical characteristics of the displays and signals, and then proceeds to try to find what they convey to, and elicit from, individuals toward which they are directed. Similarly, the study of language divides between, on the one hand, phonological and syntactical analysis, which works out the basic sound patterns and the rules governing their combination and, on the other hand, semantic analysis, which deals with meaning in both its lexical and grammatical varieties. The structural analysis takes priority in the sense that one needs to know what it is that has meaning before one can study what that meaning is. However, how the syntax is construed can be affected by what the meaning is taken to be. This is exemplified by ambiguous cases such as Chomsky has frequently cited (e.g., "They are flying planes," Chomsky, 1957, p. 87); and linguistic philosophers (e.g., Ryle, 1949) have argued a distinction between syntactic grammer and logical grammer according to which identity in the former can be misleading if taken to imply identity in the latter (e.g., "points have location but no dimension" understood as the same kind of statement as "professors have pensions but no savings"). Similarly, in cases of animal communication, there can be problems in assuming that what we perceive as the units of form correspond to what has significance in the transmission of information as far as the animals are concerned; and our discrimination of the formal divisions can be affected by what we discover the meaningful categories to be. Such discrimination could, in turn, affect the question of evolutionary origins by leading to revision of the inventory of behavior patterns for which homologues will be sought. Moreover, if the linguistic analogy were to hold to the extent of including some parallel to syntactic structure, this would be yet another dimension for comparative and evolutionary study of animal communication behavior, one that could conceivably contribute to thought about the criteria of homology in this context.

My shift to the subjunctive voice betrays that I have come to a realm of speculation still virtually unexplored. I think it worth marking the spot for possible future colonization. Some of the more persuasive reasons for this opinion come from work on seabirds.

B. Communication Behavior of Seabirds in Particular

Seabirds have a number of advantages as subjects for communication study. First, with only a few exceptions, they are colonial and monogamous, with both parents sharing in incubation of the eggs and provisioning of the chicks. There is therefore considerable conformity in circumstances prescribing occasion for a great deal of the social interaction and hence communication behavior, in the different species. Consequently, what might be found to be the case in one species can, as a rule, be readily compared with, or at least sought for in the behavior of other species. Because, according to Lack's (1968) count, the seabirds comprise 262 species, representing four orders and 13 families, there is plenty of scope for such comparison, over a range of degrees of phylogenetic relatedness.

On the other hand, there is sufficient ecological diversity to make the search for adaptive correlation worth pursuing, as Tinbergen on larids (e.g., Tinbergen, 1959) and Nelson (e.g., 1970, 1972, 1975) on sulids have demonstrated. Perhaps the major ecological contrast is between the species that breed in situations inaccessible to predators, and the species that breed where they are vulnerable to predators (Lack, 1968).

Second, being colonial, seabirds can be watched in numbers that the spacing of solitary animals precludes. Moreover, most species nest on the ground and in the open and are diurnal, which also makes for ease of observation. '

Third, the social behavior of a variety of species has been described and has received some comparative and experimental study sufficient to provide at least preliminary maps by which new research can set its courses.

The group that has been most extensively and variously studied in these ways is that of the gulls. For this reason, and because most of my own research has been on gulls, I give to them the bulk of attention in the discussion that follows.

I shall begin with the nature of the signals constituting group repertoires. Then I shall look at contexts of occurrence and sequential patterns to see what might be said about how signals are used, what they are used for, and what effects they bring about. Next, I shall comment briefly on evolutionary questions: first the matter of adaptive significance or survival value; second, the matter of evolutionary origins and evolutionary history. Finally, I shall say a little about social development. Throughout, my emphasis will be on what is problematic rather than what has been established about communication behavior of seabirds, for I believe that the study of communication is much richer in promise than in achievement.

II. SIGNAL REPERTOIRES

A. Seabirds in General

The most general description of a communication event is that it consists of a transmitter, a signal, and a receiver (cf. C. W. Morris, 1938). Signal implies "channel," which, in the context of animal communication (or "zoosemiotics," as Sebeok (1967) has suggested it be called) will be the sensory modality through which the transmission takes place. For birds in general and seabirds in particular, most communication is either visual or auditory. Tactile communication does occur in such activities as food begging, allopreening, and copulation; the fact that one can find very few instances of use of the chemical senses may be partly a consequence of their having been rarely sought. But the display postures and movements, and the vocal productions that we distinguish as calls, are what, in effect the study of communication in birds is about. At any rate, I shall deal here only with signals that are either seen or heard.

Seabirds hardly compare to some of the other groups of birds in specialization and elaboration of form and voice for communication. To put beside such glories as the tail of the peacock, the lyre of the lyrebird, and the plumes of birds of paradise, the best that the seabirds can muster are the gular sac of male frigatebirds and the faces of puffins, and even these are standouts in a predominantly plain company regimented in uniforms of black, gray, and white. No seabird can do with its voice anything approaching what the more melodious songbirds do when they sing. Indeed, to someone who has made a study of birdsong, or the display sequences of such birds as ducks, manakins, or grebes, the basic repertoires of signals of seabirds probably appear to consist of relatively simple postures, movements, and calls (Fig. 1). Despite this comparative simplicity in the forms of signal units, or perhaps because of it, the communication behavior of seabirds appears, in some cases, to be quite complex in other respects—at least this has been found to be so for gulls, which are typical of seabirds in many ways.

Fig. 1. Examples of seabird displays. (A) "Ecstatic" display by Adelie Penguins (*Pygoscelis adeliae*). (B) "Sky-pointing" by a Gannet (*Sula bassana*) about to take off from within the breeding colony. (C) Climax of "mutual display" of courting Wandering Albatrosses (*Diomedea exulans*). (D) "Gaping" displays between mated Shags (*Phalacrocorax aristotelis*). (E) Relatively unritualized threat display between Black Guillemot (*Cepphus grylle*).

B. Gulls in Particular

Although there are some earlier observations (e.g., Haviland, 1914; Herrick, 1912; Portielje, 1928; Strong, 1914), the study of communication in gulls can be considered as starting with the descriptions published during the 1930s and 1940s (Booy and Tinbergen, 1937; Deusing, 1939; Goethe, 1937, Kirkman, 1937, 1940; Steinbacher, 1938; Tinbergen, 1936a,b; Wachs, 1933). It was in such studies that many of the current names for the signal patterns were introduced. For example, "long call," "head-flagging," and "choking" are terms that come from a paper on the Laughing Gull (*Larus articilla*) by Noble and Wurm (1943). The further refinement and development of this nomenclature have been attributable mainly to Tinbergen (1953, 1959) and Moynihan (1955, 1956, 1958*a,b*, 1959*a,b*, 1962). This nomenclature labels a core repertoire of displays recognized in virtually all species studied, which includes the oblique-and-long-call, the upright posture, head-flagging, choking, head-tossing, mew call, alarm call, and copulation call. There is some variation in usage. For instance, head-flagging is now often referred to as facing-away, and I have preferred to use crooning call for what most other investigators have referred to as the mew call (e.g., Beer, 1970a,b). Nevertheless, the repertoires of the different gull species show so much correspondence with one another that judgments of homological identity are made with confidence rare in comparative ethological study. Some of the postures are illustrated in Fig. 2.

The more recent description of signal repertoires has been aided by motion picture photography, sound tape recording, videotape recording, sound spectrography, and other electronic technology. However, even with these modern techniques, the criteria for identifying the signal units remain imprecise, the ostensive definition being the basis for what agreement there is, rather than unequivocal explicit specification. Consequently, the agreement is less than total between different investigators, and the categorizing of some of the behavior is still in a state of flux, especially with regard to vocal communication. What one sees depends upon what one looks for. Differing conceptions of communication can lead to differing perceptions of what goes on.

The classic conception of Lorenz and Tinbergen, according to which animal communication is a stimulus–response system—the social signals being conceived as releasers to which innate releasing mechanisms governing specific action patterns are tuned—predisposes observation to find discrete units of communication behavior and to ignore or disregard gradations or intermediates between salient forms and differences between individuals or between situations. Similarly, most of the correlational and stochastic approaches, which came into vogue when the variability of social interaction patterns was more fully appreciated, were predicated on the atomistic assumption of a fixed set of discontinuous categories of action—the items between which the covariation could be computed, the transition probabilities estimated, the temporal relationships timed. Again, interest in what might be particular was subordinated to the attention given to finding what was general.

However, what one looks for can also be influenced by what one finds. Sometimes by chance, often without its being sought, observation and

Fig. 2. Examples of gull displays. (A) Head-tossing and facing-away of courting Laughing Gulls (*Larus atricilla*). (B) Choking between hostile Laughing Gulls. (C) Forward postures of courting Black-headed Gulls (*Larus ridibundus*). (D) Oblique-and-long-call of Black-billed Gull (*Larus bulleri*). (E) Hunched posture of fledgling Red-billed Gull (*Larus novaehollandiae scopulinus*).

experiment have uncovered facts that have forced revision of descriptive categories and conceptual frameworks. This has occurred in my work on the communication behavior of Laughing Gulls, from which I choose two examples—facing-away and long-calling—to illustrate the kinds of complexity that the more recent gull studies have revealed, as well as the interplay of perception and conception in this revelation.

C. The Facing-Away of Laughing Gulls

Facing-away or head-flagging, as originally described by Noble and Wurm (1943, p. 190) and consequently perceived by the succeeding generations of gull watchers, consisted of a gull's adopting an erect or upright posture—neck vertically extended with the bill pointing horizontally or downward to some degree—and then turning the head to face away from the other bird, repeated looking back, and then away again suggesting the flag image. Tinbergen and Moynihan (1952) reported that the side-to-side motion is missing in the display as it is performed by Black-headed Gulls (*Larus ridibundus*), which they interpreted as a further ritualization of the display's evolutionary origin, more evident in the Laughing Gull version—alternation of moving toward and moving away in a motivational situation of social ambivalence or conflict. Moynihan (1955, p.71) later remarked that Black-headed Gulls occasionally perform the looking-away movement in attitudes other than the upright posture. The consequent inappropriateness of the flag image as a general label for the display eventually led to the preference for calling it facing-away (Tinbergen, 1959).

When I came to look at Laughing Gulls I saw little to suggest a flag, but much to confirm that the facing-away movement is anything but confined to the upright posture. First in photographs and motion pictures, and then when I looked for it in the field, I found the movement to occur in nearly all the postural attitudes that have been distinguished: upright, full oblique, what Moynihan (e.g., 1962) calls the low oblique (but which might be better referred to as the horizontal posture, as it can occur in isolation from the long-call and with a distinctness suggesting that it is not just a variant of the full oblique), choking, the stooped posture in which crooning is uttered, even the attitude in which a landing gull touches down.

This distribution of facing-away over the range of other postures raised a question about how the facing-away movement should best be regarded. If one were to be guided by the atomistic conception of the communication repertoire as a set of discrete, self-contained, independent signals, then it would seem that one should recognize each of the different postural varieties of facing-away, as well as each of the postures in their "straight" forms, as a separate type of signal or display. On the other hand, to someone less inclined toward such a strictly atomistic view, there would seem to be at least as good a case for regarding the facing-away as, so to say, superimposed on the postures, in some way analogous to the attaching of prefixes or suffixes to words. Facing-away would not then be regarded as a display in and of itself, or as several types of display in and of themselves, but rather as a qualifier of display. This idea would imply that what each of the postural displays expresses or signifies is altered in the same or a similar way when facing-away is added to it. I have some evidence, admittedly far from conclusive, for the speculation that facing-away serves to sign negation for a Laughing Gull—that, for example, a gull facing-away in the upright posture is conveying that it is not hostile toward the other bird, the opposite of what is conveyed by the "straight" upright posture (Beer, 1975).

However, there is even more complication to the Laughing Gull case. Again, photographs drew my attention to details that field observation confirmed in performances of facing-away in the upright posture. In some, the position of the head and set of the feathers are such as to screen the black hood from the view of the other bird. This I call white facing-away. In other performances, the head position and feather set keep a goodly proportion of the hood, and even the white eye fringes, visible to the other bird. This I call black facing-away. In white facing-away the margin of the hood is near vertical; in black facing-away it is near horizontal. The contrasting versions are illustrated in Fig. 3. There are intermediates between these two forms, and a gull may go from one to the other without a break in the action. However, most performances, and most of the time in performance, are given to one or other of the extreme forms. Again the question is: Should we distinguish two displays in what has hitherto been regarded as one, or should we consider the variants to be end forms of a single graded signal? I have treated them as though they were discontinuous (Beer, 1975, 1976), but this is not the only possibility.

Tinbergen has used single-frame analysis of movie films to argue distinctness in the analogous case of transitions between the forward and oblique postures of Black-headed Gulls. Counts of the numbers of frames containing each position showed that the gulls hold the postures for much longer than the transitional or intermediate configurations between them. This, for Tinbergen, confirmed the reality of the forward posture and the oblique posture as distinct displays, as it showed them to be in accord with the specification of "typical intensity" (D. Morris, 1957), which he considered a condition for efficient signaling and was virtually inherent in his concept of display.

However, we must keep in mind that what can serve as a signal for an animal will depend on what can register in its perception, which may not necessarily be what strikes ours, even with the aid of technology. Marler (1969) has made the point that because of the finer temporal resolution of its ear, a bird may hear as vibrato a succession of tones alternating so rapidly that we hear them as a chord, and that the sound spectrograph can represent it either way depending on the filter setting selected. Our own experience provides an example of discrepancy between the phenomenal and the physical, which should make us hesitate in drawing inferences even from films and tape recordings. For example, the pauses we hear between words, phrases, sentences, and so forth in speech are not apparent to anything like the same degree when the sound is reproduced as an oscillogram or sound

Fig. 3. Facing-away of Laughing Gulls. (A) White facing-away in the upright posture. (B) Black facing-away in the upright and semi-upright postures.

spectrogram. This perceptual competence must be acquired in the learning of the language, for people speaking a language that we do not know seem to run all their words together. But whatever the nature of an animal's perception of the communications it receives, we run the risk of serious error if we simply assume that we share it. Again, my experience with the Laughing Gull provides case in point, this time from study of the long-call.

D. The Long-Calls of Laughing Gulls

The Laughing Gull long-call, together with the postures in which it is given, although unmistakably diagnostic of the species, has considerably more variability than one might think consistent with the rule of "typical intensity." When I came to look at sound spectrograms of the call, I found at least 15 respects in which the calls could differ from one another, and further variation if one included the postures accompanying utterance (Fig. 4). Either this variability must be irrelevant to the presumed signal function; or at least some of it must convey information that can differ between cases. The latter seemed much the more likely possibility, yet my first confirmation of it was more by accident than design.

In experimental attempts to elicit filial responses from chicks by playback of recorded calls of adults I found, contrary to initial expectation, that I could get response, from the chicks of the age I was testing, only to certain kinds of call, which included the long-call, and only if the call had been recorded from a chick's own parents. From inspection of the sound spectrograms it appeared that the individually identifying part of the long-call is in the string of short notes with which it begins, for these are constant in number and almost so in rate of repetition, duration, tonal spectrum, and frequency modulation pattern for calls of the same individual, but differ markedly between individuals (Fig. 5). The other parts of the call—the long note section and the terminal head-toss notes—vary widely in all these respects, except perhaps the tonal spectrum, in samples of calls from the same individual.

The discovery of individual characteristics and individual recognition in the communication behavior of gulls, although not something that the instinctivist, atomistic view inclined one to look for, was nevertheless hardly surprising. Indeed, Tinbergen (1953) had said that such recognition occurs in the Herring Gull (*Larus argentatus*), albeit on the basis of evidence at least some of which bears another interpretation (Beer, 1979); and Tschanz (1968) had already demonstrated individual recognition by ear in Guillemots (*Uria aalbe*) by doing playback experiments. At the time I was doing the same with Laughing Gulls, Evans (1970*a,b*) was also doing it with Ring-billed Gulls (*Larus delawarensis*) and Black-billed Gulls (*Larus bulleri*) and finding individual recognition to playback of mew calls.

Less expected was the discovery that, in addition to identifying itself individually as the sender, a long-calling Laughing Gull can specify the individual or individuals to which the signal is being sent. This I also found as a result of looking for something else. In playback tests of older chicks (2 weeks and more posthatching), in which I compared the chicks' responses to long-calls of one of their own parents and of a neighboring gull, I got strong positive response to the parent's voice in eight chicks in succession, but then tested two that did not respond at all. The contrast between these two chicks and the rest was more than could easily be dismissed as some random difference in responsiveness. I suspected that the reason for the difference might lie in the fact that for the unresponsive chicks I had used recordings

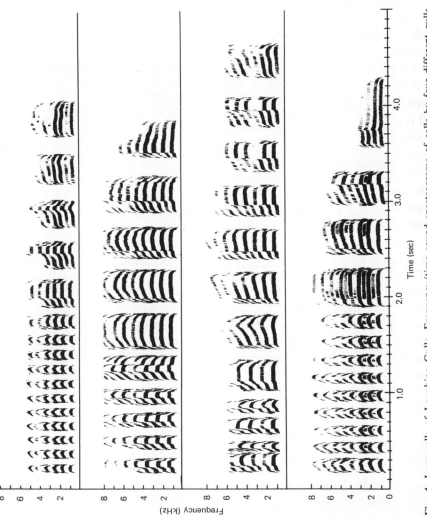

Fig. 4. Long-calls of Laughing Gulls. Frequency/time sound spectrograms of calls by four different gulls. [From Beer, (1967).]

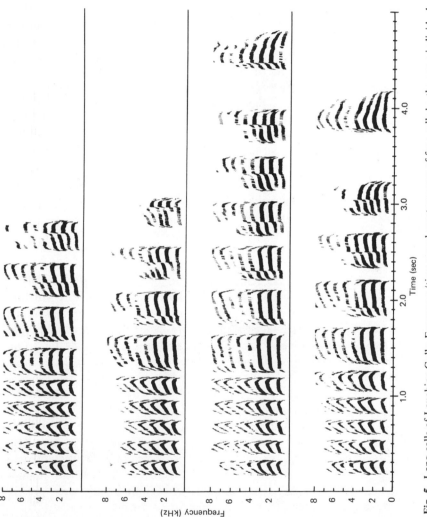

Fig. 5. Long-calls of Laughing Gulls: Frequency/time sound spectrograms of four calls by the same individual. [From Beer (1976).]

made when they were nestlings, whereas for the others I had used recordings made on or just before the day of testing. Thus suspicion was confirmed by deliberate comparison of chicks' responses to long-calls recorded from the same parent during the nestling stage and just before testing: the chicks showed significantly greater positive response to the latter than to the former.

This made me attend to the fact that a parent with 2-week-old chicks manifestly directs some of its long-calls toward its chicks and other long-calls at adults—mate, neighbors, and strangers. A parent with nestlings directs few if any long calls at them. Could it be that the older chicks respond discriminately to parental calls addressed to them? The next experiment, in which I compared chicks' responses to chick-directed long-calls and adult-directed long-calls by one of their parents, proved such discrimination to be the case: the chicks approached and called in response to the chick-directed call, but less so or not at all to the adult-directed call (Beer, 1973).

Further surprise followed when I sought for the difference by means of which the chicks were telling the two classes of call apart. Failing to find anything consistent in frequency/time sound spectrograms, I followed the example of an investigation of individual characteristics in the calls of Gannets (*Sula bassana*) (White and White, 1970) and made amplitude/time spectrograms. These did reveal a consistent difference: The short note strings of chick-directed calls began with one or two notes of slightly higher amplitude than those following, whereas in adult-directed calls the first one or two short notes were of lower amplitude than the rest, or the amplitude increased slightly with each successive note in the string (Beer, 1975, 1976) (Fig. 6). This was not where I had expected the difference to be, for I had previously noticed nothing of the sort, and there seemed more likely possibilities in the more obviously variable features of the call. Yet when I listened for the diminuendo and crescendo loudness patterns in the recordings and in the field they were clearly there in most cases.

I have dwelt on this particular case at some length because it illustrates as well as any that I know the complexities and problems with which even the initial observational and descriptive stages of the analysis of communication behavior may have to contend. It also illustrates how the animals themselves can help toward an answer to the question of what is informative to them in what we perceive as the communication behavior. In so attending to where the animals show that they draw distinctions or perceive identity in the social stimulation they receive, our efforts to arrive at a truly felicitous description of the signal repertoire or system must suspend the rule that dissociates form and function. Considerations of use and consequence intrude into the analysis of display behavior in physical terms, in a way that recalls how the functionalist psychologists found that they had to go beyond

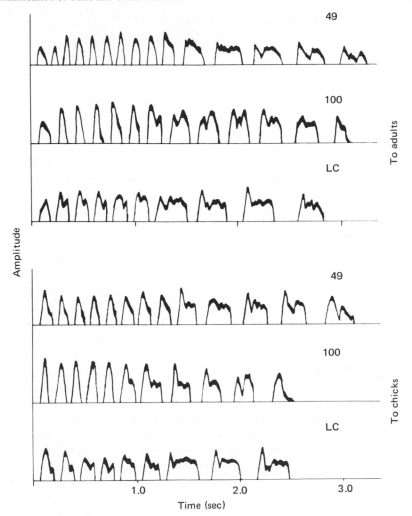

Fig. 6. Long-calls of laughing Gulls: Amplitude/time sound spectrograms of adult-addressed and chick-addressed calls by three individuals. [From Beer (1976).]

the atomistic and formal restrictions of structuralism in order to sustain their conception of mind as a product of nature.

Analysis of gull communication behavior is still in want of unequivocal criteria for settling the question: What is a display? (Beer, 1977). I believe the same to be true of other seabirds as well—indeed, of animals in general. In the meantime study of communication in action has enriched the range of possibility.

III. CONTEXT, SYNTAX, AND SEMANTICS

A. Contexts

When we turn to consideration of use and consequence of performance of communication behavior, the initial impression is that the prospect of finding there the order that we seek is even more hopeless than in the contemplation of formal characteristics. The long-call of the Laughing Gull occurs in nearly every distinguishable social context, and in a wide variety of sequential patterns with other behavior on the parts of both signaler and receiver. The same is true of some of the other putative displays as well, and almost all occur in more than one kind of context and show some variability in what appear to be their uses and effects (Beer, 1975).

Such lack of fixed order in communication behavior seems to be the rule, at least in vertebrates, and was part of the reason why the stimulus–response, or lock-and-key conception of earlier ethological communication theory gave way to the probabilistic conceptions informing the various statistical approaches, such as factor analysis and stochastic analysis. These approaches, in their turn, have been found unsatisfactory as general solutions to the problems, for they quantified the chaos without bringing the kind of order that would make clear sense of it as communication. Moreover, there were also problems of selection of parameters, class membership, and so forth, solutions to which had to be presupposed, because they could not be provided by the methods themselves; and the problem of "stationarity," which is that results obtained here and now may not agree with those obtained then and there, even when everything of supposed relevance to the situation is kept constant (Slater, 1973).

Some such cases might be explained by discovery of things that are not constant between situations. The examples I have given of individual recognition and individual address in occurrences of the Laughing Gull long-call illustrate how one could be led astray by treating all instances of what is regarded as a single type of display as the same. Another important basis for differentiation is variability of context—how an animal reacts to the same signal may differ according to the setting in which the signal arrives.

The possibility was first explicitly proposed by Haldane (1954), but Smith (1963, 1965) has the credit for putting the concept of context in animal communication into ethological circulation. G. H. Manley (1960) independently made a similar formulation on the basis of his work with Black-headed Gulls, but did not publish it.

In Smith's usage, context has a very broad reference, including anything accompanying or present with arrival of a signal, response to which is consequently affected. Thus, for example, the sex of the recipient might be part of the context, as in the case of a songbird hearing the song of a conspecific male; to another male the song means threat and so elicits agonistic response; to a female the song means invitation to courtship and so elicits sexual approach. Even my evidence of individual recognition could be construed as exemplifying the role of context in this broad sense, for the same recording elicited different response patterns from different chicks depending on their relationship to the recorded gull, family membership being a contextual variable. Closer to conventional connotations of the word are cases in which situation or circumstance affects how a signal is responded to. For example, a call given by Laughing Gulls that sounds something like *kek-kek* or *kek-kek-kek*—two or three short, clipped notes rapidly repeated—is an "alarm call" when uttered within the gullery, where it elicits the "anxiety upright posture" and flight—the reaction to approach of a potentially dangerous thing such as a person, which also stimulates the call in this situation. But outside the gullery it is given when a gull catches sight of food, and then has the effect of drawing other gulls to the vicinity, and so has been described as a "food-finding call." Such context dependent variation in response to signals is the rule for Laughing Gull communication. The long-calls, upright postures, and facings-away performed during courtship have different effects from what they have when performed during agonistic encounters. Choking has effects during boundary disputes that contrast with its effects during nest-site selection and nest relief. Crooning uttered by a parent in the vicinity of its chicks stimulates them to approach and beg; uttered in sequence with choking during border clashes, it is associated with agonistic behavior.

Some of this variability of response with context can be related to differences of form that can be distinguished in the display types, as with the two forms of facing-away in the upright posture and the numerous variants of the long-call already mentioned. But even when these are taken into account, there remain many cases of context dependent response to what appears to be morphologically the same display.

Smith argues that when this is so, the variation in response to the display arises because what elicits the response is the combination of display and context, the context being variable, but the display itself being invariant as far as what it tells about the performer is concerned. This latter, which Smith refers to as the "message" of the display, he initially supposed to be the motivational or central nervous state underlying or causing its performance (Smith, 1965, 1968). Later he revised or explicated his position to the effect that, rather than the motivation, which may plainly differ between contexts, it is the behavioral tendency that is the common denominator (1977). Thus to use his own favorite example, he finds the message of the *kt-ter* call of the Eastern Kingbird (*Tyrannus tyrannus*) to be hesitancy with regard to locomotion, for although the call occurs in at least five different situations for which motivational state cannot be the same, this behavioral correlate is present without exception. Hence, Smith categorizes the call as the "locomotory hesitance vocalization" (Smith, 1968). An illustration from the Laughing Gull might be provided by choking, which appears to signify a strong tendency to stay in place in each of the several contexts in which it occurs.

In his recent book, Smith (1977) elaborates his view that a display always has the same message and that the core of the message is always something about behavior, perhaps accompanied by subsidiary information such as individual identification. He even goes so far as to propose that we "define as displays at least the smallest formalized acts that carry consistent behavioral selection messages" (Smith, 1977, p. 411). As I have pointed out elsewhere (Beer, 1977), this is unsatisfactory as a working definition, for message–meaning analysis presupposes at least a preliminary designation of displays by other means. Of deeper theoretical significance are the questions why the message of a display should always have to be about behavior, and why it should always have to be the same. At least for some signals, such as alarm calls, especially where there are different alarm calls for different kinds of predator [e.g., Vervet Monkeys (*Cercopithecus aethiops*), Struhsaker, 1967], the simpler interpretation would seem to be that the signal has external reference, rather than that it is a sign that the animal is about to do something such as flee or hide. It may be that whenever the animal gives an alarm call we can predict what it is going to do, but if this is the same when it gives the sign for a hawk as when it gives the sign for a cat, there is still something to be explained. In any case, our being able to predict the behavior does not mean that the behavioral tendency is what is communicated by the signal to the recipient animal or animals. As Hailman (1978) has pointed out, a distinction needs to be drawn between information in the sense in which an observer gets it from quantitative analysis of behavior sequences and information in the sense of what is conveyed to recipient animals. Moreover, many of the display messages discerned by Smith are so broad and unspecific that one wonders what the point of sending them could be. In his insistence on the behavioral reference of displays, Smith may have talked himself into being false to the "interaction perspective" that he espouses (Smith, 1977), which focuses on what is actually shared in communication between the animals.

Why, then, this insistence on behavioral reference? My guess is that Lloyd Morgan's Canon, or something having similar implication in an implicit Cartesian or determinist metaphysics, gives the reason. The conception of display behavior as an involuntary concomitant of the working of the behavioral machinery, on a par with heavy breathing during hard running, assumes a "psychic faculty . . . lower in the psychological scale" (Morgan,

1894, p. 53) than the idea that display behavior might be used to refer to external objects. For the same reason, the principle that the message of a display is always the same will have greater appeal than the possibility that the same display might be used to express different things at different times through the agency of syntactic structure.

On this last point, Manley's (1960) conception of the role of context differed from Smith's. In addition to "context interpretation," which is equivalent to Smith's differences of meaning for the same display, Manley envisaged what he called "context determination"—the same display expressing different dispositions in different contexts, the differences being marked by variation in what the animal does in concert or in sequence with the display. Thus details of orientation, tilt of head, degree of lifting of the carpel joints of the wings, and whether certain specific other displays precede and follow it, distinguish a Black-headed Gull's forward posture expressing courtship tendency from one expressing hostility (Manley, 1960). The Laughing Gull's long-call, facing-away, choking, crooning, and some other displays can also be viewed in this way (Beer, 1975, 1977).

To preserve the atomistic viewpoint or sustain Smith's definition in terms of consistency of behavioral message, it could be argued that the differences of form or combination enable one to differentiate each variant as a distinct category of display. This would probably mean that our inventory of the displays of a species would be subject to continual revision as we learned to understand the communication system better, but we are having to do this anyway. Indeed, a disinterested critic might see the issue as a trivial matter of taste, a difference of preference for alternative *façons de parler* that say the same thing.

However, I think we tend to underestimate how different ways of speaking can reflect differences of conception affecting what is perceived both as factual and problematic. For example, I think it can be raised as an objection to the atomistic or splitting approach that it assumes the similarities of form between variants that it would distinguish as displays to be without significance as far as communication function is concerned. Suppose, instead, that we assume, with Smith, some common message content to attach to such similarity, but, contra Smith, that this can be detached from the impending behavior. For example, assume that the Laughing Gull's upright posture signifies hostility but that the gull performing it does not have to be hostile. Put this together with the suggestion I made earlier that white facing-away could serve as a negation sign (cf. Beer, 1975), and we can then interpret the occurrence of the upright posture in both agonistic and courtship contexts as signifying the same thing, namely hostility, but to threaten in the former and to signal lack of hostility, by coupling the posture with facing-away, in the latter. Choking and crooning can be construed along similar lines, as signifying place and the recipient's approach, respectively, the attendant behavior and context specifying what is conveyed in connection with them (Beer, 1975). However, to allow that a display could refer in this way would be all but to endow the animal with cognitive capacity of an order that goes beyond the bounds of admissible possibility for a behavioral scientist who has any desire to be taken seriously by his colleagues. So the conventional wisdom goes the other way, which could be a reason why it has so far provided so little understanding of animal communication as one supposes it must really work.

Nevertheless, dissatisfaction with this state of affairs may be helping to get a hearing for speculations involving the application of linguistic analogies to animal communication. Griffin's (1976) initiative with "cognitive ethology" is one among a number of signs of unrest with regard to Establishment doctrine about what is legitimate as explanation of animal action. At least some ethologists are now willing to reconsider whether a notion such as intention can be applied to animals, which opens the way to a more liberal attitude toward a conception of animal communication that includes concepts of use and reference, syntax and semantics. In any event, new-found complexity and features of structure in animal communication behavior appear to call for some rethinking on the subject. Two further examples from the Laughing Gull illustrate the kind of thing we have to try to deal with.

B. Hierarchical Morphology

According to what analysis of sound spectrographic evidence I have done so far, the 12 or so calls in the Laughing Gull repertoire consist of single or repeated utterance of notes drawn from a set of perhaps six types. I put the case in this tentative way because of the reason, mentioned earlier, that I cannot be sure that what I now perceive as the same or different corresponds to the gulls' sorting of the sounds, and also because there are variations, intermediates, and gradations between the categories that I distinguish in both the calls and the note types, which make me hesitate about just where to draw the lines or to draw them very rigidly. Indeed, for the notes of an individual, they differ in degree, not in kind, in the three parameters in terms of which they can be described: duration, amplitude modulation, and frequency modulation.

However, within the range of each parameter there is an unevenness of representation on the basis of which certain notes set themselves apart, to my ears at least, as constituting modal types. Such are the short notes that make up the first phase of the Laughing Gull's long-call. Spectrographically identical notes occur in several other types of call, including gackering (Burger and Beer, 1975) and the copulation call (Beer, 1976) (Fig. 7). The

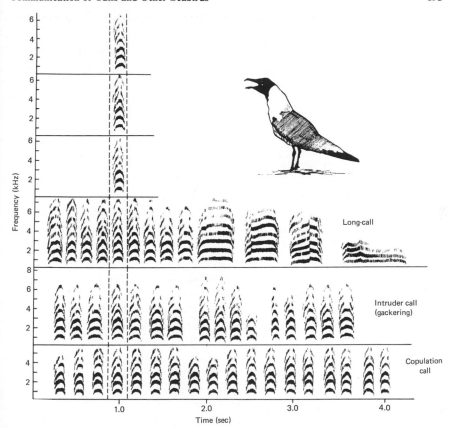

Fig. 7. Sound spectrograms illustrating similarity of note type in three different kinds of call by the same individual. [From Beer (1976).]

differences on the basis of which one can distinguish the different types of call are number of repetitions, temporal spacing of repetitions, amplitude variation, and combination with other types of note. Thus it is possible to draw an analogy with the phoneme–morpheme distinction in language—the gull vocal repertoire can be perceived as consisting of minimum units of sound which, strung together in various patterns, constitute the minimum units of sense.

A parallel analysis can be made of the elements and categories of posture and movement used for signaling. Extension of the neck, angle of the neck with respect to the body axis, angle of head with respect to neck, angle of body axis with respect to the horizontal, lifting of the carpel joints, spreading of the tail feathers, and "feather posture" (D. Morris, 1956) elsewhere on the body, and other features are formal fragments that can be varied and selected in different combinations to compose the meaningful configurations of stance and movement.

The meaningful units may in turn be combined in different ways for communicating different things. For example, the long-call of the Laughing Gull may be best construed as a compound display, or even a syntactic structure, given the wide range of its forms and uses, as well as the fact that long-note strings and head-tosses are also performed as signals on their own. But the long-call itself enters into sequential patterns with other displays in ways that suggest that syntactic order is relevant to what is expressed and conveyed in its use. In the Laughing Gull the so-called "meeting ceremony" of courtship (Tinebergen, 1959) consists of the sequence of unison long-calling with parallel, same direction orientation; facing-away in the upright or horizontal posture; then one of the birds running round the other, in the facing-away horizontal posture (presumably the "pairing charge" of Noble and Wurm, 1943), often ending with one or more head-tosses. In the Black-headed Gull the sequence is even more stereotyped: long-call, bill-up forward, upright, facing away synchronously performed by the two birds (Manley, 1960; Tinbergen, 1959). In other contexts the sequential patterning is neither so elaborate nor so rigid, but still shows differential association. For example, in the Laughing Gull, an unpaired male on territory early in the season, and a parent calling to its chicks from a distance, alternates long-calling with repeated utterance of the ke-hah call; but the former engaged in a clash with a rival, and the latter calling at chicks other than its own, will alternate long-calling with repetitions of the kow call. Variation with context in the association between crooning and choking (Beer, 1975), and the coupling of facing-away with other postures, are other examples of order for which older conceptions of animal communication behavior provide no explanation. The possibility that such syntactic complexity has semantic significance cannot be dismissed.

C. Performance Participation

The basic formula of the communication event, as transmission of a signal from a sender to a receiver via a channel, fits some forms of display interaction better or more neatly than others. Its application does not distinguish between cases in which the communication is clearly asymmetrical, from an active sender to a passive receiver; cases in which there is an alternation of the roles of sender and receiver; and cases in which there is simultaneous performance of display. The Laughing Gull presents instances of each of these patterns, and of variants within them and intermediates between them. In the first category is gackering, the display given by an incubating gull to an intruder approaching to steal nest material or attempt copulation. The latter is usually in the "anxiety upright" posture, but otherwise returns no answer to the agitation that it provokes. Alternation is exemplified by the ke-hah calling that occurs in many situations—two gulls answering one another at intervals so measured as to suggest antiphonal performance. The same impression is given by a parent and chick when they take it in turn to utter *ke-hah* in the one case, *chiz-ik* (which is almost certainly the ontogenetic antecedent of *ke-hah*) in the other (Beer, 1979). Simultaneous display of the same signal occurs with choking, and with kow calling in some of its contexts. The long call is typically a solo performance in some contexts, as, for example, in parent chick communication and in interactions between an adult and a chick not its own; but it is just as typically performed either alternately or simultaneously in other contexts. An example of the latter is the meeting ceremony, which begins with both gulls performing more or less synchronously and then alternating. Incubating neighbors often long-call in a chorus at gulls flying low overhead.

These variations in performance participation suggest that different kinds of display work in different ways, or in different ways in different situations. It has been found for choking in the nest-relief situation that the gull that chokes longest is the one most likely to get or retain possession of the nest (Herrmann, 1978); it appears that duration or persistence is pertinent to the outcome of choking bouts in other contexts as well. In contrast, duration or persistence in this sense does not appear to be relevant to use of the long-call, or the alarm call, or facing-away. Timing of response is evidently part of what conveys information in the use of ke-hah calls, especially in interactions between parents and chicks (Beer, 1979), but this is not so for kow calling or gackering.

Competence in communication thus entails more than the ability to produce the signals of the species; it also includes skill in the techniques of their use, and proficiency in working their uses. Among the latter, the conveying of messages about impending behavior is only one of a number of ends toward which a gull appears to direct display, others including to coerce, to lure, to identify, to locate, and to warn other individuals. To comprehend the communication system of a species we need to know its structure, how it is employed, and what its uses are. Our conceptions of each of these will affect what we look for in asking questions about adaptive significance, evolutionary origins, and ontogenetic development.

IV. QUESTIONS OF ADAPTIVE FUNCTION

A. Proximate Functions

I have mentioned the point that a distinction can be drawn, although not always very sharply, between proximate and ultimate functions with respect to communicating behavior; the consequences that performance of a display immediately has; and the contribution that such consequences make to reproductive success. Study of proximate function is not essentially different from study of proximate causation, except in the direction or focus of interest. Observation of ongoing behavior and experimental manipulation give the data from which inferences about proximate function can be drawn. Ultimate function can also be investigated experimentally in some cases but, more often than not, the only evidence that can be obtained comes from adaptive correlation (Beer, 1973b).

Experimental study of the proximate functions of displays has consisted of attempts to reproduce the signals artificially in natural or contrived contexts where they might be expected to elicit response or affect behavior in test birds. Playback of recordings of calls has, for example, been the means of discovery of individual recognition by ear in a number of species of seabirds, including the Manx Shearwater [*Puffinus puffinus*: (Brooke, 1978)], Guillemot [*Uria aalge*: (Tschanz, 1968)], Razorbill [*Alca torda*: (Ingold, 1973)], Common Tern [*Sterna hirundo*: (Stevenson *et al.*, 1970)], Black-billed Gull [*Larusbulleri*: (Evans, 1970a)], Ring-billed Gull [*Larcus delawarensis*: (Evans, 1970b)], and Laughing Gull (Beer, 1969, 1970a–c). One of the implications of these demonstrations is that interpretation of the meaning of a signal may require knowing the relationship between the partners in communication. An approach that scores sequential occurrences without regard to individual identity could produce data of doubtful value as biology.

My use of the playback technique also revealed that one could be misled through failure to distinguish different versions of what is assumed to be a unitary display: Laughing Gull chicks presented with playback of parental long-calls responded to chick-directed versions of the call but significantly less to adult-directed versions (Beer, 1975, 1976). Then if the meaning of a call depends on its syntactic accompaniments—other signals or signal features performed in sequence or in concert with the call—failure to take these into account in the experimental reproduction could also frustrate the effort to get at understanding of the communication system in this way.

Reproduction of visual displays is less easily achieved than reproduction of sounds, and it has been less often attempted. The most persistent and successful such endeavor with seabirds is being carried out by Stout and his colleagues in study of aggressive communication in Glaucous-winged Gulls (*Larus glaucescens*), using static and motorized models of gulls set in different postures (Galusha and Stout, 1977; Amlaner and Stout, 1978). They have obtained results in agreement with their observations of naturally occurring sequences, and confirmed a role for what can at least figuratively be referred to as lexical syntax: "a gull can, by a combination of direction of movement, posture and orientation, convey more precise and varied information than it can be [sic] any one of these display components independently" (Amlaner and Stout, 1978, p. 249). However, further deployment of this approach will probably require taking into account other features of social complexity, including that associated with individual relationships, individual recognition and the possibility of semantically significant syntactical variation.

B. Ultimate Functions

The best example of the approach of adaptive correlation to questions of ultimate function is still Esther Cullen's (1957) classic comparison of the Cliff-nesting Kittiwake (*Rissa tridactyla*) with ground-nesting gulls, such as the Black-headed Gull. She showed that many of the behavioral differences are adaptively correlated with the facts that, in contrast to ground-nesting, cliff-nesting avoids predator pressure but entails intense competition for nesting sites, little space for behavioral interaction, and, for prefledge young, the hazard of fatally falling out of the nest. For example, the greater premium on place goes with an elaboration of choking as the main territorial display and of facing-away as a means of curtailing aggressive encounters on the ledge, in the kittiwakes compared to the Black-headed Gull. In what I think is a consistent contrast with this, I found full-blown choking to be missing from the repertoire of the Black-billed Gull (Beer, 1966). This species nests on river beds subject to irregular flooding, a selection pressure favoring reduction of the period when eggs and chicks are vulnerable to destruction by flood water. Apparently as a consequence, a colony starts laying almost synchronously and almost immediately once it has selected a breeding site, and shows the closest packing of the breeding pairs to be found in any gull species. Both the speed of laying onset and the close packing require a minimum of territorial contention. Because one of the uses of choking in other species appears to be to express the highest intensity of territorial aggressiveness at close quarters (Tinbergen, 1959), its absence in the Black-billed Gull repertoire is in harmony with an apparent reduction in such territorial aggressiveness.

Also associated with the dense-nesting arrangement and speeding up of the breeding process in Black-billed Gulls may be the fact that chicks can identify their parents individually on the basis of individual differences in their mew calls (Evans, 1970a), in contrast to Laughing Gull chicks, which develop such individual recognition a day or so later, but do so on the basis of the ke-hah calls and long-calls, not mew calls (= crooning) (Beer, 1970b,c, 1973). In both species the mew calls is the only call used consistently by a parent to command the filial behavior of its chicks during the first few days posthatching. The separation of Laughing Gull nests is such that social contact outside the family through increase in a chick's mobility does not necessitate discrimination of the parents until about the end of the first week, by which time the parents have begun using the ke-hah and long-call, in addition to the mew call, to address their chicks. By contrast, the closeness of neighbors in the Black-billed Gull colony, together with the greater hatching synchrony between families compared to the Laughing Gull, apparently favors individual recognition from almost the beginning to keep filial responsiveness within the family [cf. the guillemot, where a comparable situation obtains (Tschanz, 1968)], which fits with the fact that the mew calls provide for it. Consistent with this is the fact that Ring-billed Gull chicks also individually recognize their parents by their mew calls. Nesting density and speed of development of chick mobility in this species are comparable to the figures for the Black-billed Gull (Evans, 1970b). Toward the other direction are the chicks of Franklin's Gulls (*Larus pipixcan*). Burger (1974) tested these for individual recognition by playback of recordings of long-calls and found no evidence of discrimination, even by chicks as old as 3 weeks posthatching. This correlates with the fact that, compared with the other three species mentioned, Franklin's Gull chicks spend most of their time confined to their floating nests and do not show much mobility until a relatively late age. In other words, situations calling for individual recognition by prefledge chicks of their parents seldom arise; accordingly, the ability appears not to be present. Unfortunately, Burger did not test Franklin's Gull chicks with the mew call. The preceding argument predicts, of course, that the results of such tests would also be negative. It would predict the same for the kittiwake, in which brood mobility is the most restricted of any species.

There is much variation between species, as far as communication behavior is concerned, that has yet to be interpreted in terms of adaptive correlation. For example, the Laughing Gull, and perhaps its close relative, Franklin's Gull, appear to be exceptional in the complexity and variability of their long-calls. Appreciation of this complexity and variability and of the syntactic structures and semantic uses of the communication systems of gulls and other seabirds opens up new questions and possibilities for this approach. The same is true for the other kind of evolutionary question—that having to do with historical origins and derivations.

V. DISPLAY AS DERIVED ACTIVITY

The concept of ritualization has had some of its more persuasive use in comparative studies of gulls (e.g., Tinbergen, 1959). As elsewhere, the display patterns have been interpreted as elaborations for communication function of behavior that was originally either part of the action to which the signal refers, or a "displacement activity" associated with such action (Tinbergen, 1952). Thus the upright posture is regarded as consisting of "frozen" components of the initial parts of attack movements—lifting the head to peck downward, and lifting the carpels in preparation to wing-beat—only slightly ritualized from its origin as "a mosaic of intention movements of attack, inhibited by escape tendencies" (Tinbergen, 1959, p. 47). Comparisons of form, situation of occurrence, and sequential patterning, both within and between species, supported this interpretation, according to which the upright posture is iconic of what it expresses.

Such interpretation has been more or less an ethological preserve since Whitman (1899), Heinroth (1911), and Lorenz (1941) applied a comparative morphology approach to behavior. In spite of being at the historical heart of the matter, it perhaps receives less respect than any other part of ethology. The evidence and arguments in support of its interpretations can never be conclusive. Lacking a fossil record, a clear concept of behavioral homology, and a footing for experiment, the evolutionary history of display behavior can be no more than a matter for informed conjecture (cf. Blest, 1961). Consequently, it has been dismissed by some as a sterile undertaking.

This attitude could reflect a lack of fertile imagination in the critic however. At any rate, as appreciation of the syntactic complexity of animal communication behavior deepens, the possibilities for the comparative phylogenetic approach widen. What one finds depends on what one looks for. So far the atomistic conception of communication has dictated the kinds of correspondence sought. But the syntactic structures must have evolved along with the units they order. Evidence of this has yet to be gathered.

Meantime there are apparent convergences to consider in the forms of displays. Gannets, gulls, and auks show facing-away movements; the gannet's "sky-pointing" looks like the "ecstatic" displays of penguins, the climax attitude of the "mutual display" of albatrosses, and the "erect" posture of terns (see Fig. 1); "meeting ceremonies" are common to behavior having to do with pair-bonding in seabirds in general. Variation in degrees of ritualization of putatively homologous patterns within groups give dimension to the possibilities for comparison with their analogues in other groups.

Most of the concern with behavioral phylogeny has been with the forms of posture and movement used in communication. Now that we have the means of describing and comparing the vocal signals at least as precisely as the visual, there should be increased interest in applying comparative study to the evolution of forms of utterance. For example, it is surely a point worth trying to explain in historical terms, that the long-call of Franklin's Gull has a long-note string preceding a short-long string, which is the reverse of the pattern of its closest relative, the Laughing Gull.

Again, however, there is more to be looked for than has been found. Comparative ethological study of derived activities should not be dismissed as worn out, especially in its application to seabirds. Seabirds offer opportunity for new exploration, in spite of having been a major ground for old preoccupation. However, even richer opportunity lies in the virtually uncharted region of the development of communication behavior.

VI. THE DEVELOPMENT OF COMMUNICATION BEHAVIOR

Like evolution, development is both history and process. Developmental study divides between tracing the course of emergence of later from earlier patterns, and accounting for the transitions and transformations in generative terms. As far as the communication behavior of seabirds is concerned, neither has received much attention.

Of descriptive studies, Moynihan's work (1959a) on the emergence of display patterns in gull chicks raised in captivity is the most extensive, and it has to be regarded as merely preliminary. Accounts of parent–chick interactions at different ages, and of the development of mobility, have been given from field observations of Ring-billed Gulls (Evans, 1970b; Miller and Emlen, 1975), Black-billed Gulls (Beer, 1966, Evans, 1970a), Herring Gulls (Tinbergen, 1953), Franklin's Gull (Burger, 1974), and Laughing Gulls (Beer, 1970a,b, 1979), but little systematic attention has been given to what, when, and how communication behavior makes its appearance and changes during social development.

The pattern most thoroughly studied experimentally, and from a perspective on process, is the pecking of nestling gulls, which stimulates the parents to regurgitate food. Tinbergen (1953) used presentation of models to analyze the stimulus features to which the display is tuned in nestling Herring Gulls. Hailman (1967) did the same with Laughing Gull chicks and also investigated how experience affects subsequent development. But this is only a fragment of the whole dynamic complex constituting the social development that an individual must negotiate to become a socially mature adult of its species, which entails acquiring the signal repertoire and competence in its use and comprehension.

These things do not necessarily go together. Work in progress on the Laughing Gull indicates that hand-raised birds develop all the display postures, movements and calls, but not the species-typical proficiency in when to use them and how to receive them. Depriving chicks of experience with parents until the 6th day posthatching caused retardation in their social development, which spending the next 6 days with the parents failed to make up (Beer, 1972). From field observation and experiment it is evident that there is progressive change in interactions between parents and young, including changes in the kinds of signals used and in response to them. Parents initially approach little in response to the calls of chicks; when the chicks are about 1 week old the parents almost invariably approach to chick calls if none of their chicks is with them; but by a week or so later approach is again infrequent, the parents instead using the ke-hah and long-call apparently to try to get the chicks to come to them (Beer, 1979). Chicks show a parallel shift, approaching in response to parent calls until about 14 days posthatching, and then tending more and more to call persistently in response to the parent calls, but without approaching, in a way suggesting a contest of wills over which individuals will come to which. Such experience in using signal repetition to manipulate a social situation may be crucial in the development of proficiency in communication as adults, for many kinds of display interactions between adults involve similar tactics.

Social development within a family of Laughing Gulls thus shows change in response to the same signals, in the behavior of both parents and chicks. There is also change in the kinds of signal used to get the same response. For example, a parent uses the mew call (= crooning) to get its nestlings to approach in order to feed, but as the chicks become more mobile it substitutes ke-hah calls and then adds long-calls as the signals commanding approach at a distance, reserving the mew call for when the chicks have come within physical contact distance. The chicks keep pace with this shift, transferring their approach responsiveness from mew call to ke-hah call and long-call, and becoming unresponsive to mew call alone, just before the time when the lack of individually identifying characteristics in the mew call would cause problems for keeping filial interest within the family (Beer, 1972, 1973, 1979). It is impossible as yet to say which of the two—parents or chicks—has the greater part in calling the tune for these shifts in use and response in family communication. Whatever the case, it is a reasonable supposition that these and other transitions of similar sorts are necessary steps in the passage to competence in adult social communication in this species, and probably in other gulls, and at least some other seabirds, as well.

If this conception includes conforming to the syntactic patterns, including semantically significant variations of such display forms as the long-call and facing-away, then we have more to look for than we used to think. It is my expectation that experience will prove to be involved in various and subtle ways, and that an adequate account of social development in the Laughing Gull will read more like the kind of story we associate with Piaget than either a Lorenzian account in terms of innateness or a Skinnerian account in terms of conditioning.

VII. EPILOGUE

This essay has had more to say about what might be the case than what is known to be the case. The emphasis could have gone the other way, for much more is known about seabird communication than I have chosen to include. However, the soundness of a science is less in the facts at its disposal than in the conceptions in terms of which those facts are ordered, and even differentiated. I am not alone in holding the opinion that the study of animal communication is in a flux about how to view its subject and that there may be opportunity here for imagination to bring about a new synthesis. Perhaps, therefore, the subject exists more in its possibilities than in its accomplishments—more in its doubts than in its certainties. In any event, one can make the positive point, which is the right kind on which to end, that to try to realize these possibilities and resolve these doubts the science has plenty to work on in the communication behavior of seabirds.

ACKNOWLEDGMENTS

Research referred to here on the Laughing Gull was supported by Grant No. MH 16727 from the United States Public Health Service. I thank the U.S. Fish and Wildlife Service and the New Jersey Department of Environmental Protection for permission to work in the Brigantine National Wildlife Refuge, and the Refuge Manager and his staff for their hospitality and cooperation. Figures 4–7 are reproduced with the permission of the New York Academy of Sciences, having originally appeared in *Ann. NY Acad. Sci.* **280**:413–432. This is Publication No. 332 from the Institute of Animal Behavior, Rutgers University.

REFERENCES

Amlaner, C. J., and Stout, J. F., 1978, Aggressive communication by *Larus glaucescens*. Part VI: Interactions of territory residents with a remotely controlled [sic], locomotory model, *Behaviour* **66**:223–251.

Beer, C. G., 1966, Adaptations to nesting habitat in the reproductive behaviour of the Black-billed Gull *Larus bulleri*, *Ibis* **108**:394–410.

Beer, C. G., 1969, Laughing Gull chicks: Recognition of their parents' voices, *Science* **166**:1030–1032.

Beer, C. G., 1970a, On the responses of Laughing Gull chicks to the calls of adults. I: Recognition of the voices of the parents, *Anim. Behav.* **18**:652–660.

Beer, C. G., 1970b, On the responses of Laughing Gull chicks to the calls of adults. II: Age differences and responses to different types of call, *Anim. Behav.* **18**:661–677.

Beer, C. G., 1970c, Individual recognition of voice in the social behavior of birds, *Adv. Study Behav.* **3**:27–74.

Beer, C. G., 1972, Individual recognition of voice and its development in birds, *Proc. 15th Int. Ornithol. Congr.* **1970**:339–356.

Beer, C. G., 1973a, A view of birds, *Minn. Symp. Child Psychol.* **7**:47–86.

Beer, C. G. 1973b, Species-typical behavior and ethology, in: *Comparative Psychology—a Modern Survey* (D. A. Dewsbury and D. A. Rethlingshafer, eds.), pp. 47–86, McGraw-Hill, New York.

Beer, C. G., 1975, Multiple functions and gull displays, in: *Function and Evolution in Behaviour—Essays in Honour of Professor Niko Tinbergen* (G. P. Baerends, C. G. Beer, and A. Manning, eds.), pp. 16–54, Clarendon Press, Oxford.

Beer, C. G. 1976, Some complexities in the communication behavior of gulls, *Ann. NY Acad. Sci.* **280**:413–432.

Beer, C. G., 1977, What is a display? *Am. Zool.* **17**:155–165.

Beer, C. G., 1979, Vocal communication between Laughing Gull parents and chicks, *Behaviour* **70**:118–146.

Blest, A. D., 1961, The concept of ritualization, in: *Current Problems in Animal Behaviour* (W. H. Thorpe and O. L. Zangwill, eds.), pp. 102–124, Cambridge University Press, Cambridge.

Booy, H. L., and Tinbergen, N., 1937, Nieuwe Feiten over de Sociologie van de Zilvermeeuwen, *De Levende Natuur* **41**:325–334.

Brooke, M. de L., 1978, Sexual differences in the voice and individual vocal recognition in the Manx Shearwater (*Puffinus puffinus*), *Anim. Behav.* **26**:622–629.

Burger, J. 1974, Breeding adaptations of Franklin's Gull (*Larus pipixcan*) to a marsh habitat, *Anim. Behav.* **22**:521–567.

Burger, J., and Beer, C. G., 1975, Territoriality in the Laughing Gull (*L. atricilla*), *Behaviour* **55**:301–320.

Chomsky, N., 1968, *Syntactic Structures*, Mouton, The Hague.

Cullen, E., 1957, Adaptation in the Kittiwake to cliff-nesting, *Ibis* **99**:275–302.

Deusing, M., 1939, The herring gulls of Hat Island, Wisconsin, *Wilson Bull.* **51**:170–175.

Evans, R. M., 1970a, Parental recognition and the "Mew Call" in Black-billed Gulls (*Larus bulleri*), *Auk* **87**:503–513.

Evans, R. M., 1970b, Imprinting and the control of mobility in young Ring-billed Gulls (*Larus delawarensis*), *Anim. Behav. Monogr.* **3**:193–248.

Galusha, J. G., and Stout, J. F., 1977, Aggressive communication by *Larus glaucescens*, Part IV. Experiments on visual communication, *Behaviour* **62**:222–235.

Goethe, F., 1937, Beobachtungen und Untersuchungen zur Biologie der Silbermöwe (*L.a. argentatus*) auf der Vogelinsel Memmerstand, *J. f. Ornithol.* **85**:1–119.

Griffin, D. R., 1976, *The Question of Animal Awareness*, Rockefeller University Press, New York.

Haldane, J. B. S., 1954, La signalisation animale, *Ann. Biol.* **30**:89–98.

Hailman, J. P., 1967, The ontogeny of an instinct, *Behav. Suppl.* **15**.

Hailman, J. P., 1978, Review of "The Behavior of Communicating: An Ethological Approach," by W. J. Smith, Harvard University Press, Cambridge, Mass., *Auk* **95**:771–774.

Haviland, M. D., 1914, The courtship of the Common Gull, *Larus c. canus, Br. Birds* **7**:278–280.

Heinroth, O., 1911, Beitrage zur Biologie, namentlich Ethologie und Psychologie der Anatiden, *Verh. V. Int. Ornithol. Kongr.*, Berlin, pp. 333–342.

Herrick, F. H., 1912, Organization of the gull community, *Proc. 7th Int. Zool. Congr.*, Boston **1907**:156–158.

Herrmann, J., 1977, The nest relief of Laughing Gulls: A study of mate relations during the incubation period, Ph.D. dissertation, Rutgers University.

Ingold, P., 1973, Zur lautlichen Beziehung des Elters zu seinem Kueken bei Tordalken (*Alca torda*), *Behaviour* **45**:154–190.

Kirkman, F. B., 1937, *Bird Behaviour*, Nelson, London.

Kirkman, F. B., 1940, The inner territory of the Black-headed Gull, *Br. Birds* **34**:100–104.

Kuhn, T. S., 1970 *The Structure of Scientific Revolutions, International Encyclopedia of Unified Science*, Vol. 2, No. 2, 2nd ed., University of Chicago Press, Chicago.

Lack, D., 1968, *Ecological Adaptations for Breeding in Birds*, Methuen, London.

Lorenz, K., 1941, Vergleichende Bewegungsstudien an Anatinen, *J. f. Ornithol.* **89**:194–294.

Manley, G. H., 1960, The agonistic behaviour of the Black-headed Gull, Ph.D. thesis, Bodleian Library, Oxford.

Marler, P., 1969, Tonal quality of birds sounds, in: *Bird Vocalizations* (R. A. Hinde, ed.), Cambridge University Press, Cambridge.

Miller, D. E., and Emlen, J. T., 1975, Individual chick recognition and family integrity in the Ring-billed Gull, *Behavior* **52**:124–144.

Morgan, C. L., 1894, *An Introduction to Comparative Psychology*, Murray, London.

Morris, C. W., 1938, *Foundations of the Theory of Signs, Encyclopedia of Unified Science*, Vol. 1, No. 2, University of Chicago Press, Chicago.

Morris, D., 1956, The feather postures of birds and the problem of the origin of social signals, *Behaviour* **9**:75–113.

Morris, D., 1957, "Typical intensity" and its relation to the problem of ritualisation, *Behaviour* **11**:1–12.

Moynihan, M., 1955, Some aspects of reproductive behaviour in the Black-headed Gull (*Larus r. ridibundus*) and related species, *Behav. Suppl.* **4**:1–201.

Moynihan, M., 1956, Notes on the behavior of some North American gulls. I. Aerial hostile behavior, *Behaviour* **10**:126–179.

Moynihan, M., 1958a, Notes on the behavior of some North American gulls. II. Non-aerial hostile behavior of adults, *Behaviour* **12**:95–182.

Moynihan, M., 1958b, Notes on the behavior of some North American gulls. III. Pairing behavior, *Behaviour* **13**:112–130.

Moynihan, M., 1959a, Notes on the behavior of some North American gulls. IV. The ontogeny of hostile behavior and display patterns, *Behaviour* **14**:214–239.

Moynihan, M., 1959b, A revision of the family Laridae (Aves), *Am. Mus. Novit.* **1928**:1–42.

Moynihan, M., 1962, Hostile and sexual behavior patterns of South American and Pacific Laridae, *Behav. Suppl.* **8**:1–365.

Nelson, J. B., 1970, The relationship between behaviour and ecology in the Sulidae with reference to other sea birds, *Oceanogr. Mar. Biol.* **8**:501–574.

Nelson, J. B., 1972, Evolution of the pair bond in Sulidae, *Proc. 15th Int. Ornithol. Congr.*, The Hague **1970**:371–388.

Nelson, J. B., 1975, Functional aspects of behaviour in the Sulidae, in: *Function and Evolution in Behaviour —Essays in Honour of Professor Niko Tinbergen* (G. P. Baerends, C. G. Beer, and A. Manning, eds.), Clarendon Press, Oxford.

Noble, G. K., and Wurm, M., 1943, The social behavior of the Laughing Gull, *Ann. NY Acad. Sci.* **45**:179–220.

Portielje, A. F. J., 1928, Zur Ethologie bezw. Psychologie der Silbermöwe (*Larus argentatus argentatus* Pont.), *Ardea* **17**:112–249.

Ryle, G., 1949, *The Concept of Mind*, Hutchinson, London.

Sebeok, T. A., 1967, Animal communication, *Intern. Soc. Sci. J.* **19**:88–95.

Sebeok, T. A. (ed.), 1968, *Animal Communication*, Indiana University Press, Bloomington.

Sebeok, T. A. (ed.), 1977, *How Animals Communicate*, Indiana University Press, Bloomington.

Slater, P. J. B., 1973, Describing sequences of behavior, in: *Perspectives in Ethology* (P. P. G. Bateson and P. H. Klopfer, eds.), Vol. 1, pp. 131–153, Plenum Press, New York.

Smith, J. W., 1963, Vocal communication of information in birds, *Am. Nat.* **97**:117–125.

Smith, W. J., 1965, Message, meaning and context in ethology, *Am. Nat.* **99**:404–409.

Smith, W. J., 1968, Message-meaning analysis, in: *Animal Communication* (T. A. Sebeok, ed.), pp. 44–60, Indiana University Press, Bloomington.

Smith, W. J., 1977, *The Behavior of Communicating: An Ethological Approach*, Harvard University Press, Cambridge, Mass.

Steinbacher, G., 1938, Zur Ethologie unserer einheimischen Möwenarten, *Berich. Schles. Ornith.* **23**:42–65.

Stevenson, J. G., Hutchinson, R. E., Hutchinson, J. B., Bertram, B. C. R., and Thorpe, W. H., 1970, Individual recognition by auditory cues in the Common Tern (*Sterna hirundo*), *Nature (London)* **226**:562–563.

Strong, R. M., 1914, On the habits and behavior of the Herring Gull *Larus argentatus* Pont., *Auk* **31**:22–49.

Struhsaker, T. T., 1967, Auditory communication among Vervet Monkeys (*Cercopithecus aethiops*), in: *Social Communication among Primates* (S. A. Altmann, ed.), University of Chicago Press, Chicago.

Tinbergen, N., 1936a, Waarnemingen en Proeven over de Sociologie van een Zilvermeeuwenkolonie, *De Levende Natuur* **40**:1–24; 262–280.

Tinbergen, N., 1936b, Zur Soziologie der Silbermöwe, *Larus a. argentatus* Pont., *Beitr. Fortpflanzung Biol. Vögel* **12**:89–96.

Tinbergen, N., 1952, "Derived" activities: Their causation, biological significance, origin, and emancipation during evolution, *Q. Rev. Biol.* **27**:1–32.

Tinbergen, N., 1953, *The Herring Gull's World*, Collins, London.

Tinbergen, N., 1959, Comparative studies of the behaviour of gulls (Laridae): A progress report, *Behaviour* **15**:1–70.

Tinbergen, N., 1963, On aims and methods of ethology, *Z. Tierpsychol.* **20**:410–433.

Tinbergen, N., and Moynihan, M., 1952, Head flagging in the Black-headed Gull: Its function and origin, *Br. Birds* **45**:19–22.

Tschanz, B., 1968, Trottellummen, *Z. Tierpsychol. Beiheft* **4**:1–103.

Wachs, H., 1933, Paarungsspiele als Artcharaktere, Beobachtungen an Möwen und Seeschwalben, *Zool. Anz. Suppl.* **6**:192–202.

White, S. J., and White, R. E. C., 1970, Individual voice production in gannets, *Behaviour* **37**:40–54.

Whitman, C. O., 1899, Animal behavior, in: *Biological Lectures Delivered at the Marine Biological Lab. at Woods Hole in 1898*, pp. 329–331, Marine Biological Laboratory, Woods Hole, Massachusetts.

Wilson, E. O., 1975, *Sociobiology: The New Synthesis*, Belknap Press (Harvard University), Cambridge, Mass.

Chapter 7

MECHANISMS AND ADAPTIVE VALUE OF REPRODUCTIVE SYNCHRONY IN COLONIAL SEABIRDS

Michael Gochfeld

Department of Health
Trenton, New Jersey 08625
and Division of Environmental Health
Columbia University School of Public Health
New York, New York 10032

I. INTRODUCTION

Advances in the study of the behavioral ecology of colonial birds requires understanding both temporal and spatial dispersion of reproductive activities. Coloniality itself is one aspect of spatial clustering, whereas synchronous breeding represents temporal clustering and is characteristic of many colonial species. Darling (1938) called attention to the importance of reproductive synchrony. His hypotheses, often referred to as the Darling or Fraser Darling effect, linked colony size, social stimulation, reproductive synchrony, and reduced predation into a single behavioral–ecologic–evolutionary framework. Darling hypothesized that birds in larger colonies would experience greater social stimulation which, mediated by neuroendocrine pathways, would accelerate breeding cycles, leading to earlier and more synchronous laying than in small colonies. This would be advantageous in that the time span of hatching would be reduced. Thus more chicks would hatch in shorter time periods. At the peak of hatching many chicks would be vulnerable to predation, but a predator would soon become satiated and could consume a smaller proportion of the overall crop than in a less synchronous situation where hatching was spread out over a longer period of time. Thus factors enhancing synchrony would be favored by natural selection.

Darling's model covers a broad range of phenomena, and it is not surprising that many exceptions have come to light (e.g., von Haartman, 1945; Orians, 1961; MacRoberts and MacRoberts, 1972*b*). Recently a number of authors (e.g., Veen, 1977) have examined seabird synchrony without reference to Darling's work. Some discussion of synchrony has been sidetracked by confused terminology or because various authors have emphasized different aspects of the Darling effect (e.g., colony size versus density; earliness of laying versus synchrony; overall mortality versus predation). Most importantly, there are various methods for gathering data and measuring synchrony, and these must be clarified. I believe it is valuable to recognize the importance of Darling's (1938) contribution, even when one finds that various aspects of the model do not operate in particular colonies. This chapter examines the basis on which Darling constructed his hypotheses, discusses the main features of the Darling effect, and considers the manner in which these features can be studied.

I contend that much of the difficulty in finding support for the Darling effect in birds lies in the overwhelming logistical problems inherent in testing a global theory with the resources currently available to most field biologists. Cooperative ventures that monitor the fate of several colonies provide some hope of clarifying the occurrence and role of synchrony, of elucidating the mechanisms by which it is achieved, and of determining its adaptive value. Sections II and III define important terms and discuss in detail various methods for quantifying and comparing synchrony across colonies or species. Sections IV and V discuss the importance of synchrony in understanding coloniality and the history and criticisms of the Darling effect. Sections VI and VII present original data on reproductive synchrony of Common Terns (*Sterna hirundo*) and Black Skimmers (*Rynchops niger*). Sections VIII and IX discuss the interaction of factors enhancing or reducing synchrony and the evolutionary implications of synchrony.

Colonial nesting of birds is selectively advantageous with respect to avoidance of predation and exploitation of food or restricted nest sites. It is difficult to disentangle synchrony from studies of coloniality, *per se*, yet this chapter cannot deal with the adaptiveness of coloniality (see references in Krebs, 1978). Moreover, many biological functions can be synchronized among individuals, but this chapter concerns itself only with behavior directly related to reproduction. No attempt is made here to critically review all the seabird papers that discuss synchrony or the Darling effect, nor has it been feasible to discuss in detail the important contributions to synchrony theory derived from work with nonseabirds or with other organisms. These are mentioned in passing, where appropriate. I proceed first to define certain important terms.

II. DEFINITIONS

There is confusion over the most important terms, such as colony, social stimulation, and synchrony; this section provides the definitions to be used throughout this chapter.

Colony. Although it is the basic unit in the study of colonial species, definition of a "colony" remains a relative matter influenced by the biology of the species, the homogeneity and distribution of available space, the density and dispersion of the nests, and the needs and biases of the observer. I use the term to indicate a breeding assemblage of birds in a single location. Such birds are in sufficiently close proximity to experience regular social interactions and are separated from birds in the next colony by a sufficient gap such that little or no communication occurs (at least while the birds are actually on the colony). Nelson (1966) notes that almost all seabirds have a tendency to nest close to established birds and adds (Nelson, 1970) that "a species is colonial if its typical breeding dispersion is such that each individual sees, hears or frequently interacts with many conspecifics, usually throughout breeding. . . ." This is consistent with Altman's (1965) definition of a society as an aggregation of socially intercommunicating conspecifics separated from the next by a barrier of less frequent communication.

There is great diversity in the density of colonies from Royal Terns (*Sterna maxima*) with six nests per square meter (Buckley and Buckley, 1972) to Herring Gulls (*Larus argentatus*) with one nest per 6 m². Although it will be difficult to standardize use of the term colony for different biologists or for different species or habitats, individual field workers usually develop a working definition. For example, on Long Island, New York, there are almost no ambiguities if one says that two aggregations of Common Terns separated by 1 km, are separate colonies. Closer aggregations are treated as subcolonies. Only if the aggregations are very small (a dozen pairs, for example) would one need to alter the working definition. At a distance of 1 km, birds in the air over one colony can readily see birds in the air over the next colony, but there is little direct interaction. Ambiguous cases arise with birds nesting in small groups on islands that are less than 1 km apart. Such birds may have relatively infrequent interactions; birds at one colony may respond aggressively at the appearance of a potential predator, while birds on the next island continue to incubate undisturbed. Such units would be considered separate colonies.

Not only does colony structure vary from one species to another, but a species may be colonial in one place and not in another. In seabirds, it is characteristic that birds may neglect apparently suitable breeding sites to nest close to established conspecifics, even if they must resort to relatively unsuitable sites (e.g., Ashmole, 1962). Thus coloniality implies an active process influenced by social factors, rather than a more passive aggregation determined by nest-site availability. It is with this social aspect of coloniality that this chapter is concerned.

Subcolony. Within a colony one can often identify clusters of birds that are separated from adjacent clusters by an unusued space or by a habitat discontinuity (e.g., different vegetation, a cliff, or a water gap). Such clusters may be considered subcolonies, and within a subcolony all birds are within visual and acoustic contact and interact frequently. In practice, I find that the distance between subcolonies is on the same order of magnitude as the dimensions of one subcolony. Subcolonies may be characterized by highly synchronous nesting; this is discussed in detail below.

Burger (1974a) defined subcolonies as synchronous clusters that were spatially continuous without apparent physical separation. In this case the discontinuity would be temporal rather than spatial, and I prefer to refer to these as "neighborhoods." The distinction between a subcolony and a neighborhood will not always be clear.

Neighborhood. This is a relative term referring to a cluster of birds that interact frequently with their neighbors. Characteristically birds in a neighborhood nest synchronously and are often out of phase with birds in nearby clusters (e.g., Coulson and White, 1960; Burger, 1974a).

Social Stimulation. Many behavioral and physiologic activities of birds are enhanced in frequency or intensity when in the presence of other organisms, particularly conspecifics. In many cases a specific signal elicits a particular behavior (e.g., Lehrman, 1965), whereas in other cases simple presence or activity increases the frequency of other activities. Several kinds of social stimulation can occur. Darling (1938) emphasized the role of social stimulation in synchronizing reproductive activities within colonies of birds. The following three terms refer to different kinds of social stimulation.

Social Facilitation. Tolman (1964) found that a domestic chick increased its pecking rate in the presence of another chick that was pecking. Pecking also increased, but to a lesser extent in the presence of a nonpecking chick and even increased slightly in the presence of a dead chick. I consider the former to represent social facilitation—the increase in intensity and/or frequency of a particular behavior when in the presence of another individual performing the same behavior. In discussions of synchrony, many authors have used social facilitation and social stimulation interchangeably, but future studies of mechanisms for achieving synchrony should provide more rigorous definition of the kinds of social interaction under study. Darling (1938) reported that gulls would be stimulated to engage in breeding displays in the presence of other gulls giving the same displays. This would be an example of social facilitation, a term he did not use. Darling also emphasized that the presence, the activity, and the vocalizations of the neighboring birds would stimulate and accelerate reproductive activity in general, and I would consider this an example of social stimulation in general, not of social facilitation. In this chapter, under contagious displays (Section VI) I examine social interactions in precopulatory and copulatory activities among terns. Because these two types of behavior are causally linked, I treat them under social facilitation, whereas in examining temporal association between nest-building activity and copulation among neighboring birds, I would be inclined to avoid the term social facilitation. In any case, it is important that authors define their terms clearly, no matter how broad a terminology they choose.

Social Attraction. This is another category of social stimulation. Orians (1961) and Hailman (1964) discuss the tendency of birds in breeding condition to be attracted to other birds that are breeding. This phenomenon of social attraction accounts for the growth of colonies, particularly successful colonies, and is biologically important. Social attraction is particularly important as a mechanism encouraging inexperienced, first-time breeders to settle among older, established birds, thereby improving the likelihood of their choosing a suitable nesting habitat.

Social Segregation. I introduce this term as an additional category of social stimulation. It refers to the tendency of birds in breeding condition to separate themselves from prebreeding aggregations and establish breeding groups. Such groups become synchronous neighborhoods and may be the epicenters from which colonies grow (e.g., Burger and Shisler, 1980). It could prove an important mechanism for establishing colonies and achieving synchrony, and I believe it plays an important role in certain Black Skimmer colonies (Gochfeld, 1978a, 1979) accounting for the formation of synchronous subcolonies, as well as in the Sandwich Tern colonies (*Sterna sandvicensis*) described by Veen (1977).

Synchrony. Synchrony is a measure of temporal clustering, referring to the tendency for two or more events to happen close together in time. The term is quite relative depending on one's time frame. Thus on one scale, events happening the same minute may be considered synchronous, whereas on a longer time frame one may consider events that occur on the same day as synchronous. It is most useful to compare relative synchrony. Thus given two pairs of events, pair A (A_1 and A_2) and pair B (B_1 and B_2), if the time between A_1 and A_2 is shorter than that between B_1 and B_2 one can say that pair A is more synchronous.

When we study synchrony among birds a major difficulty arises in choosing the event to study. Most studies have focused on the date of clutch initiation. However, to evaluate the Darling effect adequately, one needs to examine events preceeding clutch initiation. Burger and Shisler (1980) have actually studied synchrony of nest initiation as well as clutch initiation, in Herring Gulls.

Synchronizing. It is important to distinguish synchrony, which is a describable phenomenon, from synchronizing, which entails the mechanisms whereby any observed degree of synchrony is achieved. Factors such as climate, predation, food availability, and social stimulation can lead to relatively high synchrony. In practice it is not uncommon to find situations where synchronizing factors are prominent, but where overall synchrony is low. Thus, if a dense mass of nesting birds continually attracts new birds (social attraction), there may be a prolonged nesting season and relatively low synchrony, despite the presence of social synchronizing factors. There are also factors that reduce synchrony, and such desynchronizing factors must not be overlooked. Synchronizing factors may act proximately (e.g., adverse weather inhibiting nesting in a particular week) or ultimately (e.g., adverse weather selecting repeatedly against early nesting birds over many seasons). In both cases weather would reduce early nesting.

In distinguishing synchrony from synchronizing, it is worth clarifying that when a group of birds shows high synchrony, this is properly referred to as synchronous breeding. It may also represent synchronized breeding, but could have arisen by chance rather than through the action of synchronizing factors. This distinction is only important in clarifying that one can have high or low synchrony regardless of whether there is high or low input from synchronizing factors.

Recruitment. The addition of new individuals to a population is known as recruitment. It is used in two contexts, both extremely important in studying synchrony and the Darling effect. The role of social attraction of late-arriving birds to an active breeding colony is important, and this is one example of recruitment to a successful and established population. Predator recruitment occurs when additional predators are attracted to an abundant food source. The predator saturation by a synchronous colony, central to the Darling effect, will not operate if predator recruitment occurs.

Finally, I must reemphasize that some of these terms must be modified to suit the requirements of a particular species or study. Although uniformity of terminology is a desirable goal, it is more important that authors clearly indicate their use or understanding of terms or concepts, than that they adopt the definitions provided here. Before proceding to a discussion of the Darling Effect and synchrony, it is important to examine in detail different methods for studying and describing synchrony. This is treated in the next section.

III. SYNCHRONY METHODOLOGY

A major source of difficulty in discussing the function and causation of synchrony lies in making comparisons among studies that report data and measure synchrony in diverse ways. This section evaluates several methods, old and new, and examines their limitations. It also emphasizes that the choice of a measure depends on the goal of the study and the nature of hypotheses being tested. I make a plea for adequate presentation of data—no matter what measure one chooses—so that other workers have the opportunity to reanalyze the data presented for different comparisons.

A. Choice of Phenomenon

Almost all students of synchrony study the timing of clutch initiation (laying of the first egg in a nest). This is most convenient because each datum point is clear—an egg is present or it is not. However, because important synchronizing factors act before laying and after hatching, it is valuable, where feasible, to examine other phenomena. Burger and Shisler (1980) compare the phenology of nest initiation and clutch initiation, otherwise such comparisons are lacking.

Hatching synchrony is of particular interest with respect to the Darling effect. The point of hatching must be defined. Is it first detectable pipping, or is it presence of a chick free from the eggshell? Several studies report laying and hatching data (e.g., Vermeer, 1970). Synchronous fledging of Thick-billed Murre (*Uria lomvia*) chicks has been reported by Williams (1975). It is entirely appropriate to make any of these phenomena the subject of a study of synchrony.

B. Protracted Nesting Seasons and Renesting

Many colonial seabirds initiate a new clutch within a few weeks after losing eggs or chicks. Thus adverse factors tend to prolong the breeding season. Because these birds have already come into breeding condition and layed eggs, the factors responsible for synchrony are the adverse factors (e.g., predation, flooding) responsible for the loss of the first clutch, rather than other factors to which the birds are exposed. In most cases, such renesting should be eliminated from studies of synchrony.

Young birds nesting for the first time also tend to nest late, and like renesting birds they lay smaller clutches and smaller eggs than older birds and earlier breeders (e.g., Coulson and White, 1960). Renesting birds may stay in the same colony or move to a new one, and first-time breeders may return to natal colonies or be attracted to thriving ones. It is therefore difficult to determine (without having known-age birds) whether late nesting birds are new recruits (in which case it is clearly important to include them in synchrony studies) or renesting residents (in which case one usually excludes them). Renesting by immigrants from other colonies may be important and one may wish to include them in a synchrony study. In any case, it is important to recognize that these groups of birds exist and are basically different. Moreover, inclusion of them will decrease the overall synchrony of a colony, particularly if a span (duration) measure is used or if they make up a large proportion of the breeding birds.

An early nesting adult that loses eggs early in incubation is likely to renest during the peak period so that heavy loss of early nests (cf. Nisbet, 1975) may lead to exageration of the later peak. Several authors (e.g., Paludan, 1951) clearly indicate which nests are renests, but in many cases one must guess that late peaks include renesters, particularly in colonies where it is known that early nests have been destroyed.

One must be pragmatic and often arbitrary in eliminating late nesting when comparing synchrony among colonies. This requires a detailed knowledge of the species' local breeding phenology across several seasons. It should not be based solely on the use of a fixed percentage of the total nests, because in a bad year, renesting among Common Terns, for example, may exceed 50% of the population (Gochfeld and Ford, 1974).

In measuring synchrony, therefore, I have used several different techniques for homogenizing the data. The following techniques eliminate late outliers:

First 90%:   Calculate span of days or standard deviation
             for the first 90% of nests initiated
First 75% or first 50%:   Calculations based on the first 75% or 50%

The following techniques eliminated both the earliest and latest outliers. These measures are not symmetrically distributed around the median:

Mid-90%:   Calculate span of days or standard deviation
           for the 90% of nests initiated in the shortest
           time periods (the peak or mid-90%)
Mid-75% or mid-50%:   The same applying to the peak or mid-75% or
                      50%

These terms are used in the discussions of span and standard deviation that follow.

C. Measures of Synchrony

The measures to be discussed can be broken down into the following categories: (1) graphic, (2) fixed period, (3) span or range, (4) central tendency or standard deviation, and (5) other. The most frequently discussed measures involve the span or range of laying, but in this section I compare different methods to establish their limitations (Table I).

1. Graphic Methods

A graphic presentation is not only a useful visual format for data, but it can be used to compare synchrony among studies. Because the number of nests initiated in a day is a discontinuous variable, the data should be presented in the form of histograms. The various graphs that are most useful are shown in Fig. 1. The nest-initiation graph shows the number of nests begun on each day. The cumulative graph shows the number of nests that have been begun in the study area by the end of each day. The probability

Table I. Measures of Synchrony

Reference	Species	Synchrony measure
	Span of laying[a]	
Darling (1938)	Lesser Black-backed and Herring Gull	100% span
Coulson and White (1960)	Black-legged Kittiwake	100% span
Langham (1974)	Sandwich, Common, Roseate, Arctic Terns	100% and mid-90% span
MacRoberts and MacRoberts (1972)	Herring and Lesser Black-backed Gull	80% and 100% spans
Nelson (1976)	Great Frigatebird	100% span
Montevecchi *et al.* (1979)	Laughing Gull	Mid-75%
	Percent laying in fixed period	
Nelson (1976)	Great Frigatebird	% in 2 months
Patterson (1965)	Black-headed Gull	% in each 5 days
Veen (1977)	Sandwich Tern	% in peak 3 days
Veen (1977)	Sandwich Tern	No. in each 3 days
	Graphic methods	
Darling (1938)	Herring Gull	Right skew noted
Vermeer (1973)	Herring and Ring-billed Gulls	Cumulative % curve
	Standard deviation[b]	
Burger (1979)	Herring Gull	SD of 100%
Gochfeld (1979)	Black Skimmer	Subcolony SD
Spaans and Spaans (1975)	Herring Gull	Mean ± SD
Burger (1974a)	Franklin's Gull	Mean ± SD

[a] Orians (1961) used 100% span; Robertson (1973) used mid-50%.
[b] Horn (1970) used variance; Emlen and Demong (1975) used SD.

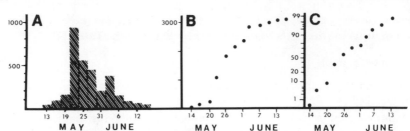

Fig. 1. Distribution of clutch initiation dates for Common Terns at Cedar Beach, New York, 1976. (A) Event histogram showing number of clutches initiated by 3-day intervals; (B) Cumulative curve showing number of nests initiated by the end of each 3-day interval; (C) probability paper graph of cumulative percent curve.

curve represents the cumulative curve on special probability paper. The latter two curves are most often presented as continuous functions. Vermeer (1973) provides additional cumulative curves. The steeper the curve, the greater the synchrony during the peak period, and the more the curve is displaced to the left, the greater the overall synchrony. Such curves allow one to quickly compare many colonies and studies. If the x axis is a percentage scale one can make comparisons among very different sizes and kinds of colonies, however, for intraspecific comparisons it is preferable to use a real-time base (e.g., number of days) for the x axis.

The initiation or event histogram provides the most familiar picture of the phenology of a colony. The range and the peakedness are immediately apparent, although only gross comparisons can be made among data sets. An additional limitation to graphic means is that when reproduced on journal pages it is often difficult to extract the raw data if one wishes to perform reanalyses. Moreover, crowding quickly reduces the comparability of cumulative or probability curves. Graphic techniques are clearly useful, but the raw-data histograms or comparable tables should be presented.

Probability paper offers some added advantages (see Sokal and Rohlf, 1969; see also Fig. 1C). In comparison with the cumulative frequency distribution, probability paper exaggerates both tails. A normal distribution would plot as a straight line on probability paper. As it is, most colonies give an approximately stright line, so that one can measure the slope directly without having to search for the straight middle portion of the cumulative curve. Probability paper tends to show very consistent synchrony among several studies of Herring Gulls (Fig. 2), and to emphasize differences one must exaggerate the x axis. The technique is not used widely in biology, but its utility should be considered, since the reduction of a distribution to a straight line has useful properties.

Skewness becomes particularly apparent on probability paper. It is apparent that the peak of laying (or median date) usually occurs early in the season, followed by a drawn-out tail of late nesting birds. In highly synchronous colonies (e.g., Veen, 1977), the median is very close to the beginning. The consistent right-skewness of reproductive event curves is apparent in Fig. 3, which compares the interval in days between the median and the 5th and 95th percentile days for various studies. Almost all studies showed fewer days between 5% and 50% than between 50% and 95% (right-skewness). Darling (1938) noted that the forepart of the curves were steeper than the right tail. In Fig. 3, normally distributed curves would lie on the diagonal.

Although limited by the need to make additional statistical comparisons, graphic techniques should remain popular and valuable adjuncts in studies of synchrony. At the least, an event histogram provides a clear description of the timing of events one has studied.

2. Fixed Time Comparisons

One can compare colonies using the percentage of nests initiated in a fixed period. One could determine what proportion of nests were initiated in

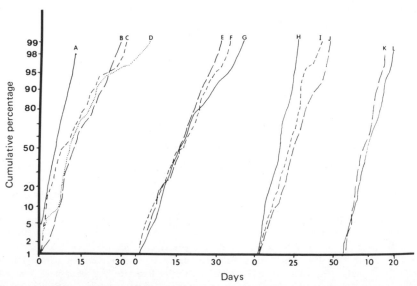

Fig. 2. Probability graphs of clutch initiation or hatching for Herring Gulls. Cumulative percent is shown on y axis; x axis is days. Clutch initiation distributions are shown for data from (A) Darling (1938); (B) 1943 data from Paludan (1951); (C) 1944 data from Paludan (1951); (D) Teeple (1977); (E) 1970 data from Davis (1975); (F) 1972 data from Davis (1975); (G) 1969 data from Davis (1975). Data from Burger (1979) are for Clam Island complex, Barnegat Bay, New Jersey; data for (H) Experimental Island; (I) Egret Island; (J) House Island; (K) 1965 hatching data; and (L) 1965 laying data from Drent (1967).

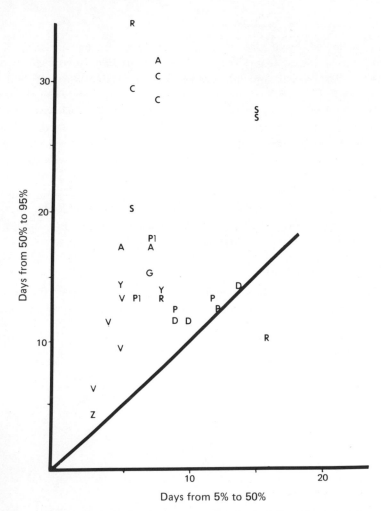

Fig. 3. Two-way plot of colonies for skewness. x Axis is interval (days) from the 5th percentile to the 50th percentile (the median) nest. y Axis is interval (days) from the 50th percentile to the 95th percentile nest. The diagonal is the line of equal intervals. Colonies above the diagonal show a right skew (protracted 2nd half of nesting season). Colonies below diagonal show a left skew (protracted first half of nesting season.) Data are from the following sources: A, Arctic Tern data from Langham (1974) (3 years); C, Common Tern data from Langham (1974) (3 years); D, Herring Gull data from Davis (1975) (3 years); G, Common Tern data from Cedar Beach, N.Y., in 1976 (M. Gochfeld, unpublished data); P, Herring Gull data from Parsons (1975); Pl, Herring Gull data from Paludan (1951); R, Roseate Tern data from Langham (1974) (3 years); S, Sandwich Tern data from Langham (1974) (3 years); V, Ring-billed and California Gull data from Vermeer (1970) (2 years); Y, Black-headed Gull data from Ytreberg (1956); Z, Cory's Shearwater data from Zino (1971). Note that almost all values show a right skew, particularly tern data reported by Langham (1974). Note also the tendency for clustering of data from each study suggesting a consistency across years for particular colonies. For discussion of aberrant Roseate Tern point, see Langham (1974).

the first 20% of the season or in the peak 20% (Table II). In many studies data are obtained at weekly intervals, and it is useful to compare proportions of nests in each interval, particularly because such data do not lend well to the standard deviation methods described below. The fixed time comparisons are particularly useful in the Tropics, where much breeding takes place throughout the year. Many papers give the numbers of birds nesting in each month, and comparisons can most conveniently be made on the basis of the percentage of birds nesting in the peak month or other period. For example, Nelson (1976) describes a colony of Great Frigatebirds (*Fregata minor*) with 90% of nests initiated in a 2-month period. Studies by Patterson (1965) and Veen (1977) examined synchrony and productivity in depth, making comparisons of success by 5- and 3-day intervals respectively. Veen also compared colonies with respect to percent of nests in peak 3-day period.

Fixed time comparisons seem particularly useful for working with grouped data. They are more useful when dealing with homogeneous samples (many species on a single island, or many colonies of one species in a state) than for comparisons among more diverse samples. The relatively low variance of the proportion of nests in the peak 20% of the season for Herring Gulls (Table II) argues for more widespread use of this method.

3. Span or Range

Darling (1938) used the time interval from initiation of the first to initiation of last clutch in a colony—the 100% span. Other authors (e.g., Langham, 1974) recognizing that a very early or late nest would distort synchrony, used narrower spans, such as the shortest period in which 90% of nests were initiated (mid-90% span). Table I indicates the different uses of span that have been published, and Table II compares different measures which I computed for published studies. In important studies on the Icteridae, Orians (1961) used the 100% span, whereas Robertson (1973) used the 50% span.

It was soon recognized that contrary to Darling's (1938) hypothesis, the span of laying was usually longer in larger colonies. The reasons for this are partly statistical (in two colonies of comparable synchrony the larger colony will have the larger range based on the normal probability curve) and partly biological (late-breeding birds may be attracted to large breeding aggregations). Coulson and White (1956) analyzed laying data for Black-legged Kittiwakes (*Rissa tridactyla*) and showed that the 100% span increased with colony size. Most criticisms of the Darling effect have arisen because larger colonies have longer 100% spans, no matter what their actual degree of synchrony.

Table II. Synchrony Measures Extracted from Studies of Herring Gull Breeding Biology

| Reference | Sample[a] type and year | Interval (days) | Number of nests | Span or range | | | | | Standard deviation | | | | | % of nests in[d] | | Kurtosis |
| | | | | 100% | First[c] | | Mid[c] | | 100% | First[c] | | Mid[c] | | First 20% | Peak 20% | |
					90%	50%	90%	50%		90%	50%	90%	50%			
Drent (1967)	1963 Lay	1	43	27	20	13	16	7	5.6	4.7	3.4	4.3	2.3	16	32	−0.25
Drent (1967)	1963 Hatch	1	47	20	19	11	15	9	5.3	4.9	2.8	4.5	2.7	19	25	−1.22
Drent (1967)	1964 Lay	1	53	24	20	11	16	7	6.1	5.1	4.1	5.1	1.9	15	41	−0.45
Drent (1967)	1964 Hatch	1	76	24	21	11	19	6	5.6	4.7	2.8	4.7	1.6	10	45	−0.63
Drent (1967)	1965 Lay	1	63	19	15	9	15	6	4.3	3.6	2.4	3.7	1.7	13	52	−0.82
Drent (1967)	1965 Hatch	1	104	16	13	8	11	6	3.6	3.2	1.5	3.0	1.8	6	42	−0.90
Darling (1938)	1937 Lay	2	65	13	11	7	10	3	3.1	2.6	1.6	2.5	1.0	13	48	−0.43
Paludan (1951)	1943 Lay	1	88	31	24	13	21	8	6.4	5.1	1.7	5.1	2.2	6	42	0.65
Paludan (1951)	1944 Lay	1	89	32	19	8	19	8	6.1	4.8	2.1	4.8	2.0	36	44	0.80
Harris (1964)		1	288	22	19	12	17	8	5.4	3.8	2.7	4.7	2.6	8	26	
Davis (1975)	1969 Lay	2	283	40	27	16	25	9	7.5	6.0	3.8	5.7	2.8	9	45	0.20
Davis (1975)	1970 Lay	2	359	37	25	18	22	9	6.4	5.3	4.2	5.2	2.7	4	45	−0.19
Davis (1975)	1972 Lay	2	210	39	25	17	25	9	7.8	6.4	4.8	6.2	2.8	10	36	−0.10
Teeple (1977)	Lay	1	33	41	23	12	23	6	8.0	5.2	3.2	5.2	1.8	6	58	−1.20
Parsons (1975)	1968 Lay	2	903	34	28	17	25	17	7.2	7.0	5.8	6.6	3.2	6	45	−0.27
Parsons (1975)	1967 Hatch	2	2339	33	24	14	21	7	6.2	5.4	3.1	5.5	2.1	8	39	−0.56
Parsons (1975)	1968 Hatch	2	903	22	19	12	17	8	7.7	7.6	4.6	6.5	3.2	11	34	−0.70
Burger (1979)[b]	House Is.	3	345	44	34	20	28	10	8.1	6.2	3.9	6.1	2.4	7	47	−0.22
Burger (1979)	Egret Is.	3	159	40	27	20	21	9	6.9	5.4	4.6	5.1	2.5	7	45	0.44
Burger (1979)	Experimental	3	88	26	21	14	17	6	5.9	4.3	3.4	4.4	1.3	14	45	−0.41
Mean:			326	29.9	22.2	13.6	19.6	7.9	6.1	5.1	3.3	4.9	2.3	9.3	40.7	−0.33
SD:			538	9.2	5.6	4.2	4.9	2.3	1.4	1.2	1.1	1.0	0.6	4.6	7.9	0.56
CV[e]:			16.5	30.8	25.2	30.9	25.0	29.1	22.9	23.5	33.3	20.4	26.1	49.5	19.4	169.7

[a] "Lay," clutch initiation data; "Hatch," day of hatching data.

[b] Data are from Burger (1979) for three islands in Clam Island Complex, Barnegat Bay, Ocean County, New Jersey for 1976 and 1977.

[c] First, calculations based on the first 90% or first 50% of the nests in a season; Mid, calculations based on the smallest time period containing 90% or 50% of the nests—this is not necessarily symmetrical around the median.

[d] Refers to the percent of all clutches which were initiated in the first 20% of the season or in the peak 20% of the season.

It should have been apparent that using the range of laying to characterize the temporal distribution of the event was as unsatisfactory as using the simple range in any statistical analysis of frequency distributions. Although several authors have reported the standard deviation (SD) of clutch initiation (e.g., Spaans and Spaans, 1975), it has not yet become a popular measure of synchrony. Horn (1970), and Emlen and Demong (1975) used variance and SD in studies of synchrony in passerines, whereas Burger (1979) and Gochfeld (1979) used SD as the primary measure of synchrony in Herring Gulls and Black Skimmers (*Rynchops niger*). These studies concluded that the range itself was not a satisfactory measure of synchrony and that measures of central tendency must be used.

When outliers are removed, however, certain span measures (e.g., first 90% or interquartile) may be useful. These are compared with other measures in Tables II and III.

4. Mean and Standard Deviation

These measures of central tendency provide data on the timing and spread of laying, both of which are germane to studies of synchrony. Where the data are not normally distributed it is often more valuable to use the median date than the mean, but in either case the SD provides a usable measure of synchrony. At times it may be desirable to substitute the coeffi-

Table III. Correlation Matrix among Selected Synchrony Measures (Based on Table II)

Method	Number of nests	Span 100%[a]	Span 90% (mid)[b]	SD 100%[a]	SD 90% (1st)[c]	SD 90% (mid)[b]	% in peak 20%	Kurtosis[a]
Number of nests	—							
100% span[a]	0.26	—						
Mid-90% span[b]	0.33	0.94	—					
100% SD[a]	0.23	0.94	0.95	—				
First 90% SD[c]	0.39	0.79	0.89	0.90	—			
Mid-90% SD[b]	0.35	0.83	0.93	0.93	0.97	—		
% in peak 20%	−0.06	−0.38	0.26	0.26	−0.01	−0.06	—	
Kurtosis[a]	0.001	0.38	0.30	0.24	0.12	0.24	0.05	—
Mean r^2	0.073	0.478	0.530	0.520	0.476	0.506	0.040	0.050
Mean partial corr.[d]	0.269	0.691	0.728	0.720	0.690	0.711	0.200	0.230

[a] Based on entire sample of nests reported.
[b] Based on 90% of clutches initiated in shortest time period (not symmetrical about median).
[c] Based on first 90% of nests initiated.
[d] Square root of mean r^2 estimates the partial intercorrelation among the variables.

cient of variation (CV = SD × 100/mean) if the SD and mean are correlated.

The following discussion considers some limitations on use of the standard deviation. It proves remarkably insensitive to colony size over a broad range of sizes, but is somewhat sensitive to use of grouped data. To demonstrate this, two hypothetical colonies were constructed with 10 nests initiated in 3 days and 30 nests initiated in 10 days. The distributions were then multiplied by 10, 100, 1,000, 10,000, and 100,000, and for each of these the SD was calculated. Figure 4 shows that SD remains constant from 10^6 down to 10^3, and increases by a small percentage for smaller samples. Thus one can mathematically compare colonies of vastly different size. The SD used in Fig. 4 is calculated from Eq. (1) below. This is the standard method used to estimate the parameter of a universe. If one wishes to describe only the variance of a particular sample, Eq. (2) is preferable, in which case there would be negligible change in SD, even at 10^2 and 10^1.

$$\text{Variance} = [\textstyle\sum X^2 - (\sum X)^2/n]/n - 1 \tag{1}$$

$$\text{Variance} = [\textstyle\sum X^2 - (\sum X)^2/n]/n \tag{2}$$

where x is the vector of dates and n is the sample size. SD = (variance)$^{1/2}$ Some calculators actually compute the variance using both methods.

Like the span, one can compute the SD based on only a portion of the nests, eliminating outliers. The rationale (based on renesting) for eliminating outliers has been presented above.

5. Using Grouped versus Ungrouped Data for Standard Deviation

In many studies, daily sampling is precluded, either for logistic reasons or to avoid excessive disturbance. How can one compare SD computed for colonies sampled on a daily basis with those sampled at 2-day, 3-day, or longer intervals? Data for Common Terns at West End Beach, New York, in 1971 were used for such comparisons. The SDs were computed for five subcolonies (n = 62 to 194 nests) using 100% of the nests in each. The values, 7.28–11.54, are represented on the y axis of Fig. 5. The data were then grouped for weekly intervals and for 3-day intervals, and various SDs were calculated and plotted against the SD for the daily values. For the weekly intervals it was assumed that all nests were initiated on the midpoint day of each week. For the 3-day intervals I used the same technique (M in Fig. 5), but I also redistributed the nests for each interval evenly over the 3 days in the interval. The SD for the "smoothed" data are shown by E in Fig. 5. Finally, I ignored the individual days and simply used the interval

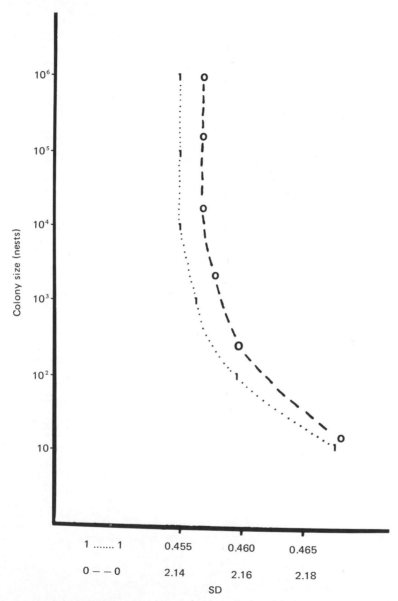

Fig. 4. Standard deviation as a function of size for colonies with the same proportional distribution of nests (see text). SD is constant over nearly four orders of magnitude, and even down to 10^1 it increased by only 2%.

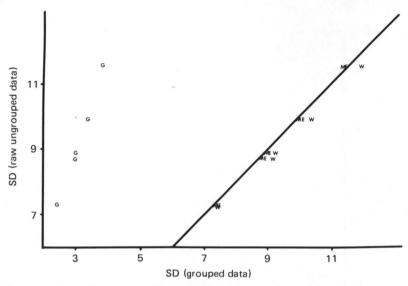

Fig. 5. Comparison of SD computed for raw data and for same data grouped in different ways, for Common Terns, at West End Beach (five subcolonies in 1971; Gochfeld, unpublished data). *y* Axis is SD for the raw ungrouped data obtained on a daily basis. *x* Axis is SD for same data grouped as follows: W, data grouped for weekly intervals with midpoint of week used for calculations (i.e., 4th day used for 1st week; 11th day used for 2nd week, . . . , 46th day used for 7th week); M,E,G, for data grouped by 3-day intervals. G uses the interval number for calculations (i.e. 1st, 2nd, . . . , interval, rather than day number); M uses the midpoint of each interval for calculations; E uses data evenly redistributed over all days in the interval. Diagonal line is locus of points where grouped SD equals ungrouped SD.

number for the SD calculations, thus the first 3-day interval was considered 1, the next 2, and so on. This has the effect of compressing the range and naturally reduces the mean and SD approximately threefold. A comparable method is discussed in Section VIII.G, where data are grouped by months, and the SDs are computed for values of 1–12.

The diagonal in Fig. 5 is the locus of points whereon the SD for grouped data equals the SD for the same data ungrouped. It is apparent that when the midpoint of a 3-day interval is used instead of the daily frequencies, the values are almost on the diagonal line. When data are evenly redistributed over 3 days in each interval, the SD is slightly increased. These two methods yield SDs that average 0.5% and 1.2% higher than the true (daily values) SD. These values lie so close to the diagonal that one should be able to use these methods interchangeably, unless one is comparing SDs which themselves have a very small variance. The values grouped by week midpoint yield a slightly higher (3.8%) SD. Values grouped by interval number yield much lower SDs and cannot be compared. Using the actual daily values provides the most accurate representation of the daily peaks and gives a lower SD, whereas smoothing the data by distributing points evenly over each day in the 3-day period gives the highest SD.

The main conclusion is that one must compare intervals of the same length. Thus if one were to compare a data set with daily values for one colony with data grouped by 3-day intervals for another colony, the data from the first should be lumped into 3-day intervals. It is apparent that intercolony differences will be reduced when grouped data are used. For many reasons, however, grouped data are desirable, and the limitations mentioned here apply to use of SD. As indicated above, fixed time comparisons may be made directly on grouped data.

6. Skewness and Kurtosis

Any comparison of frequency distributions must consider not only the mean and variance, but the shape of the distribution. Two important statistics are skewness and kurtosis, and both are important in evaluating the temporal distribution of events relevant to synchrony.

Skewness refers to the asymmetry of a curve, in which case one of the tails is drawn out and the mean and median do not coincide. This is important in examining nest initiation curves, where one sees a preponderance of nests in the early part of the seasons with a prolonged tapering off late in the season. Figure 3 illustrates that in most studies the number of days elapsed between the 5th and 50th percentile of nests initiated is shorter than between the 50th and 95th percentiles. Thus right-skewed distributions characterize nest initiation curves. Skewness can be described mathematically and tested for significance of departure from normality (Sokal and Rohlf, 1969). The sample parameter for skewness ($g1$) is calculated from the sum of the cubed deviations of each values from the mean as follows:

$$g1 = [\sum_{1}^{n}(X_i - \overline{X})^3]/[(n - 1)(n - 2)\,SD^3]$$

where $\overline{X}$ is the mean and X_i represents each value, n is the sample size, and SD is the standard deviation. IF $g1$ is negative, the distribution is skewed to the left; if $g1$ is positive, the distribution is right skewed. The value can be tested for significance (see Sokal and Rohlf, 1969).

Kurtosis refers to the presence or absence of exaggerated peaks in the distribution. If clutch initiation occurs gradually, the curve tends to be flatter than a normal distribution (platykurtic), in contrast to a highly synchronous initiation, which produces a narrow, sharp peak (leptokurtic). The sample parameter for kurtosis is $g2$, computed as follows:

$$g2 = [\sum_{1}^{n}(X_i - \bar{X})^4]/[(n - 1)(n - 2)(n - 3) SD^4] - [3(n - 1)^2]/[(n - 2)(n - 3)]$$

If $g2$ is negative the curve is platykurtic, whereas a positive $g2$ indicates a leptokurtic curve. Kurtosis, too, can be tested for significance, since for a normal distribution $g1$ and $g2$ should be zero (see Sokal and Rohlf, 1969).

Because kurtosis is a statistical description of the peakedness of a curve, it is actually a primary measure of synchrony. The more leptokurtic the curve, the more synchronous the events. However, thus far, kurtosis has not been used in studies of synchrony. In my experience the main limitation is that $g2$ is sensitive to skewness, and if a curve departs from normality, the measure of kurtosis becomes a less accurate measure of synchrony. The comparison of kurtosis ($g2$) with other synchrony measures is given in Table II. For most studies the computed kurtosis is negative, indicating platykurtosis. Empirical comparisons of curves that have been normalized or have had outlying values removed may demonstrate that this measure can be used routinely in studying synchrony. Certainly, in conjunction with the SD, kurtosis should prove useful for comparing curves obtained in single studies.

7. Miscellaneous Methods

Under this category I consider one method that has been used by Hailman (1964)—the proportions of birds at each breeding stage. In a single visit to a colony one can ascertain approximately what proportion of birds are engaged in courtship, incubation, care of preflying young, and care of fledged young. For example, Floyd (1933) noted that on a visit to Penikese Island, Massachusetts, all the young Common Terns he saw were nearly ready to fly, and he remarked on this high degree of synchrony, which was not at all typical for that species.

Hailman (1964) divided the cycle of the Swallow-tailed Gull (*Creagrus furcatus*) into three stages: (1) incubation, (2) preflight young, and (3) postflight young. A synchronous colony would have most birds at one of these stages; a less synchronous colony might have equal proportions at each stage. Testing a null hypothesis of equal frequencies at each stage, Hailman showed that for larger colonies ($N \geq 15$) the distributions showed significant synchrony. This method allows one to make approximate statements about synchrony based on brief visits and is satisfactory for comparisons among colonies of a single place or species.

8. Comparisons among Methods

It is not possible to designate a single "most valuable synchrony measure." Graphic presentations accompanied by statistical analysis are a desirable combination. By computing SD for several studies using 50%, 60%, . . . , 90%, 95%, 100% of nests, I find a breaking point between 90% and 95% in most cases. This suggests that when in doubt about the extent of renesting or the significance of late nests, one could safely use 90% of the values. However, if one is concerned about the phenomenon of synchrony itself as well as the related social factors, one may wish to focus on the peak period, looking at the mid-50% of nests, thereby eliminating early outliers as well as late ones. If one is interested in comparisons across colonies, it may be useful to eliminate only the late outliers (see Table II).

Table II compares various synchrony measures, and intercorrelations among these are shown in Table III. Certain measures, such as the proportion of nests in the first 20% of the nesting season, had low partial correlations, but for many comparisons the correlations exceeded $+0.56$ ($p <$ 0.01). The intercorrelations suggest that the 90% span, 100% SD, and 90% SD (mid) are approximately equivalent. The 100% span had a slightly lower intercorrelation (see Table III).

Kurtosis showed a low correlation with most other values, indicating that its measure of the peakedness of the curve is independent of the SD and the range. Thus it can be used in conjunction with these measures as a useful description of the temporal pattern of nesting. The other methods discussed above may find use in special cases, but at present a graphic method, coupled with an SD measure and kurtosis measure (perhaps also based on 90%), should be used for synchrony studies.

IV. SYNCHRONY AND COLONIALITY

The evolution of synchrony in colonial birds cannot be considered apart from the evolution of coloniality itself. This chapter does not examine the question of gregariousness in birds in any detail. Lack (1968), Ward and Zahavi (1973), and Krebs (1978) among many others have discussed whether predation (see Crook, 1964) or food exploitation is the main factor influencing the evolution of aggregations and coloniality. Krebs (1978, p. 309) has concluded that one can test hypotheses about the selective advantages accruing to particular types of social, antipredator, or feeding behavior, whereas it is fruitless to attempt to answer the more general question of which of these factors is the main force that shapes gregariousness.

Lack (1968, p. 135) notes that many species clump together despite widespread food availability and considers that the main advantage of such flocking lies in protection from predators. Powell (1974) and Siegfried and Underhill (1975) demonstrated experimentally that birds feeding in flocks had quicker predator detection. Munro and Bedard (1977) found that although large flocks of Common Eider (*Somateria mollissima*) ducklings were attacked by hawks more frequently than small flocks, they suffered lower mortality. Ward and Zahavi (1973) attribute a much greater range of avian assemblages to food exploitation. Krebs (1978) notes that food exploitation can be the main benefit to coliniality either if the colony happens to be close to a locally abundant food source or if groups of birds are more efficient in finding and/or capturing food than are solitary ones. In some cases colonies may arise simply as a passive aggregation of individuals in a favorable habitat wherein nest sites are in short supply, but it is probably necessary to demonstrate that such aggregations actually result from habitat shortage, as some studies (e.g., Coulson, 1971) show that such colonies are more compact than dictated by the site shortage. Snapp (1976) believed that passive aggregations were important based on studies of Barn Swallows (*Hirundo rustica*), but studies on Bank Swallows (*Riparia riparia*) implicated protection from predators (Hoogland and Sherman, 1976) and food exploitation (Emlen and Demong, 1975) in the development of synchronized colonial breeding. Nelson (1966) notes that Northern Gannets (*Sula bassana*) are limited by colony site availability, but does not suggest that this is the main reason they occur in colonies.

Lack (1968) summarized the relation of colonial breeding to flock feeding, finding that 99% of marine nidicolous, 73% of freshwater nidicolous, and 26% of all flock feeding birds nested colonially, whereas only 1% of solitary feeders and 13% of all bird species did so.

Austin (1940) contended that the main benefit to coloniality "is the greater consistency of incubation and brooding which some unknown factor engenders in a large colony." Common Terns in large, dense colonies are more vigorous in defending against human intruders than are birds in small colonies (Austin, Jr., 1933, M. Gochfeld, unpublished manuscript).

Several authors (e.g., Patterson, 1965; Tenaza, 1971) have reported reduced breeding success in birds nesting apart from colonies. However, in some cases isolated birds may fare better than their counterparts in nearby colonies. Burger (1974b) found that Brown-hooded Gulls (*L. maculipennis*) nesting in colonies suffer more predation (only 2% of 107 nests produced young), whereas solitary birds mostly escaped (81% of 16 nests produced young).

In some species the tendency to breed in a colony is very high. Ashmole (1962) reported that although many Black Noddies (*Anous tenuirostris*) in their dense colonies on Ascension Island lost their eggs because they could find only narrow unsuitable nest ledges, none colonized the numerous unoccupied ledges elsewhere on the island, and no isolated breeding pairs were seen. Because solitary birds are much more difficult for human investigators to detect than are their colonial counterparts, it is likely that for most species the tendency for solitary nesting has been underestimated. I have found that isolated nesting (one or two pairs far removed from a breeding colony) occurs annually in the Common Tern population of western Long Island. Recent coverage by helicopter has indicated that many species of colonial birds occasionally nest in isolation. The breeding success of such birds warrants comparison with that of their nearest colonial conspecifics. As Orians (1961) pointed out, merely demonstrating that isolated pairs of a colonial species are successful in breeding, does not imply that coloniality is not beneficial. It is likely that both habits are beneficial at certain times but not at others and that diversity is maintained through varying selective pressures (J. Burger, personal communication).

Factors favoring coloniality include protection against predators, enhanced utilization of resources, and enhancement of social interactions. Predator protection derives from safety in numbers and reduction of susceptible perimeter (Hamilton 1971); predator confusion (of questionable significance: Brown, 1975); quicker predator detection (Seigfried and Underhill, 1975; Hoogland and Sherman, 1976); enhanced predator mobbing (Kruuk, 1964; Palmer, 1941; Austin, 1946); and predator satiation by synchronous hatching (Darling, 1938). Resource utilization can refer to limited colony or nest sites (Lack, 1966); patchy but abundant food resources (Krebs, 1978; Ward and Zahavi, 1973); and cooperative feeding (Bartholomew, 1942). Social interactions include social stimulation leading to synchrony (Darling, 1938), decreased interference (Nelson, 1976), social attraction of inexperienced birds (Austin, 1946), and facilitation of pair formation.

Several disadvantages of coloniality arise from competition for nest sites, mates (MacRoberts, 1973), and food; from interference in reproductive activities; from attraction of predators; and from spread of parasites or diseases. With respect to predation, Tinbergen *et al.* (1967) showed that spacing is a compromise between factors leading to decreased spacing (enhanced predator mobbing) and to increased spacing (thwarting predator's search image).

The focal point of discussions of synchrony and coloniality is Darling's book (1938). To clarify discussion of synchrony it is desirable to review the types of data Darling presented (see Table IV). It is also important to recognize that significant social interactions occur among birds that are not

Table IV. Summary of Data on Gull Breeding Synchrony [Modified from Darling (1938)]

Species and year[c]	No. of breeding adults	Nests	Clutch-initiation dates			Mean clutch	Eggs laid	Eggs hatched	Productivity estimates Young fledged	
			First nest	Last nest	100% span[b]				Number	% hatched[a]
HG 1936	84–90	40	May 7	May 23	17	2.1	84	72	35	48.6
HG 1936	30–34	12	May 12	June 3	23		26	22	8	36.4[d]
HG 1936	20	7	May 18	June 12	26		16	14	3	21.4
HG 1936	4	2	May 26							
HG 1937	130–150	65	May 9	May 25	17	2.9	189	181	76	41.4
HG 1937	34–40	18	May 9							
HG 1937	16	9	May 11	June 1	22					
HG 1937	6	3	May 13	May 27	15		9	8	1	12.5
LBBG 1936	72–80	36	May 15[e]	June 7	24 (18)[e]	2.2	78	72	40–45	55–63
LBBG 1936	18	7	May 23	June 12	21		15	14	8	59.1
LBBG 1937	120	56	May 17	June 8	23	2.9	163	155	75–90	48–50
LBBG 1937	30	15	May 19	June 12	25		43	41	22	53.6

[a] % Based on number of eggs hatching.
[b] 100% span is interval in days from 1st to last nest initiation.
[c] (HG) Herring Gull, *Larus argentatus*; (LBBG) Lesser Black-backed Gull, *L. fuscus.*
[d] Heavy loss of chicks due to flooding.
[e] First next on May 15; rest delayed to May 21.

considered colonial. Indeed, Darling (1952) suggested that an important function of avian territoriality was to provide boundaries across which neighboring birds could experience mutual social stimulation.

V. DARLING'S GULL STUDIES

In 1936 and 1937, Darling (1938) studied breeding Herring and Lesser Black-backed (*L. fuscus*) Gulls on Eilean A'Chleirich (Priest Island) off the Western Highlands of Scotland (see Darling, 1944). Table IV summarizes data on size of colonies, clutch size, and synchrony (span from 1st to last nest initiation). Given the small number of colonies and the small range of sample sizes, it is remarkable that Darling developed such an all encompassing hypothesis.

Each year he found four Herring Gull colonies (2–65 nests) and two Lesser Black-backed colonies. The earliest eggs were found in the largest colonies. For Herring Gulls the span of laying was greater in small colonies (e.g., 17 days for 40 nests versus 26 days for 7 nests). In 1936 the largest colony fledged young from 49% of all the eggs that hatched, whereas a medium colony fledged 21% and the smallest fledged none. The relationships did not hold for the other species.

Darling proposed a consistent relationship between colony size and synchrony. The Darling effect links the number of birds and the total social stimulation with reproductive physiology (mediated by neuroendocrine pathways), breeding phenology, and predation. Darling suggested that birds in larger colonies would be exposed to a greater amount of visual and acoustic stimulation, which would accelerate the endocrine activity, causing late birds to breed earlier than they otherwise would. This leads to more synchronous breeding in the colony, and to synchronous hatching. Because chicks (of the gulls he studied) were vulnerable (to the predators he studied) for only about 2 weeks, Darling recognized that synchronous hatching would produce a veritable glut of vulnerable young for a short time, during which the predator could take only a fraction of those available. Thus the satiated predator would have consumed a smaller proportion of the young in a large synchronous colony than in a smaller unsynchronized one. This is summarized in Fig. 6A, wherein a predation threshold is evident, above which birds will escape predation. Nisbet (1975) provides an example of a Great-horned Owl (*Bubo virginianus*), which consumed a relatively constant biomass of tern chicks throughout the breeding season. Birds hatching their young during the peak period are more likely to have them escape predators.

Darling (1938) did not consider that in many cases predators are recruited to areas in which there is a temporary abundance of food. Also, he

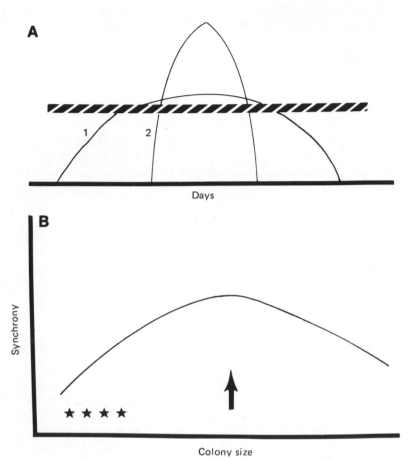

Fig. 6. (A) Curve 1, hatching distribution for a relatively asynchronous colony; curve 2, distribution for a synchronous colony of same size. Slashed line is predation threshold for a nonrecruitable predator(s). Predator can consume all individuals below the line. Only when number of individuals hatching on a day exceeds this threshold will young survive. Thus synchronous colony has more surviving young than does a less synchronous one [modified from Darling (1938)]. Curve is a model relating synchrony, predator satiation, and survival. (B) Theoretical relationship between synchrony and colony size. Synchrony should increase with size (and density) up to a certain limit beyond which social stimulation can no longer be effective (the neighborhood or subcolony). Larger aggregations may result from social attraction and may include several neighborhoods which are internally synchronous but which are not synchronized with each other. Thus SD is a biphasic function of size. Stars represent range that Darling (1938) studied. Arrow represents breaking point that Burger (1979) found to be 200–250 for Herring Gull colonies in New Jersey.

did not discuss in detail other factors that might produce synchrony or that might lead to asynchrony.

Darling's ideas received considerable attention both favorable and critical. The adequacy of Darling's data was questioned, correctly, by von Haartman (1945). However, it is clear that Darling's data illustrate how he drew his inferences, and they are not presented as statistical tests of preexisting hypotheses.

Some workers have reported that the Darling effect did not operate for a particular species (MacRoberts and MacRoberts, 1972b; Orians, 1961), whereas others have found support for one or another aspect of it (Burger, 1979; Gochfeld, 1979). The problem lies partly in the manner in which the hypotheses are presented, partly in the interplay of other factors, and partly in the logistic complexities of adequately studying the Darling effect.

A. Colony Size and Synchrony

Darling (1938) clearly proposed that the size of a breeding colony (number of pairs) would be the primary influence on the amount of social stimulation available to any pair within the colony, and that synchrony should increase (overall span from initiation of first to the last nest should decrease) as colony size increased. Many authors have subsequently found the opposite or little correlation between size and synchrony (e.g., Weidmann, 1956; Orians, 1961; Vermeer, 1963). Coulson and White (1956) reported that the span of laying increased for Black-legged Kittiwakes with increasing colony size. It is reasonable to suppose that a colony with more nests will have a longer nesting season than will a colony with fewer nests, quite apart from social stimulation and synchrony. If social stimulation is important, late breeding birds may be attracted to areas with the most stimulation causing such colonies to have longer laying seasons, without providing a clue as to the amount of synchronizing activity. Thus, as defined above, one may have a large amount of synchronizing activity with relatively low resultant synchrony.

Darling used the span of laying (first nest to last nest) as a measure of synchrony. In Section III I examined the measures and methods of studying synchrony, and concluded that using the time span is like using the range to describe a frequency distribution, an unsatisfactory measure. Unfortunately, most critics have also used range, and found, as did Coulson and White (1956) and even Darling (1938) for Lesser Black-backed Gulls, that the span increases as colony size increases (in most cases).

Veen (1977) found that Sandwich Terns in larger colonies were more synchronous than those in smaller ones. Bullough (1942) found more late-hatching eggs in small Arctic Tern (*S. paradisea*) colonies, and he attributed this to the Darling effect. In the most detailed study across colonies Burger (1979) found for Herring Gulls that synchrony (using standard deviation rather than range) increased with colony size up to about 200 pairs, beyond which point the trend reversed. Larger colonies apparently behaved like several small ones, and whatever synchronized the birds did not extend beyond a group of about 200–250 pairs. Darling (1938) provided a like interpretation of Goethe's (1937) conclusion that there were synchronized groups within his Herring Gull colony of 6000 pairs. Darling suggested that Goethe's groups behaved much like the smaller colonies Darling studied. A general model relating size and synchrony is shown is Fig. 6.

B. Density and Synchrony

It is clear from Darling's discussion and from his emphasis on social stimulation that density rather than size *per se* should be a major factor influencing synchrony, because in dense aggregations each individual will have more individuals close at hand to provide stimulation. Density is limited at the upper end by direct aggression and interference. Only in colonies of very small size (such as those studied by Darling) should there be no effect of density. In large colonies density is a key factor because it is not likely that any pair could be appreciably influenced by the activities of birds beyond its immediate neighborhood. Not only would distance decrease the social stimulation value of a display, but the presence of displaying birds closer by would mask the activities of more distance pairs. The effective social neighborhood may be limited to a circle of a very few meters [as Coulson and White, (1960) suggested] or may extend over perhaps 250 pairs (cf. Burger, 1979). In a less dense area of species the effective social neighborhood may be much larger spatially, but may contain fewer pairs.

The effects of density must be complex, must be influenced by the physiography of the colony, and must vary from species to species, depending on what aspects of the social milieu are significant for social stimulation. In some species acoustic cues may be more important than visual ones, and in a given species acoustic stimulation may play a proportionately greater role in sparse than in dense areas. A given species may nest at very different densities depending on numbers, available space, habitat, visibility, and social factors. Montevecchi *et al.* (1979) note that Laughing Gull (*Larus atricilla*) colonies in New Jersey may differ in density by a factor of 7. Results on social stimulation and synchrony obtained under one set of conditions (e.g., dense colony) may have little bearing on what goes on under other conditions (e.g., sparse colony).

For Black-headed Gulls (*L. ridibundus*), Patterson (1965) found that in denser groups, birds are more synchronous and have lower predation, but concluded that the role of density was secondary. MacRoberts and MacRoberts (1972*b*) studied Herring and Lesser Black-backed Gulls and concluded that density did not have a major effect, but their Fig. 4 seems to belie this conclusion.

Coulson and White (1956) showed that the spread of breeding was greater in older, larger colonies, but the range of density was closely related to size. This was less true in later years, but they found that the time spread of breeding was related to density. High-density areas were earlier (or *vice versa*) and large colonies had both dense–early and late–sparse areas. Because time of breeding often has a major influence on breeding success (e.g., Paludan, 1951; Parsons, 1975*a*, 1976), the effect of density on productivity, may be spurious.

Coulson and White (1960) showed that larger colonies usually have longer spans and are often cited as refuting the Darling effect. Actually they refute only the use of laying span as a measure of synchrony. They do show social stimulation and synchrony for kittiwakes, with the effect limited to a small neighborhood of about 2 m. Because birds return to the colony at different stages of readiness, Coulson and White (1960) suggest that synchronization occurs before return, but at this point they are no longer discussing synchrony *per se*. They also suggest that there may be an accumulation of stimulation over a period of years, which may account in part for enhanced success of experienced birds.

C. Timing of Breeding

Darling (1938) emphasized that large colonies would breed earlier as well as more synchronously. It would be obvious that selecting for early breeding would push the nesting season earlier and earlier each year were it not for such constraints as adverse weather or habitat conditions (Burger, 1974*a*; Burger and Shisler, 1978; Sealy, 1975), predation (Nisbet, 1975), or lack of food (e.g., Perrins, 1970).

These countervailing forces almost certainly overpower any tendency for social stimulation to lead to earlier breeding, and breeding seasonality is thus stabilized by centripetal selection. The emphasis on earliness of breeding attracted criticism, and to my mind is the least important aspect of the Darling effect.

D. Critiques of the Darling Effect

A number of authors (e.g., Klopfer and Hailman, 1965; Horn, 1970; Burger, 1979) have found support for one or more aspects of the Darling effect. Davis (1940), Armstrong (1947), Orians (1961), Hailman (1964), and Langham (1974), among others found little support for the effect. Fisher (1954) noted that age differences in different colonies would account for some of the reported effects, a conclusion emphasized more forcefully, perhaps, by Coulson and White (1956, 1960). Richdale (1951) and Austin (1945) noted comparable effects of age, while Burger (1979) concluded that her colonies were relatively homogeneous in age (based on egg size).

Orians (1961) found that larger colonies of Tricolored Blackbirds (*Agelaius tricolor*) had longer laying spans, but higher productivity than smaller colonies. He concluded that the Darling effect was at least theoretically possible. Several authors found evidence that asynchronous breeders fared worse than those at the peak [e.g., Atlantic Puffin (*Fratercula arctica*), Nettleship, 1972; Red-winged Blackbird (*Agelaius phoeniceus*), Robertson 1973; European Shag (*Phalacrocorax aristotelis*), Snow 1960; Black-headed Gull, Patterson, 1965; Bank Swallow, Emlen and Demong, 1975; storm petrels, Harris, 1969; Herring Gulls, Paludan, 1951; Parsons, 1971; Brown, 1967].

Not only do such critiques run afoul of different definitions of colony vs. subcolony, but there is no hard and fast rule for defining these aggregations. Numerous authors (e.g., Burger, 1974a; Emlen, 1952; Gochfeld, 1979; Hailman, 1964; Hall, 1970; Orians, 1961, Schaller, 1964; Ward, 1965; Hoogland and Sherman, 1976; Spurr, 1975) have found synchrony within subgroups of a colony, whereas the entire colony may be quite asynchronous.

MacRoberts and MacRoberts (1972a,b) studied Herring Gulls and Lesser Black-backed Gulls, measuring nest-density and nearest-neighbor distances on study plots. They emphasized earliness of laying (using the median date) as the most important feature of the Darling effect, whereas I regard it as much less important than synchrony. They found no significant relationship between earliness and density, nor between synchrony (100% span) and density, but the size of the aggregations in which the plots occurred was not considered. They found no significant difference between mean nearest-neighbor distance for birds laying before or after the peak, but it is not clear at what point in the season the distances were measured. This is important because late-arriving birds may establish territories between preexisting territories (filling in) and thereby alter density measures. Despite these questions, their study remains one of the most extensive attempts to examine the interplay of density laying date, and synchrony.

These and other factors will be discussed below. Alternative explanations exist for the data Darling presented (e.g., Coulson and White, 1956; Orians, 1961), for factors causing synchrony, and for functional or adaptive significance of synchrony when it occurs. Darling (1952) recognized that his model need not function in all or even most colonial species.

The evolutionary significance of the Darling effect is that it linked the behavioral aspects of nesting synchrony and its physiologic correlates with the ecological effects of predation to provide a mechanism selecting for synchronous breeding. Social facilitation of reproductive behavior mediated hormonally served as the synchronizing factor, whereas predator satiation served as the selective mechanism.

Testing the model has proven difficult. For such a global hypothesis, one must study the phenology of several colonies, evaluating social interactions and productivity. In any given season, several confounding factors may mask social facilitation and synchrony and even reverse the expected productivity. Predation is particularly difficult to study; it is rarely seen. Predators are often nocturnal and elusive and may remove their prey entirely. Darling was fortunate to work on an island where there were few predators and no opportunity of recruitment from the mainland. In most cases predators can concentrate in areas in which food becomes locally (if only temporarily abundant), and this predator recruitment would obscure the Darling Effect.

In my view the phenomenon of greatest interest in Darling's hypotheses is the role that social facilitation plays in synchronizing behavior among neighboring pairs of colonial birds and in synchronizing their reproductive physiology. In order to demonstrate that social facilitation is occurring, it is necessary to search for evidence of temporal and spatial contagiousness of certain behaviors. Southern (1974; see also Table V) provides such documentation for Ring-billed Gulls (*Larus delawarensis*), and the next section

Table V. Effect of Copulatory Wing Flagging on Neighboring Ring-billed Gulls[a]

Increased behaviors[b]	Decreased behaviors[c]
Mounting	Agonistic behaviors
Vocalizing	Walking-toward
Looking-toward	Preening
	Soliciting
	Bill pointing[c]
	Head tucked[c]
	Plumage fluffing
	Nest building[c]
	Departures and arrivals[c]

[a] After Southern (1974).
[b] All changes significant at $p < 0.05$.
[c] Behaviors did not occur during test period.

provides evidence that such contagiousness exists for the precopulatory and copulatory behavior of Common Terns.

VI. CONTAGIOUS COURTSHIP AND COPULATION DISPLAYS OF COMMON TERNS

One important aspect of synchrony is the demonstration that social facilitation actually occurs. I have studied the interactions of Common Terns to determine whether neighboring birds influence each other with respect to several types of behavior. Data were obtained during the early weeks of four breeding seasons (1973–1976) mainly on two groups of about 40 and 80 pairs of terns at West End Beach, Nassau County, New York. Southern (1974) has demonstrated that when one pair of Ring-billed Gulls was engaged in copulation accompanied by conspicuous extension and waving of the male's wings (copulatory wing flagging), the behavior of nearby birds changed (Table V). Maintenance behavior decreased, whereas courtship behavior increased. Figure 7 shows two pairs of Caspian Terns (*Hydroprogne caspia*) on Hat Island, Lake Michigan. One pair is copulating, and a nearby pair has just begun a precopulatory display. The few copulations studied appeared to be contagious (M. Gochfeld, unpublished data).

More detailed data have been obtained on Common Terns. Palmer (1941) describes in detail the courtship, precopulatory and copulatory sequences of this species. Immediately prior to copulation the male and female terns stand parallel, and as the female begins to crouch she starts to turn her body in a circle. The male circles around her, approaching slightly and assuming an erect position. Circling may continue for 30–60 sec, usually ending with a mount. The behavior is ritualized. Mounting usually results in copulation (see Cooper *et al.*, 1970). Cloacal apposition in successful copulations is accompanied by vigorous wing flapping or wing flagging, which presumably functions to allow the male to maintain balance.

I hypothesized that if social facilitation were occurring one would see a temporal clustering of displays. Moreover, those events that occurred close together in time would be close together in space. In addition, I studied whether conspicuous displays such as wing flagging might have a stimulating effect over a longer distance than do more subtle displays, such as precopulatory circling. Therefore the spatial-temporal clustering should be more obvious for circle to circle than for copulation to copulation interactions.

Fig. 7. Courting and mating Caspian Terns on edge of colony at Hat Island, Lake Michigan. When one pair began copulating, a neighboring pair began its precopulatory displaying.

The initial study involved random 30-sec observation periods in an area occupied by about 80 pairs of terns during the week prior to the peak of egg laying in 1974 and 1975.

I determined that during a mount, wing flagging occupied an average of 18.6 sec (n = 63), and a pair copulated about 7.2 times per 12-hr day (n = 4) during the week prior to clutch initiation. These values are very variable, but allow one to predict that 80 pairs should have had 578 copulations or 0.39 copulations/half-minute. At an average of 18.6 sec, there should have been 7.3 sec out of every 30 sec when copulation was occurring. Given a probability of 0.39 copulations per half-minute, the expected values for 2, 3, or 4 copulations/half-minute are 0.15. 0.06, and 0.02.

The results are shown in Table VI. The statistical tests were based on the average number of events that would have been expected if they were uniformly distributed in time. An excess of intervals with zero copulations and with two or more copulations was found. For the six samples with half-minute intervals, five yielded results consistent with temporal contagion, and all three samples using 2-min intervals provided the same result.

A more detailed study of spatiotemporal clustering was undertaken. I focused attention on the interval between one precopulatory circling display and the next, and between one copulation and the next, and also between

Table VI. Contagious Copulation Behavior among Common Terns[a]

| Interval length[b] | Total time[b] | No. of copulations observed | | No. of intervals with copulations | | | | χ^2 | df | p |
		Total	Mean/min	None	One	Two	Three or more			
0.5	15	13	0.86	23	2	3	1	9.5	2	<0.01
0.5	15	15	1.00	22	2	5	1	10.4	2	<0.01
0.5	15	43	2.86	14	5	4	7	10.7	4	<0.05
0.5	15	39	2.60	13	6	5	6	6.1	3	$0.10 < p > 0.05$
0.5	15	21	1.40	18	7	1	3	1.8	3	>0.10
0.5	15	14	0.94	23	3	1	3	9.0	2	<0.025
2.0	32	122[b]	3.81	7	2	3	15	16.0	3	<0.001
2.0	32	90[c]	2.81	11	5	4	10	8.5	3	<0.05
2.0	32	38	1.18	14	6	2	6	8.0	3	<0.05

[a] Ten observation periods of 15- or 32-min length with either ½-min or 2-min intervals were used to determine whether copulation with wing flagging (see text) tended to be temporarily clustered.

[b] Events used are precopulatory circling displays [see Palmer (1941)].

[c] Events used are mounts prior to actual copulation; abortive mounts are included. In all other samples events are copulatory wing-flagging displays.

one copulation and the next circling display. On a real-time base I recorded the intervals between displays and recorded the distance between the pairs involved.

I grouped distanced into 4-m intervals and time into 1 min. intervals. Table VII summarizes 1973 data, excluding displays where no other display followed within 4 min (17% of total). For the three types of interaction—(1) circle to circle, (2) copulation to circle, and (3) copulation to copulation—I cast data into time × distance contingency tables analyzed for independence. Because of small sample sizes, distances of 8–20 m were grouped for (1) and for (3). The hypotheses were that (1) events would be clustered in time, (2) events close in time would be close in space, and (3) copulation would be stimulating over a greater distance than circling. The opposite null hypotheses were tested.

Table VII. Temporal and Spatial Clustering of Displays by Common Terns[a,b]

Distance	1 min	2 min	3 min	4 min
Circle–circle[c]				
0–4 m	21	23	12	2
4–8 m	15	9	9	6
8–20 m	8	12	15	11
G statistic	112**	102**	39**	1.7**
Copulation–circle[d]				
0–4 m	12	9	3	0
4–8 m	15	6	4	1
8–12 m	16	4	6	1
12–16 m	8	2	5	0
16–20 m	3	4	0	0
G statistic	55**	35**	20**	1
Copulation–copulation[e]				
0–4 m	9	1	5	3
4–8 m	12	4	5	2
8–20 m	18	11	8	8
G statistic	32**	2	18**	7*

[a] Data are number of sequences occurring in certain time intervals (by minute) and at certain distances (by meters). The first event (a circle or a copulation) is at time zero and position zero. The time to the next event (either a circle or copulation) and the distance between the two pairs involved provide the data points.
[b] *Asterisks*: * $p < 0.05$; ** $p < 0.01$.
[c] Contingency table: $\chi^2 = 20.6$; $df = 6$; $p < 0.0025$.
[d] Contingency table: $\chi^2 = 10.5$; $df = 8$; $p > 0.10$.
[e] Contingency table: $\chi^2 = 4.6$; $df = 6$; $p = 0.60$.

The circle–circle data show significant dependence with a strong excess of circles occurring within 1 min and within 4 m (χ^2 = 20.6; $p <$ 0.0025). For (2) there is again an excess of close and early circles following a copulation, but overall the contingency table does not reject independence ($p > 0.10$), and the same is true for copulations following a copulation. However, for each minute column it is possible to determine whether the responding birds are randomly grouped with respect to distance from the initial displaying pair. Assuming that pairs were uniformly distributed over the study area, the expected values can be generated by the available area in each concentric ring. Using this approximation and a G test for goodness of fit, I found for (1) a heavy preponderance of close displays in the 1st, 2nd, and 3rd minutes ($p < 0.001$), but not in the 4th. For (2) there was a preponderance of near-responses at all times, whereas for (3) the near-displays occurred mainly in the 1st minute. The general reduction in G scores (Table VII) in successive minutes indicates that as more and more time elapsed, the displays were less likely to be close to the initial display.

These findings and the study by Southern (1974) indicate that breeding displays of larids can be contagious, a medium through which social stimulation could play a synchronizing role. It is important, however, to recognize that extrinsic factors may be important. Palmer (1941) and others have remarked that reproductive behavior is depressed under adverse weather conditions, and therefore a resurgence of such displays when weather improves could be misinterpreted as caused primarily by social stimulation. It is important to recognize that even in the absence of social interactions among pairs, the within-pair stimulation may suffice to induce spontaneous displays when one's neighbors are quiescent.

VII. SYNCHRONY IN BLACK SKIMMERS

The Black Skimmer breeds in dense aggregations in open habitats, nesting more compactly than necessitated by available space. Here I examine two aspects of synchrony in skimmers—(1) the timing of nesting and productivity, and (2) subcolony synchrony. The latter topic is discussed in more detail elsewhere (Gochfeld, 1979).

A. Timing of Nesting and Productivity in Black Skimmers

Data from the 1976 season have been analyzed to determine whether skimmers nesting at the peak of the season have better hatching and/or fledging success than do those nesting before or after the peak. Patterson

(1965) demonstrated such a relation for Black-headed Gulls. Numerous studies have shown that early nesting (often older birds) are likely to have larger clutches and higher productivity. Therefore it has been necessary to make comparisons with the time of the season as well as with numbers of nests.

Skimmer nests were surveyed at irregular intervals, and the number of eggs laid and clutches initiated were noted based on intervals up to 1 week early in the season and longer intervals later. More attention was paid to hatching and fledging success. The data are shown in Table VIII. It was anticipated that hatching and fledging success would decline steadily through the season, but for neither colony was there a significant decline in hatching success. At Cedar Beach, fledging success was significantly lower late in the season (r_s = +0.95, p < 0.01). The fledging success (but not hatching success) at West End Beach, New York, correlated significantly with the timing of nesting (peak nests fared better, r_s = 1.0, p < 0.001).

These data demonstrate that certain productivity measures correlate with the temporal distribution of nesting. It is often difficult to obtain percise fledging information, although it is usually simple to document hatching success. Although a higher fledging success for peak birds is directly consistent with the Darling effect, the higher hatching success requires a separate explanation. Darling (1938) did not find significant differences in hatching success among larger as opposed to smaller colonies. One can invoke Austin's (1940) observation that in large colonies social behavior engenders superior incubation performance, or Kruuk's (1964) conclusions that larger and denser aggregations are more successful at eliminating predators. The former is probably more relevant to the Black Skimmers, for which abandonment rather than predation accounts for most hatching failures.

B. Subcolony Synchrony in Black Skimmers

In the northeastern United States, Black Skimmers appear to require Common Tern colonies in which to nest (Burger and Lesser, 1978; Gochfeld, 1978a). Within large terneries the skimmers may form discrete clusters or subcolonies. If the birds are synchronized by social factors rather than by extrinsic climatic factors, one would expect that the subcolonies would show greater synchrony than would the colony as a whole, much as Hailman (1964) found for Swallow-tailed Gulls. Data on clutch initiation dates were obtained for skimmers at Cedar Beach on Long Island, New York (Gochfeld, 1976). In 1976 there were nine discrete subcolonies, whereas in 1977 13 were found. These subcolonies ranged from 5 to 36 nests. For each subcolony the SD for all nests was computed. In 1976 the colony SD was higher than all subcolony SDs, whereas in 1977 it was exceeded by only one

Table VIII. Relative Breeding Success of Peak, Prepeak, and Postpeak Black Skimmers on Long Island[a]

Dates[b]	Clutches initiated per day[c]	Eggs laid in interval	% hatched[d]	Young fledged per pair[e]	Comparison of measures[f]	r_s	p
			West End Beach				
May 27–June 3	6.37	175	84	0.8	b vs. c	0.65	NS
June 4–June 5	13.0	86	78	1.1	b vs. d	0.47	NS
June 6–June 12	9.7	222	90	0.9	b vs. e	0.65	NS
June 13–June 21	7.8	190	87	0.6	c vs. d	0.71	NS
June 22–June 27	2.2	31	64	0.3	c vs. e	1.0	<0.01
after June 27	1.4	40	55	0.2	d vs. e	0.71	NS
			Cedar Beach				
May 19–May 25	2.1	53	53	1.3	b vs. c	0.57	NS
May 26–May 29	7.0	103	86	1.4	b vs. d	0.24	NS
May 30–June 3	10.7	229	82	1.1	b vs. e	0.94	<0.01
June 4–June 5	7.5	46	74	1.2	c vs. d	0.65	<0.05
June 6–June 12	5.4	106	53	0.7	c vs. e	0.58	NS
June 13–June 14	2.5	15	67	0.6	d vs. e	0.37	NS
June 15–June 30	0.4	14	64	0.6			
July 1–July 21	0.5	27	56	0.2			

[a] Data for two Long Island, N.Y. colonies in 1976.
[b] Grouping of dates determined by field work schedule in each colony.
[c] Clutches initiated per day allows a fixed-time comparison (see Methods section).
[d] % of eggs laid during this interval that subsequently hatched.
[e] Young fledged per pair which initiated clutches in this interval.
[f] Spearman rank correlations among measures for each interval (b, date; c, clutches initiated; d, % hatched; e, young fledged per pair).

of the 13 subcolony SD's (see Table IX). Using a Bartlett's Test for Homogeneity of Variances demonstrated that for both years there were significant differences among the variances, indicating that some subcolonies were significantly more synchronous than others.

To ascertain whether the observed synchrony might have arisen at random, I tested the hypothesis that the clutch initiation dates for the 2 years were randomly distributed over the subcolonies. I simulated subcolonies by randomly assigning initiation dates to each subcolony. The dates were selected at random from the actual dates observed in the colony as a whole. Thus the computer essentially reshuffled dates at random. Acutally, 1000 simulates were constructed and for each simulation the colony SD was compared to the subcolony SDs. The colony value fell close to the median of the subcolony values. This was in marked contrast to the observed situation. In none of the 1000 simulates for 1976 did the colony value exceed 7 of the subcolony values; hence the observed rank of 1st would have arisen with a probability much lower than $p = 0.001$. The same was true for 1977. More detailed discussion of the data appears in Gochfeld (1979).

The main outcome of the study is that the observed synchrony in the subcolonies was not a random event. It sheds no light on how the synchrony was achieved and provides no basis for separating social attraction, social segregation, and social stimulation. Because the skimmers form large, dense flocks on the colony prior to nesting, it is quite likely that in this species, social stimulation occurs prior to establishment of territory, and social segregation of birds ready to breed plays an important role in the formation of subcolonies. However, the large spread of nesting dates in some large subcolonies suggests that social attraction of late-nesting skimmers is also important.

VIII. DISCUSSION OF FACTORS INFLUENCING SYNCHRONY

A. Spring Arrival and the Prelaying Period

Although most studies of synchrony have focused on dates of clutch initiation, the factors of greatest interest happen earlier in the breeding cycle, prior to laying. The following discussion presupposes that social stimulation is important.

1. Arrival

Arrival on territory involves several more or less discrete steps: (1) birds return from wintering quarters to the vicinity of the colony, (2) birds gather on or adjacent to the colony, often in flocks on loafing areas, and are present for only part of the day, (3) birds begin to occupy territories, (4)

Table IX. Standard Deviation of Laying Dates for Black Skimmer Subcolonies at Cedar Beach, New York

Year	Range of subcolony SDs	Colony SD	Rank of colony SD[a]	Subcolonies
1976	2.05–6.42	6.49	1st	9
1977	1.62–6.92	6.07	2nd	13

[a] The SD for the colony is compared to the entire vector of subcolony SDs and is given its respective rank. Thus in 1976 the colony SD was higher than all nine of the subcolony SDs, hence it ranked 1st. In 1977 the colony SD was higher than 12 of the 13 subcolony SDs, hence it ranked 2nd.

bulk of birds occupy territories, and (5) late birds fill in between existing territories.

It is not always clear whether a paper refers to arrival in the area, at the colony, or on territory. Darling (1938) noted that Priest Island Herring Gulls returned gradually over a period of weeks. On Long Island, New York, Herring Gulls may remain near the colony throughout the year (D. B. Ford, personal communication). The Northern Gannet arrives gradually at Bass Rock, but synchronously at Bonaventure Island, and the laying period may be 75 days at the former but is contracted at the latter into 45 days (Nelson, 1975).

Some species may have a fixed time requirement once on territory. For example, Baggerman *et al.* (1956) report prelaying periods of about 22 days, regardless of whether Black Terns (*Chlidonias niger*) arrived April 23 or a week later. Spurr (1975) found that in Adelie Penguins (*Pygoscelis adeliae*) laying date was not delayed in the year when their arrival had been delayed by a month as a result of ice. Burger and Shisler (1979) found that Herring Gulls arriving on Clam Island, New Jersey, in early March required 26 days to start laying, as compared with 2–3 days for birds arriving 2 months later. The latter presumably spent much time in "clubs," wherein they were exposed to social stimulation. Had they settled on territories in early May, they would have been surrounded by birds that were incubating or caring for young and would not have received stimulation from displays appropriate to the prelaying period (J. Burger, personal communication).

Vermeer (1970) reported geographic variation in duration of prelaying periods for California Gulls (*L. californicus*) (6–7 weeks in the United States as compared with 3–4 weeks in Canada), whereas Ring-billed Gulls arrived 4–5 weeks prior to laying in both areas. Initial occupation of the colony may be delayed by ice, or snow, or cold temperatures (e.g., Vermeer 1970; Paludan, 1951).

Langham (1974) found that Sandwich Terns had a prelaying period on territory of about 2 days as against Roseate (*S. dougallii*), Common, and Arctic Terns in the same colony. For the latter, social stimulation can be experienced on the breeding territories during the prelaying period, whereas for Sandwich Terns, stimulation, if relevant, occurs in prebreeding flocks. Birds ready to breed then form a subcolony together (termed social segregation: Langham, 1974; Veen, 1977).

For the most part, however, data on timing and frequency of events prior to laying are scant. It is important to consider the sequence of events in the prelaying period.

2. Prelaying Period

For a colony the prelaying period is the time between the arrival of the first birds (for a migratory species) and the laying of the first egg. For a pair it is more useful to determine the interval between territory establishment and laying. Factors influencing time of arrival (e.g., weather, snow conditions) on the one hand, and timing of laying on the other, determine the length of this period. Although it is relatively easy to determine when the first birds arrive at a colony site (if one is able to be present at the colony early in the season), it is sometimes difficult to determine when a territory is established. In Common Terns I have found (see also Palmer, 1941) that birds may be present on their future territory for only a small part of the day, whereas in some cases a territory may be used early in courtship and then apparently forsaken. Because field workers are often on a tight time schedule the early part of the breeding season is frequently ignored, and most studies of colonial birds begin early in the egg-laying period.

Unfortunately the events occurring in the prelaying period are of considerable importance in studying social factors, synchrony, and the Darling effect. Social attraction (Hailman, 1964)—the tendency of birds in breeding condition to be attracted to breeding birds—influences the settlement pattern, whereas social stimulation leading to reproductive acceleration would influence laying. During this period birds are exposed to neighbors undergoing pair formation and/or pair maintenance with abundant visual and acoustic displays.

At this stage it is usually difficult to recognize individuals (except in long-term studies), for birds are usually not yet color marked. In certain species, such as some gulls (Burger and Shisler, 1980), nest construction begins early after territory establishment, whereas in other species, such as terns, nest material is only accumulated during incubation (Palmer, 1941).

In studying the relation between density and synchrony, most authors have used density or nearest-neighbor distances between established nests, even though the events of greatest interest occurred at an earlier phase when density may be quite different. For example, many of the later-settling birds whose nests are included in the density measurements may not have been present at all at the peak of courtship and laying. Recognizing this factor, Burger and Shisler (1980) measured nearest-neighbor distances at the time that scrapes were being made in their Herring Gull colonies.

3. Epicenters and Filling In

Many studies do not focus on an entire colony, but select particular subcolonies, arbitrary study sites, or quadrats. However, colonies are clearly not homogeneous with respect to terrain, age, composition, and other factors (e.g., Coulson, 1968; Coulson and White, 1960), and one must understand the structure of the colony in terms of both temporal and spatial distribution of nesting. Patterson (1965) undertook such a study. Burger and Shisler (1980) describe in detail the growth of Herring Gull colonies from certain nuclei of early-nesting birds. Depending on both colony size and traditional factors, birds may form several nuclei and the fill in between them or may simply fill in around a single nucleus. The first nests in a colony are likely to comprise small clusters of old experienced breeders, which may form epicenters in preferred habitats. As the birds occupying these traditional sites die off, the epicenters may shift or break down, as I have found recently for old Common Tern colonies.

The importance of filling in by late arriving birds through social attraction is usually not recognized. It could, for example, account for the later median laying dates, which MacRoberts and MacRoberts (1972b) found in some of their study plots. The importance of the epicenters is that birds settling close to a center are exposed to greater stimulation than are birds nesting further away. Nelson (1976) emphasized the importance of such display nuclei for frigatebirds.

B. Synchronizing and Desynchronizing Factors

1. Synchronizing Factors

Table X summarizes factors that may in the long term or in any given year increase or decrease the degree of synchrony. Several aspects of breeding are influenced by endogenous and exogenous annual cycles of birds and their environments (see Immelmann, 1971). The annual cycle, entrained by photoperiodic influences, may be significantly modified by such subsidiary factors as weather, food availability, and social interactions. Latitudinal breeding differences relate to consistent trends in food availability or snow cover, which allow birds to breed earlier at lower latitudes. Sealy (1975) reported that late snow melting delayed and synchronized auklet nesting. Burger (1974a) found that Franklin's Gull (*L. pipixican*) breeding depended on food availability, which in turn depended on the disappearance of snow and the resurgence of cultivation. Late breakup of pack ice delayed arrival, but not egg laying, of Adelie Penguins, but the birds compensated for the shortened prelaying period by laying smaller clutches (Spurr, 1975). Onset of dry-season trade winds facilitates foraging flights of frigatebirds (Diamond, 1975) and synchronizes their nesting.

Weather factors may also act in the short term to influence nesting and egg laying. Although Paludan (1951) found no relation between egg laying and temperature, Noll (1931) found a positive relation. Palmer (1941) reported depression of courtship and copulation during foul weather, whereas Burger (1979) reports that Herring Gull nest construction decreased during rainy weather with a rebound increase the following day.

Flooding may function proximately to eliminate certain cohorts or groups of nests, promoting renesting or emigration. It may also function

Table X. Synchronizing and Desynchronizing Factors for Seabirds

Synchronizing effect	Factor	Desynchronizing effect
Endogenous cycles of birds and their environment bring them into breeding condition at a particular season (Immelmann, 1971).	Annual cycle	Cycles tend to break down in some tropical species.
Cycles are adapted to peak of food availability for egg formation or for feeding young (Perrins, 1970; Veen, 1977).	Food availability	Prolonged availability of renewable food resources favors a protracted, asynchronous nesting season.
Delay nesting by covering breeding sites (Sealy, 1975); delaying arrival (Spurr, 1975) or preventing access to food (Burger, 1974a).	Snow, ice, and flooding	Hightide cycles vary from year to year selecting for different cohorts at different seasons (J. Burger, personal communication).
Adverse weather may delay breeding with a resurgence of synchronous activity when fair weather appears.	Weather	Continued adverse weather may result in a slow "breakthrough" effect, with few nests initiated each day, and an asynchronous season.
Social segregation of birds in breeding condition (Gochfeld, 1979; Burger, 1974a; Veen, 1977) and social attraction of late birds by breeders (Hailman, 1964). Social faciliation of displays synchronizes neighbors (Darling, 1938).	Social interactions	Continous social attraction of late-arriving birds results in a long-protracted and apparently asynchronous nesting season.
Predator may select a particular cohort (Nisbet, 1975) or may eliminate most birds outside of the peak period (Darling, 1938). Antipredator mobbing may select for synchrony (Kruuk, 1964). If predator eliminates a breeding unit, renesting may be highly synchronous.	Predation	Predation may act at different times in different seasons.

ultimately by influencing habitat selection or timing of breeding. Because tide cycles vary differently from annual cycles, it is not feasible for a species to adapt its breeding schedule to avoid tidal flooding (J. Burger, personal communication).

Food availability during the egg-laying period is important and in certain species seems to exert a strong effect on timing (Perrins, 1970). Veen (1977) mentions that Sandwich Tern breeding may have been triggered by the sudden pulse of available herrings. Ultimately the timing of breeding may be adjusted so that hatching coincides with peak food availability. If there is a short peak of food available, then synchrony will be favored. If food is available over a long time and is renewable, synchrony may be disadvantageous.

In view of the important role that human disturbance plays in seabird colonies, factors that increase disturbance, such as holidays, may have an important effect. Humans not only damage seabirds directly, but facilitate predation (e.g., Kury and Gochfeld, 1975) and mortality (Gochfeld, 1973) indirectly. I have found that Least Terns, which lay early, may fledge their young prior to the 4th of July weekend, which brings hordes of beachgoers into their colonies. Later young may be captured or inadvertently chased far from their parents.

2. Desynchronizing Mechanisms

Just as selection can favor synchrony, certain factors can select for asynchrony. Such effects are apparent in the literature on asynchronous hatching (e.g., Parsons, 1972; Nisbet and Cohen, 1975), a topic that is not otherwise germane to the present discussion of synchrony. Desynchronizing factors may be evolutionary, operating as centrifugal or destabilizing selection and resulting in a large variance in the timing of reproductive cycles, or they may be ecologic, the result of proximate factors that influence different segments of a population differently, resulting in a broad time span over which reproduction is carried out.

Unpredictable Climate. If early breeders are selected against in some years and favored in others, or if peak breeding is favored some years but not others, there will remain a substantial variance in laying dates. One may prefer to view the reverse side of the coin, i.e., failure of centripetal selection (increased peak) or directional selection (earlier breeding) to operate, but it is reasonable to think in terms of a population adapted to unpredictable, harsh seasonal environments.

Recruitable Predators. If a large crop of eligible and edible young were to attract predators leading to a sudden increase in predation pressure, birds hatching before the peak (even though relatively few in number) might be less vulnerable than those hatching at the peak. If, as the supply of chicks declines (either through predation, maturation, or fewer eggs hatching) predators shift to another source of food, late-hatching chicks may also experience reduced predation pressure. Recruitable predators are quite different from those implicated in the Darling effect, which clearly implies a constant predation rate (see Fig. 6A).

Food Competition. Birds hatching at the peak may experience high food competition, and unless this is balanced by some other advantage (e.g., Emlen and Demong, 1975), it may be advantageous to spread hatching over a longer time period. This is particularly true if food is renewable and available for a long period, rather than being concentrated in a brief pulse.

C. Geographic Extent of Synchrony

Synchrony may occur at different spatial levels. At the widest level, birds breeding in the high Arctic show continentwide synchrony. They must wait for the disappearance of winter conditions and then complete reproductive activities before the arrival of adverse weather. Synchrony may occur on a regional level, but particularly a group of colonies in close proximity may show synchrony (e.g., Common Terns on Long Island: Gochfeld, 1976). By contrast, colonies close together may be asynchronous (e.g., Hailman, 1964; Burger, 1974b). Several studies have demonstrated that even very large colonies may have a high degree of synchrony throughout. An impressive example is the 375,000-pair Sooty Tern (S. fuscata) colony on Bird Island, Seychelles, wherein 75% of nests were initiated in a 9-day period (Feare, 1976). Many authors have reported that subcolonies, clusters, or neighborhoods may be synchronous, whereas the colony as a whole is not (see Table XI). The effect may be that several small, synchronous units, produce an overall normal distribution of laying [M. Gochfeld, unpublished data on Common Terns; T. R. Birkhead, personal communication for Common Murres (Uria aalge)].

Local climatic factors would be expected to produce synchrony in the given locality, but some cases cannot be so explained (e.g., Hailman, 1964; Ashmole, 1962), particularly in the tropics, so that social factors must be invoked. Considering that neighborhoods and subcolonies are defined in terms of communication and available social stimulation, one would expect synchrony to be apparent at these levels.

D. Social Factors and Synchrony

Social factors are of primary concern in the study of synchrony, even though they may not be the most obvious or important factors. It is well established that the social interactions between mates play crucial roles in bringing them into reproductive readiness and synchronizing their behavioral and physiologic cycles (Lehrman, 1965). Relevant social interactions may be classified as (1) social attraction, (2) social segregation, and (3) social stimulation. Hailman (1964) reported that Swallow-tailed Gulls in breeding condition were attracted to birds that had just begun breeding. He considered this "social attraction," and implicated it as a factor that

Table XI. Studies Reporting Synchrony on the Subcolony Level

Scientific name	Common name	Reference
Pygoscelis adeliae	Adelie Penguin	Spurr (1975)
Pelecanus erythrorhynchus	American White Pelican	Schaller (1964)
Fregata minor	Great Frigatebird	Nelson (1976)
Sula spp.	Boobies	Nelson (1975)
Diomedea irrorata	Galapagos Albatross	Harris (1973)
Larus argentatus	Herring Gull	Burger and Shisler (1980)
		Parsons (1976)
		Paynter (1949)
		Goethe (1937)
Larus fuscus	Lesser Black-backed Gull	Davis and Dunn (1976)
Larus pipixican	Franklin's Gull	Burger (1974a)
Creagrus furcatus	Swallow-tailed Gull	Hailman (1964)
Rissa tridactyla	Black-legged Kittiwake	Coulson and White (1960)
Larus ridibundus	Black-headed Gull	Patterson (1965)
Sterna sandvicensis	Sandwich Tern	Veen (1977)
Sterna fuscata	Sooty Tern	Feare (1976)
Sterna hirundo	Common Tern	Gochfeld (unpublished data)
Sterna dougallii	Roseate Tern	Gochfeld (unpublished data)
Anous tenuirostris	Black Noddy	Ashmole (1962)
Rynchops niger	Black Skimmer	Gochfeld (1979)

accounted for synchrony within colonies, whereas there was little synchrony between colonies. I believe that social attraction is of overwhelming importance in seabird breeding biology for a least two reasons: (1) young birds breeding for the first time may learn of appropriate breeding sites if they are attracted to an area where conspecifics are nesting; (2) birds that have failed at a nesting site may recognize more suitable sites by the presence of apparently successful breeders. The social attractiveness of a colony may continue over a period of months so that it will continually attract new breeders. Thus the very social factor that might synchronize new arrivals will lead to a prolongation of the breeding season for the colony as a whole and thereby to a reduction in the apparent synchrony. Thus methods must be employed that will allow one to separate these conflicting features.

I define "social segregation" as the settlement together of a group of birds in a like breeding condition. I suggest (Gochfeld, 1979) that from a large prebreeding flock of Black Skimmers, a group of birds that are ready to breed might segregate itself and form a synchronous subcolony. Several days later another comparable group might form another subcolony. This has only been apparent in certain seasons and is clearly not happening in other years. A comparable phenomenon has been suggested for Franklin's Gulls (Burger, 1974a) and Sandwich Terns (Veen, 1977).

Social Stimulation and social facilitation involve the direct effect of behavioral interactions on the breeding of neighbors. Social stimulation is not only important within a mated pair, but in territorial and colonial species as well. Darling (1952) suggested that a major function of territorial behavior is to provide social stimulation for the participants. Demonstration of social facilitation (see Tolman, 1964) involves demonstrating that when one bird performs a particular behavior, there is an increased likelihood that neighboring birds will display the same behavior, and that this tendency is influenced by the behavior, and that this tendency is influenced by the behavior of the first bird and not by some outside event, such as the approach of a predator (see above).

Social stimulation is an more general term, and in the context of the Darling effect it means that birds performing any of several reproductive-type displays will stimulate nearby birds to perform similar displays. Southern (1974) showed that copulatory wing flagging by Ring-billed Gulls leads to a significant increase in the amount of breeding-type activities and a concomitant reduction in maintenance activities (see Table V). In this chapter I provide comparable evidence for contagiousness of copulation in Common Terns (see above).

One must distinguish proximate synchronizing factors, which I would consider social interactions, short-term weather conditions, and immediate food availability, from ultimate factors, such as the annual cycle and the selective effect of food availability on the timing of the reproductive cycle. I use these terms differently from the way Immelmann (1971) uses them.

Nelson (1970, p. 521) distinguishes synchrony from seasonality. Seasonality is imposed on temperate species by annual changes in climate and food availability, and secondarily imposes a certain synchrony on temperate populations. This synchronized effect increases with latitude. In a more constant environment one may have high synchrony (imposed by pulses of food availability) or nearly aseasonal breeding. This is considered in more detail below.

1. Density and Synchrony

Social stimulation will be greater in dense areas than in sparse areas. Patterson (1965) found a relation between density and success and believed it involved the greater synchrony in dense areas. Burger (1974a) found that visibility was a key factor influencing nest spacing in Franklin's Gulls and in other gull species as well (Burger, 1977). Colonial birds apparently pack themselves more closely together than is dictated by available space. This increases the frequency of aggressive and other interuptive interactions, and beyond a certain point it facilitates predation as well (cf. Tinbergen *et al.*, 1967). The advantages of dense nesting include antipredator behavior and perhaps enhanced social stimulation. Kruuk (1964) emphasizes enhanced antipredator mobbing by Black-headed Gulls in denser areas.

One must expect density to be a compromise between a bird's tolerance of other birds and its willingness to chase them off over a long distance. As a season progresses, either this willingness changes, or the persistence of intruders may increase, so that density changes in the course of a season (e.g., Coulson and White, 1960; Burger and Shisler, (1980).

Because the distance over which social stimulation can be effective is finite, one would expect greater social stimulation in a dense assemblage than in a sparse one. Similarly, the spatial extent of synchrony can be variable.

2. Synchrony and Interference

Next to competition for space or food, physical interference in reproductive activities is the most frequently mentioned disadvantage of coloniality. In dense-nesting assemblages aggression from neighboring birds frequently interrupts the routines of courtship, copulation, incubation, and feeding of young. Interference during copulation is particularly common. In gull colonies cannibalism by neighbors or specialists is frequently mentioned (e.g., Parsons, 1975b), whereas in tern colonies food kleptoparasitism may be important.

Herring Gulls out of phase with others may be important predators (J. Burger, personal communication). Nelson (1976) found that late males displaying in the dense nesting assemblages that form around display sites, are a frequent cause of nest, egg, and chick destruction in frigatebirds.

The role of synchrony and interference merits additional study. It is not at all intuitive that asynchrony would enhance interference with copulation, for example. One might imagine that a neighbor ready to copulate might be more likely to interrupt an incipient copulation than would an incubating neighbor. In that case the advantage might lie in simply being later than one's neighbors, an advantage offset by the reduced productivity of late breeders.

E. Synchrony and Predation

Predation is a universal problem affecting seabird colonies, and avoidance of predator-prone areas is a prime factor in nest-site selection. Food exploitation and predation are often presented as the two major factors that shape the evolution of coloniality. Predation can operate by selecting for safe colony sites or by promoting antipredator behavior. The Darling effect hypothesizes an added benefit of predator satiation, whereby a large pulse of chicks may be more than a predator can handle, ensuring that some will survive.

1. Social Aspects of Antipredator Aggression

Kruuk (1964) provides an oft-cited report on Black-headed Gull defense against predators. Different predators elicited different combinations of attach, fleeing, and other behavioral elements. He found that gulls, crows, mustelids, and hedgehogs were deterred by group mobbing, whereas foxes attacking on dark nights were not. He found effectiveness of mobbing closely related to density.

Williamson (1949) found that colonial Parasitic Jaegers (*Stercorarius parasiticus*) vigorously attacked human intruders, whereas solitary ones did not; he attributed this difference to social stimulation. Lemmetyinen (1971) found that solitary and colonial Common and Arctic terns were equal in frequency of attack, and effectively deterred predators. He provides an excellent discussion and literature review (Lemmetyinen, 1971) on the role of colony protection by aggression, which was criticized by von Haartman (1945) and favored by Kruuk (1964), Tinbergen *et al.* (1967), and Lack (1968). For an overall review of the behavioral factors influencing predation see Curio (1976).

The lack of aggressiveness of solitary jaegers might be the result of fear or use of other protective devices, such as distraction displays associated with cryptic nests. The aggressiveness of isolated terns, which Lemmetyinen (1971) suggested might be attributable to their general lack of opportunity to release aggression, may be atypical in that the colonies he worked in were relatively sparse. Austin (1946) did find that small Common Tern colonies have lower aggression rates than large ones, whereas after population crashes even large aggressive colonies show a diminution in attack frequency (Austin, 1946).

For most species, detailed comparisons of aggression rates among areas of differing density or synchrony remain to be done. Several authors (e.g., Lemmetyinen, 1971; Drent, 1967) have noted an increase in antipredator aggression during incubation, with a peak occurring just prior to hatching. In a synchronous colony, with more birds at the same stage with respect to hatching, the mobbing effect can be enhanced, whereas earlier or late hatching chicks will not benefit from the aggressive protection.

2. Predator Satiation: Myth or Reality

The key to the Darling effect involves the saturation of a predator by a synchronous hatch of young. Darling (1938) emphasized that a predator taking a relatively constant number of young chicks would take a higher proprotion of young prior to and after the hatching peak. At the peak there would be more vulnerable young than the predator could handle. There is an important assumption that the predator is nonrecruitable, for the Darling effect can not operate if additional predators are attracted by the abundance of young at the peak. Nisbet (1975) provided an example of this, noting that a Great-horned Owl took a relatively constant biomass of young Common Tern chicks, removing a decreasing number of birds each night as more larger chicks became available. The biomass taken remained constant despite a 100-fold increase in available biomass

Aside from Nisbet's report, accounts of predator satiation include feral cats preying on Sooty Terns (Ashmole, 1963), Carrion Crows on Black-headed Gull chicks (Patterson, 1965), Herring Gulls cannibalizing conspecifics (Parsons, 1971), and Short-eared Owls (*Asio flammeus*) on young Swallow-tailed Gulls (Snow and Snow, 1968). Williams (1975) reported that synchronous fledging of young murres swamped their gull predators.

Examples from other groups of birds are noteworthy. Elgood and Ward (1963) noted that a snake might prey continuously on weaverbirds in a single colony, and if hatching were synchronous this single, nonrecruitable predator would be swamped. The synchronous roost departures of birds and bats and synchronized calving in ungulates and marine mammals are further examples of predator satiation (Wilson, 1975). Synchronous metamorphosis of anurans appears to reduce snake predation during this vulnerable period (Arnold and Wassersug, 1978). Even more dramatic examples include the synchronous emergences of insects, particularly the periodic cicadas and mast fruiting of many tropical tree species, which appears to satiate their insect predators (e.g., Janzen, 1971).

It is apparent that predator satiation can occur for many organisms, but seabirds do not provide particularly good examples. The roles of synchrony influencing predation, and *vice versa*, are thus more complicated than Darling (1938) envisioned, and factors such as aggressive defense are often more important and conspicuous than is predator satiation.

F. Confounding Variables

Up to this point it has been apparent that studies of timing and synchrony involve considerations of colony size, habitat, density, predation, and other factors. The central theme of reproductive adaptations of which synchrony may be one, is that productivity must be maximized, at least over the breeding lifetime. Demonstrating increased productivity is the central problem. Patterson (1965) found that fledging success was higher for peak gulls than for early or late-nesting birds. Table XII reviews evidence for enhanced success of peak birds. This may involve advantages resulting from synchrony or from extrinsic factors (i.e., weather, predation, food), which

Table XII. Advantages of Nesting during the Peak Period

Species	Mechanism	Reference
Adelie Penguin	Improved incubation and decreased desertion	Spurr (1975)
Herring Gull	Reduced cannibalism	Parsons (1976)
Herring Gull	Higher hatching due to adverse conditions early in season	Parsons (1975)
Herring Gull	Hatching success increased through season regardless of peak	Harris (1964)
Black-headed Gull	Reduced predation	Patterson (1965)
Sandwich Tern	Larger clutch and reduced chick and egg loss	Veen (1977)
Sooty Tern	Reduced pecking and predation	Feare (1976)
Bank Swallows	Enhances learning of feeding areas	Emlen and Demong (1975)

favor nesting by the majority of birds at that time. It will be difficult to distinguish cause from effect.

Success is influenced by age of the male and female, experience of the individual and its partner, previous experience with the same partner, size of eggs and clutch, incubation quality, position of nest, and ability to obtain food and protect against predators. Any or all of these may depend on the ability or propensity of certain individuals to respond to social stimulation of neighbors.

During the past two decades researchers have begun to consider the interplay of multiple factors influencing productivity. Productivity is highly variable for any species (e.g., Austin, Jr., 1933) between colonies and years. The interactions of the above mentioned factors may be complex, and the kinds of interactions may vary from species to species and from place to place. The combinations of interactions are astronomical in number, and the effects on productivity may be difficult to assess.

A review of the literature on the interaction of such factors is beyond the scope of this chapter. Suffice it to say that interactions involving density and earliness seem to be of primary importance. Coulson (1968) has stressed the differences between central and edge birds and Patterson (1965) and Tenaza (1971) showed that edge birds fared worse with respect to predation. However, the nature of the edge must vary from colony to colony, and it is not apparent that the geographic center of a colony necessarily behaves as a biological center in the sense of Coulson (1968). Montevecchi (1977) and Burger and Lesser (1978) found heavier predation in the center than on the edges of salt marsh gull and tern colonies.

A large number of studies have shown that older, more experienced breeders are more productive, and this criticism was leveled at Darling (1938). It was suggested that colony differences were caused not primarily by size but by the fact that small colonies usually were predominantly comprised of young birds, hence would be predicted to have low productivity.

It is important to realize that complex interactions influence timing and synchrony and productivity. It is hardly likely that any study could consider even a majority of the influential factors, but studies can be designed to hold some of the factors constant.

G. Synchrony in Tropical Seabirds

This chapter cannot consider tropical seabird seasons in detail. The main annual control of seasonality is lacking in the tropics, but with respect to oceanic currents, trade winds, food availability, and precipitation they are far from aseasonal. For many species breeding seasons are readily detectable, whereas some species breed with less than annual periodicities. Schreiber and Ashmole (1970) describe breeding seasons for 18 species on Christmas Island. The data for most species are not quantitative, but for almost all at least one distinct peak is noted. The data may be summarized as follows: four species with one short laying period, five species with a long laying season, four species that lay in all months and show a single peak; four species that lay in all months and show two peaks, and one species with two short laying seasons 6 months apart (Sooty Tern). The latter species is unique in showing highly synchronous breeding peaks often without an annual cycle. The cycle may be 6 months on Christmas Island and 9.7 months on Ascension Island (Chapin, 1954). Schreiber and Ashmole (1970) conclude that the relatively aseasonal species must be synchronized by social factors and that social attraction may play an important role.

In some cases, despite lack of seasonal constraints, tropical species may show extremely high synchrony (e.g., Sooty Tern: Feare, 1976). The methodology for studying tropical synchrony requires some modification. It will be different for seasonal and for aseasonal species. One approach is to define a particular period (e.g., 1 or 2 months) and note the percentage of the annual number of nests initiated during that period. It would also be feasible to use SD, comparing number of nests initiated for each month among species or islands.

Synchronizing factors in the tropics remain to be determined for many species. Nelson (1976) believed that direct social stimulation of displaying males was important in frigatebirds. He also suggests (Nelson, 1974) that in some way seasonally high productivity of an upwelling has selected for seasonal breeding in Christmas Island birds. Additional factors, such as dry-season trade winds (Diamond, 1975) or monsoons, can synchronize tropical breeding. Ashmole (1962) discusses this matter in detail.

Table XIII summarizes synchrony for several species of tropical seabirds. No attempt has been made to exhaust the literature. Rather the data suggest some of the comparisons that can be made. It is apparent that the SD is constrained by the limited sample size (12 months) and ranges up to 4.0. One can see that there is considerable variation in the synchrony for Red-billed Tropicbirds, which may be annual or geographic. Dorward's (1962) data for two species of boobies at two locations each, suggest that the White Booby (*S. dactylatra*) is more synchronous than is the Brown Booby (*S. leucogaster*), but the difference is not marked if one considers only the proportion nesting in the peak month. Data for the Black Noddy show remarkable similarity in timing for two seasons (Ashmole, 1962).

There was a highly significant negative correlation between SD and the proportion of nests in either the peak 1, 2, or 3 months. The correlation between SD and the peak 2 months was the highest. The average SD for the 12 tropical data sets was 2.25 ± 1.21, which was substantially higher than for temperate zone studies (0.29 ± 0.18; $p < 0.01$). It is important to distinguish between the SD computed on a daily basis and SD computed on a monthly basis. The latter, suitable for comparing tropical studies, would be extremely gross for comparing temperate ones.

Certain tropical species, such as the Sooty Tern (Feare, 1976; W. B. Robertson, personal communication), show such extremely high synchrony and seasonality, that one must examine them separately from species such as the tropicbirds, which breed almost every month of the year. The Black Noddy appears intermediate since Ashmole (1962) reports nests initiated over a 4-7 month period.

It would, however, be feasible to examine clutch initiation on a daily basis, for small assemblages of tropical seabirds, in order to compare the relative synchrony among populations. Hailman's (1964) study of the Swallow-tailed Gull provides an example for which the daily distribution of nesting could have been used to compare the colonies.

H. Other Aspects of Synchrony

Two important areas that lie outside the scope of this chapter are (1) synchrony of breeding of colonial passerine birds and (2) synchrony of other activities, such as feeding. No attempt has been made to review these areas, but some significant studies are mentioned here.

Some of the most important studies of avian synchrony have been accomplished with non-seabirds. The Quelea (*Quelea quelea*), an African weaverbird (Ploceinae), forms the largest aggregations of any passerine, nesting in colonies of more than 100,000 pairs. An individual colony is

Table XIII. Examples of Synchrony in Tropical Seabirds

Species	Reference	Total nests	Mean month[a]	SD[b]	Maximum %			Nesting[c] months
					1 mo.	2 mo.	3 mo.	
Yellow-billed Tropicbird	Stonehouse (1962)	712	June 14	3.49	10	19	28	12
Red-billed Tropicbird	Stonehouse (1962)	291	July 27	2.76	18	33	46	12
Red-billed Tropicbird	Schreiber and Ashmole (1970)	233	Aug 12	3.61	27	51	70	11
Red-billed Tropicbird	Fleet (1974)	3397	June 21	1.93	21	39	56	12
Red-billed Tropicbird	Fleet (1974)	3465	May 15	1.23	55	75	86	12
White Booby	Dorward (1962)	649	June 24	1.11	55	74	89	10
White Booby	Dorward (1962)	75	June 9	1.38	37	71	94	8
Brown Booby	Dorward (1962)	345	July 1	3.70	29	38	50	11
Brown Booby	Dorward (1962)	133	Mar 18	2.39	47	49	50	7
Christmas Island Shearwater	Schreiber and Ashmole (1970)[c]	124	July 28	3.88	27	36	50	12
Black Noddy	Ashmole (1962)	111	Apr 10	0.90	55	85	95	4
Black Noddy	Ashmole (1962)	200	Apr 13	0.61	50	96	99	4
Common Tern[d]	Gochfeld (unpublished data)	650	May 25	0.26	80	98	100	3

[a] Mean month calculation is based on all nests being initiated on the 15th of the month: Correlation coefficients for SD vs. peak 1 month: −0.76; SD vs. peak 2 months: −0.86; SD vs. peak 3 months: −0.85.
[b] Standard deviation calculated based on number of nests each month.
[c] Nesting months is number of months in which nests were initiated.
[d] Data on a Temperate species has been added for comparison.

highly synchronous, although there may be low synchrony over a region. Ward (1972) found colonies formed by birds in the same stage of readiness (social segregation), and such synchrony could have significant antipredator benefits, as Darling (1938) suggested through predator satiation (see Elgood and Ward, 1963).

Hall (1970) found that two species of weaverbirds showed synchrony, which arose through a common response to onset of rains plus social stimulation, and colonies had to attain a certain size before males actually secured mates. Perhaps the most impressive demonstrations of a Darling effect in birds comes from aviary studies of the Village Weaver (*Ploceus cucullatus*) by Collias *et al.* (1971*a,b*), who found that larger colonies nested earlier and showed better nest building and nest acceptance.

Colonial blackbirds (Icteridae) have also shed light on synchrony. Studies by Orians (1961), Horn (1970), and Robertson (1973) are important. Orians (1961) concluded that it would be difficult to adduce evidence on productivity for sufficient colonies in comparable habitats to test the Darling effect adequately. For swallows, Snapp (1976) found no evidence for social stimulation as an adaptation; Hoogland and Sherman (1976) suggested important antipredator functions; Emlen and Demong (1975) emphasized the role of social learning of feeding areas. In 10 loose colonies of Carrion Crows (*Corvus corone*) synchrony (based on 100% span of laying dates was inversely correlated with density, and in the denser groups predation rates were inversely correlated with synchrony (Yom Tov, 1975). Social facilitation plays an important role in vocal communication of birds and, for example, appears to be characteristic of the Pampas Meadowlark (*Sturnella defilippii*) where the rate of song output per male increases with group size; these songs play a role in territoriality and in attracting females (Gochfeld, 1978*b*).

Important work on synchrony in other groups of vertebrates is discussed by Wilson (1975) and in this chapter (see Section VIII.E.2). Invertebrates provide even more impressive examples in their swarming behavior. Perhaps the most important examples are represented by tropical plants, wherein mast fruiting appears to be a specific adaptation for avoiding predation (Janzen, 1971).

A variety of activities other than egg laying may be synchronous, e.g., migratory departures (Bowman and Whitman, 1972), diving of feeding seabirds (Siegfried *et at.*, 1975; Gochfeld, manuscript in preparation), panics of gulls and terns (Marples and Marples, 1934), and roost departures (ffrench, 1967). Ward and Zahavi (1973) and Krebs (1978) emphasize the role of avian aggregations and related behaviors in conveying information about food availability, and direct cooperation in feeding (e.g., Bartholomew, 1942) may be a factor in selecting for synchrony.

IX. EVOLUTIONARY IMPLICATIONS OF SYNCHRONY

The discussion of the role of synchrony may be divided into the following questions:

1. Does synchrony occur (i.e., are some species, populations, or colonies consistently more synchronous than others)? This is mainly a methodologic question.
2. What factor(s) produce the observed synchrony in the short term (year of study)? This is mainly an ecological question.
3. Is social stimulation (e.g., contagiousness of displays) occurring? In what manner does it contribute to synchrony? This is mainly a dual behavioral and physiologic question.
4. What is the impact of synchrony on the natural history of the species? Can an adaptive function be identified? This is mainly an evolutionary question.

In any given study of a colonial species it may not be feasible to address more than just one of these questions, but one must not lose sight of the complex components of the problem. Much criticism of the Darling effect arises from the fact that numerous factors influence the timing of reproduction, which may not be consistent across years or across colonies.

A. Does Synchrony Occur?

Because synchrony is a relative term, this question is meaningless; it must therefore be rephrased to emphasize that one group of birds may be more synchronous than another. Numerous studies (see Table XI) have reported that synchrony occurs in certain size groups, indicating that the temporal clustering of breeding is more than one would expect if specific synchronizing factors were not operative. It is safe to conclude that in certain cases synchrony is occurring and requires study and explanation.

B. What Factors Produce Synchrony?

Social factors contribute strongly to synchrony. Social segregation involves the formation of nesting units by birds that are at the same stage of the reproductive cycle. It implies that whatever has led to the synchrony (whether it be social behavior or coincidence of annual cycles) occurs prior to establishment of territories. Social attraction involves the attraction of birds ready to breed to a group that is already breeding. It may result in the acceleration of the cycles of late-arriving birds, which may enhance synchrony. However, protracted attraction of birds will cause an overall reduction of apparent synchrony. Social facilitation may play an important role as Darling (1938) suggested, but one cannot assume, *a priori* that it must be working.

Extrinsic factors, such as weather and habitat conditions, may play important roles in synchronizing reproduction. If birds at varying stages of readiness are required to wait a long time for favorable conditions, all may be ready to breed at the same time when conditions permit. Such high degrees of synchrony may occur in the absence of social stimulation.

C. Is Social Stimulation Occurring?

Two studies (Southern, 1974; also the work outlined in this chapter) provide direct evidence that contagiousness of breeding displays occurs. This provides a mechanism for synchronizing behavior and reproductive cycles among neighboring birds.

D. What Is the Adaptive Value of Synchrony?

Darling (1938) emphasized that predator satiation would occur if many chicks hatched at once. This is likely to be true with a large colony and a nonrecruitable predator. In many cases, it will not operate, however. Other antipredator devices, such as living in a large colony and mounting effective defensive attacks, may be more important. Because the intensity of attacks waxes and wanes systematically during the breeding cycle (e.g., Lemmetyinen, 1971), it is advantageous for birds to breed synchronously.

In future studies it will be important to consider items 1, 2, and 3 below, and to answer the appropriate questions. The basic question is whether synchrony has evolved as a specific adaptation or as an important corollary to other aspects of reproductive timing, or whether it is simply a mathematical construct. There are several possibilities:

1. Synchrony is only apparent, a passive outcome of many birds nesting together:
   a. Is reproductive timing simply a result of the annual cycle entrained by daylength and modified by food availability (e.g., Snapp, 1973)?
   b. Do climatic conditions (e.g., snow, flooding) render breeding grounds uninhabitable for awhile (Burger, 1974a,b; Burger and Shisler, 1978; Beer, 1966)?

2. Early nesting is adaptive, and synchrony is a passive consequence.
    a. Do predators take more late-nesting adults or young?
    b. Does adverse autumn weather favor earlier nesting?
    c. Do older, more experienced birds breed earlier and fare better?
    d. Is it important to enhance opportunities for renesting?
    e. Are early-nesting birds successful for unknown reasons?
    f. Is social facilitation accelerating breeding cycles?
3. Synchrony has been evolved as an adaptation.
    a. Does peak hatching period correspond to peak food availability?
    b. Does social facilitation synchronize breeding among neighboring birds?
    c. Does synchronous hatching lead to predator satiation?
    d. Do predators remove both early- and late-nesting adults, their eggs and/or young (centripetal selection)?
    e. Do birds hatching at peak have better chance of learning feeding areas (e.g., Emlen and Demong, 1975)?

As synchrony is a relative term, it is more valuable to describe it mathematically (see Section III) than to determine whether it is occurring. It is useful, also, to attempt to identify what synchronizing and desynchronizing factors are operating in the short term (season of study) and in the long term (evolutionarily). It is important to determine the reproductive outcome of birds breeding with varying degrees of synchrony, particularly when the confounding variables (see Section VIII.F above) can be estimated. Finally, it is important to remember that synchrony can be advantageous for a species in one place or season and disadvantageous elsewhere. Moreover, synchrony can be advantageous for different reasons for different populations.

This chapter has provided definitions and methods that should be considered in studies of reproductive behavior and ecology of colonial birds. Synchrony should be examined in the context of those factors that produce it and those that influence reproductive success. Even when synchrony is adaptive for a species over the long term, it may prove maladaptive for a particular season. Thus desynchronizing factors operate, often in an unpredictable fashion, and these factors may themselves be adaptive, assuring that the peak of hatching is not so sharp that the entire population fails in the face of adverse conditions. Colonies may benefit from not putting all their eggs into a single time "basket."

High degrees of synchrony, when they occur, are dramatic and tempt evolutionary interpretation. Darling's (1938) model, which integrates behavior, physiology, ecology, and evolution, remains worthy of testing, even though it may never operate in its entirety. Certain aspects of the Darling effect may prevail and should be sought. It is always possible to find something correlated with synchrony, but cause and effect will prove difficult to unravel when dealing with the timing of reproduction in colonial birds.

ACKNOWLEDGMENTS

My field studies of terns and skimmers have been supported by the Frank M. Chapman Fund, the Society of Sigma Xi, and by the U.S. Environmental Protection Agency. I appreciate the field assistance of L. Farber, D. B. Ford, D. J. Gochfeld, R. Gochfeld, J. Grafton, R. O. Paxton, D. Riska, C. Safina, and others. Through the years Dr. J. Burger has provided much stimulating discussion on the topic of synchrony, and I thank her for allowing me to write about a subject so dear to her. Without her devotion to synchrony, and to the author, this chapter would have been completed much more swiftly or not at all.

REFERENCES

Altman, S. A., 1965, Sociobiology of Rhesus Monkeys: II. Stochastics of social communication, *J. Theoret. Biol.* **8**:490–522.

Armstrong, E. A., 1947, *Bird Display and Behaviour*; Lindsay Drummond, London.

Arnold, S. J., and Wassersug, R. J., 1978, Differential predation on metamorphic anurans by Garter Snakes (*Thamnophis*): Social behavior as a possible defense, *Ecology* **59**:1014–1022.

Ashmole, N. P., 1962, The Black Noddy *Anous tenuirostris* on Ascension Island, *Ibis* **103b**:253–273

Ashmole, N. P., 1963, The biology of the Wideawake or Sooty Tern *Sterna fuscata* on Ascension Island, *Ibis* **103b**:297–364.

Austin, O. L., 1940, Some aspects of individual distribution in the Cape Code tern colonies, *Bird-Banding* **11**:155–169.

Austin, O. L., 1945, The role of longevity in successful breeding of the Common Tern, *Bird-Banding* **16**:21–28.

Austin, O. L., 1946, The status of the Cape Cod terns in 1944: A behaviour study, *Bird-Banding* **17**:10–27.

Austin, O. L., Jr., 1933, The status of Cape Cod terns in 1933, *Bird-Banding* **4**:190–198.

Baggerman, B., Baerends, G. P., Heikens, H. S., and Mook, J. H., 1956, Observations on the behaviour of the Black Tern, *Chlidonias n. niger* (L) in the breeding area, *Ardea* **44**:1–71.

Bartholomew, G. A., 1942, The fishing activity of Double-crested Cormorants on San Francisco Bay, *Condor* **44**:13–21.

Beer, C. G., 1966, Adaptations to nesting habitat in the reproductive behaviour of the Black-billed Gull (*Larus bulleri*), *Ibis* **108**:394–410.

Bowman, M. C., and Whitman, S. L., 1972, Simultaneous migration of Sandhill Cranes in Florida, *Auk* **89**:668.

Brown, J. L., 1975, *The Evolution of Behavior*, Norton, New York.

Brown, R. G. B., 1967, Breeding success and population growth in a colony of Herring and Lesser Black-backed Gulls, *Larus argentatus* and *L. fuscus*, *Ibis* **109**:503–515.

Buckley, F. G., and Buckley, P. A., 1972, The breeding ecology of Royal Terns *Sterna* (*Thalasseus*) *maxima maxima*, *Ibis* **114**:344–359.

Bullough, W. A., 1942, Observations on the colonies of the Arctic Tern (*Sterna macrura* Naumann) on the Farne Islands, *Proc. Zool. Soc. London* **112**:1–12.

Burger, J., 1974a, Breeding adaptations of Franklin's Gull (*Larus pipixican*) to a marsh habitat, *Anim. Behav.* **22**:521–567.

Burger, J., 1974b, Breeding biology and ecology of the Brown-hooded Gull in Argentina, *Auk* **91**:601–613.

Burger, J., 1977, Role of visibility in nesting behavior of *Larus* gulls, *J. Comp. Physiol. Psychol.* **91**:1347–1358.

Burger, J., 1979, Colony size: A test for breeding synchrony in Herring Gull (*Larus argentatus*) colonies, *Auk* **96**:694–703.

Burger, J., and Lesser, F., 1978, Selection of colony sites and nest sites by Common Terns, *Sterna hirundo* in Ocean County, N.J., *Ibis* **120**:433–449.

Burger, J., and Shisler, J. K., 1978, Nest site selection and competitive interactions of herring and laughing gulls in New Jersey, *Auk* **95**:252–266.

Burger, J., and Shisler, J. K., 1979, The immediate effects of ditching a salt marsh on nesting Herring Gulls, *Larus argentatus*, *Biol. Conserv.* **15**:85–103.

Burger, J., and Shisler, J. K., 1980, The process of colony formation among Herring Gulls *Larus argentatus* in New Jersey, *Ibis* **122**(in press).

Chapin, J. P., 1954, The calendar of Wideawake Fair, *Auk* **71**:1–15.

Collias, N. E., Victoria, J. K., and Shallenberger, R. J., 1971a, Social facilitation in weaverbirds: Importance of colony size, *Ecology* **52**:823–828.

Collias, N. E., Brandman, M., Victoria, J. K., Kiff, L. F., and Rischer, C. E., 1971b, Social facilitation in weaverbirds: effects of varying the sex ratio, *Ecology* **52**:829–836.

Cooper, D., Hays, H., and Pessino, C., 1970, Breeding of the Common and Roseate Terns on Great Gull Island, *Proc. Linn. Soc. N.Y.* **71**:83–104.

Coulson, J. C., 1968, Differences in the quality of birds nesting in the centre and on the edges of a colony, *Nature (London)* **217**:478–479.

Coulson, J. C., 1971, Competition for breeding sites causing segregation and reduced young production in colonial animals, *Proc. Adv. Study Inst. Dynamics Numbers* **1970**:257–266.

Coulson, J. C., and White, E., 1956, A study of colonies of the Kittiwake, *Rissa tridactyla* (L), *Ibis* **98**:63–79.

Coulson, J. C., and White, E., 1960, The effect of age and density of breeding birds on the time of breeding of the kittiwake *Rissa tridactyla*, *Ibis* **102**:71–86.

Crook, J. H., 1964, The evolution of social organisation and visual communication in the weaver birds (Ploceinae), *Behav. Suppl.* **10**:1–178;

Curio, E., 1976. *The Ethology of Predation*, Springer-Verlag, New York.

Darling, F. F., 1938, *Bird Flocks and the Breeding Cycle*, Cambridge University Press, Cambridge.

Darling, F. F., 1944, *Island Years*, G. Bell and Sons, Ltd., London.

Darling, F. F., 1952, Social behavior and survival, *Wilson Bull.* **69**:183–191.

Davis, D. E., 1940, Social nesting habits of the Smooth-billed Ani, *Auk* **57**:179–218.

Davis, J. W. F., 1975, Age, egg-size and breeding success in the Herring Gull *Larus argentatus*, *Ibis* **117**:460–473.

Davis, J. W. F., and Dunn, E. K., 1976, Intraspecific predation and colonial breeding in Lesser Black-backed Gulls *Larus fuscus*, *Ibis* **118**:65–77.

Diamond, A. W., 1975, Biology and behaviour of frigatebirds *Fregata* spp. on Aldabra Atoll, *Ibis* **117**:302–323.

Dorward, D. F., 1962, Comparative biology of the White Booby and Brown Booby, *Sula* spp. at Ascension, *Ibis* **103b**:174–220.

Drent, R. H., 1967, Functional aspects of incubation in the Herring Gull, *Behav. Suppl.* **17**:1–132.

Elgood, J. H., and Ward, P., 1963, A snake attack upon a weaverbird colony: Possible significance of synchronous breeding activity, *Bull. Br. Ornithol. Club* **83**:71–73.

Emlen, J. T., Jr., 1952, Social behavior in nesting Cliff Swallows, *Condor* **54**:177–199.

Emlen, S. T., and Demong, N. J., 1975, Adaptive significance of synchronized breeding in a colonial bird: A new hypothesis, *Science* **88**:1029–1031.

Feare, C. J., 1976, The breeding of the Sooty Tern *Sterna fuscata* in the Seychelles and the effects of experimental removal of its eggs, *J. Zool. London.* **179**:317–360.

ffrench, R. P., 1967, The Dickcissel on its wintering grounds in Trinidad, *Living Bird* **6**:123–140.

Fisher, J., 1954, Evolution and bird sociality, in: *Evolution as a Process* (J. S. Huxley, A. C. Hardy, and E. B. Ford, eds.). pp. 71–83, Unwin, London.

Fleet, R. R., 1974, The Red-tailed Tropicbird on Kure Atoll, *Am. Ornithol. Union Monogr.* **16**:1–64.

Floyd, C. B., 1933, Further notes from Penikese Island terns, *Bird-Banding* **4**:200–202.

Gochfeld, M., 1973, Effect of artefact pollution on the viability of seabird colonies on Long Island, New York, *Environ. Pollu.* **5**:1–5.

Gochfeld, M., 1976, Waterbird colonies of Long Island, New York. 3. Cedar Beach Ternery, *Kingbird* **26**:63–80.

Gochfeld, M., 1978*a*, Colony and nest site selection by Black Skimmers, *Proc. 1st Conf. Colonial Waterbird Gp.* **1**:78–90.

Gochfeld, M., 1978*b*, Social facilitation of singing: Group size and flight song rates in the Pampas Meadowlark, *Ibis* **120**:338–339.

Gochfeld, M., 1979, Breeding synchrony in Black Skimmers: Colony vs. subcolonies, *Proc. 2nd Conf. Colonial Waterbird Gp.* **2**:171–177.

Gochfeld, M., Synchronous diving by rafts of waterfowl, (manuscript in preparation).

Gochfeld, M., and Ford, D. B., 1974, Reproductive success in Common Tern colonies near Jones Beach, Long Island, New York, in 1972: A hurricane year, *Proc. Linn. Soc. N.Y.* **82**:63–76.

Goethe, F., 1937, Beobactungen und Untersuchungen zur Biologie der Silbermowe (*Larus a. argentatus* Pontopp.) auf der Vogelinsel Memmersand, *J. f. Ornithol.* **85**:1–119.

Hailman, J. P., 1964, Breeding synchrony in the equatorial Swallow-tailed Gull, *Am. Nat.* **98**:79–83.

Hall, J. R., 1970, Synchrony and social stimulation in colonies of the Black-headed Weaver *Ploceus cucullatus* and Vieillot's Black Weaver *Melanopteryx nigerrimus*, *Ibis* **112**:93–104.

Hamilton, W. D., 1971, Geometry for the selfish herd, *J. Theoret. Biol.* **31**:295–311.

Harris, M. P., 1964, Aspects of the breeding biology of the gulls *Larus argentatus*, *L. fucus* and *L. marinus*, *Ibis* **106**:432–456.

Harris, M. P., 1969, The biology of storm petrels in the Galapagos Islands, *Proc. Calif. Acad. Sci.* **37**:95–166.

Hoogland, J. L., and Sherman, P. W., 1976, Advantages and disadvantages of Bank Swallow (*Riparia riparia*) coloniality, *Ecol. Monogr.* **46**:33–56.

Horn, H. S., 1970, Social behavior of nesting Brewer's Blackbirds, *Condor* **72**:15–23.

Immelmann, K., 1971, Ecological aspects of periodic reproduction, in: *Avian Biology* (D. S. Farner and J. R. King, eds.), pp. 342–391, Academic Press, New York.

Janzen, D. H., 1971, Seed predation by animals, *Annu. Rev. Ecol. Syst.* **2**:465–492.

Klopfer, P., and Hailman, J. P., 1965, Habitat selection in birds, in: *Advances in the Study of Animal Behavior* (D. S. Lehrman, R. A. Hinde, and E. Shaw, eds.), pp. 279–303, Academic Press, New York.

Krebs, J. R., 1978, Colonial nesting in birds, with special reference to the Ciconiiformes, in: *Wading Birds* (A. Sprunt, J. C. Ogden, and S. Winkler, eds.), pp. 299–314, National Audubon Society, New York.

Kruuk, H., 1964, Predators and anti-predator behaviour of the Black-headed Gull (*Larus ridibundus*), *Behav. Suppl.* **11**:1–129.

Kury, C. R., and Gochfeld, M., 1975, Human interference and gull predation in Cormorant colonies, *Biol. Conserv.* **8**:23–34.

Lack, D., 1966, *Population Studies of Birds*, Clarendon Press, London.

Lack, D., 1968, *Ecological Adaptations for Breeding in Birds*, Methuen, London.

Langham, N. P. E., 1974, Comparative breeding biology of the Sandwich Tern, *Auk* **91**:255–277.

Lehrman, D. S., 1965, Interaction between internal and external environments in the regulation of the reproductive cycle of the Ring Dove, in: *Sex and Behavior* (F. A. Beach, ed.), Wiley, New York.

Lemmetyinen, R., 1971, Nest defence behaviour of Common and Arctic Terns and its effects on the success achieved by predators, *Orn. Fenn.* **48**:13–24.

MacRoberts, M. H., 1973, Extramarital courting in Lesser Black-backed and Herring Gulls, *Zeits. f. Tierpsychol.* **32**:62–74.

MacRoberts, M. H., and MacRoberts, B. R., 1972a, The relationship between laying data and incubation period in Herring and Lesser Black-backed Gulls, *Ibis* **114**:93–97.

MacRoberts, M. H., and MacRoberts, B. R., 1972b, Social stimulation of reproduction in Herring and Lesser Black-backed Gulls, *Ibis* **114**:495–506.

Marples, G., and Marples, A., 1934, *Sea Terns or Sea Swallows*, Country Life, Ltd., London.

Montevecchi, W. A., 1977, Predation in a salt marsh Laughing Gull colony, *Auk* **94**:583–585.

Montevecchi, W. A., Impekoven, M., Segre-Terkel, A., and Beer, C. G., 1979, The seasonal timing and dispersion of egg-laying among Laughing Gulls *Larus atricilla*, *Ibis* **121**:337–344.

Munro, J., and Bedard, J., 1977, Gull predation and creching behaviour in the Common Eider, *J. Anim. Ecol.* **46**:799–810.

Nelson, J. B., 1966, Population dynamics of the Gannet (*Sula bassana*) at the Bass Rock with comparative information from other Sulidae, *J. Anim. Ecol.* **35**:443–470.

Nelson, J. B., 1970, The relationship between behaviour and ecology in the Sulidae with reference to other seabirds, *Oceanograph. Marine Biol. Annu. Rev.* **8**:501–574.

Nelson, J. B., 1974, The biology of the seabirds of the Indian Ocean, Christmas Island, *J. Marine Biol. Assoc. India* **14**:643–662.

Nelson, J. B., 1975, Functional aspects of behaviour in the Sulidae, in: *Function and Evolution in Behaviour* (G. Baerends, C. Beer, and A. Manning, eds.), Clarendon Press, Oxford.

Nelson, J. B., 1976, The breeding biology of frigatebirds—A comparative review, *Living Bird* **14**:113–155.

Nettleship, D. N., 1972, Breeding success of the Common Puffin (*Fratercula arctica*, L.) on different habitats of Great Island, Newfoundland, *Ecol. Monogr.* **42**:239–268.

Nisbet, I. C. T., 1975, Selective effects of predation in a tern colony, *Condor* **77**:221–226.

Nisbet, I. C. T., and Cohen, M. E., 1975, Asynchronous hatching in Common and Roseate Terns, *Sterna hirundo* and *S. dougallii, Ibis* **117:**314–379.

Noll, H., 1931, Neue biologische Beobachtungen an Lachmoven (*Larus argentatus*), *Beitr. Fortpfl. Vogel* **7:**7–9.

Orians, G. H., 1961, Social stimulation within blackbird colonies, *Condor* **63:**330–337.

Palmer, R. S., 1941, A behavior study of the Common Tern (*Sterna* hirundo hirundo L.), *Proc. Boston Soc. Nat. Hist.* **42:**1–119.

Paludan, K., 1951, Contributions to the breeding biology of *Larus argentatus* and *Larus fuscus, Dansk Naturhist. Foren. Viddensk. Meddel.* **114:**1–121.

Parsons, J., 1971, Cannibalism in Herring Gulls, *Br. Birds* **64:**528–537.

Parsons, J., 1972, Eggs size, laying date and incubation period in the Herring Gull, *Ibis* **114:**536–541.

Parsons, J., 1975*a*, Seasonal variation in the breeding success of the Herring Gull: An experimental approach, *J. Anim. Ecol.* **44:**553–573.

Parsons, J., 1975*b*, Asynchronous hatching and chick mortality in the Herring Gull *Larus argentatus, Ibis* **117:**517–520.

Parsons, J., 1976, Nesting density and breeding success in the Herring Gull *Larus argentatus, Ibis* **118:**537–546.

Patterson, I. J., 1965, Timing and spacing of broods in the Black-headed Gull *Larus ridibundus, Ibis* **107:**433–459.

Paynter, R. A., Jr., 1949, Clutch-size and the egg and chick mortality of Kent Island Herring Gulls, *Ecology* **30:**146–166.

Perrins, C. M., 1970, The timing of birds' breeding seasons, *Ibis* **112:**242–255.

Powell, G. V. N., 1974, Experimental analysis of the social value of flocking by Starlings (*Sturnus vulgaris*) in relation to predation and foraging, *Anim. Behav.* **22:**501–505.

Richdale, L. E., 1951, *Sexual Behaviour of Penguins*, University of Kansas Press, Lawrence.

Robertson, R. J., 1973, Optimal niche space of the Red-winged Blackbirds: Spatial and temporal patterns of nesting activity and success, *Ecology* **54:**1085–1093.

Schaller, G. B., 1964, Breeding behavior of the White Pelican at Yellowstone Lake, Wyoming, *Condor* **66:**3–23.

Schreiber, R. W., and Ashmole, N. P., 1970, Sea-bird breeding seasons on Christmas Island, Pacific Ocean, *Ibis* **112:**363–394.

Sealy, S. G., 1975, Influence of snow on egg-laying in auklets, *Auk* **92:**528–538.

Siegfried, W. R., and Underhill, L. G., 1975, Flocking as an anti-predator strategy in doves, *Anim. Behav.* **23:**504–508.

Siegfried, W. R., Frost, P. G. H., Kinahan, J. B., and Cooper, J., 1975, Social behaviour of Jackass Penguins at Sea, *Zool. Africana* **10:**87–100.

Snapp, B. D. M., 1976, Colonial breeding in the Barn Swallow (*Hirundo rustica*) and its adaptive significance, *Condor* **78:**471–480.

Snow, B. K., 1960, The breeding biology of the Shag (*Phalacrocorax aristotelis*) on the island of Lundy, Bristol Channel, *Ibis* **102:**554–575.

Snow, B. K., and Snow, D. W., 1968, Behavior of the Swallow-tailed Gull of the Galapagos, *Condor* **70:**252–264.

Sokal, R. R., and Rohlf, F. J., 1969, *Biometry*, Freeman, San Francisco.

Southern, W. E., 1974, Copulatory wing-flagging: A synchronizing stimulus for nesting Ring-billed Gulls, *Bird-Banding* **45:**210–216.

Spaans, M. J., and Spaans, A. L., 1975, Enkele gegevens over de broedbiologie ban de Zilvermeeuw *Larus argentatus* op Terschelling, *Limosa* **48:**1–39.

Spurr, E. B., 1975, Breeding of the Adelie Penguin *Pygoscelis adeliae* at Cape Bird, *Ibis* **117:**324–338.

Stonehouse, B., 1962, The tropic birds (Genus *Phaethon*) of Ascension Island, *Ibis* **103b**:474–479.

Teeple, S. M., 1977, Reproductive success of Herring Gulls nesting on Brothers Island, Lake Ontario, in 1973, *Can. Field Nat.* **91**:148–157.

Tenaza, R., 1971, Behaviour and nesting success relative to nest location in Adelie Penguins (*Pygoscelis adeliae*), *Condor* **73**:81–92.

Tinbergen, N., Impekoven, M., and Franck, D., 1967, An experiment on spacing-out as a defense against predation, *Behaviour* **28**:307–321.

Tolman, C. W., 1964, Social faciliation of feeding behaviour in the domestic chick, *Anim. Behav.* **12**:245–251.

Veen, J., 1977, Functional and causal aspects of nest distribution in colonies of the Sandwich Tern (*Sterna s. sandvicensis* Lath.), *Behav. Suppl.* **20**:1–193.

Vermeer, K., 1963, The breeding ecology of the Glaucous-winged Gull (*Larus glaucescens*) on Mandarte Island, B. C., *Occas. Papers Br. Columbia Provincial Mus.* **13**:1–104.

Vermeer, K., 1970, Breeding biology of California and Ring-billed Gulls: A study of ecological adaptation to the inland habitat, *Can. Wildl. Service Rep. Ser.* **12**.

Vermeer, K., 1973, Comparison of egg-laying chronology of Herring and Ring-billed Gulls at Kawinaw, Manitoba, *Can. Field Nat.* **87**:306–308.

von Haartman, L., 1945, Zur biologie der Wasser- und Ufervogel in Scharenmeer Sudwest-Finnlands, *Acta Zool. Fenn.* **44**:1–120.

Ward, P., 1965, The breeding biology of the Black-faced Dioch (*Quelea quelea*) in Nigeria, *Ibis* **107**:326–349.

Ward, P., 1972, Synchronisation of the annual cycle within populations of *Quelea quelea* in East Africa, *Proc. XV Int. Ornithol. Congr.*, pp. 702–703.

Ward, P., and Zahavi, A., 1973, The importance of certain assemblages of birds as "information centres" for food finding, *Ibis* **115**:517–534.

Weidmann, U., 1956, Observations and experiments on egg laying in the Black-headed Gull (*L. ridibundus* Linn.), *Br. J. Anim. Behav.* **4**:150–161.

Williams, A. J. 1975, Guillemot fledging and predation on Bear Island, *Orn. Scand.* **6**:117–124.

Williamson, K., 1949, The distraction behaviour of the Arctic Skua, *Ibis* **91**:307–313.

Wilson, E. O., 1975, *Sociobiology*, Harvard University Press, Cambridge.

Yom Tov, Y., 1975, Synchronization of breeding and intraspecific interference in the carrion crow, *Auk* **92**:778–785.

Zino, P. A., 1971, The breeding of Cory's Shearwater *Calonectris diomedea* on the Salvage Islands, *Ibis* **113**:212–217.

Chapter 8

DEVELOPMENT OF BEHAVIOR IN SEABIRDS: AN ECOLOGICAL PERSPECTIVE

Roger M. Evans

Department of Zoology
University of Manitoba
Winnipeg, Canada R3T 2N2

I. INTRODUCTION

Young seabirds are decidedly terrestrial creatures that spend all or a large part of their prefledging developmental period on or near the surface of the ground, cliffs, or other firm substrates. Consequently, the developing seabird faces problems in adaptation to its natal environment that are common to a wide range of birds, including species whose entire life cycle is restricted to terrestrial habitats. Despite their early terrestrial existence, however, young seabirds are also irrevocably linked to the primarily marine ecology of their parents, whose selection of breeding habitat and foraging behavior imposes strong selective constraints on the developmental adaptations of the young (Lack, 1968). The interaction of these terrestrial and marine influences provides a unique blend of selective forces that underlie the diversity and richness of developmental patterns found in seabirds today. This review is an attempt to describe some of this diversity, and where possible to relate it to the particular ecological conditions to which the individuals of particular species are adapted.

Historically, studies of behavior development have been strongly oriented towards causal or mechanistic processes (Hinde, 1970; Gottlieb, 1976). In seabirds, a similar causal emphasis has often been present, e.g., in the now-classic analyses of the stimulus parameters eliciting pecking in young Herring Gulls and other larids (Tinbergen, 1953; Hailman, 1967; Quine and Cullen, 1964). Another traditional approach to developmental studies, perhaps best exemplified by the work of Nice (1962) and Moynihan

(1959a) on the development of behavior in Ring-billed Gulls and Franklin's Gulls, involves a detailed examination of behavioral changes from hatching onward. A blending of this chronological approach with causal analyses is particularly well exemplified by recent studies of behavior embryology (Gottlieb, 1973).

Although there is unquestioned merit in causal, chronological, and embryological studies of behavior development, it is also true that developing organisms are ecologically dependent organisms that must adapt to conditions within their natal environment if they are to survive, develop more or less the normal behavior patterns of their species, and ultimately reproduce their genes. This ecological, or functional, approach to behavior development has long been present in studies of seabirds, but has commonly been contained within larger works on behavior or ecology (e.g., Tinbergen, 1953; Sladen, 1958; Tuck, 1960; Lockley, 1961, 1962; Snow, 1963; Nelson, 1978; and many others). Recently, more attention has been directed at the ecological and functional aspects of behavior ontogeny as a topic worthy of investigation in its own right, in areas as diverse as the development of mammalian play (Bekoff, 1972) and early behavior of hatchling lizards (Stamps, 1978). While preparing the present review, it has been my intention to draw on all aspects of behavior development in seabirds, including the causal, chronological, embryological and ecological, but to emphasize the latter, while simultaneously attempting to place the others into an ecological perspective.

A list of seabird species mentioned in the text is contained in the Appendix.

II. LIFE HISTORY AND DEVELOPMENTAL PATTERNS

The main life history attributes of seabirds and their primary ecological correlates have been reviewed by Lack (1968). As a group, seabirds tend to be long-lived, with low intrinsic reproductive rates due to the combined effects of small clutch sizes and delayed reproductive maturity. These traits are characteristic of the life history pattern of a wide range of so-called K-selected species (Horn, 1978) and are thought to have evolved in situations where individuals are faced with severe limitation of resources in populations that have grown to and stabilized at or near the carrying capacity (K) of their particular environment.

In seabirds, it has been convincingly argued that food is the resource that is most commonly limited in supply (Ashmole, 1963; Lack 1968;

Nelson, 1975, 1978). Under these conditions, the evolution of small brood sizes combined with increased parental investment (Trivers, 1974) for those young that are produced, as well as generally retarded growth and developmental rates of the young, constitute four developmentally relevant traits that are thought to have been favored as adaptations enhancing the survival of the young until they develop sufficiently to cope with their severe environment.

A. Precocial and Altricial Development

Precocial young are those which leave the nest soon after hatching and go to their food supplies, either with or without their parents (Nice, 1962). Semiprecocial species have been defined by Nice (1962) as species that are able to locomote soon after hatching, but that normally stay in the vicinity of the nest where they are fed by their parents. Altricial species are hatched virtually naked and in an immature state of development. Semialtricial species are similar to altricial species, except they are fully covered with natal down at the time of hatching, as are precocial and semiprecocial species.

The adaptive significance of precocial or altricial developmental modes apparently has not been incorporated into any general theory of life-history strategies, but the ecological parameters involved may be quite similar. As discussed by Lack (1968) and Case (1978), the fully precocial mode of development represents an important adaptation for species that utilize a food resource that can be exploited most efficiently by young that accompany a parent to the food supply, thereby reducing energy costs incurred by parents that transport food to a localized nest site. In seabirds, four species of alcids (Xantus', Craveri's, Ancient, and Japanese Murrelets) are fully precocial (Sealy, 1973). The young of at least one, and probably all of these four species are not fed at the nest before departing for the feeding grounds at 1–2 days of age (Sealy, 1976).

In all other seabirds, food is brought to the nestlings, which remain at or near the nest for all or a portion of the prefledging period. Reduced precocity of the young in these species is usually interpreted as an adaptation to foraging strategies in which the normal food of the young is too difficult for the young to reach or else too difficult for them to catch or handle even if it were accessible (Fjeldsa, 1977; Case, 1978). If it is not adaptive in such instances to take the young to the food, then a more sedentary early life, at or near the nest site, constitutes a means of reducing energy expended on locomotion by the young, and makes it easy for a foraging parent to locate its young when the parent returns with food. It has also been established that there is a trend towards smaller eggs in species that are less precocial (Lack, 1968; Sealy, 1975). According to Case (1978), this effectively shifts the main load of parental investment in offspring away from the egg formation stage toward the subsequent incubation and posthatch stages where both parents can participate fully in the reproductive effort.

A semiprecocial mode of development appears to represent a balance between conditions favoring reduced movements of the young, as discussed above, and other conditions favoring an early development of locomotion in the vicinity of the nest site or territory (Evans, 1977a). In most gulls and terns, the semiprecocial young are cryptic in color and where terrain permits move quickly away from the nest and crouch and hide (most species) or run in mobile creches (Royal Tern: Buckley and Buckley, 1972a; Black-billed Gulls: R. M. Evans, personal observation) when disturbed. Most gulls and terns are inland or inshore feeders that typically breed on sites that are more exposed to predators than are those of the more pelagic forms of seabirds (Lack, 1968). The early development of mobility in the semiprecocial young of the gulls and terns thus represents an important antipredator adaptation (see Kruuk, 1964). Similar considerations apply to the cryptic semiprecocial young of the skuas. An ability to move to shade to avoid heat stress is also important in several ground-nesting species (Howell and Bartholomew, 1962; Dawson et al., 1976).

The semiprecocial alcids exhibit highly variable mobility patterns. The Razorbill and the two species of murres leave the nest and accompany a parent to the feeding grounds at 18–25 days of age, well before they can fly (Tuck, 1960; Greenwood, 1964). Sealy (1973) has suggested that this pattern represents a situation intermediate between the precocial and other semiprecocial alcids. The adaptive significance of this intermediate pattern in the Razorbill and murres again appears to be related to the most efficient means of exploiting food resources, in this case by moving the young to the food as soon as they are able to safely reach the sea and locomote efficiently in the marine environment (Lack, 1968; Sealy, 1973; Birkhead, 1977). The situation for the more sedentary semiprecocial alcids is less clear. Most of these species nest in protected crevices or burrows (reviewed in Sealy, 1973), where mobility is restricted and less necessary as an antipredator strategem than in the more open-nesting larids discussed above. Indeed, movement out to the entrance of a burrow can expose otherwise protected young to predators (Nettleship, 1972). Under these conditions, it might be expected that many alcids would exhibit a convergence toward a more semialtricial mode of development, and some have been so classed by at least one investigator (Cody, 1973).

A primary distinction between altricial and semialtricial development is the presence of natal down at the time of hatching in semialtricial species (Nice, 1962). The adaptive significance of down appears likely to be related to the advantages of an early onset of temperature regulation which occurs earlier in most down-covered young than in naked altricial young (Table I). The importance of down for temperature regulation is further suggested by the close correlation in time between the onset of temperature regulation and the growth of a dense coat of natal down at about 2–3 weeks in the altricial pelecaniformes (Bartholomew *et al.*, 1953; Bartholomew and Dawson, 1954; Nelson, 1978). The advantages of an early onset of temperature regulation have not been investigated in detail in seabirds, but it has been noted that the onset of temperature regulation can free the parents from the need to brood their young (Farner and Serventy, 1959) and so

Table I. Functional Onset of Temperature Regulation (Thermogenesis) in Relation to Mode of Development in Young Seabirds

Order and species	Mode of development	Onset of temperature regulation	Source
Sphenisciformes			
Adelie Penguin	Semialtricial	10–12 days	Goldsmith and Sladen (1961)
Procellariiformes			
Black-footed Albatross	Semialtricial	1 day	Howell and Bartholomew (1961)
Laysan Albatross	Semialtricial	1 day	Rice and Kenyon (1962)
Slender-billed Shearwater	Semialtricial	1 day	Farner and Serventy (1959)
Giant Petrel	Semialtricial	18–26 days	Conroy (1972)
Leach's Petrel	Semialtricial	4–6 days	S. W. Harris (1974)
Pelecaniformes			
White Pelican	Altricial	2–3 weeks	Bartholomew *et al.* (1953)
Brown Pelican	Altricial	2–3 weeks	Bartholomew and Dawson (1954)
Masked Booby	Altricial	2–3 weeks	Bartholomew (1966)
Charadriiformes			
McCormick Skua	Semiprecocial	2–3 days	Spellerberg (1969)
Herring Gull	Semiprecocial	1 day	Dunn (1976)
California Gull	Semiprecocial	1–2 days	Behle and Goates (1957)
Ring-billed Gull	Semiprecocial	1 day	Dawson *et al.* (1976)
Laughing Gull	Semiprecocial	1 day	Dawson *et al.* (1972)
Kittiwake	Semiprecocial	1 day	Maunder and Threlfall (1972)
Common Tern	Semiprecocial	1–2 days	Lecroy and Collins (1972)
Roseate Tern	Semiprecocial	1–2 days	Lecroy and Collins (1972)
Common Murre	Semiprecocial	4–5 days	Johnson and West (1975)
Thick-billed Murre	Semiprecocial	4–5 days	Johnson and West (1975)
Pigeon Guillemot	Semiprecocial	1 day	Drent (1965)
Ancient Murrelet	Precocial	1 day	Sealy (1976)

permit them to leave the young alone while the parents forage for food (Conroy, 1972). Although the need to guard chicks from other hazards may preclude the parents leaving the young in some species, it seems likely that an early development of temperature regulation to permit parents to gather scarce food constitutes one of the main adaptive aspects of a semialtricial, as opposed to a purely altricial, mode of development.

In the Giant Petrel and the Adelie Penguin, the onset of temperature regulation and the time that parents leave their young unattended and forage for food are both delayed relative to most other semialtricial species (Table I). Whether the delay in onset of temperature regulation is a manifestation of a generally greater degree of altriciality in these species is not certain. It is relevant, however, that in at least one species of penguin, the King Penguin, the newly-hatched young are essentially naked, like an altricial species (Stonehouse, 1960).

B. Developmental Stages

As in embryological studies, it is sometimes useful to set out a series of developmental categories or stages through which the behavior of a young bird can be expected to pass as it matures. The most detailed early set of developmental stages for birds was described by Nice (1964) in her studies of the Song Sparrow (*Melospiza melodia*) and applied to a wide range of precocial and semiprecocial species, including gulls and terns (Nice, 1962). This scheme was subsequently applied to Adelie Penguins (Spurr, 1975) and, in modified form, to Ring-billed Gulls (Evans, 1970a). The main features of these stages are outlined in Table II, along with a separate but not mutually exclusive set of stages presented for the Adelie Penguin by Sladen (1958).

The developmental stages proposed by Nice (column 1 of Table II) are based on the sequential development of various motor coordinations, beginning with those associated with nutrition (Stage I), followed by the gradual maturation of comfort movements (Stages II and III), then the development of locomotor abilities and nest leaving (Stage IV), followed by the onset of aggression and finally flight (Stage V) (column 2 of Table II).

Although the stages of development described by Nice appear to have applicability across a wide range of species, it is important to note that because the onset of various stages was defined along different behavior parameters (nutrition, comfort movements, locomotion, aggression), the order in which the defining events occur is not necessarily uniform for all species. One important exception amongst the seabirds concerns the stage at which aggression begins. While aggression commonly occurs after the onset of locomotor abilities and nest leaving as outlined by Nice (1962, 1964),

Table II. Stages in the Development of Young Seabirds

Developmental stages (Nice, 1962)	Characteristics (Nice, 1962)	Adelie Penguins (Spurr, 1975)	Corresponding stages in brood mobility (Evans, 1970a)	Alternative stages for penguins (Sladen, 1958)
I. Post or late embryonic	Nutritional coordinations	Brooded by parent, begs	In nest ("incipient mobility")	Chicks guarded in or near the nest ("guard stage")
II. Preliminary	Comfort movements begin	Some comfort movements	In nest ("Incipient mobility")	Chicks guarded in or near the nest ("guard stage")
III. Transition	Maturation of comfort movements	Stands, comfort movements common, pants	In nest ("Incipient mobility")	Chicks guarded in or near the nest ("guard stage")
IV. Locomotory	Leaving nest	Avoidance reactions, moves on territory, pecks siblings	Moves on territory ("restricted mobility")	Chicks guarded in or near the nest ("guard stage")
V. Socialization	Aggression, flight	May leave colony, creching, aggression, feeding chases	Moves away from territory ("extended mobility")	Chicks form creches off territory ("creche stage")

aggression between siblings can arise while the young are still confined to the nest (kittiwake: Cullen, 1957) or even before the young are capable of walking (White Boobies, Brown Boobies: Nelson, 1978; White Pelican: Knopf, 1975; Johnson and Sloan, 1978; R. M. Evans personal observation). In such instances it becomes difficult if not impossible to assign the developing young to particular stages on the basis of criteria involving both aggression and locomotor abilities. Use of a single behavior dimension, such as increasing degrees of brood mobility, or functional categories like "guard" and "creche" stages (Table II, last two columns) appear less likely to lead to ambiguities of the sort described above for boobies and pelicans.

In her discussion of developmental stages in precocial birds, Nice (1962) pointed out that her Stages I–III represent periods of immobility in the young, followed by a period of relative mobility in Stage IV, and full activity in Stage V. These mobility characteristics were also recognized by Spurr (1975) for the Adelie Penguin, and formed the basis for the three stages of mobility recognized by Evans (1970a) for Ring-billed Gulls. In each instance, the three stages of mobility correlate closely in time with (1) the time that the young are restricted to the nest, followed by (2) limited movements outside the nest within the home territory and, finally, (3) more extensive movements away from the home territory. These three stages in the development of brood mobility thus provide a unifying theme amongst the various schemes listed in Table II. They have, in addition, the merit of being readily identifiable for diverse species under most field conditions and have important implications in relation to adaptations to nest sites, parent young interactions, and individual recognition, as discussed in the following sections.

III. DEVELOPMENTAL ADAPTATIONS TO PHYSICAL HAZARDS AT THE NEST

The nest site and its immediate surroundings constitute an important focal point of activity for the developing young of all but the most mobile species of seabirds. Where physical barriers to brood mobility are present, as in the many species of petrels, alcids, and some penguins that place their nests in burrows or crevices, few if any specialized adaptations to the nest site have been identified. The situation is quite different in species that place their nests in more hazardous situations, such as on open cliff ledges, trees, marshes, or unstable river beds. In these species, adaptations in the young to the physical hazards present at the nest site can constitute an important part of their overall developmental strategy. Since the classic study of cliff-nesting adaptations in the kittiwake by Cullen (1957), the behavior of adults and young at these more hazardous nest sites have been examined for several seabird species (Table III). The list of studies in Table III is illustrative rather than exhaustive, but will provide entry into the main literature for the major groups involved.

A. Marsh Nests

Several species of gulls and terns nest in marsh habitats (Table III; also Burger, 1974a) where the young face problems of wetting and water logging if they leave the nest. As shown for the Franklin's Gull and Brown-hooded

Table III. A Selection of Studies of Seabirds Breeding at Nest Sites in Potentially Hazardous Locations

	Nest site			
Order and species	Exposed cliff	Tree	Other[a]	Source
Procellariiformes				
Black-browed Albatross	X			Tickell and Pinder (1972, 1975)
Grey-headed Albatross	X			Tickell and Pinder (1972, 1975)
Pelecaniformes				
Brown Pelican		X		Schreiber (1977)
Pink-backed Pelican		X		Burke and Brown (1970)
Peruvian Booby	X			Nelson (1978)
Atlantic Gannet	X			Nelson (1967a, 1978)
Shag	X			Snow (1960, 1963)
Darter		X		Vestjens (1975)
Frigate Birds		X		Nelson (1975)
Charadriiformes				
Glaucous Gull	X			Smith (1966)
Black-billed Gull			R	Beer (1966a)
Franklin's Gull			M	Burger (1974a)
Kittiwake	X			Cullen (1957) McLannahan (1973)
Swallow-tailed Gull	X			Hailman (1965)
Black Tern			M	Cuthbert (1954)
Brown Noddie		X		Dorward and Ashmole (1963)
Black Noddie	X	X		Cullen and Ashmole (1963)
Fairy Tern		X		Howell (1978)
Common Murre	X			Tschanz and Hirsbrunner-Scharf (1975)

[a] (M) Marsh; (R) nests on unstable river beds.

Gull (Burger, 1974a,b), unmolested young raised under these conditions tend to remain localized on their nest platforms for at least the first 3–4 weeks after hatching, even though they are capable of moving quickly away in response to intruders. In young Franklin's Gulls, Burger (1974a) has shown that the young exhibit a strong tendency to walk upwards on an inclined plane, a response that should facilitate return up to the nest from the water should they be frightened away from it.

Another physical hazard faced by marsh nesters arises from the action of waves, which can wash away nests and render the site unsuitable for occupancy, as found for the Forster's Tern (McNicholl, 1971) and Black Tern (Bergman et al., 1970). It is apparently not known whether the young of these species exhibit any special adaptations to cope with unstable conditions in the marsh, but they do on occasion make use of their early locomotor capacity, characteristic of the family laridae, to move to other brooding sites within the marsh (McNicholl, 1971; Cuthbert, 1954).

B. Nests on Unstable Riverbeds

Potential loss of the nest site also occurs in ground-nesting gulls and terns which make use of river beds subject to late spring flooding, as described for the Black-billed Gull (Stead, 1932; Beer, 1966a). The young of this species exhibit a tendency to leave the nest site and territory and become somewhat nomadic, in loose creches that form in the vicinity of the colony (Beer, 1966a; Evans, 1970b). It is possible that the early mobility of this species may represent an important developmental adaptation to their river-bed nest sites should these become flooded.

C. Tree Nests

Because of the danger of falling from the nest, young reared in tree nests commonly exhibit a lower level of mobility than do their ground-nesting relatives. Exceptions are present, as in the tree-nesting Bonaparte's Gull, where the young leave the nest as early as 2–3 days after hatching (Jehl and Smith, 1970). Less extreme movements away from the nest when the young approach fledging is characteristic of several other arboreal species (references in Table III), but the young of these species usually remain closely associated with the nest at earlier ages.

The particular behavior patterns of arboreal young that are responsible for their low level of mobility apparently have not been studied in detail for any species. There is, however, evidence that levels of locomotor activity may be depressed in arboreal species, as shown by Watson (1908), who observed appreciably less locomotor activity and mobility in laboratory-reared young of the Brown Noddie than in the young of the ground-nesting Sooty Tern. It would be useful to determine if a similar depression in early locomotor activity is present in other tree-nesting seabird species (cf. cliff nesters, Section III.D).

Another behavior that may reduce the danger of falling from tree nests is restraint in food begging, as noted in frigatebirds and Abbott's Booby by Nelson (1975, 1978). However, Nelson (1978) also noted that begging restraint in sulids correlates well with plentiful food supplies, and so may not be primarily an adaptation to the danger of falling. The fact that the Red-footed Booby is commonly short of food and shows a greater level of begging than any other sulid despite its tree-nesting habit (Nelson, 1978) lends support to the latter interpretation. A reduction in the intensity of prefledging practice flights, as identified in the Abbott's Booby constitutes a more likely adaptation to tree nesting in this species (Nelson, 1978). Adaptations enabling arboreal young to cling tightly to small tree limbs or twigs by as early as the first day after hatching have also been described, in the Fairy Tern (Dorward, 1963; Howell, 1978).

D. Exposed Nests on Cliffs

Young reared on narrow cliff ledges must, of necessity, restrict their activities to the nest or its immediate vicinity if they are to avoid falling. Behavior patterns of the young that are thought to reduce the likelihood that the young will fall from open cliff ledges are listed in Table IV for the kittiwake (after Cullen, 1957; McLannahan, 1973) and Common Murre (after Tschanz and Hirsbrunner-Scharf, 1975; Wehrlin, 1977), two species which have been studied in some detail. It is noteworthy that the tendency to avoid light, which is strongly developed in the Common Murre, was not clearly present in young kittiwakes tested by McLannahan (1973), while a reduced tendency to beg for food apparently has not been reported as a possible cliff adaptation in the Common Murre. Young Common Murres do exhibit a reduced tendency to initiate feeding, but this difference has been interpreted as a means of reducing potential food losses to neighbors rather than as an adaptation to cliff ledges *per se* (Tschanz and Hirsbrunner-Scharf, 1975). The reduced or delayed tendency to participate in prefledging flight practice in the kittiwake (Table IV), is understandably absent from the common murre, due to its unusual habit of leaving the breeding cliff for the open sea well before flight ability matures.

It should be noted that the kittiwake and Common Murre are not particularly closely related, both probably representing specialized forms

Table IV. Behavior Patterns That Reduce the Likelihood of Falling from Exposed Cliff Ledges in Young Kittiwakes and Common Murres

Behavior	Kittiwake	Common Murre
Reduced tendency to stand or wander	[a,b]	[c]
Faces or is attracted to wall	[a,b]	[c,d]
Avoids light	Not found[b]	[c,d]
Avoids cliff edge	[b]	[d]
Restrained begging	[a,b]	?
Reduced or delayed flight practice	[a,b]	Leaves ledge before flight matures[c,d]

[a] Cullen (1957).
[b] McLannahan (1973).
[c] Tschanz and Hirsbrunner-Scharf (1975).
[d] Wehrlin (1977).

within their respective families (Cullen, 1957; Bedard, 1969). The similarities that are present in the way the young of these two species have adapted to cliff-ledge nest sites are therefore probably examples of convergence. Other cliff nesting seabirds that have been tested also show evidence of convergence with the kittiwake and Common Murre. Thus, a reduced tendency to run when disturbed is present in the young of several cliff-nesting species, including the Fulmar (Duffey, 1951), the Atlantic Gannet (Nelson, 1967a), the Swallow-tailed Gull (Hailman, 1965), the Black Noddy (Cullen and Ashmole, 1963), the Glaucous, Thayers, and Iceland Gulls (Smith, 1966), and cliff-reared young Herring Gulls (McLannahan, 1973). In gannets and Swallow-tailed Gulls, young reared in cliff nests tend to turn toward the rock wall (Nelson, 1966; Hailman, 1965). Young gannets also show restraint when begging for food (Nelson, 1967a). A tendency to delay or reduce practice flights until the wings have matured, just prior to fledging, is also present in Swallow-tailed Gulls (Hailman, 1965), Atlantic Gannets (Nelson, 1978), and Black Noddies (Cullen and Ashmole, 1963).

Avoidance of cliff edges has apparently been found in all cliff-nesting species that have been tested (e.g., Emlen, 1964; Hailman, 1965, 1968; McLannahan, 1973). Avoidance of a cliff edge is, however, also characteristic of many other vertebrate species that do not breed on cliffs (Walk and Gibson, 1961), including ground-nesting (Hailman, 1968) and marsh-nesting (Burger, 1974a) gulls, and so represents more than just an adaptation specific to cliff nesting. Notwithstanding the widespread occurrence of this response, it is significant to note that actual withdrawal from an abyss is particularly well developed in young kittiwakes, who respond to both visual and tactile stimuli from ledges (McLannahan, 1973). Taken together, the data therefore suggest that although cliff-edge responses are widespread, they can be especially strongly selected for in species that normally place their exposed nests on narrow cliff ledges.

Herring Gulls, which are usually ground nesters but sometimes cliff nesters, have provided useful insights into the role of early experience in the ontogeny of cliff-edge responses. In an early study with this species, Emlen (1964) examined cliff-edge responses of young from both types of nest site, and found a greater reluctance to jump from an elevated platform in young from cliff nests. By switching eggs between the two types of nest, he was able to demonstrate that differences in early experience, rather than in genetics, were responsible for the observed results. In a more detailed laboratory study, McLannahan (1973) verified that early rearing experience on a cliff can strengthen the cliff avoiding tendency in young Herring Gulls. Interestingly, McLannahan found that the cliff edge responses of young kittiwakes were affected only marginally by the physical characteristics of the rearing site. These results suggest a much less plastic ontogeny in the more highly specialized kittiwake, as compared with the more generalized, opportunistic Herring Gull.

IV. THE SOCIAL ENVIRONMENT

With the exception of the four species of fully precocial alcids (Section II.A), the newly hatched and developing young of all seabirds are fed and often brooded or guarded at or near the nest site by their parents. Because of this early reliance on parental care, parents no doubt constitute the most important aspect of the social environment for the young of most if not all seabirds. Ocurring less generally, but nevertheless of importance where they are present, are relationships with siblings in species having multiple-young broods and, in most colonial species, relationships with adults or young of other nearby family units. These three aspects of the social environment, parents, siblings, and other adults or young, are considered sequentially below.

A. Interactions with Parents

1. Care Soliciting by the Young

Young seabirds resemble their terrestrial counterparts in that they produce begging or "distress" signals that can be interpreted as attempts by the young to manipulate the parent (Dawkins and Krebs, 1978) to ensure that the young receive an appropriate level of parental investment in the form of brooding and feeding. In many species, vocalizations by the young are known to begin before hatching, during the pipped egg stage. In the Common Murre (Tschanz, 1965, 1968) and Laughing Gull (Impekoven, 1973), calls produced by pipped eggs have been observed to induce parents to call back to the young. Adult Black-headed Gulls (Beer, 1966*b*) and Ring-billed Gulls (Emlen and Miller, 1969) have been observed actually attempting to feed vocalizing pipped eggs. These observations have prompted suggestions that the calls of the young are important in the change from incubation to care of the young after hatching (see Beer, 1966*b*; Impekoven, 1973). The probable importance of vocalizations from pipped eggs in eliciting parental behavior is also suggested by the strong development of such calling in the young of otherwise relatively quiet species, such as the White Pelican (Schaller, 1964; R. M. Evans, personal observation). Experiments in which new young have been substituted for unpipped eggs (Beer, 1966*b*; Emlen and Miller, 1969; Kepler, 1969) have shown that prehatching exposure of adults to chick calls is not necessary for the onset of parental behavior in the species examined, but such results do not rule out the possibility that prehatch calls may have some facilitating effect on parental care (see Impekoven, 1973). Further experimental studies of the role of embryonic or neonatal vocalizations in facilitating the switch from incubation to brooding and chick-feeding behavior in seabirds are required.

The occurrence of vocalizations by the young when begging for food is widespread amongst seabirds, as is the tendency of the young to peck at or in the vicinity of the parent's bill. Pecking at the parent's bill has been studied especially for young gulls and terns (Tinbergen, 1953; Cullen, 1962; Quine and Cullen, 1964; Hailman, 1967; Nystrom, 1972), and has also been noted in all other orders of seabirds (Allan, 1962; Schaller, 1964; Nelson, 1966, 1967*b*; Brown and Urban, 1969; Burke and Brown, 1970; Fisher, 1970; Conroy, 1972; Spurr, 1975). In gulls, there is experimental evidence (Henderson, 1975) that both pecking and vocalizations increase as a function of deprivation time ("hunger") of the young. Long-term quantitative studies of begging are rare, but in at least one species, the Atlantic Gannet, the ratio of begging to feeding has been found to correlate strongly with chick age (Chapter 9). The Red-tailed Tropicbirds are unusual amongst seabirds in that the young gape in the fashion of many song birds rather than peck at the parent's bill (Howell and Bartholomew, 1969).

Although both pecking (most seabirds) and vocalizations (all?) of the young are correlated with obtaining food from the parent, evidence that both of these behaviors actually have a functional role in causing the parent to bring forth food is difficult to obtain (Nelson, 1966). One recent study

(Miller and Conover, 1979) that has attempted to answer this question experimentally is of particular relevance. By blindfolding young Ring-billed Gull chicks, Miller and Conover were able to deprive the young of the ability to peck at the parent's bill, although the young could still vocalize normally. Another group of young were surgically devocalized, but able to beg by pecking at the parent's bill. When devocalized chicks were presented to adults whose own were 1–3 days of age, feeding of the devocalized chicks was depressed significantly compared to sham-operated or unoperated controls. In contrast, the blindfolded chicks, who could call but not peck at the parent's bill, were offered food at rates equivalent to normal controls. The difference between devocalized and blindfolded chicks disappeared when tests were conducted with parents that had up to 3 days' experience feeding normal young. The results therefore suggest that for Ring-billed Gulls, at least, begging for food by vocalizing is more effective than pecking at the parent's bill during the initiation of feeding by a parent soon after the young hatch. As discussed by Miller and Conover (1979), the main function of the bill pecking response in gulls may be to direct the feeding activities of the young toward the source of food and thereby facilitate obtaining food when it is offered by the parent.

Chick vocalizations have been observed to stimulate parental feeding in other seabirds (e.g., Stonehouse, 1960), and might be expected to be especially important for young that do not open their eyes until some time after hatching (e.g., some penguins: Warham, 1974, 1975; shags: Snow, 1966), during nighttime feedings (Diamond, 1973) or for young hatched in burrows where visual cues may be greatly reduced or lacking (Thorensen, 1964).

In species whose young may wander away from the nest site, as in ground-nesting penguins, albatrosses, skuas, gulls, terns, and some others, a parent returning to the nest or territory to feed its young may initiate the feeding process by calling to the young, which typically respond vocally and approach the parent or home territory for feeding (Tinbergen, 1953; Sladen, 1958; Penney, 1968; Evans, 1970a; Fisher, 1970; Nelson, 1978). By placing young Laughing Gulls inside opaque-walled pens adjacent to a nest site, Beer (1979) has shown that vocalizations given by the young in response to their calling parent serve to attract the parent to the youngster. Consequently, a parent and its young are readily united for purposes of feeding even when the young are well away from the nest and hidden in vegetation.

2. Responses to Adult Alarm Calls

According to Tinbergen (1953), young Herring Gulls in the pipped egg stage respond to adult alarm calls by becoming silent. Upon hatching, a similar inhibition of activities in response to adult alarm calls has also been reported in gulls (Tinbergen, 1953; Tinbergen and Falkus, 1970; Smith and Diem, 1972; Impekoven, 1976). Within 1–2 days of hatching, the response begins to shift to a pattern characterised by rapid dispersal to a hiding place followed by a silent, immobile crouch (Tinbergen, 1953; Smith, 1966; Tinbergen and Falkus, 1970; Burger, 1974a; Impekoven, 1976). This behavior has usually been interpreted as an important antipredator adaptation (Kruuk, 1964). Similar behavior is common in many ground nesting skuas and terns (Palmer, 1941; Stonehouse, 1956; Cullen, 1960; Spellerberg, 1971a). In older young, more extended movements out onto adjacent water (e.g., gulls, terns, pelicans: R. M. Evans, personal observations) or into creches (Section IV.C.2) may also occur, but in these cases it is usually not clear whether the young are responding to adult alarm signals or to the physical presence of the observer or other disturbance.

Within dense colonies, it is usual for a predator or human intruder to evoke alarm calls from many adults, hence it is often not clear whether responses elicited by adult alarm calls are best considered as examples of interactions between young and their parents, or between young and other adults. In at least one species—the Laughing Gull—there is evidence (Beer, 1973) that the combined calls of the alarmed colony are much more effective than the alarm calls of a single adult. Comparisons between species which breed in dense colonies, like the Laughing Gull, with species that nest farther apart or in isolation, such as the Bonaparte's gull, should prove instructive.

3. Parent–Young Conflict

According to the theoretical formulations of parental investment developed by Trivers (1974), there normally will come a time during development when the young would improve its inclusive fitness by obtaining more parental care than the parent is selected to provide. Circumstantial evidence for such parent–young conflict is widespread within seabird colonies, as exemplified by young that beg extensively from parents that refuse to provide additional food (Warham, 1963; Young, 1963a; Schaller, 1964; Penney, 1968; Nelson, 1967b, 1969; Hunt, 1972; Spurr, 1975; Elston et al., 1977). Brief starvation periods prior to fledging, reported for some Procellariiformes (Rowan, 1952; Lockley, 1961; Tickell, 1960; Richdale, 1963; Harris, 1966; Round and Swan, 1977) are also indicative of possible parent–young conflict. However, prefledging starvation periods are much less prevalent than once imagined (Richdale, 1951), and there are many instances reported where parents have returned to a nest site after the young have departed (Richdale, 1951; Warham, 1963; Drent, 1965; Manuwal, 1974; Harris, 1976).

B. Interactions between Siblings

Seabirds as a group are characterized by small clutch sizes, with many species laying only one egg per clutch, (all Procellariiformes, and some species within the Sphenisciformes, Pelecaniformes, and Charadriiformes. Reviewed in Lack, 1968). Where two or more young are present, however, biologically important sibling interactions are possible and commonly occur.

Multiple-young broods of seabirds are characterized by an asynchronous hatch within the brood, younger offspring hatching 1–2 days (Adelie Penguin: Spurr, 1975; many gulls and terns: Barth, 1955, Schreiber, 1970; Howell et al., 1974), up to several days (e.g., boobies: Kepler, 1969; Simmons, 1967) after the first. As formulated by Lack (1954), asynchronous hatching within a brood appears to be a mechanism for selective brood reduction in times of food scarcity, the older chick typically, although not invariably, winning in the competition for food with its younger and weaker sibling. It has been suggested recently (Werschkful and Jackson, 1979) that sibling competition of this sort may select strongly for more rapid nestling development. As a corollary, lack of sibling competition in seabirds having only one young at a time may be conducive to the evolution of slow developmental rates.

Several studies have documented the occurrence of overt antagonism within broods. In the kittiwake, Cullen (1957) noted that apparently peaceful siblings immediately began to fight when the parent arrived with food. In this species, the onset of begging by the younger sibling induced the older sibling to attack. Only when the older sibling was fed was the younger able to get food. Sibling aggression is less commonly observed in ground-nesting gulls than in the kittiwake (Cullen, 1957), but it is common in some ground-nesting seabirds, notably the skuas and some members of the Pelecaniformes.

Sibling aggression in the South Polar Skua has been described by several authors (Young, 1963b; Spellerberg, 1971a; Procter, 1975). These studies have documented a high incidence of severe aggression and fratricide resulting from a tendency of the (usually) older and larger chick to harass and drive away its younger sibling. As described by Procter (1975), a dominance relationship between siblings develops soon after hatching, which can result in the dominant chick being avoided by the subordinate, which, as a result, commonly moves away from the protection of the parents and becomes more susceptible to predation or other mortality factors. By depriving young of food, Procter (1975) was able to show that food deprivation was primarily responsible for stimulating aggression. Where food is more abundant, sibling aggression is reduced or absent and broods of two are more commonly reared (Young, 1978; Trillmich, 1978).

Sibling aggressive interactions showing a striking convergence with those of the skuas occur in those members of the Pelecaniformes where the young hatched from two-egg clutches are faced with food shortages. In pelicans, severe harassment of the younger chick by its older sibling can result in the production of lacerations, attempts to escape from the nest, greatly reduced growth rates and frequently death (Din and Eltringham, 1974; Johnson and Sloan, 1978; R. M. Evans, personal observation). In the White Pelican, Knopf (1975) found that only 10% of a sample of 515 nests produced two young. In the same species, Johnson and Sloan, (1978) found that the younger sibling died as a direct result of physical abuse by the older sibling in more than 90% of observed nests containing two young. In the White Booby, the situation is even more extreme, more than one chick per brood virtually never surviving (Nelson, 1967b).

As shown by Kepler (1969) for the Blue-faced (= White) Booby, the young that survives is often (22% of sample) from the second egg laid, owing to failure of the first egg to hatch or to early death of the first young. Moreover, a young was fledged significantly more often from two-egg clutches than from one-egg clutches, thus indicating the laying of two eggs can represent an adaptive strategy for the parents despite the high incidence of sibling aggression and fratricide. A recent theoretical account of fratricide (O'Connor, 1978) shows that in extreme situations, the inclusive fitness of the victim itself can be enhanced by brood reduction.

It should not be assumed from the above accounts that all interactions between seabird siblings are necessarily aggressive in nature. On the contrary, siblings pecking at each others bill tips may facilitate the onset of feeding soon after hatching in young gulls (Hailman, 1961), and aggregative responses in some species may be strongly developed (Evans, 1970a). Even in White Pelicans, naked young during their first week tend to creep together and huddle, presumably for warmth, when placed on a table top at room temperature (R. M. Evans, personal observation). The extension of such early aggregative responses to the formation of creches is considered further below (Section IV.C.2).

C. Interactions with Neighbors and Other Conspecifics

In colonial species where nests are not hidden from one another behind ledges, in crevices or in burrows, interactions with neighbors and other adults or young constitute an important aspect of the social environment of the developing young. In some species, where other adults constitute a significant source of direct mortality or predation, as in the Gannet and Herring Gull and its close relatives, the presence of neighbors can have profound implications for the developing young. Some of these implications are considered next, followed by an examination of the development of creching behavior.

1. Adapting to the Presence of Dangerous Neighbors

The pressures that can be exerted on a developing young by its neighbors are perhaps seen in most striking relief in the Atlantic Gannet, where young that leave their tiny (2.3 pairs/m^2) territory run a high risk of being pecked and killed by neighbors (Nelson, 1978). This situation may be ameliorated somewhat by the continual guarding of the nest site and young by at least one parent, although as a consequence of this protective behavior the threat posed by a chick's neighbors is also continuously present. Faced with these severe social conditions, young Atlantic Gannets have adapted in what appears to be the most effective way possible under the circumstances—they normally do not wander from home, even when nests are placed on relatively large, flat substrates (Nelson, 1978).

The situation faced by young Herring Gulls and most other large ground-nesting gulls is somewhat different, but no less dangerous. Territories of Herring Gulls (approx 0.1 pair/m^2, Parsons, 1971) are appreciably larger than for the gannet, the young are cryptic in color, and the territory is usually well vegetated. However, these protective advantages to the young are offset by the highly developed tendencies of adults to attack and often kill and even eat young that trespass on their territories (Dutcher and Bailey, 1903; Ward, 1906; Herrick, 1912; Deusing, 1939; Paludan, 1951; Tinbergen, 1953; Harris, 1964; Brown, 1967; Parsons, 1971; Hunt, 1972). Cannibalism may be especially prevalent amongst adults that have lost their own broods (Davis and Dunn, 1976), although it is by no means limited to such adults. The danger to the young is increased as a result of their tendency to disperse when alarmed (Section IV.A.2), and as shown for the closely related Glaucous-winged Gull (Hunt and McLoon, 1975), by their tendency to become more active when hungry. Unlike the Atlantic Gannet, young Herring Gulls cannot adapt to such an environment by remaining at the nest, as that would render them highly conspicuous to potential predators and cannibalistic adults. Their main adaptation to dangerous neighbors appears to be a tendency to remain localized on their parents' territory (Tinbergen, 1953), combined with a markedly reduced tendency to approach parental attraction calls of their own species (Evans, 1973, 1979; Section V.B).

Instances of infanticide have been reported for many other seabirds, especially in the larids during and after periods of disturbance by humans (Austin, 1929; Marples and Marples, 1934; Pettingill, 1939; Emlen, 1956; Behle and Goates, 1957; Tinbergen and Falkus, 1970; Vermeer, 1970;

Feare, 1976; Hunt and Hunt, 1976; Conover *et al.*, 1980), and instances of cannibalism have been reported in the McCormick Skua (Spellerberg, 1971*b*). Milder hostility in the form of threats or nonlethal pecks are a more common response of adults to wandering, trespassing young in all orders of seabirds (Stonehouse, 1960; Dorward, 1962; Snow, 1963; Warham, 1963; Schaller, 1964; Nelson, 1967*a*; Penney, 1968; Brown and Urban, 1969; Burke and Brown, 1970: Fisher, 1970; Harris, 1973; Spurr, 1975; Tschanz and Hirsbrunner-Scharf, 1975). Even when not inflicting serious damage, such aggression by the adults is presumably sufficient to discourage adoptions. Food robbing by foreign young has also been reported (e.g., Elston *et al.*, 1977), and this behavior would presumably also be inhibited by aggressive acts directed at the young.

Responses by the young to attacks by adults are highly variable, and include turning or running away (many species), attempting to get under the adults as reported for the densely nesting Common Murre (Tschanz and Hirsbrunner-Scharf, 1975), or some form of appeasement posture, such as bill hiding (Cullen, 1957; Moynihan, 1959*a*; Dorward, 1962; Cullen and Ashmole, 1963; Nelson, 1967*a*; Fisher, 1970; Howell *et al.*, 1974). In addition, the young of many species soon develop some form of aggressive response (Nice, 1962) that can, as the young grow older, develop into an effective means of individual or territorial defence (Duffy, 1951; Cullen and Ashmole, 1963; Snow, 1963; Beer, 1966*a*; Nelson, 1967*b*, 1969; Dinsmore, 1972; Spurr, 1975; Howell, 1978).

2. Creches

Creches are characteristic of several colonial ground-nesting penguins (Richdale, 1951; Pettingill, 1960), but are by no means restricted to this group. The young of all ground-nesting pelicans apparently exhibit creching behavior (Schaller, 1964; Brown and Urban, 1969; Vestjens, 1977), and even in the tree-nesting Pink-backed Pelican the young from adjacent nests within a given tree may form into small groups when conditions permit (Burke and Brown, 1970). Creching is also characteristic of the Royal Tern (Buckley and Buckley, 1972*a*), some populations of the Sandwitch Tern (Smith, 1975), other large crested terns (reviewed in Buckley and Buckley, 1972*a*), and is present in at least two species of gull, the Black-billed Gull (Beer, 1966*a*; R. M. Evans, personal observation) and the Slender-billed Gull (Isenmann, 1976). Aggregations of prefledging young have also been noted in other densely nesting species, especially during times of disturbance (Nelson, 1966; Evans, 1970*a*; Fisher, 1970: Harris, 1973; Howell *et al.*, 1974), but it is questionable whether the latter instances represent true creches.

The age at which creching begins is highly variable, both within and between species, and ranges from 2–3 days after hatching in the highly mobile

Royal Terns to 2–4 weeks in most semialtricial and altricial species (Table V). In the virtually altricial King Penguin, the onset of creching can be delayed until about the 6th week after hatching (Stonehouse, 1960). Creching clearly can not begin until the young are sufficiently mobile to enable them to aggregate, and this, combined with delayed temperature regulation (Table I), may account for the general tendency for creching to be delayed in altricial species (Table V). Density of cover and extent of disturbance may also influence the timing and extent of creching (Smith, 1975).

According to most authors (see Yeates, 1975; Buckley and Buckley, 1972a), creching represents an adaptation to reduce predation, the grouping of the young rendering them less susceptible to attack than they would be if they remained scattered on their nesting territories. Creching as a means of temperature regulation has also been discussed (Brown and Urban, 1969; Yeates, 1975). Whether the immediate benefit derived from creching be primarily thermoregulation, protection from predators, or both, it is evident that it provides a potential alternative to continuous brooding or guarding by the parents, thereby freeing them for other activities, such as foraging (Stonehouse, 1953; Buckley and Buckley, 1972a). It is noteworthy in this regard that creching species are either flightless (penguins) or else forage relatively far from the breeding colony, as in pelicans (Brown and Urban, 1969; Burke and Brown, 1970; Johnson and Sloan, 1978), Royal Terns (Erwin, 1977) and Slender-billed Gulls (Isenmann, 1976). It seems reasonable to expect that in all these situations the parents would benefit from an increase in foraging time. If so, then it follows that creching, like burrow nesting and an early onset of temperature regulation in semialtricial young (Section II.A), can usefully be viewed as an adaptation serving to free the parents from incubation or guard duties so that they can devote more time to foraging. This interpretation of the ultimate adaptive significance of creching appears to be consistent with existing data, but requires further testing.

V. IMPRINTING AND THE DEVELOPMENT OF SPECIES IDENTIFICATION

A. Imprinting

The study of imprinting to parents or young companions has received little attention in seabirds, due perhaps to the lack of overt following of the parent to food sources in most species. Possibly the earliest account of an imprinting process in seabirds was that of Watson (1908) in his studies of Noddy and Sooty Terns. He noted that group-reared young noddies tended

Table V. The Onset of Creching Behavior in Relation to Developmental Mode in Three Orders of Seabirds

Order and species	Mode of development	Onset of creching	Source
Sphenisciformes			
King Penguin	Altricial (?)	6th week	Stonehouse (1960)
Adelie Penguin	Semialtricial	28–31 days	Sladen (1958)
		17–32 days	Taylor (1962)
		21–25 days	Spurr (1975)
Rockhopper Penguin	Semialtricial	13–16 days	Pettingill (1960)
		19–23 days	Warham (1963)
Fiordland-crested Penguin	Semialtricial	21–28 days	Warham (1974)
Erect-crested Penguin	Semialtricial	~3 weeks	Warham (1972)
Pelecaniformes			
Great White Pelican	Altricial	25–30 days	Brown and Urban (1969)
White Pelican	Altricial	3–4 weeks	Behle (1958)
		26–30 days	Schaller (1964)
Charadriiformes			
Black-billed Gull	Semiprecocial	~2 weeks	Beer (1966a)
Royal Tern	Semiprecocial	2–3 days	Buckley and Buckley (1972a)

to aggregate at an age (3–4 days) when they would show hostile and fear reactions to strange objects. Subsequently, Kirkman (1937) noted that initial approaches to him by Black-headed Gulls of up to 12-hr age became replaced with fear responses soon thereafter. It has been observed that human keepers of young gulls are readily approached (Strong, 1923; R. M. Evans, personal observation).

In the above examples, it is not clear whether simple exposure to a visual stimulus was sufficient to produce attachments and discriminations between the familiar and novel objects, or whether conditioning arising from feeding was involved. Because parental feeding is a normal correlate of parent–young interactions in all semiprecocial and altricial seabirds, it might be supposed that they would not exhibit true imprinting, which is usually considered to arise only from simple exposure, in the absence of overt reinforcement with food or other reward (Bateson, 1966; Sluckin, 1972; Hess, 1973).

I tested the imprinting hypothesis for Ring-billed Gulls by exposing them to siblings, parents, and inanimate objects in the absence of associated feedings (Evans, 1970a, 1975a). In all cases, initial approach responses were maintained or facilitated (*sensu* Gottlieb, 1976) to the familiar stimuli, while novel objects began to elicit avoidance or withdrawal responses, in a manner consistent with imprinting in the more widely studied chickens and ducks. Evidence for a "sensitive" or "critical" period was also obtained—exposure to others was highly effective in producing approaches if exposure occurred during the first 3 days after hatching, but became progressively less effective thereafter, and was completely ineffective, under the conditions of the test, by 6 days of age. A critical period terminating after the end of 3 days is longer than in precocial species (cf. Bateson, 1966; Sluckin, 1972), and may reflect the slower rate of development in the semiprecocial Ring-billed Gull. In both groups, the critical period appears to terminate at or about the time of nest leaving (cf. Bjarvall, 1967; Evans, 1970a), and so may function to ensure that the young are able to distinguish between conspecifics and other similar species or potential predators by the time the young become mobile.

B. Species Identification by Voice

The interpretation that visual imprinting is a mechanism for species identification at or soon after the time of nest leaving has been complicated in recent years by the finding (Gottlieb, 1971) that in some ducks and domestic chickens, the young are able to respond selectively to species-typical maternal cells upon hatching, even when deprived of prior exposure to such stimuli ("parentally naive" young).

A strong ability to discriminate between parental mew calls of their own and other species has also been found in parentally naive Ring-billed Gulls and Herring Gulls (Evans, 1973, 1975b). Young Ring-billed Gulls exhibit a pattern similar to that of young precocial species, in that they approach more strongly to the mew call of their own species than to the equivalent call of the Herring Gull or Black-billed Gull. They also vocalize more strongly to the call of their own species, as expected from the functional correlation between approach and vocal begging responses.

Young Herring Gulls also exhibit a highly significant ability to discriminate between the calls of their own species and that of the Ring-billed Gull and Black-billed Gull, but they differ from the other species so far studied in that they were found to approach and vocalize less, rather than more, to the calls of their own species. This seemingly anomalous result is readily repeatable (Evans, 1979), and is still present even when young Herring Gulls are exposed to the conspecific call during the pipped egg stage (Evans, 1973).

Within a breeding colony, young Herring Gulls show apparently normal levels of approach and vocal responses to their mewing parents at feeding time (Tinbergen, 1953; Brian Knudsen, personal communication), so evidently some form of experience gained under natural rearing conditions facilitates responses to parental mew calls in this species. The tendency for parents to call before feeding the young suggests that food reinforcement could be involved. I examined this possibility in a recent study (Evans, 1979) in which young Herring Gulls were fed in the presence of a conspecific mew call. Tests of their responses to mew calls of both Ring-billed Gulls and Herring Gulls were conducted before (1 day old) and after (7 days old) food training. Results showed that the initial preference for the foreign call was reversed as a result of training with food. Responses to the conspecific call were facilitated (approaches) or maintained (calls), while responses to the other call declined. Evidently food training can influence responses to species-typical calls in this species, and so may play a role in the normal development of species identification. A similar conditioning to species-typical stimuli is suggested by the results of crossfostering experiments with Arctic Terns and Common Terns (Busse, 1977) and other crossfostering experiments (Harris, 1970; Harris et al., 1978), which have resulted in mates actually being selected from the foster species.

Although the evidence reviewed above indicates that young Herring Gulls are able to develop a preference for the mew call of their own species by least as early as 1 week of age, their failure to do so before training appears to be unique amongst species examined to date (Gottlieb, 1971), and poses questions as to the adaptive significance, if any, of such a developmental pattern. Two separate issues can be identified: (1) why do the young respond at such low levels to conspecific mew calls, and (2) why is the Herring Gull mew call relatively less attractive than the equivalent call of the Ring-billed Gull or Black-billed Gull?

As suggested elsewhere (Evans, 1979), the low levels of responding to adult mew calls by young Herring Gulls may be an adaptation to the severe conditions of infanticide and cannibalism found within colonies of this species (Section IV.C.1). A reduced tendency to respond to mew calls should reduce the likelihood of a youngster being accidentally lured away from the relative safety of its own parents' territory, especially at an early age before the onset of functionally effective individual recognition of its own parents. In addition, Herring Gulls exhibit a high degree of nest-site tenacity, both adults and young tending to remain localized on or near the territory while the brood is young (Tinbergen, 1953). Such localization means that the young are not as likely to be strongly selected to follow and run after their parents as are more mobile species. Ring-billed Gulls and Black-billed Gulls, in contrast, are less subject to infanticide and cannibalism (Vermeer, 1970; Beer, 1966a; R. M. Evans, personal observation), and are also more mobile than Herring Gulls (Evans, 1970a,b), hence a stronger approach and vocal responsiveness to parental calls is more likely to have been favored, much as in more fully precocial species.

The mew calls employed by Ring-billed, Herring, and Black-billed Gulls are strongly divergent (illustrated in Evans, 1973), and may also have been subject to selection pressures similar to those outlined above for their young. Thus, the tendency for broods of the Ring-billed and Black-billed Gull to leave the territory at an early age, sometimes forming large aggregations (Evans, 1970a; Beer, 1966a), would be expected to favor the evolution of a highly attractive parental mew call as well as a strong response to it by the young. A similar pressure does not seem likely in the more sedentary Herring Gull. Mew calls of Herring Gulls may, however, have been subject to other, counterselection pressures that have caused them to become relatively less attractive. According to most interpretations of gull displays (Moynihan, 1959b; Tinbergen, 1959; Stout et al., 1969; Beer, 1975), mew calls are thought to perform multiple functions during the breeding season. As described by Tinbergen (1959), one such function of mew calls is to act as a threat, or "distance increasing" display. If mew calls can sometimes function as distance increasing threats, it seems reasonable to expect a corresponding tendency toward a reduction in mew-call attractiveness in Herring Gulls, which characteristically defend relatively large territories (Parsons, 1971). Ring-billed and Black-billed Gulls both have small territories (mean inter-nest distance less than 1 m, Evans, 1970a,b), hence any such tendency toward reduced attractiveness in their mew calls should be less than in Herring Gulls.

It is clear from the results reviewed in this section that imprinting and food reinforcement combined with an apparently innate capacity to identify species-typical parental calls constitute a complex of mechanisms potentially capable of ensuring appropriate recognition of conspecifics under normal rearing conditions. However, only a few species have as yet been studied, and the extent to which the results described here for gulls can be generalized to other seabird groups is not known. Moreover, the apparently atypical developmental pattern exhibited by young Herring Gulls indicates that generalization even between fairly closely related species must be undertaken with caution. This area constitutes a fertile field for future ecologically oriented studies of behavior development in seabirds.

VI. FAMILY MAINTENANCE AND RECOGNITION

According to current evolutionary theory (Williams, 1966; Hamilton, 1972; Wilson, 1975; Krebs and Davies, 1978), natural selection favors individuals that behave so as to maximize their own inclusive fitness. Because adult seabirds are on average more closely related to their own offspring than to the offspring of their neighbors, it is to be expected that mechanisms favoring the maintenance of family units will be strongly selected for, especially during those periods when parental investment is large and resources scarce, as when parents are feeding dependent young.

In seabirds, the initial mechanism favoring family maintenance is simply localization of broods at or near the nest. Subsequently, in species where dependent young become mobile, family maintenance is favored by selective responsiveness to the home nest site and to members of the family unit. Important responses among family members include calling and approaching their own young by parents, parental rejection of foreign young, approaches and begging by chicks to their own parents, withdrawal, silence, and sometimes aggression directed at foreign adults by young, and approaches between siblings. Selective responses between members of a family unit usually, but not necessarily always, involve individual recognition by at least one member of the interacting pair (Section VI.B). These various behaviors facilitating family maintenance are considered separately below, but it should be noted that they are often not mutually exclusive and may act together.

A. Brood Localization at or near the Nest

Immediately after hatching, contact between young seabirds and their parents is readily maintained by strong nest-site attachments by the parents combined with poorly developed locomotor capacities of the young. In some species, as in the cliff-nesting kittiwake discussed earlier (Section III.D), the young continue to stay in or close to the nest, where they are easily located by their parents at least until fledging.

In ground-nesting species, where physical barriers to brood mobility are reduced or absent, a tendency for the parent to direct its parental behavior towards young that are in or near the nest may still be an effective means of maintaining family unity during an initial posthatch period when the young normally restrict their activities to the parents' territory. The responses of a parent to young on the territory soon after hatching have recently been examined for the Ring-billed Gull by Conover et al. (1980). In this species, Conover et al. (1980) found that parents with young chicks of their own show more parental care towards chicks on their territory when chicks are (1) similar in size to the parent's own, (2) located close to the parent's nest, (3) located away from the nests of neighbors, and (4) located near vegetated areas that the parent's own brood had likely used in the past. Parental behavior was also greater (5) if the parent's own chicks were not in the nest. To this list could perhaps be added (6) the occurrence of normal behavior on the part of the offspring (Conover et al., 1980; Beer, 1979).

An important adaptation that can facilitate the localization of the brood at or near the nest is an ability on the part of the young to develop learned attachments to the nest site or adjacent area. Evidence for site attachment by the young during the rearing period is unfortunately difficult to obtain under natural conditions, due to the possibility that remaining at a nest may reflect nothing more than a tendency not to be mobile, as in altricial or semialtricial young. Moreover, even if movements away from the nest have occurred, return to the site may be prompted by selective responses to the parents or siblings (Section VI.C) rather than to stimuli from the site itself (Noseworthy and Lien, 1976). Clear evidence for site attachment based on field observations is, however, provided by the well-documented tendency of young Adelie Penguins to leave the creche and move to their natal territory where they are fed by the returning parent (Sladen, 1958). That such an ability is independent of the presence of the parent on the territory has been documented by playback of parental calls from locations outside the colony in the absence of the parents (Penney, 1968) and by instances in which the young return to the territory before the parent arrives with food (Sladen, 1958; Spurr, 1975). Young Sooty Terns (Dinsmore, 1972) and Laysan Albatrosses (Fisher, 1970) are also able to return to their home territory from nearby hiding or loafing places in the apparent absence of a parent.

Several training experiments have been conducted that demonstrate an early development of site attachments in seabirds. Watson (1908) trained eight young Sooty Terns, beginning on their 5th day, to run up a short inclined plane to reach a box where food was provided at mealtime. This response, which is somewhat analogous to the reunion of adults and young near the nest site at feeding times as described by the same author (and see Dinsmore, 1972), was learned by all experimental young by their 8th day. In Herring Gulls, Tinbergen (1953) has suggested that young rapidly become conditioned to the natal territory, and select particular hiding places on it when the colony is disturbed. An ability for experimentally displaced colony-reared young Herring Gulls to hide in the same type of vegetation (grass, herb, or shrub) as that which predominated around their nest (Noseworthy et al., 1973; Noseworthy and Lien, 1976) provides experimental evidence for an attachment to familiar habitat cues in this species.

In Ring-billed Gulls, young soon learned to go directly to a particular hiding place when trained away from the colony, and also demonstrated a significant, albeit weak, preference for one of two artificial habitats in which they were reared (Evans, 1970a). A strong preference for a distinctively patterned artificial substrate on which it had been fed has been documented for the Common Murre (Tschanz, 1968). Young Common Murres also learned to recognize local and surrounding features of artificial ledges on which they were reared (Wehrlin, 1977). Young Atlantic Gannets, which tend to return to or near their natal territory to breed (philopatry), may also be assumed to develop strong site attachments as youngsters (Nelson, 1978).

B. Recognition of Offspring by Parents

The tendency for adults to reject foreign young that trespass or are experimentally placed in their nest or on their territory has sometimes been taken as evidence that the rejecting parent is able to individually recognize its own young. As a corollary, a failure to reject foreign young might be taken as evidence that individual recognition of offspring by the parent is absent. Recent experimental evidence indicates that both of these assumptions may be invalid.

The assumption that rejection of foreign chicks indicates an ability of a parent to individually recognize its own young fails to take into account the possibility that such rejection is based entirely on the behavior of the young, which may itself be able to distinguish between parents and foreign adults and respond differently to each. Experimental evidence that behavior, rather than the individual identity, of the young can influence acceptance of a young by an adult has been obtained recently for Laughing Gulls (Beer, 1979).

Using playback techniques, Beer (1979) examined the ability of adult Laughing Gulls to respond selectively to vocalizations of their young at an age (6–8 days) when adults of other ground-nesting species have been shown to reject foreign young (citations in Section VI.D.) Under the conditions of Beer's test, parents failed to respond selectively to the calls of their own young. In a subsequent experiment, young belonging to a parent and a foreign young were placed in small opaque cages, one on either side of the nest. When the parent returned to the territory and called, its own young vocalized in return, the parent oriented and approached the calling chick, and so, in effect, responded selectively to the calls of its own chick even though apparently unable to individually recognize the chick's voice, as demonstrated in the first experiment. Selective responsiveness by the adult was thus presumably based on the occurrence of normal vocal behavior by the chick, rather than on individual recognition by the parent. Although these experiments do not rule out the possibility that the parents may have been able to individually recognize their own chicks by visual cues, they do show that behavior of the young may be important, and cannot be ignored unless explicitly ruled out by the design of the test.

The assumption that a failure to reject foreign young indicates a lack of recognition of its young by a parent has also been challenged by recent experimental evidence. In a study of the Caspian Tern, Shugart (1977) made use of a sensitive two-choice situation to test for chick recognition before the time that adults of this species normally reject foreign young placed in their nest. Shugart removed the original nest and replaced it with two other nests, placed 30 cm on either side of the original site. A chick of the home parent was then placed into one of the new nests, and a foreign chick of the same age and color phase was placed into the other. Parents returning to these nests selected their own offspring significantly more than the foreign chick when their own young were only 2–3 days old.

Similar results have been obtained for other species. For example, in the Fairy Tern, where the young normally remains close to or on the (tree) nest site for many days, rejection of foreign young was absent until at least 11 (probably more) days of age. In a special roof-nesting situation, where it was possible for the young to move away from the nest, parents went directly to their own 3-day-old young and tended them while ignoring others nearby (Howell, 1978). The particular type of nest site used by the parents has also been found to be relevant in Herring Gulls. In this species, adults from typical ground nests are known to reject foreign young when their young are about five days old (Tinbergen, 1953). In contrast, cliff-nesting Herring Gulls have recently been found to accept foreign young when the parents' own young are more than 1 week of age (von Rautenfeld, 1978; see also Harris, 1964).

The above-described experiments with Caspian Terns, Fairy Terns, and Herring Gulls show clearly that it is not valid to infer an inability to recognize merely because adults fail to reject foreign chicks. This does not, however, detract from the important implications of chick rejection data when assessing the probable ecological and adaptive significance of parent–young recognition—that adults reject potentially mobile foreign young but accept foreign young at an age when nest-site or mobility constraints normally render brood mixing unlikely constitutes strong circumstantial evidence that the functional onset of recognition is an important mechanism to ensure family unity in mobile broods.

C. Recognition of Parents by Young

When parent–young interactions are viewed from the perspective of the young, there is no compelling reason to assume that the best interests of the young are always served by restricting its begging and care soliciting behavior exclusively to its biological parent. On the contrary, provided a youngster suffers no ill treatment, natural selection would be expected to favor chicks that from time to time obtain care from nonparents, especially during times of need as when food supplies are less than optimal. It is therefore not surprising that the seabird literature contains many accounts of apparently indiscriminate begging and care soliciting by young reared in dense colonies, where the potential for such opportunities abound. In at least some instances, the young are successful in obtaining food from neighbors (Sladen, 1958; Nelson, 1966; Dinsmore, 1972; Howell et al., 1974; Feare, 1976; Elston et al., 1977), and in extreme instances, young are able to gain adoption by foster parents (Palmer, 1941; Brown *et al.*, 1967; Howell *et al.*, 1974; Hunt and Hunt, 1975; Spurr, 1975; Feare, 1976; Conover *et al.*, 1980).

Despite instances of indiscriminate care soliciting by the young, such behavior is apparently not usually the most efficient strategy for a chick to adopt. One important reason for this is the strong counterstrategy of rejecting strange chicks commonly employed by the adults (Section IV.C.1). In addition, it seems likely that on average, chicks that are overly enthusiastic about soliciting care from nonparents would often tend to become attached to other broods which could soon swell to numbers above the optimum that the (foster) parent could effectively feed. There is evidence for many seabird species that young reared in broods of artificially inflated size grow more slowly or survive less well than do young in broods of normal size (Rice and Kenyon, 1962; Harris, 1966; Norman and Gottsch, 1969; Nettleship, 1972; Diamond, 1973; Perrins *et al.*, 1973; Jarvis, 1974; Lloyd, 1977; see also Nelson, 1978), hence the creation of larger than normal broods as a result of indiscriminate begging and care soliciting by the young should be selected against in the young as well as in the parents. An ability to respond selectively to their own parents should therefore be favored in the young of all species for which there is a potential for a significant degree of misdirected begging and care soliciting by the young, a condition present wherever dependent young are sufficiently mobile to bring them into direct contact with their neighbors.

Selective responsiveness based on individual recognition of parental calls by the young has now been documented in several species of colonial seabirds (reviewed in Beer, 1970; Spurr, 1975; Evans, 1977*a*; Table VI). Two basic playback methods have been used, employing either (1) a successive presentation of one or more parental calls and an equivalent series of calls from a foreign or neighboring adult, or else (2) a simultaneous presentation of the two call series. Where both methods have been used (Evans, 1970*a*), individual recognition was demonstrable at a younger age by the simultaneous presentation method, indicating that it is a more sensitive, although perhaps thereby less rigorous, test method. Both simultaneous and successive calling by a parent and foreign adult may occur in breeding colonies (Evans, personal observation), hence the two test methods appear equally valid.

Responses of the young to their own parents voice include vocal begging and approaches to the sound source and, in some species (e.g., Adelie Penguin: Penney, 1968), movement to the home territory. All these responses facilitate reunion of the chick with its parent, and so tend to ensure that parental investment is directed selectively toward the parent's own offspring.

D. Time of Onset of Parent–Young Recognition

It has commonly been suggested that the onset of parent–young recognition should occur at least as early as the age at which broods begin to move about to a degree that significantly increases the risk of brood mixing (e.g., Davies and Carrick, 1962; Evans, 1970a,b, 1977a; Miller and Emlen, 1975). To further test this generalization, in this section I examined the relationships between (1) the onset of parent–young recognition, (2) the age at which the young normally leave the nest, and (3) the age at which the young normally leave the vicinity of the home territory. Results are listed in Table VI.

The criteria used to assess parent–young recognition are listed at the bottom of Table VI and indicated for each species in the body of the table. Criteria based on selective responses to own parents or young, experiments involving playback of recorded voices, and rejection of foreign young provide the best evidence that individual recognition was involved, although in the case of rejection of foreign young it is not clear whether recognition is by the parent or by the young (Section VI.B). Aggression by the young directed at foreign adults may also be based on individual recognition, but differences in the behavior of family members and foreign individuals when reacting to the nest site may also have provided relevant cues. For many of the studies listed in Table VI, determining whether a brood was on or off the territory, and selecting age ranges for recognition and mobility, involved a certain degree of interpretation on my part. Where the margin of error seemed potentially large, I have put a question mark after the estimated age value.

From the evidence in Table VI, parent–young recognition apparently develops most rapidly in semiprecocial species, especially where the young begin to vacate the nest at an early age, as in the Common Murre, Crested Tern, and Royal Tern. Recognition is also present soon after hatching in most of the other semiprecocial larid species examined. In the altricial and semialtricial species listed in Table VI, there is a delay in the onset of both nest leaving and recognition, particularly in the gannet, shag, and the two albatrosses. Although only one species of penguin and two Pelecaniformes are listed, it is probable that they reflect the general situation within their respective orders. Details of the onset of recognition have rarely been determined for penguins and Pelecaniformes, but the existence of recognition at least by the onset of extensive brood mobility or creche formation has been suggested or documented for several species in both groups [penguins: reviewed in Spurr (1975); pelicans: Schaller (1964); Brown and Urban (1969)].

Examination of the relationship between the onset of recognition and

Table VI. The Onset of Parent–Young Recognition in Relation to the Age at Which Broods Leave Their Nest and Vicinity of Their Territory

Order and species	Onset of recognition and type of evidence[a]	Broods begin to leave nest	Broods begin to leave vicinity of home territory	Source
Sphenisciformes				
Adelie Penguin	8–17 days (R)	11–15 days	21–25 days	Thompson and Emlen (1968); Spurr (1975)
Procellariiformes				
Wandering Albatross	56–140 days (C,R)	? – 150 days	263–303 days	Tickell and Pinder (1972); Tickell (1968)
Laysan Albatross	42–49 days (R)	? – 60 days	165 days (mean)	Rice and Kenyon (1962)
Pelecaniformes				
Atlantic Gannet	~60 days (F)	90 days (mean)	90 days (mean)	White (1971); Nelson (1978)
Shag	29 days (R)	25–30+ days	28–33+ days	Snow (1963)
Charadriiformes				
Larinae (gulls)				
Herring Gull	By 5 days (R) (on flat land; see text)	1 day	By 10 days	Tinbergen (1953)
Ring-billed Gull	3–5 days (P)	2–6 days	3–9+ days	Evans (1970a)
Black-billed Gull	2–4 days (P)	2–4 days	2–5 days	Beer (1966a); Evans (1970b, and unpubl. notes)

Laughing Gull	1–8 days (P)	3–8 days	12?–35 days	Beer (1970); Hahn (1977)
Franklin's Gull	9–14 days (R) 16 days (Y)	3–10? days	>28 days	Burger (1974a)
Sterninae (terns)				
Common Tern	By 2–4 days (P) By 5 days (R)	2–5 days	By 12 days	Palmer (1941); Stevenson et al. (1970)
Arctic Tern	2 days (D)	1–2 days	2–3 days	Pettingill (1939); Busse and Busse (1977)
Sooty Tern	3–5 days (R)	<3 days	3–4 days	Watson (1908); Dinsmore (1972)
Royal Tern	1–2 days (C)	1 day	2–3 days	Buckley and Buckley 1972a,b
Crested Tern	1–2 days (R)	By 3 days	By 3(?) days	Davies and Carrick (1962)
Caspian Tern	By 2–3 days (C,R) 2–9 days (R)	?	6 days	Shugart (1977)
Fairy Tern	3 days (C) (if off nest, see text)	3 days (if on flat surface)	70–96 days	Dorward (1963) Howell (1978)
Alcidae (murres)				
Common Murre	<1 day (D,P)	1 day	18–25 days	Tuck (1960); Tschanz (1968)
Razorbill	5(?)–10 days	?	16–23 days	Ingold (1973)

[a] (C) Parents respond selectively to own young; (P) playback experiments to young; (Y) young attack foreign adults; (D) young respond selectively to own parents; (R) rejection of foreign young by parents.

the two measures of mobility in Table VI indicates, first, that recognition usually develops before, and often appreciably before, the age at which the young normally begin to leave the vicinity of the home territory. Development of recognition well before the onset of such extended mobility is especially evident for species in which the young sometimes leave the nest at a relatively early age but then stay nearby for some time before fledging, as in the two albatrosses, the Fairy Tern, Common Murre, and some of the gulls (especially if undisturbed). The onset of recognition usually precedes, rather than coincides with, the onset of extended brood mobility.

A closer coincidence appears to exist between the onset of recognition and the time of nest leaving (compare second and third columns of Table VI). Excluding the four species for which I did not obtain an estimate of the onset of nest leaving, the onset of recognition occurred before nest leaving in at least four and probably five of the species listed in Table VI; recognition developed only after nest leaving in at least five and possibly six species, while both occurred on the same day in at least four species. When interpreting these data, it is important to note that ages of nest leaving are probably biased downwards due to the almost inevitable disturbances caused by the human investigator (many authors), whereas the development and use of more sensitive test methods may well lower the age at which recognition is known to occur in some species.

The results listed in Table VI can perhaps best be interpreted as indicating that parent–young recognition will almost certainly develop before the onset of brood movements away from the home territory, at least in species such as those listed in Table VI where parental care continues subsequent to vacation of the territory (see the next section for situations involving possible lack of recognition). The data for recognition in relation to departure from the home territory are thus in accord with the interpretation that recognition should develop at least as early as the onset of extensive brood mobility.

Interpretation of the close correlation between nest leaving and the onset of recognition is more equivocal, because of the several instances in which recognition has not been documented at the time of nest leaving. Closer examination of the circumstances involved in these species suggests that they do not in fact constitute significant exceptions to the relationship between potential brood mixing and mobility. This is particularly true for the Franklin's Gull, which is the only species listed in which nest leaving is shown as occurring well before the onset of recognition. In this marsh-nesting species, under relatively undisturbed conditions, the young do not normally leave the floating platform on which their nests are placed until fledging, and so run little risk of brood mixing as a result of leaving the nest cup prior to the onset of recognition (Burger, 1974a). Similar considerations are likely to apply in the other, less striking, instances where nest leaving sometimes precedes the onset of recognition. In the Ring-billed Gull, for example, the young are closely attended by the parents during temporary feeding excursions away from the nest during the first few days after hatching, and even when alarmed the young do not normally run more than a few centimeters when they first begin to leave the nest (Evans, 1970a). The potential for brood mixing is correspondingly small until mobility increases up to, or beyond, the margins of the home territory, by which time recognition is normally present (for similar conclusions see Beer, 1970; Nelson, 1978).

Although the above account has emphasized the probable importance of brood mobility and potential brood mixing as the main functional correlates of the onset of parent–young recognition, other selection pressures may also be involved. One important additional possibility is suggested by data from the black noddy (Cullen and Ashmole, 1963). In this species, cliff nest sites may be in short supply and heavily competed for by breeding adults, who show an unusually high degree of nest site tenacity throughout the year. The adults forage far away from the colony, and begin to leave the chick unattended within a few days of hatching. In the absence of the parents, other adults may attempt to land and claim the ledge but are threatened and driven off by the young. Under these circumstances, it seems evident that an ability on the part of the young to recognize its own parents and thereby selectively direct threats and pecks at foreign adults would be adaptive. Whether recognition is based on individual or generalized behavioral cues in this species is apparently not known, and warrants further detailed examination. Although generally less striking than in the Black Noddy, the importance of parent–young recognition to permit efficient defence of the home nest or territory by the young may also be relevant for other species in which this behavior occurs (e.g., Duffy, 1951; Snow, 1963; Nelson, 1967b, 1969, 1978; Dinsmore, 1972; Spurr, 1975; Howell, 1978).

In conclusion, the evidence to date suggests that where it occurs, the onset of individual parent–young recognition tends to correlate, albeit loosely, with the age at which interactions between the brood and foreign young or adults begin to endanger family autonomy. Depending on the circumstances, relevant interactions may arise either as a result of mobility of the brood or because of encroachment by others. The importance of brood mobility has been widely documented; encroachment by others does occur and seems important for some species and may be of potential importance for many, but requires further study.

E. Species That May Lack Parent–Young Recognition

In the kittiwake, where nests are typically placed on small ledges to which the young are restricted until fledging, chicks put into the nests of foster parents were not rejected (Cullen, 1957), even though adults of this species have been shown to have a highly developed capacity to individually recognize the voices of their mates (Wooller, 1978). In more recent play-back experiments with kittiwakes, Woller (personal communication) has confirmed an apparent inability of young kittiwakes to respond selectively to the voices of their own parents. The cliff nest site noramlly inhibits brood mixing in the kittiwake, young do not normally defend the ledge (Cullen, 1957), and parents do not normally feed their young away from the nest ledge after fledging (Coulson, 1974), so there seems to be no compelling ecological reason why individual parent–young recognition should be manifest in this species.

Failure to reject foreign young that normally do not leave the nest site until fledging and the termination of parental care has also been found in a few other species, including the cliff-nesting Atlantic Gannet (Nelson, 1966), the Black-browed and Grey-headed Albatrosses (Tickell and Pinder, 1972), the burrow-nesting Manx Shearwater (Brooke, 1978), and possibly the Red-tailed Tropicbird (Howell and Bartholomew, 1969; Fleet, 1974). Richdale (1951) has suggested that lack of parent–young recognition may be charac-teristic of petrels that feed their young only in the burrow.

Caution is needed in interpreting these negative results since more sensitive methods or test situations involving different circumstances could produce different conclusions, but the weight of the evidence to date appears to favor the view that parent–young recognition may fail to be manifest at all where it is not ecologically required to ensure appropriate dispensation of parental investment. The situation in the Atlantic Gannet, where the young apparently recognize their parents, but the parents do not exhibit rejection of foreign young (Nelson, 1978) is difficult to interpret, and merits further study.

F. Recognition of Siblings

The possibility that young are able to respond selectively to their sibl-ings has received less study then parent–young recognition, but there is evi-dence that sibling recognition does occur in species having broods of two or more young (Evans, 1970a; Noseworthy and Lien, 1976). In Ring-billed Gulls, colony-reared young released on an open area away from the colony formed aggregations with both siblings and nonsiblings, but nevertheless exhibited a significant tendency to approach siblings selectively when tested at 4–6 days of age (Evans, 1970a). When tested under the confines of a test runway, colony-reared young responded selectively to their siblings at 4, but not at 3 days of age. Laboratory reared young also selectively approached their rearing companions, when tested at 4.5 days of age (Evans, 1970a). In mobile broods of Black-billed Gulls, I have similarly noted instances in which individually marked siblings moved about together within a loose creche in the apparent absence of their parents (Evans, unpublished notes). Also suggestive of sibling recognition are observations of young threatening or attacking foreign young, described for several seabird species (Cullen and Ashmole, 1963; Snow, 1963; Warham, 1963; Schaller, 1964; Nelson, 1966, 1967b, 1969; Dinsmore, 1972; Spurr, 1975; Howell, 1978).

The adaptive significance of sibling recognition has received even less direct attention than have attempts to document the phenomenon. However, recent theoretical advances in the understanding of the role of kin selection in the evolution of social behavior (Hamilton, 1972) suggest that sibling recognition may be of fundamental importance in instances where it permits cooperation or selective sharing of excess resources and parental investment with a sibling rather than with an unrelated offspring. In mobile young, for example, close spatial adherence between siblings based on individual recognition should favor the efficient deployment of parental care for all the young of a brood, a result that should enhance the inclusive fitness of the young provided resources are not in too limited a supply (O'Connor, 1978). In broods localized at a nest or other site, sibling recognition should facilitate the ability of young to recognize and deter encroachment by foreign young should the latter attempt to gain access to food or other form of care delivered by the home parent.

G. Developmental Mechanisms

As shown by the classic studies of Tschanz (1965, 1968), young of the common murre may begin to learn the characteristics of their own parents voice as a result of being exposed to it during the pipped egg stage. The tendency for the parent to call in association with shuffling or turning the eggs appears to facilitate the ability of the young to learn at this time (Clements and Lien, 1976). In the Laughing Gull, prenatal exposure to sound also has an effect on the development of auditory discriminations, but only when tested before hatching (Impekoven and Gold, 1973). After hatching, the responsiveness of Laughing Gulls and Herring Gulls may be enhanced as a result of prenatal exposure to adult calls, but selective responses to the familiar call have not been shown (Impekoven and Gold, 1973; Evans, 1973). Adults of other species also vocalize over their pipped eggs (e.g., Evans, 1970b; Fisher, 1970; Tinbergen and Falkus, 1970), but it remains to be shown what effect, if any, this experience has on the development of individual recognition.

After hatching, visual imprinting and reinforcement from parental feeding become available as potential mechanisms mediating the development of individual recognition. In Ring-billed Gulls, I recently (Evans, 1980) examined the effects of feeding the young in the presence of one of two mew calls recorded from different individual parents. The young readily learned to discriminate between the two calls, vocalizing (begging) and approaching more often in response to the familiar, food-reinforced call.

That visual imprinting stimuli constitute a basis for individual recognition has apparently not been demonstrated in seabirds. Visual imprinting stimuli may be indirectly involved in the development of individual recognition, however, because of the reinforcing effects of visual imprinting stimuli on the learning of parental calls, as shown for Ring-billed Gulls (Evans, 1977b). Direct proof that this learning mechanism actually functions in a natural setting when parents call to their young is lacking, but it is relevant that discriminations were formed under laboratory conditions when two separate training methods specifically designed to mimic aspects of the natural situation were used.

The training methods involving reinforcement from visual imprinting stimuli that were found to produce significant preferences for a particular adult call in Ring-billed Gulls can, for simplicity, be referred to as "parent–stranger" and "parent–neighbor" procedures. These represent two of the most common types of individual recognition problems to which a young seabird is likely to be exposed within the breeding colony. In the first training method young were trained by exposing them to a familiar visual imprinting stimulus combined with one adult mew call (representing exposure to "parent" stimuli at the nest), alternating with periods of silence with no active visual imprinting stimulus (representing an absence of early exposure to a "stranger"). in the "parent–neighbor" method, young were similarly exposed to "parent" stimuli, but this was alternated with periods of reduced exposure to a second Ring-billed Gull mew call (representing reduced exposure to a "neighbor"). When young gulls trained by either of these procedures were later subjected to successive discrimination tests involving the two Ring-bill mew calls in the absence of active visual simuli, they demonstrated a greater responsiveness to the "parent" call in each case (Evans, 1977b).

Developmental mechanisms responsible for the development of individual recognition have as yet been studied in only a few species, but the data so far obtained suggest that reinforcement from parental feedings, combined with reinforcement from the sight of a familiar, imprinted parent, and sometimes embryonic exposure to parental calls, combine to ensure a functionally adequate degree of parental recognition by the young. Apparently nothing is known about the way in which parents become conditioned to their own offspring.

VII. SUMMARY

The behavior of young seabirds is adapted to a combination of ecological factors associated with their physical and social rearing environments. Small brood size, extensive parental care, and slow developmental rates, characteristic of K-selected species, are common and correlate in most species with a difficulty in obtaining food.

Four species of murrelets (Alcidae) are fully precocial and leave the nest for their marine feeding grounds soon after hatching. All other seabird young are fed at or near the nest, usually until about the time of fledging. Semiprecocial young of most gulls and terns are able to move away from the nest to seek shelter or protection from predators. Altricial young of most Pelecaniformes (e.g., pelicans, gannets, boobies) are less mobile. They are almost naked when hatched, and are therefore brooded in the nest at least until down grows out and temperature regulation develops. The semialtricial young of the penguins and Procellariiformes (e.g., albatrosses, petrels) are down-covered at hatching and have an early onset of thermoregulation, which appears to be an adaptation allowing an early cessation of brooding and hence a greater proportion of time spent foraging by the parents. In all species, development can usefully be related to three stages of brood mobility: (1) young at the nest, (2) young on the home territory, and (3) young beyond the vicinity of the home territory.

Major hazards associated with the physical characteristics of the nest in some seabirds include the danger of falling from cliff or tree nests, and getting wet and waterlogged at marsh nest sites. Young at such sites are typically less mobile than ground-nesting species. young kittiwakes and Common Murres, which have been studied intensively, exhibit other important convergent adaptations to their cliff nest sites, including facing the cliff wall and avoiding cliff edges. There is experimental evidence from Herring Gulls that early rearing experience at cliff nest sites can enhance the ability of the young to avoid falling over cliff edges.

The main elements in the social environment include parents and, where present, siblings and neighbors. Vocal begging for food from parents develops early and is often present in the pipped egg stage. Recent experiments with Ring-billed Gulls indicate that for newly hatched young, begging for food by calling is significantly more effective than pecking at a parent's bill. Young of many species commonly continue to beg from parents that are unwilling to provide food, suggesting a widespread occurrence of parent–young conflict. Sibling conflict is also present, especially in skuas, pelicans, and some boobies, where the aggressive behavior of one young commonly leads to the death of the other, usually younger, member of the brood. Such fratricide appears to be an adaptation to insufficient food which may, in extreme situations, enhance the inclusive fitness of the victim as well as the survivors.

In colonial species, neighbors constitute an important, and often dangerous, aspect of the developmental environment. Young Atlantic Gannets and Herring Gulls have adapted to potentially lethal attacks from hostile neighbors by remaining closely associated with their own, parent-protected, territory. Sublethal pecks at wandering young are of widespread occurrence amongst ground-nesting seabirds; such pecks presumably function to ensure that parental investment is not wasted on unrelated young. Young may also threaten or attack other young that encroach on their territory. Where territorial attachment is reduced, mobile young often form creches, as in many species of ground-nesting penguins, pelicans, and some gulls and terns. Creches appear to provide an alternative to persistent brooding and guarding, and thus may be an adaptation freeing the parents to devote more time to foraging.

Imprinting has been documented in the Ring-billed Gull. Parentally naive young Ring-billed Gulls also approach and beg selectively in response to the parental mew call of their own species. Young Herring Gulls can also identify species typical mew calls, but are unusual in that they approach and beg less to the conspecific call than to the analogous call of the Ring-billed and Black-billed Gulls. A recent experiment has shown that with subsequent exposure to the conspecific call during feeding, young Herring Gulls soon learn to prefer it. The initially low level of response to the conspecific call in this species may be an adaptation reducing the risk that they will approach and be killed by hostile neighbors. The greater attractiveness of Ring-billed and Black-billed Gull calls correlates with, and seems likely to be an adaptation to, the greater degree of brood mobility in those species. Imprinting and species identification by voice have been studied in too few seabird species to warrant further generalizations at this time.

Because parents are more closely related to their offspring than to other young in a colony, mechanisms that favor the maintenance of family unity are common. Initially, family unity is ensured by low levels of mobility in the young, combined with parental tendencies to respond selectively to young that are on their own territories. In species in which the broods become mobile, parent–young recognition soon develops, as shown experimentally by the rejection of foreign young by parents, and by selective responses to parents by the young. Recent experimental studies have shown that the behavior of the young, which may itself recognize the parent, can constitute a cue used by parents when responding to their young. It has also been found that parents that fail to reject a foreign young may nevertheless be able to recognize their own young when presented with an appropriate choice test. A survey of parent–young recognition in seabirds indicates that where it occurs, recognition almost always develops before, and often well before, the young leave the vicinity of the home territory; recognition usually develops at about the time of nest leaving. An ability to defend the nest against intruders, as exemplified by the Black Noddy Tern, may favor an ability to recognize individuals prior to the onset of brood mobility in some species.

Some cliff-nesting species, such as the kittiwake, Atlantic Gannet and the Black-browed and Grey-headed Albatrosses, as well as some burrow-nesting petrels, do not normally appear to exhibit any manifestations of parent–young recognition. In these species, dependent young stay at or near the nest, and the young are not fed away from the nest after fledging, hence there appears to be no compelling ecological need for individual recognition between parents and their young.

Recognition of siblings has received less attention than parent–young recognition. Sibling recognition has been documented in gulls, and may be of adaptive importance as a means of facilitating efficient deployment of parental investment.

Learning of a parent's voice by young in the pipped egg stage has been demonstrated in the Common Murre, but such learning is apparently not of widespread importance as a mechanism for the development of individual recognition in seabirds. Recent experiments with Ring-billed Gulls have demonstrated that posthatch learning of individal mew calls can arise as a result of reinforcement provided by visually imprinted stimuli and by food. Apparently nothing is yet known about the way parents become conditioned to their own young.

APPENDIX: SPECIES OF SEABIRDS MENTIONED IN TEXT, LISTED BY ORDER AND FAMILY

Order: Sphenisciformes
  Family: Spheniscidae (penguins)
    Adelie Penguin, *Pygoscelis adeliae*
    Erect-crested Penguin, *Eudyptes sclateri*
    Fiordland-crested Penguin, *E. pachyrhynchus*
    King Penguin, *Aptenodytes patagonica*
    Rockhopper Penguin, *E. crestatus*
Order: Procellariiformes
  Family: Diomedeidae (albatrosses)
    Black-browed Albatross, *Diomedea melanophris*
    Black-footed Albatross, *D. nigripes*
    Grey-headed Albatross, *D. chrysostoma*

Laysan Albatross, *D. immutabilis*
Wandering Albatross, *D. exulans*
Family: Procellariidae (petrels, shearwaters)
Fulmar, *Fulmarus glacialis*
Giant Petrel, *Macronectes giganteus*
Manx Shearwater, *Puffinus puffinus*
Slender-billed Shearwater, *P. tenuirostris*
Family: Hydrobatidae (storm petrels)
Leache's Petrel, *Oceanodroma leucorhoa*
Order: Pelecaniformes
Family: Phaethontidae (tropicbirds)
Red-tailed Tropicbird, *Phaethon rubricauda*
Family: Pelecanidae (pelicans)
Brown Pelican, *Pelecanus occidentalis*
Great White Pelican, *P. onocrotalus*
Pink-backed Pelican, *P. rufescens*
White Pelican, *P. erythrorhynchos*
Family: Sulidae (gannets, boobies)
Abbott's Booby, *Sula abbotti*
Atlantic Gannet, *Morus bassanus*
Blue-faced Booby (see White Booby)
Blue-footed Booby, *Sula nebouxii*
Brown Booby, *S. leucogaster*
Gannet (see Atlantic Gannet)
Masked Booby (see White Booby)
Peruvian Booby, *S. variegata*
Red-footed Booby, *S. sula*
White Booby, *S. dactylatra*
Family: Phalacrocoracidae (Shags, Cormorants)
Shag, *Phalacrocorax aristotelis*
Family: Anhingidae (darters)
Darter, *Anhinga melanogaster*
Family:Frigatidae (frigatebirds)
Order: Charadriiformes
Family: Stercorariidae (skuas)
McCormick Skua, *Catharacta maccormicki*
South Polar Skua (see McCormick Skua)
Family: Laridae (gulls, terns)
(Subfamily: Larinae, gulls)
Black-billed Gull, *Larus bulleri*
Black-headed Gull, *L. ridibundus*
Bonaparte's Gull, *L. philadelphia*
Brown-hooded Gull, *L. maculipennis*
California Gull, *L. californicus*

Franklin's Gull, *L. pipixcan*
Glaucous Gull, *L. hyperboreus*
Glaucous-winged Gull, *L. glaucescens*
Herring Gull, *L. argentatus*
Iceland Gull, *L. glaucoides*
Kittiwake, *Rissa tridactyla*
Laughing Gull, *L. atricilla*
Ring-billed Gull, *L. delawarensis*
Slender-billed Gull, *L. genei*
Swallow-tailed Gull, *L. furcatus*
Thayer's Gull, *L. thayeri*
(Subfamily: Sterninae, terns)
Arctic Tern, *Sterna paradisaea*
Black Noddie, *A. tenuirostris*
Black Tern, *Chilodonias niger*
Brown Noddie, *Anous stolidus*
Caspian Tern, *Sterna caspia*
Common Tern, *S. hirundo*
Crested Tern, *S. bergii*
Fairy Tern, *Gygis alba*
Forster's Tern, *Sterna forsteri*
Roseate Tern, *S. dougallii*
Royal Tern, *S. maximus*
Sandwitch Tern, *S. sandvicensis*
Sooty Tern, *S. fuscata*
Family: Alcidae (murres, auks, puffins)
Ancient Murrelet, *Synthliboramphus antiquum*
Common Murre, *Uria aalge*
Craveri's Murrelet, *Endomychura craveri*
Guillemot (see Common Murre)
Japanese Murrelet, *Synthliboramphus wumizusume*
Pigeon Guillemot, *Cepphus columba*
Razorbill, *Alca torda*
Thick-billed Murre, *Uria lomvia*
Xantus' Murrelet, *Endomychura hypoleuca*

REFERENCES

Allan, R. G., 1962, The Madeiran Storm Petrel *Oceanodroma castro*, *Ibis* **103b**:274–295.
Ashmole, N. P., 1963, The regulation of numbers of tropical oceanic birds, *Ibis* **103b**:458–473.
Austin, O. L., Jr., 1929, Contributions to the knowledge of the Cape Cod *Sterninae*, *Bull. NE Bird-Banding Assoc.* **5**:123–140

Barth, E. K., 1955, Egg-laying, incubation and hatching of the Common Gull (*Larus canus*), *Ibis* **97**:222–239.

Bartholomew, G. A., 1966, The role of behavior in the temperature regulation of the Masked Booby, *Condor* **68**:523–535.

Bartholomew, G. A., Jr., Dawson, W. R., and O'Neill, E. J., 1953, A field study of temperature regulation in young White Pelicans, *Pelecanus erythrorhynchos*, *Ecology* **34**:554–560.

Bartholomew, G. A., Jr., and Dawson, W. R., 1954, Temperature regulation in young pelicans, herons, and gulls, *Ecology* **35**:466–472.

Bateson, P. P. G., 1966, The characteristics and context of imprinting, *Biol. Rev.* **41**:177–220.

Bédard, J., 1969, Adaptive radiation in Alcidae, *Ibis* **111**:189–198.

Beer, C. G., 1966a, Adaptations to nesting habitat in the reproductive behaviour of the Black-billed Gull *Larus bulleri*, *Ibis* **108**:394–410.

Beer, C. G., 1966b, Incubation and nest-building behaviour of Black-headed Gulls. V: The post-hatching period *Behaviour* **26**:189–214.

Beer, C. G., 1970, Individual recognition of voice in the social behavior of birds, in: *Advances in the Study of Behavior* (D. S. Lehrman, R. A. Hinde, and E. Shaw, eds.), Vol. 3, pp. 27–74, Academic Press, New York.

Beer, C. G., 1973, A view of birds, *Minn. Symp. Child Psychol.* **7**:47–86.

Beer, C. G., 1975, Multiple functions and gull displays, in: *Function and Evolution in Behaviour* (G. P. Baerends, C. G. Beer, and A. Manning, eds.), pp. 16–54, Clarendon Press, Oxford.

Beer, C. G., 1979, Vocal communication between Laughing Gull parents and chicks, *Behaviour* **70**:118–146.

Behle, W. H., 1958, *The Bird Life of Great Salt Lake*, University of Utah Press, Salt Lake City.

Behle, W. H., and Goates, W. A., 1957, Breeding biology of the California Gull, *Condor* **59**:235–246.

Bekoff, M., 1972, The development of social interaction, play, and metacommunication in mammals: An ethological perspective, *Q. Rev. Biol.* **47**:412–434.

Bergman, R. D., Swain, P., and Weller, M. W., 1970, A comparative study of nesting Forster's and Black Terns, *Wilson Bull.* **82**:435–444.

Birkhead, T. R., 1977, Adaptive significance of the nestling period of Guillemots *Uria aalge*, *Ibis* **119**:544–549.

Bjärvall, A., 1967, The critical period and the interval between hatching and exodus in mallard ducklings, *Behaviour* **23**:141–148.

Brooke, M. de L., 1978, Sexual differences in the voice and individual vocal recognition in Manx Shearwater (*Puffinus puffinus*), *Anim. Behav.* **26**:622–629.

Brown, L. H., and Urban, E. K., 1969, The breeding biology of the Great White Pelican *Pelecanus onocrotalus roseus* at Lake Shala, Ethiopia, *Ibis* **111**:199–237.

Brown, R. G. B., 1967, Breeding success and population growth in a colony of Herring and lesser Black-backed Gulls *Larus argentatus* and *L. fuscus*, *Ibis* **109**:502–515.

Brown, R. G. B., Blurton Jones, N. G., and Hussell, D. J. T., 1967, The breeding behaviour of Sabine's Gull, *Xema sabini*, *Behaviour* **28**:110–140.

Buckley, F. G., and Buckley, P. A., 1972a, The breeding ecology of Royal Terns *Sterna (Thalasseus) maxima mamima*, *Ibis* **114**:344–359.

Buckley, P. A., and Buckley, F. G., 1972b, Individual egg and chick recognition by adult Royal Terns (*Sterna maxima maxima*), *Anim. Behav.* **20**:457–462.

Burger, J., 1974a, Breeding adaptations of Franklin's Gull (*Larus pipixcan*) to a marsh habitat, *Anim. Behav.* **22**:521–567.

Burger, J., 1974b, Breeding biology and ecology of the Brown-hooded Gull in Argentina, *Auk*, **91**:601–613.

Burke, V. E. M., and Brown, L. H., 1970, Observations on the breeding of the Pink-backed Pelican *Pelecanus rufescens*, *Ibis* **112**:499–512.

Busse, von K., 1977, Prägungsbedingte akustische Arterkennungsfähigkeit der Küken der Flubseeschwalben und Küstenseeschwalben *Sterna hirundo* L. und *S. paradisaea* Pont. *Z. Tierpsychol.* **44**:154–161.

Busse, von K., and Busse, K., 1977, Prägungsbedingte Bindung von Küstenseeschwalbenküken (*Sterna paradisaea* Pont.) an die Eltern und ihre Fähigkeit, sie an der Stimme zu erkennen, *Z. Tierpsychol* **43**:287–294.

Case, T. J., 1978, On the evolution and adaptive significance of postnatal growth rates in the terrestrial vertebrates, *Q. Rev. Biol.* **53**:243–282.

Clements, M., and Lien, J., 1976, Paired rotation and auditory stimulation of Common Murre *Uria aalge aalge* embryos and its post hatch effect, *Behav. Biol.* **17**:417–423.

Cody, M. L., 1973, Coexistence, coevolution and convergent evolution in seabird communities, *Ecology*, **54**:31–44.

Conover, M. R., Klopfer, F. D., and Miller, D. E., 1980, Stimulus features of chicks and other factors evoking parental protective behaviour in Ring-billed Gulls, *Anim. Behav.* (in press).

Conroy, J. W. H., 1972, Ecological aspects of the biology of the Giant Petrel, *Macronectes giganteus* (Gmelin), in the maritime antarctic, *Br. Antarctic Survey Sci. Rep.* **75**:1–74.

Coulson, J. C., 1974, Kittiwake, in: *The Seabirds of Britain and Ireland* (S. Cramp, W. R. P. Bourne, and D. Saunders, eds.), pp. 134–141, Collins, London.

Cullen, E., 1957, Adaptations in the kittiwake to cliff nesting, *Ibis* **99**:275–302.

Cullen, J. M., 1960, Some adaptations in the nesting behaviour of terns, *Proc. Int. Ornith. Congr.*, Helsinki **12**:153–157.

Cullen, J. M., 1962, The pecking response of young Wideawake Terns *Sterna fuscata*, *Ibis* **103b**:162–173.

Cullen, J. M., and Ashmole, N. P., 1963, The Black Noddy *Anous tenuirostris* on Ascension Island, Part 2. Behaviour, *Ibis* **103b**:423–446.

Cuthbert, N. L., 1954, A nesting study of the Black Tern in Michigan, *Auk* **71**:36–63.

Davies, S. J. J. F., and Carrick, R., 1962, On the ability of Crested Terns, *Sterna bergii*, to recognize their own chicks, *Aust. J. Zool.* **10**:171–177.

Davis, J. W. F., and Dunn, E. K., 1976, Intraspecific predation and colonial breeding in lesser Black-backed Gulls, *Larus fuscus*, *Ibis* **118**:65–77.

Dawkins, R., and Krebs, J. R., 1978, Animal signals: Information or manipulation? in: *Behavioural Ecology, an Evolutionary Approach* (J. R. Krebs and N. B. Davies, eds.), pp. 282–309, Sinauer Associates, Sunderland.

Dawson, W. R., Bennett, A. F., and Hudson, J. W., 1976, Metabolism and thermoregulaton in hatchling Ring-billed Gulls, *Condor* **78**:49–60.

Dawson, W. R., Hudson, J. W., and Hill, R. W., 1972, Temperature regulation in newly Hatched Laughing Gulls (*Larus atricilla*), *Condor* **74**:177–184.

Deusing, M., 1939, The Herring Gulls of Hat Island, Wisconsin, *Wilson Bull.* **51**:170–175.

Diamond, A. W., 1973, Notes on the breeding biology and behavior of the magnificent frigatebird, *Condor* **75**:200–209.

Din, N. A., and Eltringham, S. K., 1974, Breeding of the Pink-backed Pelican *Pelecanus rufescens* in Rwenzori National Park, Uganda, *Ibis* **116**:477–493.

Dinsmore, J. J., 1972, Sooty Tern behavior, *Bull. Fla. State Mus. Biol. Sci.* **16**:129–179.

Dorward, D. F., 1962, Behaviour of boobies *Sula* spp., *Ibis* **103b**:221–234.

Dorward, D. F., 1963, The Fairy Tern *Gygis alba* on Ascension Island, *Ibis* **103b**:365–378.

Dorward, D. F., and Ashmole, N. P., 1963, Notes on the biology of the Brown Noddy *Anous stolidus* on Ascension Island, *Ibis* **103b**:447–457.

Drent, R. H., 1965, Breeding biology of the Pigeon Guillemot, *Cepphus columba*, *Ardea* **53**:99–160.

Duffey, E., 1951, Field studies on the Fulmar *Fulmarus glacialis*, *Ibis* **93**:237–245.

Dunn, E. H., 1976, The development of endothermy and existence energy expenditure in Herring Gull chicks, *Condor* **78**:493–498.

Dutcher, W., and Baily, W. L., 1903, A contribution to the life history of the Herring Gull (*Larus argentatus*) in the United States, *Auk* **20**:417–431.

Elston, S. F., Rymal, C. D., and Southern, W. E., 1977, Intraspecific kleptoparasitism in breeding Ring-billed Gulls, *Proc. Conf. Colonial Waterbird Group* **1977**:102–109.

Emlen, J. T., Jr., 1956, Juvenile mortality in a Ring-billed Gull colony, *Wilson Bull.* **68**:232–238.

Emlen, J. T., Jr., 1964, Determinants of cliff edge and escape responses in Herring Gull chicks in nature, *Behaviour* **22**:1–15.

Emlen, J. T., and Miller, D. E., 1969, Pace-setting mechanisms of the nesting cycle in the Ring-billed Gull, *Behaviour* **33**:237–261.

Erwin, R. M., 1977, Foraging and breeding adaptations to different food regimes in three seabirds: The Common Tern, *Sterna hirundo*, Royal Tern, *Sterna maxima*, and Black Skimmer, *Rynchops niger*, *Ecology* **58**:389–397.

Evans, R. M., 1970a, Imprinting and mobility in young Ring-billed Gulls, *Larus delawarensis*, *Anim. Behav. Monog.* **3**:193–248.

Evans, R. M., 1970b, Parental recognition and the "mew call" in Black-billed Gulls (*Larus bulleri*), *Auk* **87**:503–513.

Evans, R. M., 1973, Differential responsiveness of young Ring-billed Gulls and Herring Gulls to adult vocalizations of their own and other species, *Can. J. Zool.* **51**:759–770.

Evans, R. M., 1975a, Responsiveness to adult mew calls in young Ring-billed Gulls (*Larus delawarensis*): effects of exposure to calls alone and in the presence of visual imprinting stimuli, *Can. J. Zool.* **53**:953–959.

Evans, R. M., 1975b, Responsiveness of young herring gulls to adult "mew" calls, *Auk* **92**:140–143.

Evans, R. M., 1977a, Semi-precocial development in gulls and terns (Laridae), *Marine Sciences Res. Lab. Tech. Rep.* **20**:213–239.

Evans, R. M., 1977b, Auditory discrimination-learning in young Ring-billed Gulls (*Larus delawarensis*), *Anim. Behav.* **25**:140–146.

Evans, R. M., 1979, Responsiveness of young herring gulls to stimuli from their own and other species: Effects of training with food, *Can. J. Zool.* **57**:1452–1457.

Evans, R. M., 1980, Development of individual call recognition in young Ring-billed Gulls (*Larus delawarensis*): An effect of feeding, *Anim. Behav.* (in press).

Farner, D. S., and Serventy, D. L., 1959, Body temperature and the ontogeny of thermoregulation in the Slender-billed Shearwater, *Condor* **61**:426–433.

Feare, C. J., 1976, The breeding of the Sooty Tern *Sterna fuscata* in the Seychelles and the effects of experimental removal of its eggs, *J. Zool. London* **179**:317–360.

Fisher, M. L., 1970, *The Albatross of Midway Island*, Southern Illinois University Press, Carbondale.

Fjeldsa, J., 1977, *Guide to the Young of European Precocial Birds*, Skarv Nature Publ., Strandgarden.

Fleet, R. R., 1974, The red-tailed tropicbird on Kure Atoll, *Am. Ornithol. Union Monog.* **16**:1–64.

Goldsmith, R., and Sladen, W. J. L., 1961, Temperature regulation of some Antarctic penguins, *J. Physiol.* **157**:251–262.

Gottlieb, G., 1971, *Development of Species Identification in Birds*, University of Chicago Press, Chicago.

Gottlieb, G., 1973, *Behavioral Embryology*, Academic Press, New York.

Gottlieb, G., 1976, The roles of experience in the development of behavior and the nervous system, in: *Neural and Behavioral Specificity* (G. Gottlieb, ed.), pp. 25–54, Academic Press, New York.

Greenwood, J., 1964, The fledging of the Guillemot *Uria aalge* with notes on the Razorbill *Alca torda*, *Ibis* **106**:469–481.

Hahn, D. C., 1977, Parent–offspring relations in the Laughing Gull (*Larus atricilla*), *American Ornithologists' Union Abstracts* 1977, No. 115.

Hailman, J. P., 1961, Why do gull chicks peck at visually contrasting spots? A suggestion concerning social learning of food-discrimination, *Am. Nat.* **95**:245–247.

Hailman, J. P., 1965, Cliff-nesting adaptations of the Galapagos Swallow-tailed Gull, *Wilson Bull.* **77**:346–362.

Hailman, J. P., 1967, The ontogeny of an instinct, *Behav. Suppl.* **15**:1–159.

Hailman, J. P., 1968, Visual-cliff responses of newly-hatched chicks of the Laughing Gull *Larus Atricilla*, *Ibis* **110**:197–200.

Hamilton, W. D., 1972, Altruism and related phenomena, mainly in social insects, *Annu. Rev. Ecol. Syst.* **3**:193–232.

Harris, M. P., 1964, Aspects of the breeding biology of the gulls *Larus argentatus, L. fuscus* and *L. marinus*, *Ibis* **106**:432–456.

Harris, M. P., 1966, Breeding biology of the Manx Shearwater, *Puffinus puffinus*, *Ibis* **108**:17–33.

Harris, M. P., 1970, Abnormal migration and hybridization of *Larus argentatus* and *L. fuscus* after interspecies fostering experiments, *Ibis* **112**:488–498.

Harris, M. P., 1973, The biology of the Waved Albatross *Diomedea irrorata* of Hood Island, Galapagos, *Ibis* **115**:483–510.

Harris, M. P., 1976, Lack of a "desertion period" in the nestling life of the Puffin *Fratercula arctica*, *Ibis* **118**:115–118.

Harris, M. P., Morley, C., and Green, G. H., 1978, Hybridization of Herring and Lesser Black-backed Gulls in Britain, *Bird Study* **25**:161–166.

Harris, S. W., 1974, Status, chronology, and ecology of nesting storm petrels in Northwestern California, *Condor* **76**:249–261.

Henderson, B. A., 1975, Role of the chick's begging behavior in the regulation of parental feeding behavior of *Larus glaucescens*, *Condor* **77**:488–492.

Herrick, F. H., 1912, Organization of the gull community, *Proc. Int. Zool. Congr.* **7**:156–158.

Hess, E. H., 1973, *Imprinting*, Van Nostrand Reinhold Co., New York.

Hinde, R. A., 1970, *Animal Behaviour*, 2nd ed., McGraw-Hill, New York.

Horn, H. S., 1978, Optimal tactics of reproduction and life-history, in: *Behavioural Ecology, and Evolutionary Approach* (J. R. Krebs and N. B. Davies, eds.), pp. 411–429, Sinauer Associates, Sunderland.

Howell, T. R., 1978, Ecology and reproductive behavior of the Grey Gull of Chile and of the Red-tailed Tropicbird and White Tern of Midway Island, *Natl. Geog. Soc. Res. Rep.* **1969**:251–284.

Howell, T. R., and Bartholomew, G. A., 1961, Temperature regulation in Laysan and Black-footed Albatrosses, *Condor*, **63**:185–197.

Howell, T. R., and Bartholomew, G. A., 1962, Temperature regulation in the Sooty Tern *Sterna fuscata*, *Ibis* **104**:98–105.

Howell, T. R., and Bartholomew, G. A., 1969, Experiments on nesting behavior of the Red-tailed Tropicbird, *Phaethon rubricauda*, *Condor* **71**:113–119.

Howell, T. R., Araya, B., and Millie, W. R., 1974, Breeding biology of the Grey Gull, *Larus modestus*, *Univ. Calif. Publ. Zool.* **104**:1–57.

Hunt, G. L., Jr., 1972, Influence of food distribution and human disturbance on the reproductive success of Herring Gulls, *Ecology*, **53**:1051–1061.

Hunt, G. L., Jr., and Hunt, M. W., 1975, Reproductive ecology of the Western Gull: The importance of nest spacing, *Auk* **92**:270–279.

Hunt, G. L., Jr., and Hunt, M. W., 1976, Gull chick survival: The significance of growth rates, timing of breeding and territory size, *Ecology* **57**:62–75.

Hunt, G. L., Jr., and McLoon, S. C., 1975, Activity patterns of gull chicks in relation to feeding by parents: Their potential significance for density-dependent mortality, *Auk* **92**:523–527.

Impekoven, M., 1973, The response of incubating Laughing Gulls (*Larus atricilla* L.) to calls of hatching chicks, *Behaviour* **46**:94–113.

Impekoven, M., 1976, Responses of Laughing Gull chicks (*Larus atricilla*) to parental attraction- and alarm- calls, and effects of prenatal auditory experience on the responsiveness to such calls, *Behaviour* **56**:250–278.

Impekoven, M., and Gold, P. S., 1973, Prenatal origins of parent–young interactions in birds: A naturalistic approach, in: *Behavioral Embryology* (G. Gottlieb, ed.), pp. 325–356, Academic Press, New York.

Ingold, P., 1973, Zur Lautlichen Beziehung des Elters zu seinem Kueken bei Tordalken (*Alca torda*), *Behaviour* **45**:154–190.

Isenmann, P., 1976, Contribution à l'étude de la biologie de la reproduction et de l'étho-écologie du goéland railleur. *Larus genei*, *Ardea* **64**:48–61.

Jarvis, M. J. F., 1974, The ecological significance of clutch size in the South African Gannet (*Sula capensis* (Lichtenstein)), *J. Anim. Ecol.* **43**:1–17.

Jehl, J. R., Jr., and Smith, B. A., 1970, *Birds of the Churchill Region, Manitoba*, Special Publ. No. 1, Manitoba Museum of Man and Nature, Winnipeg.

Johnson, S. R., and West, G. C., 1975, Growth and development of heat regulation in nestlings, and metabolism of adult Common and Thick-billed Murres, *Ornis Scand.* **6**:109–115.

Johnson, R. F., Jr., and Sloan, N. F., 1978, While pelican production and survival of young at Chase Lake National Wildlife Refuge, North Dakota, *Wilson Bull.* **90**:346–352.

Kepler, C. B., 1969, Breeding biology of the Blue-faced Booby, *Sula dactylatra personata* on Green Island, Kure Atoll. *Publ. Nuttall Ornithol. Club*, **8**:1–97.

Kirkman, F. B., 1937, *Bird Behaviour* T. Nelson and Sons, London.

Knopf, F. L., 1975, Spatial and temporal aspects of colonial nesting of the White Pelican, *Pelecanus erythrorhynchos*, Ph.D. thesis, Utah State University, Logan.

Krebs, J. R., and Davies, N. B., 1978, *Behavioural Ecology, an Evolutionary Approach*, Sinauer Associates, Sunderland.

Kruuk, H., 1964, Predators and anti-predator behaviour of the Black-headed Gull (*Larus ridibundus L.*), *Behav. Suppl.* **11**:1–129.

Lack, D., 1954, *The Natural Regulation of Animal Numbers*, Clarendon Press, Oxford.

Lack, D., 1968, *Ecological Adaptations for Breeding in Birds*, Methuen, London.

LeCroy, M., and Collins, C. T., 1972, Growth and survival of Roseate and Common Tern chicks, *Auk* **89**:595–611.

Lloyd, C. S., 1977, The ability of the Razorbill *Alca torda* to raise an additional chick to fledging, *Ornis Scand.* **8**:155–159.

Lockley, R. M., 1961, *Shearwaters*, Doubleday, New York.

Lockley, R. M., 1962, *Puffins*, Doubleday, New York.

Manuwal, D. A., 1974, The natural history of Cassin's Auklet (*Ptychoramphus aleuticus*), *Condor* **76**:421–431.

Marples, G., and Marples, A., 1934, *Sea Terns or Sea Swallows*, Country Life, Ltd., London.

Maunder, J. E., and Threlfall, W., 1972, The breeding biology of the Black-legged Kittiwake in Newfoundland, *Auk* **89**:789–816.

McLannahan, H. M. C., 1973, Some aspects of the ontogeny of cliff nesting behaviour of the Kittiwake (*Rissa tridactyla*) and the Herring Gull (*Larus argentatus*), *Behaviour* **44**:36–88.

McNicholl, M. K., 1971, The breeding biology and ecology of Forster's Tern (*Sterna forsteri*) at Delta, Manitoba, M.Sc. thesis, University of Manitoba, Winnipeg.

Miller, D. E., and Conover, M. R., 1979, Differential effects of chick vocalizations and bill-pecking on parental behavior in the Ring-billed Gull, *Auk*. **96**:284–295.

Miller, D. E., and Emlen, J. T., Jr., 1975, Individual chick recognition and family integrity in the Ring-billed Gull, *Behaviour* **52**:124–144.

Moynihan, M., 1959*a*, Notes on the behavior of some North American gulls. IV. The ontogeny of hostile behavior and display patterns, *Behaviour* **14**:214–239.

Moynihan, M., 1959*b*, A revision of the family Laridae (Aves), *Am. Mus. Novitates No.* **1928**:1–42.

Nelson, J. B., 1966, The behaviour of the young gannet, *Br. Birds* **59**:393–419.

Nelson, J. B., 1967*a*, Colonial and cliff nesting in the gannet, *Ardea* **55**:60–90.

Nelson, J. B., 1967*b*, The breeding behaviour of the White Booby *Sula dactylatra*, *Ibis* **109**:194–231.

Nelson, J. B., 1969, The breeding behaviour of the Red-footed Booby *Sula sula*, *Ibis* **111**:357–385.

Nelson, J. B., 1975, The breeding biology of frigatebirds—A comparative review, *Living Bird* **14**:113–156.

Nelson, J. B., 1978, *The Sulidae, Gannets and Boobies*, Oxford University Press, Oxford.

Nettleship, D. N., 1972, Breeding success of the Common Puffin (*Fratercula arctica* L.) on different habitats at Great Island, Newfoundland, *Ecol. Monog.* **42**:239–268.

Nice, M. M., 1962, Development of behavior in precocial birds, *Trans. Linn. Soc. N.Y.* **8**:1–211.

Nice, M. M., 1964, *Studies in the Life History of the Song Sparrow*, Vol. II, Dover, New York.

Norman, F. I., and Gottsch, M. D., 1969, Artificial twinning in the Short-tailed Shearwater *Puffinus tenuirostris*, *Ibis* **111**:391–393.

Noseworthy, C. M., and Lien, J., 1976, Ontogeny of nesting habitat recognition and preference in neonatal Herring Gull chicks, *Larus argentatus* Pontoppidan, *Anim. Behav.* **24**:637–651.

Noseworthy, C., Stoker, S., and Lien, J., 1973, Habitat preferences in Herring Gull chicks, *Auk* **90**:193–194.

Nyström, M., 1972, On the quantification of pecking responses in Young Gulls (*Larus argentatus*), *Z. Tierpsychol.* **30**:36–44.

O'Connor, R. J., 1978, Brood reduction in birds: Selection for fratricide, infanticide and suicide? *Anim. Behav.* **26**:79–96.

Palmer, R. S., 1941, A behavior study of the Common Tern (*Sterna hirundo hirundo* L.), *Proc. Boston Soc. Nat. Hist.* **42**:1–119.

Paludan, K., 1951, Contributions to the breeding biology of *Larus argentatus* and *Larus fuscus*, *Vidensk. Medd. fra Dansk. Naturh. Foren.* **114**:1–128.

Parsons, J., 1971, Cannibalism in Herring Gulls, *Br. Birds* **64**:528–537.

Penney, R. L., 1968, Territorial and social behavior in the Adelie Penguin, in: *Antarctic Bird Studies* (O. L. Austin, Jr., ed.), pp. 83–131, American Geophysical Union Publ. No. 1686.

Perrins, C. M., Harris, M. P., and Britton, C. K., 1973, Survival of Manx Shearwaters *Puffinus puffinus*, *Ibis* **115**:535–548.

Pettingill, O. S., Jr., 1939, History of one hundred nests of Arctic Tern, *Auk* **56**:420–428.

Pettingill, O. S., Jr., 1960, Creche behavior and individual recognition in a colony of Rock-hopper Penguins, *Wilson Bull.* **72**:213–221.

Procter, D. L. C., 1975, The problem of chick loss in the South Polar Skua *Catharacta maccormicki*, *Ibis* **117**:452–459.

Quine, D. A., and Cullen, J. M., 1964, The pecking response of young Arctic Terns *Sterna macrura* and the adaptiveness of the "releasing mechanism," *Ibis* **106**:145–173.

Rautenfeld, Von D. B. von, 1978, Bemerkungen zur Austauschbarkeit von Küken der Silbermöwe (*Larus argentatus*) nach der ersten Lebenswoche, *Z. Tierpsychol.* **47**:180–181.

Richdale, L. E., 1951, *Sexual Behavior in Penguins*, University of Kansas Press, Lawrence.

Richdale, L. E., 1963, Biology of the Sooty Shearwater *Puffinus griseus*, *Proc. Zool. Soc. London* **141**:1–117.

Rice, D. W., and Kenyon, K. W., 1962, Breeding cycles and behavior of Laysan and Black-footed Albatrosses, *Auk* **79**:517–567.

Round, P. D., and Swann, R. L., 1977, Aspects of the breeding of Cory's Shearwater *Calonectris diomedea* in Crete, *Ibis* **119**:350–353.

Rowan, M. K., 1952, The Greater Shearwater *Puffinus gravis* at its breeding grounds, *Ibis* **94**:97–121.

Schaller, G. B., 1964, Breeding behavior of the White Pelican at Yellowstone Lake, Wyoming, *Condor* **66**:3–23.

Schreiber, R. W., 1970, Breeding biology of Western Gulls (*Larus occidentalis*) on San Nicolas Island, California, 1968, *Condor* **72**:133–140.

Schreiber, R. W., 1977, Maintenance behavior and communication in the Brown Pelican, *Am. Ornith. Union Monog.* **22**:1–78.

Sealy, S. G., 1973, Adaptive significance of post-hatching developmental patterns and growth rates in the Alcidae, *Ornis Scand.* **4**:113–121.

Sealy, S. G., 1975, Egg size of murrelets, *Condor* **77**:500–501.

Sealy, S. G., 1976, Biology of nesting Ancient Murrelets, *Condor* **78**:294–306.

Shugart, G. W., 1977, The development of chick recognition by adult Caspian Terns, *Proc. Conf. Colonial Waterbird Gp* **1977**:110–117.

Simmons, K. E. L., 1967, Ecological adaptations in the life history of the Brown Booby at Ascension Island, *Living Bird* **6**:187–212.

Sladen, W. J. L., 1958, The pygoscelid penguins, *Falkland Islands Dependencies Survey Sci. Rep.* **17**:1–97.

Sluckin, W., 1972, *Imprinting and Early Learning*, 2nd ed., Methuen, London.

Smith, A. J. M., 1975, Studies of breeding Sandwich Terns, *Br. Birds* **68**:142–156.

Smith, N. G., 1966, Adaptations to cliff-nesting in some Arctic Gulls (*Larus*), *Ibis* **108**:68–83.

Smith, J. E., and Diem, K. L., 1972, Growth and development of young California Gulls (*Larus californicus*), *Condor* **74**:462–470.

Snow, B., 1960, The breeding biology of the Shag *Phalacrocorax aristotelis* on the island of Lundy, Bristol Channel, *Ibis* **102**:554–575.

Snow, B. K., 1963, The behaviour of the Shag, *Br. Birds* **56**:77–103; 164–186.

Snow, B. K., 1966, Observations on the behaviour and ecology of the flightless Cormorant *Nannopterum harrisi*, *Ibis* **108**:265–280.

Spellerberg, I. F., 1969, Incubation temperatures and thermoregulation in the McCormick Skua, *Condor* **71**:59–67.

Spellerberg, I. F., 1971a, Breeding behaviour of the McCormick Skua *Catharacta maccormicki* in Antarctica, *Ardea* **59**:189–230.

Spellerberg, I. F., 1971b, Aspects of McCormick Skua breeding biology, *Ibis* **113**:357–363.

Spurr, E. B., 1975, Behavior of the Adelie Penguin chick, *Condor* **77**:272–280.

Stamps, J. A., 1978, A field study of the ontogeny of social behavior in the lizard *Anolis aeneus*, *Behaviour* **66**:1–31.

Stead, E. F., 1932, *The Life Histories of New Zealand Birds*, Search Publ. Co., London.

Stevenson, J. G., Hutchison, R. E., Hutchison, J. B., Bertram, B. C. R., and Thorpe, W. H., 1970, Individual recognition by auditory cues in the Common Tern (*Sterna hirundo*), *Nature (London)* **226**:562–563.

Stonehouse, B., 1953, The Emperor Penguin *Aptenldytes forsteri* Gray. I. Breeding behaviour and development, *Falkland Island Dependencies Survey Sci. Rep.* **6**:1–33.

Stonehouse, B., 1956, The Brown Skua *Catharacta skua lonnbergi* (Mathews) of South Georgia, *Falkland Islands Dependencies Survey Sci. Rep.* **14**:1–25.

Stonehouse, B., 1960, The King Penguin *Aptenodytes patagonica* of South Georgia I. Breeding behaviour and development, *Falkland Islands Dependencies Survey Sci. Rep.* **23**:1–81.

Stout, J. F., Wilcox, C. R., and Creitz, L. E., 1969, Aggressive communication by *Larus glaucescens* Part I. Sound communication, *Behaviour* **34**:29–41.

Strong, R. M., 1923, Further observations on the habits and behavior of the Herring Gull, *Auk* **40**:609–621.

Taylor, R. H., 1962, The Adelie Penguin *Pygoscelis adeliae* at Cape Royds, *Ibis* **104**:176–204.

Thompson, D. H., and Emlen, J. T., 1968, Parent–chick individual recognition in the Adelie Penguin, *Antarctic J.* **3**:132.

Thoresen, A. C., 1964, The breeding behavior of the Cassin Auklet, *Condor* **66**:456–476.

Tickell, W. L. N., 1960, Chick feeding in the Wandering Albatross *Diomedea exulans* Linnaeus, *Nature (London)* **165**:116–117.

Tickell, W. L. N., 1968, The biology of the great albatrosses, *Diomedea exulans* and *Diomedea epomophora*, in: *Antarctic Bird Studies* (O. L. Austin, Jr., ed.), pp. 1–55, American Geophysical Union Publ. No. 1686.

Tickell, W. L. N., and Pinder, R., 1972, Chick recognition by albatrosses, *Ibis* **114**:543–548.

Tickell, W. L. N., and Pinder, R., 1975, Breeding biology of the Black-browed Albatross *Diomedea melanophris* and Grey-headed Albatross *D. chrysostoma* at Bird Island, South Georgia, *Ibis* **117**:433–451.

Tinbergen, N., 1953, *The Herring Gull's World*, Collins, London.

Tinbergen, N., 1959, Comparative studies of the behaviour of gulls (Laridae): A progress report, *Behaviour* **15**:1–70.

Tinbergen, N., and Falkus, H., 1970, *Signals for Survival*, Clarendon Press, Oxford.

Trillmich, F., 1978, Feeding territories and breeding success of the South Polar Skuas, *Auk* **95**:23–33.

Trivers, R. L., 1974, Parent–offspring conflict, *Am. Zool.* **14**:249–264.

Tschanz, B., 1965, Beobachtungen und Experimente zur Entstehung der "persönlichen" Beziehung zwischen Jungvogel und Eltern bei Trottellummen, *Verhandl. Schweiz. Naturforsch. Gesellschaft* **1964**:211–216.

Tschanz, B., 1968, Trottellummen, Die Entstehung der persönlichen Beziehungen zwischen Jungvogel und Eltern, *Z. Tierpsychol. Beiheft* **4**:1–100.

Tschanz, B., and Hirsbrunner-Scharf, M., 1975, Adaptations to colony life on cliff ledges: A comparative study of guillemot and razor bill chicks, in: *Function and Evolution in Behaviour* (G. P. Baerends, C. G. Beer, and A. Manning, eds.), pp. 358–380, Clarendon Press, Oxford.

Tuck, L. M., 1960, The murres, their distribution, populations and biology, a study of the genus *Uria, Can. Wildl. Service Rep. Ser.* **1**:1–260.

Vermeer, K., 1970, Breeding biology of California and ring-billed gulls: A study of ecological adaptation to the inland habitat, *Can. Wildl. Service Rep. Ser.* **12**:1–52.

Vestjens, W. J. M., 1975, Breeding behaviour of the darter at Lake Cowal, NSW, *Emu* **75**:121–131.

Vestjens, W. J. M., 1977, Breeding behaviour and ecology of the Australian Pelican, *Pelecanus conspicillatus*, in New South Wales, *Aust. Wildl. Res.* **4**:37–58.

Walk, R. D., and Gibson, E. J., 1961, A comparative and analytical study of visual depth perception, *Psychol. Monog.* **75**(15):1–44.

Ward, H. L., 1906, Why do herring gulls kill their young? *Science* **24**:593–594.

Warham, J., 1963, The rockhopper penguin, *Eudyptes chrysocome*, at Macquarie Island, *Auk* **80**:229–256.

Warham, J., 1972, Aspects of the biology of the Erect-crested Penguin *Eudyptes sclateri*, *Ardea* **60**:145–184.

Warham, J., 1974, The Fiordland Crested penguin *Eudyptes pachyrhynchus*, *Ibis* **116**:1–27.

Warham, J., 1975, The crested penguins, in: *The Biology of Penguins* (B. Stonehouse, ed.), pp. 189–269, University Park Press, London.

Watson, J. B., 1908, The behavior of Noddy and Sooty Terns, *Carnegie Inst. Wash. Publ. No.* **103**:187–255.

Wehrlin, von J., 1977, Verhaltensanpassungen junger Trottellummen (*Uria aalge aalge* Pont.) ans Felsklippen-und Koloniebrüten, *Z. Tierpsychol.* **44**:45–79.

Werschkul, D. F., and Jackson, J. A., 1979, Sibling competition and avian growth rates, *Ibis* **121**:97–102.

White, S. J., 1971, Selective responsiveness by the Gannet (*Sula bassana*) to played-back calls, *Anim. Behav.* **19**:125–131.

Williams, G. C., 1966, *Adaptation and Natural Selection*, Princeton University Press, Princeton, N. J.

Wilson, E. O., 1975, *Sociobiology, the New Synthesis*, Harvard University Press, Cambridge, Mass.

Wooller, R. D., 1978, Individual vocal recognition in the Kittiwake Gull, *Rissa tridactyla* (L.), *Z. Tierpsychol.* **48**:68–86.

Yeates, G. W., 1975, Microclimate, climate and breeding success in Antarctic Penguins, in: *The Biology of Penguins* (B. Stonehouse, ed.), pp. 397–409, University Park Press, London.

Young, E. C., 1963a, Feeding habits of the South Polar Skua *Catharacta maccormicki*, *Ibis* **105**:301–318.

Young, E. C., 1963b, The breeding behaviour of the South Polar Skua *Catharacta maccormicki*, *Ibis* **105**:203–233.

Young, E. C., 1978, Behavioural ecology of *lonnbergi* skuas in relation to environment on the Chatham Islands, New Zealand, *N. Z. J. Zool.* **5**:401–416.

Chapter 9

PARENTAL INVESTMENTS BY SEABIRDS AT THE BREEDING AREA WITH EMPHASIS ON NORTHERN GANNETS, *Morus bassanus*

W. A. Montevecchi

Department of Psychology
Memorial University of Newfoundland
St. John's, Newfoundland
Canada A1B 3X9

and

J. M. Porter*

Department of Biology
Acadia University
Wolfville, Nova Scotia
Canada B0P 1X0

I. INTRODUCTION

Present studies indicate that marine birds bond monogamously (e.g., Richdale, 1951; Austin and Austin, 1956; Coulson, 1966; cf. Chapter 5), are relatively long-lived, and generally exhibit delayed maturity (Lack, 1968). With the exception of murres (Tuck, 1961), razorbills, and some murrelets (Sealy, 1973), seabird hatchlings exhibit altricial through semiprecocial modes of development. All seabirds, except for some murrelets, are nidiculous during a considerable portion of posthatch development and therefore require extensive parental care at the breeding area. Parents provide offspring with body heat and food, and depending on the species, protection from weather and predators, and opportunities for learning (Nice, 1962).

* *Present address*: Department of Zoology, University of Manitoba, Winnipeg, Manitoba, Canada R3T 2N2.

Major reviews of avian parental behavior are for the most part based either on compilations of general breeding biology information (e.g., Kendeigh, 1952; Skutch, 1975) or on physiological analyses of parental behavior patterns (e.g., Lehrman, 1961; Silver, 1978). In this chapter parental behavior is viewed in an evolutionary context as an interactive process between adults and developing offspring whereby optimal breeding success is achieved. The ecological parameters of food resources, predator pressures, and habitat features shape interrelated patterns of mating systems, egg and clutch sizes, developmental modes and parental care strategies. A unified complex of adaptation emerges, variations being interpretable in terms of life-history design.

Direct energy transfer from parent to offspring through egg production, incubation, brooding, and feeding, as well as nesting behavior is given focal attention. owing to the scattered nature of the subject material a complete survey of parental behavior by seabirds has not been attempted, although comparisons among these birds are made throughout. Prenatal (Impekoven, 1976) and postfledging (Chapter 10) parent–offspring interactions are considered elsewhere. We begin by examining some of the relationships between developmental mode and parental investment followed by an outline of the basic types of parental-feeding and chick-begging behaviors shown by seabirds. The relative parental investments of Northern Gannet mates (*Morus bassanus*) are then treated extensively.

II. MODE OF DEVELOPMENT AND PARENTAL INVESTMENT

It is impossible to understand diversity in avian parental care strategies without considering the physical and behavioral capacities of hatchlings. Precocial and semiprecocial chicks are actually or potentially mobile, capable of temperature regulation soon after hatching, and the former may forage at an early age. Altricial chicks are sedentary, unable to maintain body temperatures initially and more dependent on parental care (Nice, 1962; Sealy, 1973; Evans, 1977; Ricklefs, 1979). Different developmental modes necessitate different parental activities. For instance, after hatching altricial young require prolonged brooding by parents (Dunn, 1975b), whereas the mobility of precocials make guiding and chick recognition more important parental concerns (Nice, 1962; Lack, 1968; Evans, 1977).

The eggs of seabirds that engage in extensive parental care are most often relatively small with low caloric content. The proportionally smallest eggs are produced by the largest seabirds whose chicks develop altricially (penguins, pelicans, sulids, cormorants). Emperor Penguins (*Aptenodytes fosteri*), King Penguins (*A. patagonicus*), Gentoo Penguins (*Pygoscelis papua*), and Double-crested Cormorants (*Phalacrocorax carbo*) lay eggs that average about 2% or less of adult weight (calculated from Lack, 1967, 1968). The proportionally heaviest eggs are laid by seabirds whose chicks develop semialtricially or semiprecocially (shearwaters, Storm Petrels, Diving Petrels, terns, alcids). The eggs of White-throated (*Nesofregatta albigularis*) and Blue-gray Noddies (*Procelsterna caerula*), Wilson's (*Oceanites oceanicus*), and White-faced (*Pelagodroma marina*) Storm Petrels vary at 25–30% adult weight (calculated from Lack, 1968). The energy densities of eggs are higher for birds that show more precocial development and lay large eggs with large yolks and high lipid contents (Ricklefs, 1977; Ar and Yom-Tov, 1978; Ricklefs and Montevecchi, 1979).

The clutch sizes of marine birds are not strongly associated with precocity of development. For example, single egg clutches are produced not only by altricial frigatebirds, but also by the alcids that are intermediate between precocial and semiprecocial development (Sealy, 1973). By comparison, multiple egg clutches are laid by some of the altricial Pelecaniformes, most semiprecocial gulls and terns, and the precocial alcids (see Table I).

Precocial and semiprecocial Charadriiformes begin to develop homeothermy before hatching and can maintain steady body temperatures in the face of environmental fluctuation soon after hatching (e.g., Drent, 1970, Dunn, 1976*b*; Ricklefs, 1979). Altricial seabirds develop homeothermy less rapidly, but faster than do small altricial passerines (Ricklefs, 1974; Dunn, 1975*b*, 1976*a*). The earlier acquisition of temperature regulation by precocial chicks may limit the amount of assimilated energy available for growth (Ricklefs, 1979; see Case, 1978*a*). Time spent brooding may compete with foraging time and may limit feeding rates and chick success (e.g., Langham, 1972).

Providing food is the most direct transfer of energy from parent to hatched offspring. Self-feeding precocial offspring free parents from this provisioning, and nidifugous chicks that are not self-feeding may reduce parental energy expenditure associated with transporting food to a fixed site.

Owing to high proportions of embryonic tissue at hatching and to large, well-developed digestive systems, altricial hatchlings grow more rapidly than do similarly sized precocials that have greater proportions of mature tissue (Ricklefs, 1973, 1979). Semiprecocials also grow rapidly, being more like altricials than precocials in this respect (Ricklefs, 1973, 1979). Compressed growth rates have been considered adaptive in minimizing vulnerability to predators and in allowing departure from temperate nesting areas before deteriorating fall weather conditions set in

Table I. Proportionate Egg and Clutch Weights, Clutch Sizes, and Developmental Modes of Marine Birds[a]

Order family	Common name	Egg/adult wt (%)	Clutch/adult wt (%)	Clutch size	Mode of development
Sphenisciformes (all)					
Spheniscidae	Penguins	3.0	6.0	2	Semialtricial
Procellariiformes (all)					
Diomedeidae	Albatrosses	8.0	8.0	1	Semialtricial
Procellariidae	Fulmars, shearwaters, large petrels	15.1	15.1	1	Semialtricial
Hydrobatidae	Storm petrels	25.2	25.2	1	Semialtricial
Pelecanoididae	Diving petrels	15.3	15.3	1	Semialtricial (?)
Pelecaniformes (some)					
Phaethontidae	Tropicbirds	11.0	11.0	1	Semialtricial
Pelecanidae	Brown Pelican	2.6	5.2–7.8	2–3	Altricial
Sulidae	Gannets, boobies	4.2	3.5–14.0	1–4	Altricial
Phalacrocoracidae	Cormorants	2.4	4.8–12.0	2–5	Altricial
Fregatidae	Frigatebirds	6.0	6.0	1	Altricial
Charadriiformes (some)					
Stercorariidae	Skuas, jaegers	9.6	19.2	2	Semiprecocial
Laridae					
Larinae	Gulls	10.8	21.6–32.4	2–3	Semiprecocial
Sterninae	Terns	19.9	19.9–59.7	1–3	Semiprecocial
Rynchopidae	Skimmers	—	—	2–4	Semiprecocial
Alcidae	Auks, murres, dovekies, guillemots, puffins	14.5	9.6–47.4	1–2	Precocial, semiprecocial

[a] Adapted and calculated from Nice (1962), Palmer (1962), Lack (1968), Sealy (1975), and Nelson (1978b).

(Lack, 1967, 1968; Case, 1978*b*; Nelson, 1978*a,b*; cf. Ricklefs, 1979). Slower-growing young that need less energy per unit time may permit larger brood sizes (Ricklefs, 1979).

Developmental modes are strongly interrelated with feeding ecology. Self-feeding precocial species are restricted to land birds that utilize foods such as seeds, vegetation, and insect larvae, which they require little strength or skill to obtain (Ricklefs, 1974; Case, 1978*a*). Most precocial birds are, in fact, primary consumers (Ar and Yom-Tov, 1978). Birds specialized to feed on prey that must be actively hunted and pursued tend to develop altricially (Ricklefs, 1974). This relationship between food characteristics and developmental mode suggests why seabirds, with the exception of the plankton feeding, precocial murrelets, exhibit less than full precocial development.

III. PARENTAL FEEDING AND FOOD BEGGING

Parent–offspring feeding interactions are species typical and may be shared by all members of a family or order. Variations in the form and intensity of these patterns occur ontogenetically. Environmental influences that shape these interactions have produced phylogenetic divergences.

A. Types of Interactions

Three distinct parental feeding techniques are found among seabirds: (1) offering food carried in the bill, (2) regurgitating food that may be held in the bill or dropped on the ground, and (3) regurgitating food directly to offspring. A characteristic style of begging is associated with each feeding method.

1. Offering Food Carried in the Bill and Noncontact Begging

Most terns and alcids transport fish in their bills and offer them directly to offspring. Parents using this technique often call loudly just before landing with food (Hardy, 1957) or feeding (Tschanz and Hirsbrunner-Scharf, 1975). Such communication is widespread in seabirds and often serves to signal and locate mates and offspring. When a tern chick does not take the fish from a incoming parent, the adult often flies up, circles, and relands; this behavior frequently induces a chick to feed (e.g., Hardy, 1957). Rapid feeding by chicks may be motivated by sibling rivalry in terns and neighbor competition in murres, as well as parental tendencies to take back and eat fish that are not devoured quickly (Quine and Cullen, 1964; Tschanz and Hirsbrunner-Scharf, 1975). Common Terns (*Sterna hirundo*) and razorbills (*Alca torda*) appear quite passive when chicks take fish from them (Tschanz and Hirsbrunner-Scharf, 1975; A. Storey, personal communication). By comparison, murres are more active when feeding young, holding them close to their body beneath unfolded wings while calling loudly and repeatedly.

In some seabirds using this parental feeding method, males may feed females before laying and during incubation (Nisbet, 1973, 1977); feedings between mates are rare in the posthatch period. In Common Terns and Mediterranean Gulls (*Larus melanocephalus*), an attending parent (female?) may take a fish from an incoming mate (male?); the chick may then take the fish, or the parent (female?) may eat it (Borudulina, 1966; A. Storey, personal communication).

Murres carry fish lengthwise in the bill, whereas terns, Razorbills, and puffins carry them crosswise. Murres always and terns virtually always deliver a single fish per feeding visit (Tuck and Squires, 1955; Hays *et al.*, 1973; I. C. T. Nisbet, personal communication), whereas Razorbills and puffins deliver one or more (Bédard, 1969; Arnason and Grant, 1978). The former seabirds tend to feed offspring more frequently do than the latter.

Chicks of parents that deliver food in the bill do not usually beg by direct contact, and generally display less elaborate begging behavior compared to chicks whose parents regurgitate food (see below). Tern chicks run to a parent with food and may call, wing raise, and crouch, but usually do not peck the parent's bill (Hardy, 1957; A. Storey, personal communication; cf. Nice, 1962; Quine and Cullen, 1964). Among alcids, begging probably does not involve much bill contact, although Razorbill chicks peck fish from the parent's bill (Tschanz and Hirsbrunner-Scharf, 1975). The begging shown by the chicks being considered here is probably instrumental in obtaining a meal by conveying information that they will eat the food that the parent has to offer. When parents transport food in the bill, chicks receive immediate visual information about the possiblity of feeding. Quite a different situation seems to hold, at least early in the nestling period, for chicks whose parents deliver internally transported food. Furthermore, when a parent carries food in the bill there is probably less need for intimate coordination between adult and chick during food transfer (see below), and therefore elaborate begging/feeding interactions may not have been selected.

2. Regurgitating Food That May Be Held in the Bill or Dropped on the Ground and Contact Begging

Regurgitation of partially digested food for chicks is predominant in gulls (Hailman, 1967). Parents may offer regurgitated food in the bill or drop it on the ground in front of chicks. Both types of presentation are often used by the same individual over the nestling cycle (see below; Appendix E in Hailman, 1967; Henderson, 1975; Kirkham, 1977). When chicks do not peck at food regurgitated on the ground, the parent lowers its bill to the food, often stimulating the chicks to eat. If chicks still do not respond, the parent picks up the food, holds it briefly, and then swallows it (Hailman, 1967; Evans, 1977). On returning to the nest or feeding site, parents often call, and this may induce neonates to beg and attract older chicks that may be wandering or hiding nearby (e.g., Evans, 1970; Henderson, 1975). In rare instances, females attending chicks may be fed by incoming males (e.g., Henderson, 1975). Female Herring Gulls (*Larus argentatus*) have been observed taking food from mates, then feeding chicks about an hour later (J. Burger, personal communication).

Young gull chicks peck at the tips or marks near the tips of parents' bills (Tinbergen and Perdeck, 1953; Collias and Collias, 1957; Weidman and Weidman, 1958; Hailman, 1967; Tinbergen and Falkus, 1969). The "pecking signs" on the bills of many gulls reflect the necessity of viewing parent–offspring begging/feeding interactions as an integrated, adaptive complex. Chicks often call while pecking, and this probably enhances efforts to secure food (Hailman, 1967).

Soon after hatching there is less physical contact in begging/feeding interactions. Growing gull chicks develop head pumping and head tossing displays, which accompanied by new vocalizations, replace bill pecking as the primary begging response. Bill pecking may, however, terminate a sequence of head tossing and pumping. Larger nestlings are perhaps more physically threatening to parents, and may fare best by pecking at parents very little, even then only following an extended period of ("submissive") displaying. Begging again becomes more assertive and physical among older nestlings that may be as large or larger than parents. These chicks often pull food from the parent's throat (Hailman, 1967; Tinbergen and Falkus, 1969; cf. Trivers, 1974). The more intricate begging of ground nesting gull chicks as compared to terns and alcids, may be used to gain information about the possibility of and to time a parental feed, as well as to prime parental regurgitation.

3. Regurgitating Food Directly to Offspring and Subtle Contact Begging

These behaviors are exhibited almost exclusively by seabirds that develop altricially—Sphenisciformes, Procellariiformes, and Pelecaniformes, though also by Black-legged Kittiwakes (*Rissa tridactyla*), Wideawake Terns (*Sterna fuscata*), and Black Noddies (*Anous tenuirostris*) (Watson, 1908; Richdale, 1943; 1951; Witherby *et al.*, 1947; Cullen, 1957; Quine and Cullen, 1964). Some albatrosses provide offspring with internally produced fluids along with collected food (Rice and Kenyon, 1962). Procellariiformes are the most developed in this regard and feed offspring concentrated, well-processed stomach oil (Matthews, 1949; Imber, 1976). Owing to altricial development, parents must take the complete initiative in feeding neonates; activities such as preening, prodding, and lifting on the webs of the feet, help to coordinate begging/feeding interactions. The internal transport and consequent digestion of parental food must aid the absorption of nutrients by newly hatched chicks (see Section IV.C.4).

Following neonatal development, chicks commonly touch or nibble the parent's bill, frequently in the gular area (see Allan, 1962). Chicks often call from beneath the parent and push up on the parent's ventrum (e.g., gannets). Leach's Storm Petrel (*Oceanodroma leucorhoa*) chicks vocalize when touched on the head (J. Lien, personal communication). Older penguin chicks are exceptional with regard to the association between direct regurgitation feeding and subtle begging activity and peck vigorously at the bills and flanks of parents before a feed (Warham, 1975), although younger ones do not beg so actively.

Similar to ground-nesting gulls, chicks fed directly by regurgitation must at least early in the nestling period interact with parents to determine if they have food and to induce regurgitation. Begging and parental activity bring about a precise coordination that is needed during food transfer. For gannets and perhaps other altricial seabirds, food-begging bouts come to be intimately coupled with parental stimulation during early development. A parent can also terminate begging activity by settling over a chick's head.

B. Development

At hatching, adults show a gradual transition from incubation to more direct parent care, such as brooding and feeding (Beer, 1966). Calls from pipping eggs may help promote this transition (Impekoven, 1973), and Black-headed Gulls (*Larus ridibundus*) and Northern Gannets have been observed regurgitating food onto pipping eggs (Beer, 1966; Nelson, 1978a,b). Vocal exchanges between parents and young during feeding, brooding, and preening lead to the development and strengthening of family bonds and individual recognition. Tactile stimulation provided by neonates, especially pushing up behavior common in many nestlings (Lehrman, 1955; Hailman, 1967; Impekoven, 1976), also promotes the early establishment and maintenance of parental care.

Owing to development and changing food demands, the process of food transfer between parent and offspring often changes over the nestling period. Parents initiate food transfer with hatchlings, particularly altricial ones that have little motor ability or energy reserve. Altricial young quickly develop the musculature and behavior needed to interact more actively with parents. All young regardless of developmental mode become progressively more assertive in soliciting food (e.g., Dorward, 1962). Parental initiative in feeding interactions tends to wane as resistance to the assertions of growing chicks increases (e.g., Hailman, 1967; Tinbergen and Falkus, 1969; Kirkham, 1977; Nelson, 1978b). Such developmental trends have been interpreted as indicators of changing strategies related to parent–offspring conflict (Trivers, 1974). Different seabird species use only one of the basic parental feeding techniques.

Among seabirds that do not constantly attend young the development of homeothermy affords parents more potential foraging time and thus opportunity to feed chicks at higher rates (e.g., Langham, 1972). The ontogenetic elaboration of vocal and begging displays, such as those of some larids, allows the chicks of more precocial and more nidifugous seabirds to communicate begging information over greater distances. This communication is useful for the maintenance of the parental care of mobile chicks (e.g., Evans, 1970) and possibly for the maintenance of parent–offspring attachments after fledging (see Chapter 10 and Burger, in press).

C. Environmental Influences

Environmental pressures related to kleptoparasitism, foraging area and nest sites are considered briefly here. The transport of ingested food to offspring aids in the avoidance of kleptoparasitism (see Borudulina, 1966; Arnason and Grant, 1978). Harassing frigatebirds, jaegers, and skuas can, however, induce vomiting in seabirds that carry food internally (e.g., Nelson, 1978b). Yet even so, ingestion imposes greater difficulties for potential kleptoparasites. Risks of kleptoparasitism may be further reduced by nocturnal activity.

Internal food transport might also be expected to enchance flight efficiency and permit more distant foraging without food spoilage (Ashmole, 1971). Procellariiformes that remove water from foods and deliver stomach oil to offspring may reap both advantages. Transporting food internally may be expected to be more common among pelagic foragers, but this behavior is not well associated with offshore fishing habits. Many alcids forage offshore but carry fish in the bill, whereas many sulids and larids that fish close to shore regurgitate food to offspring. Penguins feed inshore and ingest food for young, although some species, such as Adelies (*Progoscelis adeliae*), travel long distances once out of water (Lack, 1968). It could also be more difficult to swim underwater with fish in the bill. The consistency, size, and amount of prey carried also bear on transport methods.

Parental-feeding and chick-begging methods are to some degree adaptively correlated with nest-site features. Unlike the chicks of ground nesting gulls, young kittiwakes are fed by direct regurgitation, a technique well suited for nestlings on narrow cliff ledges (Cullen, 1957). Compared with sulids and larids that nest on level ground, kittiwake and gannet chicks beg less actively (Cullen, 1957; Nelson, 1967). Vigorous begging is strongly selected against in cliff-nesting seabirds. Such selection, also to be expected among tree-nesting sulids, is evident in the mild begging of Abbott's Booby (*Sula abbotti*) chicks (Nelson 1978a). As always, striking exceptions are not easily interpretable: Red-footed Boobies (*Sula sula*) nest in trees, yet their offspring exhibit the most frenzied begging actions of any sulid (Nelson, 1978b).

Nesting environments also influence other aspects of parent–young interactions. For instance, the calls of petrels returning to burrows may lead to the responsiveness of chicks (e.g., Allan, 1962). Most burrow and crevice nesters have single offspring, and as there is no sibling rivalry and parents are certain of a chick's location in such circumstances, begging may be expected to be a relatively uncomplicated matter, although methods of parental feeding could apparently require a high degree of parent–young coordination to offset this. Nestling periods tend to be longer in most burrow-nesting seabirds, and very little is known about their parental behavior.

IV. PARENTAL INVESTMENTS BY NORTHERN GANNETS

Northern Gannets are large, sexually monomorphic, long-lived seabirds that first breed at 4–7 years (Fig. 1). Both sexes share in all aspects of nesting and parental behavior. They are specialized feeders, well known for their spectacular plunge dives from heights of up to 30 m. Gannets are opportunistic foragers and may feed in waters adjacent to breeding areas or well out at sea. They nest colonially in 34 locations on relatively inaccessible islands, stacks, and coastal cliffs in the North Atlantic (Nelson, 1978a,b); six colonies are in North America (Nettleship, 1976).

Gannets arrive at Canadian breeding areas 6–8 weeks before laying single-egg clutches. Incubation requires about 6 weeks, after which chicks are cared for at the nest site for 13 weeks. Hatching and fledging success are about 85% and 95%, respectively (Nelson, 1978a,b; Montevecchi and Kirkham, 1980; cf. Poulin, 1968). When chicks leave the nest, the parent–offspring bond is severed, and postfledging mortality is estimated to be around 65% during the first year (Nelson, 1978a).

There is considerable seasonal and individual variation in the relative parental investments of gannet mates. Here we attempt to assess the average behavioral and physiological involvements of mates from colony occupation until departure to wintering areas.

Fig. 1. Pair of sexually monomorphic Northern Gannets.

A. Methods

1. Study Sites

Research was done in each of the three Newfoundland colonies: Funk Island (49°46′N, 53°11′W), Baccalieu Island (48°07′N, 52°47′W), and Cape St. Mary's (46°50′N, 54°12′W); and on Great Bird Rock (47°50′N, 61°09′W) in the Magdalen Islands, Quebec (Fig. 2). Descriptions of these gannetries can be found elsewhere (Tuck, 1961; Nettleship, 1976; Porter, 1978; Montevecchi *et al.*, 1980). Data were collected periodically on 11 nests at Cape St. Mary's May–October 1977; on nine nests on Baccalieu Island May–August 1977; on 11 and 12 nests on Funk Island in August 1978 and 1979, respectively; and on 13 nests on Great Bird Rock in June and July, 1979. Chicks in study areas hatched relatively synchronously, although study nests on Funk contained nestlings ranging from 4–7 weeks in age in 1978 and 1–4 weeks in 1979.

2. Data Collection

Data were collected (at Cape St. Mary's unless otherwise indicated) in daytime watches of 1–16 hr during incubation (Inc; Cape St. Mary's, Bac-

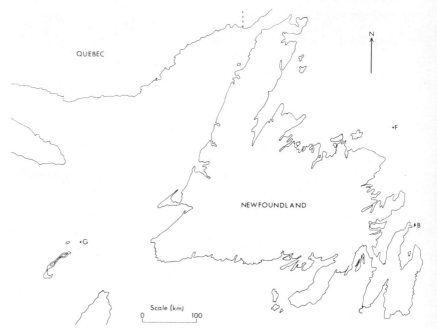

Fig. 2. Location of the three gannet colonies: F, Funk Island; B, Baccalieu Island; C, Cape St. Mary's; G, Great Bird Rock.

calieu, Great Bird Rock) and these posthatch periods: week 1 (PH1; Cape St. Mary's, Funk 1979, Great Bird Rock); week 4 (PH4; Cape St. Mary's, Great Bird Rock); week 5 (PH5; Funk 1978); week 7 (PH7; Baccalieu); week 8 (PH8); and week 13 (PH13). A total of 4356 nest observation hours were monitored. Data collected included (1) diurnal attendance shifts at the nest including the time (a) one or both parents were present and (b) males and females were present; (2) mate change-overs (i.e., when one parent departed from the nest or nest site following the arrival of the mate); (3) begging bouts (i.e., sequences separated by pauses of 1 sec or more), directed to males and females; (4) parental feeds (i.e., whenever a chick fully inserted its head deep into a parent's throat, and the adult showed regurgitation activity followed by the chick's swallowing) by males and females; and (5) latencies of begging bouts and parental feeds following parental arrivals at nest sites.

Samples of fresh eggs, different-aged chicks, and regurgitated feeds were collected for weighing, measuring, analysis of organic composition, and determination of caloric contents. Both adults and chicks often regurgitated food when approached on the nest, providing the basis for food sampling.

3. Aging Chicks and Sexing Adults

Aging Technique. Chicks were aged on the basis of laying, hatching, and fledging dates; and on extrapolations from growth data (Nelson, 1964; Poulin, 1968). Ages are assumed accurate to within about 5 days.

Sexing Adults. Mates were sexed behaviorally and morphologically. During copulation the male stands on the female's back and bites firmly on the nape of her neck (Fig. 3; see also Nelson, 1978a,b). Whenever a bird arrives at the nest site, mates inevitably engage in a greeting display that involves billing (Fig. 4) during which one bird (either the incoming or attending mate) usually makes motions to or actually bites the other's nape. Birds that performed this behavior were judged to be males. These judgments were consistent with other sex differences, such as the nearly exclusive collection of nest material by males early in the season (Nelson, 1978a,b). In extended observations of more than 50 pairs we noted one incubating pair in which both mates consistently engaged in vigorous neck biting and both frequently collected nest material. By late July the golden head coloration of females fades markedly and is much whiter than the males,

Fig. 3. Copulation. Note the neck biting.

especially on the forehead. This divergence increased throughout the remainder of the nesting cycle, and birds could be sexed quickly and easily (see also Nelson, 1978a,b). Individuals were often recognized by scars, dirt, or oil on the plumage, feather patterns, and so forth.

B. Results

1. Egg Size and Composition

Gannets lay relatively small eggs (114.1 ± 1.7 g; Ricklefs and Montevecchi, 1979) equal to about 3.5% of adult weight (3263 ± 91 g). The percentages of yolk and of lipid in the yolk, (Fig. 5), and the energy density (0.9 kcal · g^{-1}) are the lowest yet found in an avian egg.

The average total energy in a gannet egg is 105.0 kcal (Ricklefs and Montevecchi, 1979). Assuming a 25% cost of egg production (Ricklefs, 1974), about 141 kcal are needed to produce an egg.

Fig. 4. Billing display following the arrival of a mate.

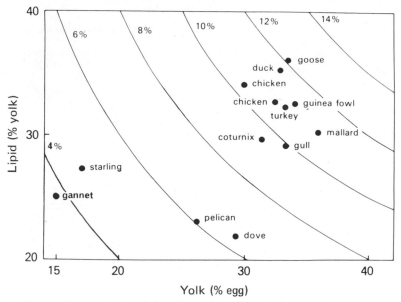

Fig. 5. Scatter diagram of the relative size of the yolk and percentage of lipid in the yolk of birds' eggs. Diagonal curves represent different levels of lipid as a percentage of the whole egg, defined by the relationship: lipid/egg = (yolk/egg) × (lipid/yolk). [From Ricklefs (1977).]

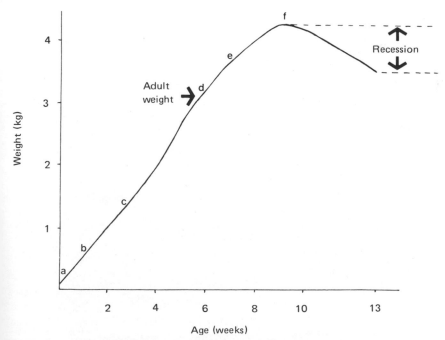

Fig. 6. Growth of gannet chicks. Based on data collected from chicks in captivity and on Poulin (1968) and Nelson (1978*b*). Points a–f on curve correspond to chicks in Fig. 7.

Fig. 7. Different-aged gannet chicks: (A) day 1: just hatched. (B) day 8: eyes open, some down growth. (C) 3 weeks: head up, covered with down. (D) 5 weeks: erect posture, thick down. (E) 7

weeks: growth of scapulars, primaries, and retrices. (F) 9 weeks: feather growth has replaced down.

Fig. 7. (*Continued*)

2. Growth and Development

The altricial chicks hatch weighing about 80 g, with eyes shut, virtually no down, and little motor or thermoregulatory capacity. A single
hatchling that we have analyzed was made up of 87% water and hatched
with a yolk reserve of 17% of its body weight, the yolk sac being comprised
of 80% water and 20% protein (Ricklefs and Montevecchi, manuscript in
preparation). By 3–4 weeks chicks weigh 1000–1500 g (Fig. 6), and lipid

Fig. 7. (*Continued*)

deposits make up about 10% of the body weight (Ricklefs and Montevecchi, manuscript in preparation). By this age, chicks have grown a thick coat of fluffy, white down (Fig. 7C) and have developed homeothermic capacity (Fig. 8; Harvey, 1979). When exposed to low ambient temperatures, gannet chicks younger than 10 days very rarely shiver and then only slightly, chicks 10–15 days of age posthatch usually shiver somewhat more, while those 15 days and older shiver vigorously (Montevecchi, manuscript in preparation). Down-covered chicks also have lower conductance rates, i.e., loss of heat to surroundings, than younger chicks (Fig. 9). By 8–10 weeks of age chicks weigh up to 4500 g (over 30% lipid, Ricklefs and Montevecchi, manuscript in preparation) and are 50% heavier than parents. Feather growth is rapid during the 2nd and 3rd months, and chicks take on a dark appearance, as down is replaced by feathers (Fig. 7F). Chicks show a slight weight recession during the last month of the nestling period, but still fledge with fat reserves at weights well above adult levels (Fig. 7; Nelson, 1964, 1978a,b; Ricklefs and Montevecchi, manuscript in preparation).

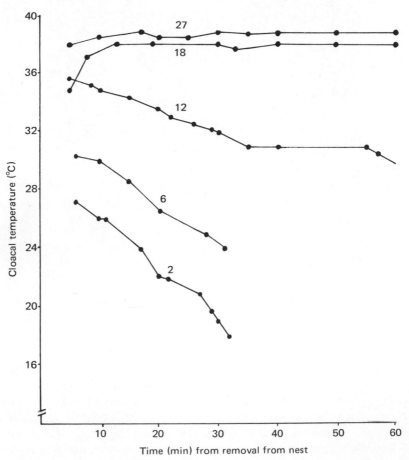

Fig. 8. Change in cloacal temperatures of five different-aged (2, 6, 12, 18, 27 days) gannet chicks removed from nests and exposed to temperatures of 12°–13°C.

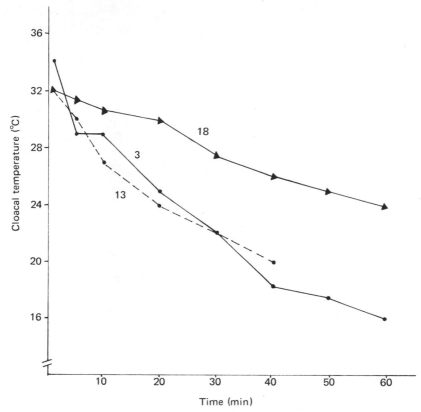

Fig. 9. Conductance of 3-day (no down), 13-day (some down), and 18-day (downy) dead gannet chicks as measured by decrease in cloacal temperature during exposure to ambient temperatures below 15°C.

3. Nutrition for Chicks

Based on regurgitations collected at nests in Newfoundland, the main foods are in decreasing order of importance: mackerel (*Scomber scombrus*), capelin (*Mallotus villosus*), squid (*Ilex illecebrosus*), and in some years Atlantic saury (*Scomberexos saurus*), herring (*Clupea harengus*), and small cod (*Gadus morhua*). There are marked seasonal fluctuations in the utilization of these prey (Fig. 10). Mackerel, sand launce (*Ammodytes hexapterus*), and squid (*Loligo spp.*) were the most common foods in the colony on Great Bird Rock.

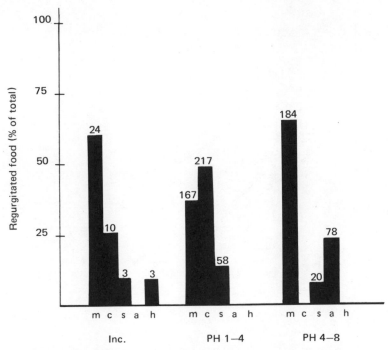

Fig. 10. Percentages of different foods in regurgitations of gannets on Baccalieu and Funk Islands during incubation and the 1st and 2nd months of the nestling cycle. Absolute numbers given at the tops of columns. Abbreviations: m, mackerel; c, capelin; s, squid; a, Atlantic saury; h, herring.

Table II. Nutritional Characteristics of the Major Foods Fed to Northern Gannet Chicks

Food	Metabolizable energy (kcal/g[a])	Percentage composition				Protein index[b]
		Water	Protein	Fat	Ash	
Mackerel	2.45	62.3	16.2	18.6	2.0	6.61
Capelin	1.08	77.1	11.9	6.0	2.1	11.02
Squid	0.97	78.2	15.9	3.0	1.8	16.39

[a] Wet weights (assumes 100% digestibility): protein, 4.3 kcal/g.; fat, 9.0 kcal/g.
[b] Percent protein content divided by metabolizable energy (Ricklefs, 1974).

Table II shows the organic compositons and caloric contents of the major foods fed to chicks. Owing to high lipid content, mackerel is the most energy rich food. Capelin and squid have higher protein indexes and substantially lower lipid levels than mackerel. An average ($\pm$ SE) feed (regurgitation) of mackerel (316.3 $\pm$ 15.6 g) around midnestling period is substantially larger than one of either capelin (130.6 $\pm$ 52.0 g) or squid (137.1 $\pm$ 15.8 g). A mackerel feed contains about 775 kcal, which is more than five times that in an average feed of capelin (141 kcal) or squid (133 kcal).

4. Parental Attendance and Change-Overs

Females incubate significantly more than males and account for 74% (23,192 min) of the total time (31,471 min) a sexed mate was observed settled on an egg. Throughout the nestling period chicks are continuously attended by at least one parent. Overall, parents are together at the nest site an average of 11% of the time. Both parents are at the nest about 8% of the time from hatching until 2 months posthatch, whereas they are together approximately 20% of the time toward the end of the nestling period (Fig. 11). Pairs were present at 12.5% (58/465) of the nests checked at dusk.

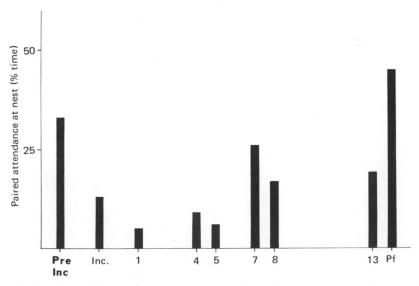

Fig. 11. Percentage of time gannet parents were together at the nest over the nesting cycle. Pf, Postfledging period.

Table III. Average (± SE) Diurnal Nest Shifts[a] (min) by Male and Female Gannets over the Nestling Cycle

Developmental stage	Male and female	n	Male	Female[b]
PH1	424.8 ± 95.4	13	362.3 ± 132.6	497.8 ± 143.8
PH4	222.4 ± 22.0	97	170.6 ± 22.8	302.9 ± 40.7*
PH5	240.0 ± 16.1	160	198.2 ± 16.7	294.6 ± 32.7*
PH7	139.6 ± 21.1	73	85.6 ± 15.7	226.3 ± 49.3*
PH8	204.9 ± 16.0	108	203.6 ± 23.1	206.6 ± 21.7
PH1–8	217.2 ± 9.7	451	175.3 ± 10.8	270.6 ± 13.9*
N			263	188

[a] Shifts bounded by change-overs, very early onsets or very late terminations of watches. This procedure eliminated the brief shifts.
[b] Values followed by asterisk (*): Student's t-test of means of male vs. female data: $p \# 0.05$.

Based on data collected in 1979 on Great Bird Rock, the mean (± SE) diurnal incubation shift is just over 6 hr (377.4 ± 51.3 min), and, after removal of the briefest stints of less than 60 min, is about 7 hr (432.9 ± 53.1 min). Diurnal chick attendance shifts by parents average (± SE) 217.1 ± 9.7 min, decreasing from about 7 hr in PH1 to about 2.3 hr in PH7 and PH8. On Funk Island and Great Bird Rock in 1979, parental attendance shifts with young 1–4 weeks posthatch, averaged about 7 hr (425.8 ± 46.1 min) for males and about 8 hr (492.0 ± 32.9 min) for females. Parental shifts by females during most periods are significantly longer than those of males (Table III). Males make more visits to the nest site than do females

Table IV. Parental Visits by Males and Females to the Nest Site and Change-Overs

Developmental stage	No. nest visits		Visits that led to change-overs (%)	
	Male	Female[a]	Male	Female[a]
INC	126	35*	22	60*
PH1	47	10*	41	100*
PH4	195	76*	32	78*
PH5	85	50*	42	71*
PH7	84	30*	21	70*
PH8	108	49*	25	61*
PH13	28	15*	39	60
Postfledge (PH14)	33	18*	36	50
Total:	706	283*	30	69*

[a] Values followed by asterisk (*): $p_s < 0.05$ (χ^2 tests based on 50:50 expectancy).

(Table IV). Many of these visits do not result in change-overs, and early in the season involve the presentation of nest material to the female.

Figure 12 shows the precentages of daily incubation and chick attendance times put in by males and females. Based on a 50:50 expectancy, females spend significantly more time at the nest in PH1–8; subsequently males are at the nest site more.

After dark and before first light there is very little activity in the colony and birds do not enter or leave the nesting area. A marked increase in movement and an influx of arriving adults begins around dawn. On the basis of data from 10 all-day watches, change-over rates are relatively constant through the day (frequencies in the first and last intervals are extrapolated for watches that began after 0530 or ended before 2130 hr): 22% (66/293) occur between 0530 and 0930 hr, 26% (77/293) occur between 0930 and 1330 hr, 25% (72/293) occur between 1330 and 1730 hr, and 27% (78/293) occur between 1730 and 2130 hr. At 220 nest checks at dusk, about 50% (111) of the single parents were males.

5. Begging and Parental Feeding

The chick's begging behavior changes markedly as motor coordination develops in early life (Figs. 7 and 13). At about 1 week of age, chicks orient

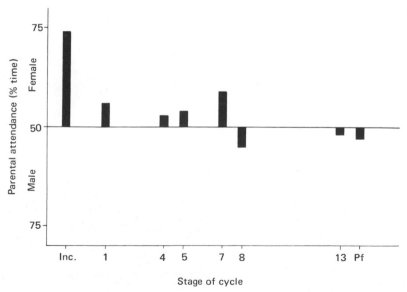

Fig. 12. Percentages of time male and female gannet parents were in attendance at the nest over the nesting period.

Fig. 13. Begging (A) and feeding (B) of a gannet chick about 3 weeks of age.

Table V. Diurnal Variation[a] (Percentage of Occurrence) of Begging Bouts and Parental Feedings during Different 4-hr Intervals of 10 All-Day Watches[b]

Developmental stage[c]		Time of day			
		0530–0930 hr	0930–1330 hr	1330–1730 hr	1730–2130 hr
PH1	Begs[d]	19% (38)	34% (66)	18% (35)	29% (57)
	Feeds[d]				
PH4	Begs	25% (124)	15% (74)	23% (113)	38% (187)
	Feeds	17% (22)	8% (10)	28% (60)	47% (60)
PH5	Begs	16% (172)	27% (297)	32% (249)	24% (265)
	Feeds	19% (21)	29% (32)	29% (32)	22% (24)
PH7	Begs	18% (70)	16% (64)	33% (129)	32% (126)
	Feeds	17% (6)	22% (8)	47% (17)	14% (5)
PH8	Begs	24% (188)	26% (202)	27% (208)	22% (172)
	Feeds	24% (13)	24% (13)	22% (12)	30% (16)
PH1–8	Begs*	20% (592)	24% (703)	28% (834)	27% (807)
	Feeds*	19% (62)	19% (63)	29% (96)	32% (105)

[a] Frequencies given in parentheses. Frequencies of first and last intervals extrapolated whenever watches began after 0530 hr or ended before 2130 hr.
[b] Single watches for PH1 and PH7, two watches for PH4 and PH8, and four watches for PH5.
[c] Asterisk (*): $p_s < 0.001$ (χ^2 tests based on 25% expectancy in each interval).
[d] Data describe parent–young interactions in PH1.

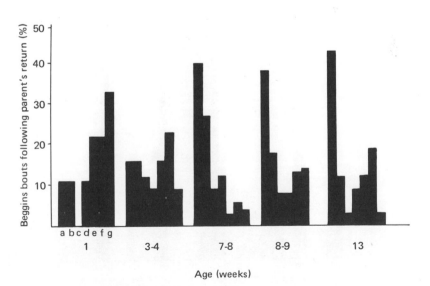

Fig. 14. Percentage of begging bouts occurring in seven intervals—a, 0–5; b, 6–10; c, 11–15; d, 16–30; e, 31–60; f, 61–120; g, 120 min)—following the arrival of a parent at the nest over the nestling cycle.

begging responses to the parent's bill. More begging bouts and parental feeds occur in the afternoon and evening than in the morning. The diurnal variation of these activities, however, is not striking (Table V); this also holds when males and females are considered separately.

Neonates and adults interact most, well after a parent arrives at the nest. At about 1 month posthatch, the begging behavior of chicks is relatively evenly distributed in time with reference to parent arrivals, whereas chicks 2 months and older initiate begging and beg most, soon after a parent returns (Fig. 14). A similar temporal distribution is evident for parental feeding (Fig. 15).

Begging does not always result in feeding. Furthermore, entries into the parent's bill do not always result in food transfer. Food transfer occurred in 84% (660) of the instances (784) in which chicks (1–7 weeks of age) were observed from close range inserting their heads into a parent's bill. Chicks show the greatest frequency of begging bouts at about 1 month of age (Table VI). Parental feeding frequency decreases over the nestling period and drops markedly just before fledging (Table VI). The ratio of begging bouts to parental feeds increases as the chicks grow older, showing a sharp rise in their last week at the nest (Fig. 16).

Owing to seasonal fluctuations in the exploitation of different prey (Fig. 10), the average (± SE) size of a parental regurgitation is smaller in

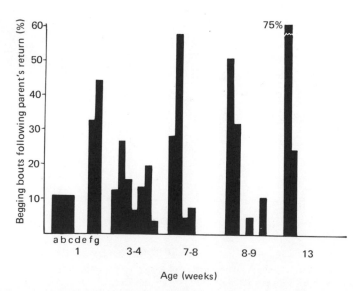

Fig. 15. Percentage of feeds occurring in seven intervals, as in Fig. 14, following the arrival of a parent at the nest over the nestling cycle.

Table VI. Average (± SE) Frequency of Begging Bouts and Parental Feeds per Chick per Day over the Nestling Cycle

Developmental stage	Begs	Feeds
PH1	26.3 ± 8.9	5.6 ± 1.5
PH4	39.5 ± 9.4	4.9 ± 0.9
PH5	30.1 ± 4.9	3.0 ± 0.4
PH7	18.3 ± 4.1	2.0 ± 0.5
PH8	16.4 ± 4.1	1.2 ± 0.3
PH13	22.4 ± 10.2	0.4 ± 0.3

the neonatal period during July (216.5 ± 30.0 g), compared to August (359.6 ± 24.7 g; $t = 3.24$, $n = 50$, $p < 0.005$), when a greater proportion of mackerel are taken. On Funk Island in August 1978, the sizes of the regurgitations of breeding adults (279.8 ± 12.4 g) were larger than those of immature plumaged birds in roosting areas (245.9 ± 12.3 g $t = 1.81$, $n = 198$, $p < 0.10$), although the percentage of different foods regurgitated were the same in the two age classes (Kirkham and Montevecchi, manuscript in preparation). Males feed neonatal chicks more than females, whereas females feed older chicks more than males (Table VII).

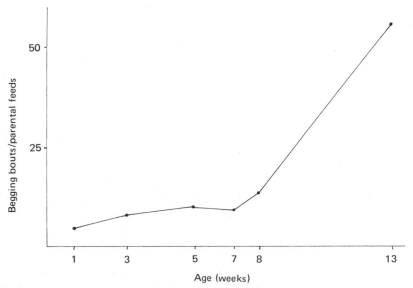

Fig. 16. Ratio of average frequency begging bouts to parental feed over the nestling cycle.

Table VII. Parental Feeds by Male and Female
Northern Gannets

Developmental stage	Male	Female
PH1	22	10
PH4[a]	167	135
PH5	54	68
PH7	10	15
PH8	12	23
PH13[b]	1	2
PH1–13	266	253

[a] Based on 50 : 50 expectancy between frequency of
male and female feeds (PH1–4); $\chi^2 = 5.80$, $p < 0.05$.
[b] PH5–13; $\chi^2 = 5.19$, $p < 0.05$.

C. Discussion

1. Prelaying and Egg Investments

Like other large seabirds, gannets lay proportionately small eggs (Lack, 1968; Ricklefs, 1977; Ar and Yom-Tov, 1978). Gannet eggs have small yolks, low lipid levels, and low-energy densities, as is characteristic of birds showing altricial development. In this context, gannet eggs fall at the extreme altricial end of the spectrum (Ricklefs, 1977; Ricklefs and Montevecchi, 1979). Small investments in eggs are associated with extensive postnatal parental care, and both tendencies are linked to altricial development. As illustrated in Fig. 5, two clusters of data are evident—the birds in one grouping (gannets, pelicans, doves, and starlings) develop altricially, whereas those in the other, with the exception of the semiprecocial gulls, develop precocially.

In addition to egg production by the female, both sexes make substantial prelaying investments in site-defense and nesting behavior. During the breeding season gannets are the most agonistic of the seabirds (Nelson, 1978a,b). Male gannets, like many other male seabirds, arrive at breeding areas before females and establish or reestablish nest sites by engaging in extensive aggressive behavior, an activity that can be very demanding (Nelson, 1978a,b).

Males also collect virtually all the nest material. Perhaps to reduce costs and risks associated with ovum development, females collect little, but do most of the nest building. Males weigh, on average, 168 g less than females during incubation, whereas both sexes have equivalent weights dur-

ing the posthatch period (Nelson, 1978*a*). Nelson contends that the prelaying period is the most stressful part of the breeding cycle for the male. For
kittiwakes, Coulson and Wooller (1976) report higher mortality of males
compared to females early in the breeding season (see also Nelson, 1978*a*;
Ryder, 1979*a*).

A gannet's egg, including production, costs about 141 kcal (Ricklefs
and Montevecchi, 1979), whereas a 168-g depletion of fat lost by males is
equivalent to a loss of 1500 kcal. We interpret this to indicate that up to the
onset of incubation the female Northern Gannet's reproductive investment
does not exceed the male's.

2. Growth, Development, and Nutrition

The rapid growth rates of Northern, Cape (*Morus capensis*), and Australasian (*M. serrator*) Gannets, compared to those of tropical boobies, are
attributed to the abundant food flushes in cold ocean water, and longer
daylengths that allow more time for foraging. It is necessary for gannets to
fledge before the onset of deteriorating fall weather in temperate regions
(Lack, 1967, 1968; Ricklefs, 1973; Nelson, 1978*a,b*).

Gannet hatchlings, like other altricial young, appear to have high water
levels in tissues and low energy densities (see Ricklefs, 1974; 1979). The
Double-crested Cormorant, another Pelecaniforme, has a similar yolk
reserve at hatchling: 16% (Dunn, 1973 cited in Ricklefs, 1974) compared to
17% for the gannet. These percentages are much higher than those of small
altricial passerines (Ricklefs, 1974).

Gannet hatchlings are continuously brooded on the parent's large,
webbed feet for 3 weeks (during which time they become homeothermic)
before the period of most rapid growth from 4–7 weeks (Harvey, 1979).
Skeletal muscle growth (especially the leg) that permits heat production
through shivering, decreasing body surface to volume ratios, and decreasing
conductance levels are important determinants of the rate of homeothermic
development (see also Ricklefs 1974, 1979). The absence of shivering in
chilled neonatal gannet chicks reflects a lack of substantial leg-muscle
development. The capacity to shiver develops gradually during the first 3
weeks, and it generates from the leg and rump regions of chicks; wing
shivering dependent on the maturation of pectoral muscle is not developed
by this age. Shivering and muscle activity may be the only means of
thermogenesis in birds (Steen and Enger, 1957; West, 1965; Aulie, 1976).
The conductances of gannet chicks sacrificed at different points in the neonatal period also decrease as a function of age, and temperature regulation
may simply be a consequence of other important developmental processes
(Ricklefs, 1974). By freeing caloric allotments that might otherwise be chan-
neled into temperature regulation, the delay of homeothermy in altricial,

compared to precocial birds, makes more assimilated energy available for growth (Ricklefs, 1979). More rapid onsets of temperature regulation allow parents more time to forage and may reduce the risk of predation at the nest for some parents (see Dunn, 1975*b*). Double-crested Cormorants are homeothermic by 14–15 days (Dunn, 1975*a*), and Masked Boobies (*Sula dactylatra*), like gannets, can thermoregulate by about 18 days (Bartholomew, 1966). Large altricial species can thermoregulate earlier in development than can small ones; the growth of thick down and large body size facilitates temperature regulation (Ricklefs, 1974). Razorbill and murre chicks are fairly homeothermic at hatching and are downy and capable of vigorous shivering (Montevecchi, unpublished data; see also Johnson and West, 1975). Gannet neonates can also gular flutter rapidly in response to slight temperature increases, a behavior that helps insure that intense solar radiation levels in nesting areas do not lead to hyperthermia (see also Howell and Bartholomew, 1962; Bartholomew, 1966).

Parental feeds are larger for older chicks (see also Palmer, 1941; Hawksley, 1957; cf. Harris, 1966). The seasonal variations in the foods delivered to gannet chicks reflect the timing of inshore migrations of different prey. There are also differences among colonies in the proportions of different foods brought to offspring (Kirkham, 1980; see Nelson, 1978*a*,*b*; Section IV.C.3). In Newfoundland, gannet chicks older than 5 weeks are fed primarily mackerel. The high lipid content of mackerel (see also Dingle, 1976) complements the fat deposition of older chicks. An 8-week chick that weighed about 25% more than an adult showed a sharp weight loss when switched from a diet of capelin to squid (a food with a lower lipid level; Table II; Harvey, 1979). The high-protein indexes of capelin, a common food of neonates, may complement rapid structural growth in early development. Analyses of the organic composition of different-aged chicks will aid in our understanding of the apportionment of nutrients during growth (Ricklefs and Montevecchi, manuscript in preparation). On the basis of the food eaten by two captive gannet chicks that developed normally and whose weights were comparable to those of chicks reared by parents, it can be estimated that chicks consume about 24,176 g of fish worth 45,545 kcal over a 13-week period in the nest.

Following a peak at 8–10 weeks, chick weights decrease somewhat, although fledglings depart heavier than adults. This recession is attributable to many factors: decreased parental feeding rates (Table V), decreased feeding motivation of older chicks (see also Nelson, 1978*a*), increased energy and water requirements for feather growth (see Ricklefs, 1968*b*), and increased activity (mostly wing flapping). Many shearwater (*Puffinus tenuirostris*, *P. gravis*, *P. griseus*, *P. puffinus*) nestlings at maximum weight are heavier than parents and are rarely fed, if at all, during the last 2 weeks before fledging (Harris, 1966; Lack, 1968). It is common among birds that forage for long periods of time, such as seabirds, swifts, and swallows, for nestling weights to exceed those of adults. The fat reserves of nestlings and fledglings are useful in the face of reduced parental feeding rates or after independence (Harris, 1966; Lack 1967, 1968; Ricklefs 1968a,b). Gannet and Brown Pelican (*Pelecanus occidentalis*) chicks can go up to 2 weeks without food and still develop normally (Schreiber, 1976; Nelson, 1978a). Gannet fledglings need time to develop plunge-diving fishing skills, and fat reserves aid in the transition to independence (Nelson, 1978a). Seabird nestlings, such as those of frigatebirds and tropical sulids, that do not grow heavier than adults are fed by parents for extended periods after departure from the breeding area (Lack, 1968; Dunn, 1975a; Chapter 10).

3. Parental Attendance and Foraging Time

Owing to frequent, intense agonistic interactions among nesting neighbors, gannet chicks are constantly attended by at least one parent throughout the nestling period. Parental attendance limits the time available for foraging, but this does not seem to cause difficulties for gannets. Parents spend substantial time together at the nest, indicating that they are not overly taxed in providing sufficient food for themselves and offspring. The percentage of time mates are paired at the nest is highest with older chicks that place greater feeding demands on parents. The high feeding rates of neonates are not well associated with adult arrivals at the nest, which may impose greater foraging demands early in the nestling cycle, resulting in parents being together at the nest less often than later in the season. Even though neonates are provided with small feeds, parents probably cannot delay the assimilation of food they hold, and must replenish supplies frequently (Nelson, 1978a,b).

In Contrast to Lack's (1954, 1966) theory that clutch size is optimally adjusted to the maximum number of offspring parents can raise (see also Ashmole, 1963; Pearson, 1968), gannets, like many seabirds, are not pushed too close to their foraging limitations in rearing a chick. Nelson (1978b) reports two cases in which single males whose mates had died, successfully fledged offspring (see also Nisbet et al., 1978). Nelson (1964, 1978a,b) reports that gannets can increase parental feeding rates and successfully fledge experimental broods of two chicks. Jarvis (1974), however, questions whether the parents of twins actually gain a reproductive advantage, as the slightly lighter weight of twinned nestlings compared to singletons might lower postfledging survival. Jarvis (1974) finds no reproductive advantage due to artificial twinning in Cape Gannets, although Nelson (1978b) points out ecological differences that might account for the Northern Gannets' ability to rear twins and the Cape Gannets' lack of it.

Some gulls are also exceptions to Lack's clutch size hypothesis, because they can rear larger than normal broods (e.g., Vermeer, 1963; Harris and Plumb, 1965; Pearson, 1968; Haymes and Morris, 1977). By comparison with gulls, Procellariiformes, gannets, and Alcidae—birds that usually forage farther from shore and for more patchily distributed food—cannot rear larger than normal broods (Harris, 1966, 1978; Lack, 1968; Lloyd, 1978).

The incubation and parental attendance shifts of gannets in North America are substantially shorter than those of gannets on Bass Rock and Ailsa Craig, Scotland, although they are similar to those in the British Colony of Bempton (Nelson, 1978a,b). The shift durations reported for the British birds include nocturnal time, whereas ours do not. After taking nighttime activity into account, Newfoundland birds are still absent from the nest for shorter periods than gannets in Scottish colonies. This suggests that it might be easier for gannets to forage and fledge two-chick broods in Newfoundland. Even among North American colonies there are differences —gannets on Great Bird Rock put in longer attendance shifts than do gannets in Newfoundland. Cape Gannets put in long incubation and chick attendance shifts comparable to those of Northern Gannets in Scotland (Jarvis, 1971). Systematic comparisons of the distribution and abundance of food resources around different colonies and in different years are needed. From one colony to another, gannets utilize different foods and varied proportions of similar foods (Nelson, 1978a,b; Kirkham, 1980). Air and ocean temperatures are lower in the Western North Atlantic, possibly favoring more intense brooding and greater food abundance (flushes), whereas longer daylengths at British colonies allow gannets more potential foraging time.

Gannets can fly approximately 80 km/hr or more (Montevecchi, unpublished data; Barlee, 1956 cited in Nelson, 1978b). Assuming that hunting birds capture prey soon after a school is located and then return fairly directly to the nest, it can be estimated that gannets forage within about 120 km of colonies in Newfoundland. Longer trips, perhaps up to 500 km, may be required around colonies in Great Britain (Nelson, 1978a,b) and in the Gulf of St. Lawrence. The seasonal decrease in parental attendance shifts in Newfoundland is consistent with observations that gannets forage closer to shore as prey (capelin, mackerel, saury, squid) move toward the coastline. It may be more difficult for gannets to obtain small prey (capelin) that they exploit early in the nestling period. The constant occurrence of change-overs during the day suggests there are no daily or tidal fluctuations in prey availability (see also Tuck, 1961; Drent, 1965; Slater, 1976).

Females brood significantly more than males, possibly a continuation of the female's greater role in incubation. By comparison, Nelson (1978a,b) finds that males incubate more and attend chicks more than females.

Nelson's time budgets include nocturnal time at the nest. Our observations indicate that either mate is equally likely to remain alone at the nest at night, and thus night shifts are unlikely to account for the different male/female attendance ratios found in western and eastern North Atlantic colonies. Our study and Nelson's (1978 *a,b*) both indicate that males attend older chicks more than females do. More detailed comparisons are needed. Male Manx Shearwaters incubate more than females (Harris, 1966), whereas female Least Terns (*Sterna albifrons*) and Common Terns brood more than males (Hardy, 1957; I. C. T. Nisbet and A. Storey, personal communications), and male Crested penguins (*Eudyptes* spp.) do virtually all the brooding (Warham, 1975). Male Common Terns and Great Black-backed Gulls (*Larus marinus*) forage farther and exploit different prey than females (A. Storey, personal communication; Belopolskii, 1961). It would be interesting to relate these male/female investment patterns to ecological factors.

After the chick fledges, mates spend significantly more time together at the nest site, and courtship and territorial behavior increase substantially, apparently strengthening the pair bond and attachment to the nest site. Activity in the colony in the postfledging period is reminiscent of that before laying, and as in the prelaying period, the male may play a greater role in nest-site defense (Nelson, 1978*a,b*). Males probably have higher levels of circulating sex hormones at the end of the season, as indicated by the deep, golden coloration of their head plumage compared to the whiter heads of females.

4. Begging and Parental Feeding

Slightly more begging and feeding occurs in the afternoon and evening. Gannets in Scotland (Nelson, 1978*b*) and New Zealand (Wodzicki and McMeekan, 1947) show similar diurnal patterns. Such activity may help ensure that chicks do not become hungry at night. We have observed parents feeding young during fading light an hour after sunset. Because gannets can regurgitate several hours after they return to the nest, it is possible that parents returning to the nest late in the evening may feed chicks after dark.

Begging and feeding become more tightly associated with parental arrivals at the nest as chicks grow older. Chicks may learn that incoming adults bring food (see Hailman, 1967) and may beg more effectively with age; parents, in turn, may be more motivated to deliver sooner the large feeds they bring to older chicks. Whatever the mechanisms, the small feeds and the delay between parental arrival and feeding of neonates ensure that food is predigested by adults before transfer to newly hatched chicks.

Begging behavior, especially that of young chicks, is closely associated with and often initiated by parental activities, such as the vigorous preening and prodding of chicks. Such interactive behavior between parent and young serves to coordinate the activity of the two during food transfer. Parents also thwart the begging activity of chicks by settling on them.

Younger chicks are fed more often than are older chicks. Low parental feeding rates just prior to fledging are reflected in the weight recession of chicks. The feeding rates reported here are in fair agreement with rates of "feeding bouts" found by Nelson (1978a,b), although the gannets studied in Scotland engage in many more food transfers between parent and offspring. It is unlikely that the amount of food delivered to offspring differs substantially in these locations, as growth rates and fledging ages are similar. The substantially longer foraging trips of gannets in Scotland probably result in greater digestion, which in turn requires parents to make more deliveries per unit of food transferred. It is not possible to compare seasonal feeding trends between the two locations, as Nelson's findings are based on a single 48-hr watch of 27 different-aged chicks. More detailed intercolony comparisons of the gannets' feeding ecology are needed, particularly among colonies with different environmental and climatic conditions.

Gannets carry large quantities of food that are more than sufficient for young chicks. The direct regurgitation feeding method helps ensure that feed size increases while feeding rates decrease with chick age. Among birds that transport food in the bill, parental feeding rates might be more likely to increase over the nestling period (e.g., Tuck, 1961).

Begging does not always elicit parental feeding, and a chick is not always fed when it inserts its head into the parent's bill (see also Nelson, 1978a,b). As they grow older, chicks beg more for a feed; a trend that may help explain why older chicks are usually fed soon after a parent arrives at the nest but may also reflect a developing parent–offspring conflict with regard to food transfer (Trivers, 1974). Feed sizes increase over the nestling period but are relatively stable after week 5, when chicks exceed adult weight and take large feeds, and when mackerel is the predominant food. It would be informative to determine the proportions of unsuccessful begging–feeding interactions over the nestling period (see Wortis, 1969; Trivers, 1974).

Our data do not allow for speculation about possible selective prey utilization by gannets. Limited speculation may be possible by comparing seasonal fluctuations in the proportions of different foods taken with comparable fisheries statistics, as the major prey of gannets are commercially important. Consistent with the idea that delayed maturity allows specialized feeders time to develop efficient foraging skills that enable them to obtain sufficient food for themselves and offspring (e.g., Lack, 1968; Orians, 1969; Recher and Recher, 1969; cf. Nelson, 1977, 1978a), the regurgitations of immature, nonbreeding gannets smaller than those of breeding adults (Kirkham, 1980).

During the neonatal period young chicks are brooded more by females and fed more by males that are absent from the nest more. Overall, females probably deliver more food energy to chicks, as they do most of the feeding of older chicks that take much larger feeds. A similar situation may hold for Blue-footed Boobies (Nelson, 1978b). The longer attendance shifts of female Northern Gannets suggests that they may forage more efficiently than males. As they visit the nest more frequently, it is not likely that males forage further than females. While away from the nest site males may loaf more, although in view of their stronger tenacity to and higher level of agonistic behavior at nest sites, this seems unlikely. Telemetric studies could aid in the study of the off-duty activities of parents.

5. Parental Contributions by Males and Females

Females do more direct parental care (i.e., incubating, brooding, attending, and food-energy provisioning of chicks,) than males. Males invest more in the establishment and possession of nest sites. The greater parental care by females fits with Trivers' (1972) contention that owing to differential gametic commitments by the sexes, the female should invest more in offspring than the male even in monogamous systems. The male's reproductive investments, such as those in the prelaying period that appear to easily offset the small egg investments of female gannets are not, however, accounted for in Trivers' scheme.

Among marine birds, female terns, gulls, and alcids make the greatest energetic commitments to eggs (Table I), and in many of these species males engage in extensive feeding of mates before and after laying. Male Black-headed Gulls (*Larus ridibundus*) may feed females from a month before laying until after hatching (Kirkman, 1937). In some species males provide females with substantial (perhaps all) nutrition before egg laying (Watson, 1908; Nisbet, 1973, 1977; see also Borudulina, 1966; Brown, 1967). In what may perhaps be thought of as a ritualized analogy to courtship feeding, the males of some species (e.g., the gannet) collect most of the nest material and present it to the female. For some seabirds it may be energetically inefficient and risky for females that are heavier (owing to egg development) to forage or collect nest material in areas distant from the nest site (see Nisbet, 1977). In general, female investments in eggs may have often been greatly overestimated relative to male parental investments (see also Gladstone, 1979).

It would be informative to relate the variation in the sharing of parental activities between mates to fitness (see Coulson, 1966), but fledging successs among gannets is so high as to preclude a thorough study of this except on a very long term basis. At 30 study nests for which we have success data, only a single failure has been recorded. The chick died at 31 days of unknown causes, although probably because of exposure during a sustained wet spell.

The apparent ease with which gannets rear offspring allows them substantial flexibility in the sharing of parental activity between mates (see also Wallman, *et al.*, 1979). Although it was not dealt with here, study of the variation of mate relationships among different pairs will be crucial for a full understanding of "species-typical" parental strategies. Analysis of the sharing of parentally related behavior throughout the breeding cycle, especially in successively monogamous seabirds, provide interesting ground for further investigation. Purely selfish interactions need not be expected between mates that bond over successive cycles (Lazarus and Inglis, 1978).

ACKNOWLEDGMENTS

J. Burger, P. Cottrell, I. R. Kirkham, S. G. Sealy, A. Storey and L. M. Tuck provided helpful comments on the manuscript. G. Coombes, L. Grimmer, B. D. Harvey, I. R. Kirkham, and R. Purchase contributed greatly to the field work. R. E. Ricklefs allowed modification of his figure (Fig. 5) on the egg composition of different birds and provided information on growth in gannets, and B. D. Harvey permitted reproduction of his figure (Fig. 8) on homeothermy. The Canadian Coast Guard and Department of Fisheries, provided transportation to and from Funk Island (special thanks are due to Captain P. Grandy) and housing on Baccalieu Island. The Wildlife Division of the Newfoundland Department of Tourism provided housing at Cape St. Mary's and along with the Government du Quebec and Canadian Wildlife Service allowed us to study the gannets. The lightkeepers at Cape St. Mary's (M. Careen, S. Careen, and C. Warren) and on Baccalieu (E. Blundon, R. Hyde, R. Noonan, P. Rice, C. Riggs, and L. Walsh) helped us in many ways. The late Dr. Leslie M. Tuck, through his scholarship and research, his example, courage, and support, provided us with invaluable inspiration and opportunity in our work with the gannets. Research was financed by the Atlantic Provinces Inter-university Committee on the Sciences, Memorial University of Newfoundland, and the Natural Sciences and Engineering Research Council of Canada, Grant No. A0687. To each of these individuals and institutions we remain indebted.

REFERENCES

Allan, R. G., 1962, The Madeiran Storm Petrel *Oceanodroma castro*, *Ibis* **103b**:274–296.
Ar, A., and Yom-Tov, Y., 1978, The evolution of parental care, *Evolution* **32**:655–669.

Arnason, E., and Grant, P. R., 1978, The significanc of kleptoparasitism during the breeding season in a colony of Arctic Skuas *Stercorarius parasiticus* in Iceland, *Ibis* **120**:38–54.

Ashmole, N. P., 1963, The regulation of numbers of tropical oceanic birds, *Ibis* **103b**:458–473.

Ashmole, N. P., 1971, Seabird ecology and the marine environment, in: *Avian Biology* (D. S. Farner and J. R. King, eds), Vol. 1, pp. 223–286, Academic Press, New York.

Aulie, A., 1976, The pectoral muscles and the development of thermoregulation in chicks of Willow Ptarmigan (*Lagopus lagopus*), *Comp. Biochem. Physiol.* **53A**:343–346.

Austin, O. L., and Austin, O. L., Jr., 1956, Some demographic aspects of the Cape Cod population of Common Terns (*Sterna hirundo*), *Bird-Banding* **27**:55–66.

Barlee, J., 1956, Flying for business and pleasure, *Shell Aviation News*, pp. 2–9.

Bartholomew, G. A., 1966, The role of behavior in the temperature regulation of the Masked Booby, *Condor* **68**:523–535.

Bédard, J., 1969, Adaptive radiation in Alcidae, *Ibis* **111**:189–198.

Beer, C. G., 1966, Incubation and nest-building behavior of Black headed Gulls. V: The post-hatching period, *Behavior* **26**:189–214.

Belopolskii, L. O., 1961, *Ecology of Sea Colony Birds of the Barents Sea*, Israel Program for Scientific Translations, Jerusalem.

Bent, A. C., 1921, *Life Histories of North American Gulls and Terns*, Smithsonian Institute U.S. National Museum Bulletin 113.

Borudulina, T. L., 1966, Biological and economic importance of gulls and terns of southern USSR water bodies, *Ecol. Morphol. Birds Mann. Natl. Sci. Fnd.* (transl.), Washington, D.C.

Brown, R. B. G., 1967, Courtship behaviour in the Lesser Black-backed Gull, *Larus fuscus*, *Behavior* **29**:122–153.

Burger, J., in press, On becoming independent in Herring Gulls: Parent–young conflict. *Am. Nat.*

Case, T. J., 1978a, Endothermy and parental care in the terrestrial vertebrates, *Am. Nat.* **112**:861–874.

Case, T. J., 1978b, On the evolution and adaptive significance of postnatal growth rates in the terrestrial vertebrates, *Q. Rev. Biol.* **53**:243–282.

Collias, E. C., and Collias, N. E., 1957, The response of chicks of the Franklin's Gull to parental bill-color, *Auk* **74**:371–375.

Coulson, J. C., 1966, The influence of the pair bond and age on the breeding biology of the Kittiwake Gull (*Rissa tridactyla*), *J. Anim. Ecol.* **35**:269–279.

Coulson, J. C., and Wooller, R. D., 1976, Differential survival rates among breeding Kittiwake Gulls *Rissa tridactyla* (L.), *J. Anim. Ecol.* **45**:205–213.

Cullen, E., 1957, Adaptations in the Kittiwake to cliff-nesting, *Ibis* **99**:275–302.

Dingle, J. R., 1976, Technology of Mackerel Fishery, Bibliography and Survey of Literature, Information Canada, Ottawa.

Dorward, D. F., 1962, Behaviour of boobies, *Sula* spp., *Ibis* **103b**:221–234.

Drent, R. H., 1965, Breeding biology of the Pigeon Guillemot *Cepphus columba*, *Ardea* **53**:100–160.

Drent, R. H., 1970, Functional aspects of incubation in the Herring Gull, in: *The Herring Gull and Its Egg* (G. P. Baerends and R. H. Drent, eds.), *Behav. Suppl.* **17**:1–132.

Dunn, E. H., 1973, Energy allocation of nestling Double-crested Cormorants, Ph.D. Thesis, University of Michigan, Ann Arbor.

Dunn, E. H., 1975a, Growth, body components and energy content of nestling Double-crested Cormorants, *Condor* **77**:431–438.

Dunn, E. H., 1975b, The timing of endothermy in the development of altrical birds, *Condor* **77**:288–293.

Dunn, E. H., 1976a, Development of endothermy and existence energy expenditure of nestling Double-crested Cormorants, *Condor* **78**:350–356.

Dunn, E. H., 1976*b*, The development of endothermy and existence energy expenditure in Herring Gull chicks, *Condor* **78**:493–498.

Evans, R. M., 1970, Imprinting and mobility in young Ring-billed Gulls, *Larus delawarensis*, *Anim. Behav. Monogr.* **3**:193–248.

Evans, R. M., 1977, Semi-precocial development in gulls and terns (Laridae), *Proc. NE Reg. Mtg. Anim. Behav. Soc., St. John's, Newfoundland*, pp. 213–239.

Gladstone, D., 1979, Promiscuity in monogamous colonial birds, *Am. Nat.* **114**:545–557.

Hailman, J. P., 1967, The ontogeny of an instinct, *Behav. Suppl.* **15**:1–159.

Hardy, J. W., 1957, The Least Tern in the Mississippi Valley, *Publ. Mus. Mich. State Univ. Biol. Ser.* **1**:1–60.

Harris, M. P., 1966, Breeding biology of Manx Shearwaters (*Puffinus puffinus*), *Ibis* **108**:17–33.

Harris, M. P., 1978, Supplementary feeding of young puffins *Fratercula arctica*, *J. Anim. Ecol.* **47**:15–23.

Harris, M. P., and Plubm, W. J., 1965, Experiments on the ability of Herring Gulls *Larus argentatus* and Lesser Black-backed Gulls *L. fuscus* to raise larger than normal broods, *Ibis* **197**:256–257.

Harvey, B. D., 1979, Growth, food consumption, and homeothermy of nestling Gannets (*Morus bassanus*), B.Sc. thesis, University of New Brunswick, Fredericton.

Hawksley, O., 1957, Ecology of a breeding population of Arctic Terns, *Bird-Banding* **28**:57–92.

Haymes, G. T., and Morris, R. D., 1977, Brood size manipulations in Herring Gulls, *Can. J. Zool.* **55**:1762–1766.

Hays, H., Dunn, E., and Poole, A., 1973, Common, Arctic, Roseate, and Sandwich Terns carrying multiple fish, *Wilson Bull.* **85**:233–236.

Henderson, B. A., 1975, Role of the chicks' begging behavior in the regulation of parental feeding behavior of *Larus glaucescens*, *Condor* **77**:488–492.

Howell, T. R., and Bartholomew, G. A., 1962, Temperature regulation in the Red-tailed Tropic Bird and the Red-footed Booby, *Condor* **64**:6–18.

Imber, M. J., 1976, The origin of petrel stomach oils—A review, *Condor* **78**:366–369.

Impekoven, M., 1973, The response of incubating Laughing Gulls *Larus atricilla* to calls of hatching chicks, *Behavior* **46**:94–113.

Impekoven, M., 1976, Prenatal parent–young interactions in birds and their long-term effects, in: *Advances in the Study of Behavior* (J. S. Rosenblatt, R. A. Hinde, E. Shaw, and C. G. Beer, eds.), Vol. 7, pp. 201–253, Academic Press, New York.

Jarvis, M. J. F., 1971, Ethology and ecology of the Southern African Gannet *Sula capensis*, Ph.D. thesis, University of Cape Town, South Africa.

Jarvis, M. J. F., 1974, The ecological significance of clutch size in the South African Gannet (*Sula capensis* Lichtenstein), *J. Anim. Ecol.*, **43**:1–17.

Johnson, S. R., and West, G. C., 1975, Growth and development of heat regulation in nestlings, and metabolism of adult Common and Thick-billed Murres, *Ornis Scand.* **6**:109–115.

Kendeigh, S. C., 1952, Parental care of its evolution in birds, *Ill. Biol. Mongr.* **22**:1–356.

Kirkham, I. R., 1977, The feeding ecology and behaviour of Ring-billed Gull (*Larus delawarensis*) chicks, B.Sc. honours thesis, Brock University, St. Catharine's, Ontario.

Kirkham, I. R., 1980, Nestling development, energetics and parental investment in Northern Gannets (*Morus bassanus*), M.Sc. thesis, Memorial University of Newfoundland, St. John's, Newfoundland.

Kirkman, F. B., 1937, *Bird Behaviour*, T. Nelson and Sons, London.

Lack, D., 1954, *The Natural Regulation of Animal Numbers*, Oxford University Press, Oxford.

Lack, D., 1966, *Population Studies of Birds*, Oxford University Press, Oxford.

Lack, D., 1967, Interrelationships in breeding adaptations as shown by marine birds, *Proc. Int. Ornithol. Congr.* **14**:3–42.

Lack, D., 1968, *Ecological Adaptations for Breeding in Birds*, Methuen, London.

Langham, N. P. E., 1972, Chick survival in terns *Sterna* spp. with particular reference to the Common Tern, *J. Anim. Ecol.* **41**:385–395.

Lazarus, J., and Inglis, I. R., 1978, The breeding behaviour of the Pink-footed Goose: Parental care and vigilant behaviour during the fledging period, *Behaviour* **65**:62–88.

Lehrman, D. S., 1955, The physiological basis of parental feeding behaviour in the Ring Dove *Streptopelia risoria, Behavior* **7**:214–286.

Lehrman, D. S., 1961, Hormonal regulation of parental behavior in birds and infrahuman mammals, in: *Sex and Internal Secretions* (W. C. Young, ed.), Vol. II, pp. 1268–1382, Williams & Wilkins, Baltimore, Md.

Lloyd, C. S., 1978, The ability of the Razorbill *Alca torda* to raise an additional chick to fledging, *Ornis Scand.* **8**:155–159.

Matthews, L. H., 1949, The origin of stomach oil in the petrel, with comparative observations on the avain proventriculus, *Ibis* **91**:373–393.

Montevecchi, W. A., and Kirkham, I. R., 1980, Breeding success of Northern Gannets at two Newfoundland colonies, *Ibis* (submitted).

Montevecchi, W. A., Kirkham, I. R., Purchase, R., and Harvey, B. D., 1980, Colonies of Northern Gannets in Newfoundland, *Osprey* **11**:2–8.

Nelson, J. B., 1964, Factors influencing clutch size and chick growth in the North Atlantic Gannet *Sula bassana, Ibis* **106**:63–77.

Nelson, J. B., 1967, Colonial and cliff nesting in the gannet, compared with other Sulidae and the Kittiwake, *Ardea* **55**:60–90.

Nelson, J. B., 1977, Some relationships between food and breeding in the marine Pelecaniformes, in: *Evolutionary Ecology* (B. Stonehouse and C. Perrins, eds.), pp. 77–87, University Park Press, London.

Nelson, J. B., 1978a, *The Gannet*, Buteo Books, Vermillion, North Dakota.

Nelson, J. B., 1978b, *The Sulidae: Gannets and Boobies*, Oxford University Press, Oxford.

Nettleship, D. N., 1976, Gannets in North America: Present numbers and recent population changes, *Wilson Bull.* **88**:300–313.

Nice, M. M., 1962, Development of behavior in precocial birds. *Trans. Linn. Soc. N.Y.* **8**:1–212.

Nisbet, I. C. T., 1973, Courtship-feeding, egg-size and breeding success in Common Terns, *Nature (London)* **241**:141–142.

Nisbet, I. C. T., 1977, Courtship-feeding and clutch size in Common Terns *Sterna hirundo*, in: *Evolutionary Ecology* (B. Stonehouse and C. Perrins, eds.), pp. 101–109, University Park Press, London.

Nisbet, I. C. T., Wilson, K. J., and Broad, W. A., 1978, Common Terns raise young after death of their mates, *Condor* **80**:106–109.

Orians, G. H., 1969, Age and hunting success in the Brown Pelican (*Pelecanus occidentalis*), *Anim. Behav.* **17**:316–319.

Palmer, R. S., 1941, A behavior study of the Common Tern (*Sterna hirundo hirundo* L.), *Proc. Boston Soc. Nat. Hist.* **42**:1–119.

Palmer, R. S., 1962, *Handbook of North American Birds*, Yale University Press, New Haven, Conn.

Pearson, T. H., 1968, The feeding biology of seabird species breeding on the Farne Islands, Northumberland, *J. Anim. Ecol.* **37**:521–552.

Porter, J. M., 1978, The parental time investment and feeding behaviour of the North Atlantic Gannet [*Morus bassanus* (L.)] at Cape St. Mary's Newfoundland, B.Sc. honours thesis, Acadia University, Wolfville, Nova Scotia.

Poulin, J-M., 1968, Réproduction du Fou de Bassan (*Sula bassana*) l'Île Bonaventure, Quebec (Perspective Écologique), M.Sc. thesis, Laval University, Quebec.

Quine, D. A., and Cullen, J. M., 1964, The pecking response of young Arctic Terns and the adaptiveness of the "releasing mechanism," *Ibis* **106**:145–173.

Recher, H. R., and Recher, J. A., 1969, Comparative foraging efficiency of adult and immature Little Blue Herons (*Florida caerulea*), *Anim. Behav.* **17**:320–322.

Rice, R. W., and Kenyon, K. W., 1962, Breeding cycles and behavior of Laysan and Black-footed Albatrosses, *Auk* **79**:517–567.

Richdale, L. E., 1943, The Kuoka or Diving Petrel, *Trans. R. Soc. NZ* **75**:42–53.

Richdale, L. W., 1951, *Sexual Behavior in Penguins*, University of Kansas Press, Lawrence.

Ricklefs, R. E., 1968a, Patterns of growth in birds, *Ibis* **110**:419–451.

Ricklefs, R. E., 1968b, Weight recession in nestling birds, *Auk* **85**:30–35.

Ricklefs, R. E., 1973, Patterns of growth in birds. II. Growth rate and mode of development, *Ibis* **115**:177–201.

Ricklefs, R. E., 1974, Energetics of reproduction in birds, in: *Avian Energetics* (R. A. Paynter, Jr., ed.), pp. 152–297, Nuttal Ornithology Club, No. 15, Cambridge, Mass.

Ricklefs, R. E., 1977, Composition of eggs of several bird species, *Auk* **94**:350–356.

Ricklefs, R. E., 1979, Patterns of growth in Birds. V. A comparative study of development in the Starling, Common Tern, and Japanese Quail, *Auk* **96**:10–30.

Ricklefs, R. E., and Montevecchi, W. A., 1979, Size, organic composition and caloric content of Northern Gannet (*Morus bassanus*) eggs, *Comp. Physiol. Biochem.* **64A**:161–165.

Ryder, J. P., 1979, Possible origins and adaptive value of female–female pairing in gulls, in: *Proceedings of the Colonial Waterbird Group Meeting, 1978* (W. E. Southern, ed), University of Northern Illinois, DeKalb.

Schreiber, R. W., 1976, Growth and development of nestling Brown Pelicans, *Bird-Banding* **47**:19–39.

Sealy, S. G., 1973, Adaptive significance of post-hatching developmental patterns and growth rates in the Alcidae, *Ornis Scand.* **4**:113–121.

Sealy, S. G., 1975, Egg size of murrelets, *Condor* **77**:500–501.

Silver, R., 1978, The parental behavior of ring doves, *Am. Sci.* **66**:209–215.

Skutch, A. F., 1975, *Parent Birds and Their Young*, University of Texas Press, Austin.

Slater, P. J. B., 1976, Tidal rhythm in a seabird, *Nature (London)* **264**:636–638.

Steen, J., and Enger, P. S., 1957, Muscular heat production in pigeons during exposure to cold, *Am. J. Physiol.* **191**:157–158.

Tinbergen, N., and Perdeck, A. C., 1950, On the stimulus situation releasing the begging response in the newly hatched Herring Gull chick (*Larus argentatus* Pont.), *Behaviour* **3**:1–39.

Tinbergen, N., and Falkus, H., 1969, *Signals for Survival*, Clarendon Press, Oxford.

Trivers, R. L., 1972, Parental investment and sexual selection, in: *Sexual Selection and the Descent of Man 1871–1971* (B. Campbell, ed), pp. 136–179, Aldine, Chicago.

Trivers, R. L., 1974, Parent-offspring conflict, *Am. Zool.* **14**:249–264.

Tschanz, B., and Hirsbrunner-Scharf, M., 1975, Adaptions to colony life on cliff ledges: a comparative study of guillemot and razorbill chicks, in: *Function and Evolution of Behavior* (G. P. Baerends, C. G. Beer, and A. Manning, eds.), pp. 358–380, Oxford University Press, London.

Tuck, L. M., 1961, *The Murres*, Canadian Wildlife Service, Rep. Ser. No. 1.

Tuck, L. M., and Squires, H. I., 1955, Food and feeding habits of Brunnich's Murre (*Uria lomvia lomvia*) on Akpatok Island, *J. Fish. Res. Board Canada* **12**:781–792.

Vermeer, K., 1963, The breeding ecology of the Glaucous-winged Gull (*Larus glaucescens*) on Mandarte Island, *B.C. Occas. Pap. B.C. Prov. Mus.* **13**:1–104.

Wallman, J., Grabon, M., and Silver, R., 1979, What determines the pattern of sharing of incubation and brooding in Ring Doves? *J. Comp. Physiol. Psychol.* **93**:481–492.

Warham, J., 1975, The Crested Penguins, in: *The Biology of Penguins* (B. Stonehouse, ed.), pp. 189–269, MacMillan, London.

Watson, J. B., 1908, The behavior of Noddy and Sooty Terns. *Carnegie Inst. Wash. Publ.* **103**:187–255.

Weidman, R. M., and Weidman, U., 1958, An analysis of the stimulus situation releasing food-begging in the Black-headed Gull, *Anim. Behav.* **6**:114.

West, G. C., 1965, Shivering and heat production in birds, *Physiol. Zool.* **38**:111–120.

Witherby, H. F., Jourdain, F. C. R., Ticehurst, N. F., and Tucker, B. W., 1947, *The Handbook of British Birds*, Vol. 5, Witherby, London.

Wodzicki, K. A., and McMeekan, C. P., 1947, The Gannet on Cape Kidnappers, *Trans. R. Soc. N.Z.* **76**:429–452.

Wortis, R. P., 1969, The transition from dependent to independent feeding in the young Ring Dove, *Anim. Behav. Mongr.* **2**:1–54.

Chapter 10

THE TRANSITION TO INDEPENDENCE AND POSTFLEDGING PARENTAL CARE IN SEABIRDS

Joanna Burger

Department of Biology
Livingston College
and Center for Coastal and Environmental Studies
Rutgers University
New Brunswick, New Jersey 08903

I. INTRODUCTION

Behavior is one component of adaptation to the environment; thus examinations of particular kinds of behavior provide insights into the evolution of species. Comparing behaviors among animals provides a framework for generating ideas about how animals adapt. In birds, parental care involves considerable time and energy, and thus provides a good opportunity to examine the diversity, mechanisms, and adaptative significance of behavior. Parental care refers to all behaviors that contribute to the maintenance and survival of their young. Most references to parental care in the literature are limited to the early phase of care of the nestlings, whereas a few studies examine care up until "fledging." In its broadest sense, parental care includes any behavior that contributes to the survival of the young once they have departed from the nest and are partially independent of their parents.

Any consideration of parental care must include the costs and benefits of that care to both the parents and the young. These costs and benefits must project beyond the immediate breeding season to future seasons of the adults, as well as merely survival for the young. These considerations prompted Lack (1954), Williams (1966), and Ricklefs (1973, 1977) to discuss the evolution of reproductive strategies in birds. Any reproductive effort should be a compromise between the benefits (such as increased fecundity) and the costs (such as increased adult mortality) of the effort. Presumably increased parental care per offspring increases the survival prospects of that young while potentially decreasing the long-term survival of the parent. This led Trivers (1972), Parker and MacNair (1978), and others to comment on the evolutionary conflict of parents with their offspring about precisely how much parental care each offspring should receive. Whereas evidence exists that parents providing the greatest care increase the survival of their young (Nisbet, 1973; Jarvis, 1974; Harris, 1978), we often take it for granted that such care decreases the adult's survival. Examples of parents deserting young do exist (Harris, 1969; Burger, 1974; Boersma, 1977), and presumably such desertions may relate to increasing adult survival.

Parental care is often discussed as being either prefledging or postfledging. However, a more useful concept is that such care is a continium from the intense care including brooding, protection, and feeding required by a newly hatched bird to the stage when the young are fully independent and living completely apart from their parents. Clearly, particular developmental stages can be delineated along the way: reaching a peak nestling weight, being fully feathered, making a first flight, making sustained flights, leaving the nest, finding some food, and eventually obtaining all their own food. These stages do not always occur in the same order, and some stages are never reached (e.g., flight in penguins). It is unclear which of these landmarks deserves the title of fledging. Authors often discuss prefledging behavior as if everyone knows exactly what that means. However, it is difficult to determine when fledging occurs for penguins that will never fly, for those petrels that remain in the nest after parents have departed, or for those alcids that leave the colony when only half-grown to travel to foraging grounds with their parents.

Fledging has been variously defined, and it is helpful to examine the common usages. Thomson (1964) stated that when the process of acquisition of true feathers by a young bird is completed, it is fledged. Coulson and White (1958) considered fledging to be the time when the young finally vacated the nest. Maunder and Threlfall (1972) defined fledging as the vacating of the nest for more than four consecutive days. Others avoid the issue by defining an age when they consider the bird to be fledged, and this age may bear no relationship to either flight or independence (see Burger, 1978; Murphy, 1979; Tremblay and Ellison, 1979). Schwartz (1966) defined fledging as the age of first flight. It is the last definition that seems to be most common in the literature. Even deciding on the time of the first flight is difficult, because some workers literally mean the first short flight, whereas others require a longer flight, and still others require sustained flight. In reality, the term fledging is not useful unless it is carefully defined in each study for each bird. I am using fledging here *only* to mean the age at which the bird first flies (short flight). I will give ages for developmental stages such as leaving the nest, flying, or leaving the colony.

Despite the ambiguities of the term "fledging," for all groups of birds there is a clear period when the young are completely dependent for food and protection, and this period (normally referred to as the prefledging period) is not the main topic of this chapter, but has been considered in the preceding chapter by W. A. Montevecchi and J. M. Porter. This chapter concerns itself with the transition from complete dependence (in terms of protection and food requirements) to independence in seabirds. It thus includes departure from the nest and colony, flight behavior, and prolonged (but decreasing) parental care.

The transition from dependence to independence is very variable from family to family, and from species to species within families. Major issues in this transition include provisioning by parents, protection by parents, and learning foraging areas and techniques by young, Protection from inclement weather (rain, snow, cold, or floods) is usually required by small chicks and is no longer necessary when young reach adult size and are fully feathered. Predator protection is usually a function dispensed with long before the transition period, because once young approach adult size they generally can take care of themselves. It is of interest to consider the relationship between growth and independence. Ricklefs (1973), in examining growth rates in birds, stated that increased growth rate reduces the period of vulnerability of the young. This is true only for vulnerability to predators, and then only if size alone can protect them when they otherwise would not fly. In gannets, for example, the parents protect the young until they leave the colony (Nelson, 1970). Many birds become independent when they have achieved maximum size. However, for many seabirds this is not the case, and young may be fully grown (overgrown in the case of petrels) and yet be vulnerable to predators or starvation when they cannot leave to find their own food. There is little direct information on young learning foraging areas or techniques (but see Ashmole, 1971; LeCroy, 1972), and so this is discussed only in passing. This chapter, then, largely examines the role of parents in provisioning young seabirds from complete dependence to complete independence.

The seas provide abundant food for many species of birds during the breeding season as well as throughout the rest of the year. Seabirds, however, generally include those species that obtain their food from the sea during the breeding season, and typically live on or near the sea outside of the breeding season. Many seabirds (more than three-quarters) have evolved the habit of hunting far from their breeding grounds (Nelson, 1970). Seabirds congregate where food is concentrated along convergent "fronts" or current "rips," which are contact points between unlike water masses where zooplankton concentrate at the surface (Cromwell, 1953; Sealy,

1973*c*). This led Murphy (1936) to suggest that seabirds could be used to assay characteristics of the ocean. Ashmole and Ashmole (1968) attempted this procedure around Christmas Island.

Offshore foraging increases the area for food gathering, but it also increases the time interval between return trips to the breeding island. This constraint has implications for clutch size, incubation bout lengths, and feeding intervals. Presumably, increasing the size of the foraging area reduces the effect of ephemeral or erratic food supplies (Lack, 1968; Nelson, 1970). These feeding location constraints thus allow seabirds to be classified as inshore (as far as the edge of the continental shelf) or offshore (beyond these limits) feeders.

Lack (1967, 1968) has suggested that inshore feeders have clutches of two or three and that they have shorter fledging times than do offshore feeders. Offshore feeders generally start to breed at a later age, have clutches of only one, have longer incubation bouts and feeding intervals, and have longer fledging periods than do inshore feeders. Thus offshore seabird young can often survive long periods of starvation (Lack, 1968; Schreiber and Ashmole, 1970). Presumably delayed maturity results when birds require considerable periods of time to perfect the hunting technique, learn the foraging areas, and become capable of rearing their own young without undue stress to themselves (Nelson, 1970). Differences in foraging abilities as a function of age have been found for every seabird examined (see Section II). Such differences in foraging ability certainly suggest that parental care during the transition period when young first learn the techniques and feeding areas would be adaptive.

There are some 270 species of seabirds in four orders of 15 families (Palmer, 1962; Thomson, 1964; Lockley, 1974). Hereafter, all mention of the Procellariidae should be taken also to apply to the Hydrobatidae and Pelecanoididae unless otherwise specified. The taxa considered in this chapter are listed in Table I.

Few papers discuss postfledging parental care. Several handbooks (such as those by Bent, 1919, 1921, 1922; Witherby *et al.*, 1941) do not include this type of information, although Palmer (1962) and Belopolskii (1957) include it where possible. Nonetheless, I was able to find information for a third of the species (Table II). I included information in Table II only when it was precise, dealt with marked or recognizable individuals, and was complete. Other information is included in the family description column. I have not included all those references that state that young were fed for a "few days," or for "some time," as it was difficult to decide what a few days was. Wherever reliable sources contradicted each other, I listed the oldest age that chicks were fed by parents, because clearly variations might exist with respect to population size, food sources, and predation pressures. All these data can be considered conservative estimates. That is, some indi-

Table I. Family and Common Names of Seabirds Considered in This Chapter

Order	Family	Common names	No. of species
Spenisciformes	Speniscidae	Penguins	15
Procellariiformes	Diomedeidae	Albatrosses	13
	Procellariidae	Fulmars, petrels, shearwaters	60
	Hydrobatidae	Storm petrels	18
	Pelecanoididae	Diving petrels	5
Pelecaniformes	Phaethontidae	Tropicbirds	3
	Pelecanidae	Pelicans	6
	Sulidae	Gannets, boobies	9
	Phalacrocoracidae	Cormorants	35
	Anhingidae	Darters	4
	Fregatidae	Frigatebirds	5
Charadriiformes	Stercorariidae	Skuas	4
	Laridae	Gulls, terns, noddies	39
	Rynchopidae	Skimmers	3
	Alcidae	Auks	22

viduals are fed for at least this period. It is hoped that it will serve as a basis for extending our information on this phase of breeding biology.

In general, some seabirds exhibit no parental care beyond fledging (flying), some species exhibit parental feeding for a month or less, and others exhibit extended care for periods of over a month (Fig. 1). There is no parental care in the Spheniscidae and Procellariiformes once they leave the colony. However, parental behavior varies in that penguins leave and are abandoned when only half-grown, whereas petrels are abandoned when well above adult weight. Members of the Alcidae exhibit no parental care after young fly, but the location of the care prior to flying varies considerably (some are fed at the nest site, and some are fed far from the breeding colonies). Postfledging care is not known to extend beyond a month in some members of Diomedeidae and Phaethontidae, and in all members of the Stercorariidae, Anhingidae, Rynchopidae, and Pelecanidae. The care varies from only a few days to almost a month. Extended parental care of a month to several months occurs in members of the Sulidae, Fregatidae, Laridae, and Phalacrocoracidae.

Information on the transition period is sketchy for some families, contradictory in some, and fairly complete in others. Data from marked, known-age birds is particularly lacking from terns, and to a lesser extent from gulls. Observations during the winter months, and ambiguous statements (i.e., "for some weeks") suggest extended parental care but more data are needed. At the other extreme, data from members of the Sulidae

Table II. Developmental Stages in Seabird Chicks[a]

Common name and scientific name	Clutch size	Incubation period	Incubation bouts	Time on nest (longest usual)	Age that chicks first fly	Age chicks leave colony	Period of intense care of young	Pre-fledging care	Post-fledging care	Maximum total days of care of young	Sources
Spheniciformes											
Spheniscidae											
Rockhopper Penguin (*Eudyptes chrysocome*)	2	33	10–19	67–71		67–71	26	67–71	0	67–71	Warham (1963, 1975)
Fiordland Penguin (*Eudyptes pachyrhynchus*)	2	34	5–10	75		75	21	75	0	75	Warham (1974)
Adelie Penguin (*Pygoscelis adeliae*)	2	34	7–18	56		56	22	56	0	56	Taylor (1962)
King Penguin (*Aptenodytes patagonica*)	1	54	4–15	342		390	37	390	0	390	Stonehouse (1956)
Yellow-eyed Penguin (*Megadyptes antipodes*)	2	43	–	118		97–118		106	0	106	Lack (1966)
Little Penguin (*Eudyptula minor*)	2	36	+	58		58	14	58	0	58	Reilly and Balmford (1975)
Procellariiformes											
Diomedeidae											
Laysan Albatross (*Diomedea immutabilis*)	1	64	2–21	165	145	165	17	165	Several weeks	165	Rice and Kenyon (1962)
Black-footed Albatross (*Diomedea nigripes*)	1	66	1–18	180	140	180	19	140	40	180	Rice and Kenyon (1962); Palmer (1962); Miller (1940)
Waved Albatross (*Diomedea irrorata*)	1	60	19–22	178	167	178	Several weeks	167	11	178	Harris (1973)

Species											Reference
Royal Albatross (*Diomedea epomophora*)	1	79	—	236	236	236	42	236	0	236	Richdale (1954); Tickell (1968)
Wandering Albatross (*Diomedea exulans*)	1	78	14–21	278	278	278	35	278	0	278	Serventy et al. (1971); Carrick et al. (1960)
Procellariidae, Hydrobatidae, Pelecanoididae											
Northern Fulmar (*Fulmarus glacialis*)	1	55–57	4–5	51	46–51	51	14	51	0	51	Palmer (1962); Cramp and Simmons (1977); Bent (1922)
Flesh-footed Shearwater (*Puffinus carneipes*)	1	—	—	92	92	92	2–3	90	0	90	Warham (1958)
Great Shearwater (*Puffinus gravis*)	1	52	—	105	84	105	—	70	0	70	Rowan (1952); Palmer (1962)
Sooty Shearwater (*Puffinus griseus*)	1	56	1–8	96	96	96	—	85	0	85	Richdale (1963)
Short-tailed Shearwater (*Puffinus tenuirostris*)	1	53	10–16	108	94	109	—	85	0	85	Marshall and Serventy (1956); Serventy (1966); Palmer (1962)
Manx Shearwater (*Puffinus puffinus*)	1	51	3–26	76	70	76	Few days	60	0	60	Harris (1966, 1969); Cramp and Simmons (1977)
Little Shearwater (*Puffinus assimilis*)	1	52–58	1	75	70–75	75	—	64	0	64	Glauert (1946); Cramp and Simmons (1977)
Audubon's Shearwater (*Puffinus lherminieri*)	1	51	—	80	70–80	80	—	80	0	80	Snow (1965); Palmer (1962); Harris (1969)
Frigate Petrel (*Pelagodroma marina*)	1	55	3–5	67	52–67	67	—	67	0	67	Palmer (1962)
Dark-rumped Petrel (*Pterodroma phaeopygia*)	1	50–54	12	112	109–112	112	—	102	0	102	Harris (1970a)

(Continued)

Table II. (*Continued*)

Common name and scientific name	Clutch size	Incubation period	Incubation bouts	Time on nest (longest usual)	Age that chicks first fly	Age chicks leave colony	Period of intense care of young	Pre-fledging care	Post-fledging care	Maximum total days of care of young	Sources
Leach's Petrel (*Oceanodroma leucorhoa*)	1	42	1–11	72	63–70	72	5	68	0	68	Wilbur (1968); Cramp and Simmons (1977)
Wilson's Petrel (*Oceanites oceanicus*)	1	43	1	52	52	52	—	52	0	52	Fisher and Lockley (1954); Roberts (1934–1937)
Diving Petrel (*Pelecanoides urinatrix*)	1	48	—	57	57	57	14	49	0	49	Richdale (1943a, 1945)
Giant Petrel (*Macronectes giganteus*)	1	59	1–25	117	102–117	117	18	117	0	117	Warham (1962, 1966); Austin (1961)
Black-capped Petrel (*Pterodroma hasitata*)	1	53	8–14	100	90–100	100	2	96	0	96	Palmer (1962)
White-faced Storm Petrel (*Pelagodroma marina*)	1	49–56	3–9	67	57	67	0	67	0	67	Richdale (1943b)
Black-bellied Storm Petrel (*Fregetta tropica*)	1	38–44	3	71	65–71	71	—	71	0	71	Beck and Brown (1971)
Storm Petrel (*Hydrobates pelagicus*)	1	41	1–5	73	73	73	7	71	0	71	Cramp and Simmons (1977); Davis (1947)
Grey-faced Petrel (*Pterodroma macroptera*)	1	53–57	1–20	118	110	118	2	118	0	118	Imber (1976); Warham (1956)
Pelecaniformes											
Phaethontidae											
Red-billed Tropicbird (*Phaethon aethereus*)	1	42–44	2–5	100	80	100	—	90	0	90	Stonehouse (1962); Thomson (1964)

Species										References
Red-tailed Tropicbird (*Phaethon rubricauda*)	1	44	4–16	105	75–84	105	17	75	30	105	Fleet (1974); Gibson-Hill (1949)
Yellow-billed Tropicbird (*Phaethon lepturus*)	1	41	3–4	85	75	85	—	85	0	85	Stonehouse (1962)
Pelecanidae											
Pink-backed Pelican (*Pelecanus rufescens*)	2	34	0.3	76	70–75	96	12	75	21	96	Burke and Brown (1970); Din and Eltringham (1974)
White Pelican (*Pelecanus erythrorhynchos*)	2	29	1	30–80	60	80	7	60	20+	80+	Schaller (1964); Palmer (1962); Hall (1925)
Sulidae											
Red-footed Booby (*Sula sula*)	1	46	3	220	91–133	320	Few days	130	190	320	Nelson (1969a,b, 1970); Verner (1961)
Blue-faced Booby (*Sula dactylatra*)	1–2	43	1	270	115	270	20	115	156	271	Nelson (1967a, 1970); Kepler (1969)
Brown Booby (*Sula leucogaster*)	2	43	0.5	245	98–119	413	21	119	118–259	377	Dorward (1962a,b); Simmons (1967); Nelson (1970)
Gannet (*Sula bassana*)	1	44	1	97	84–97	97	—	97	0	97	Nelson (1966, 1970)
Abbott's Booby (*Sula abbotti*)	1	56	3	266	168	266	—	168	180	348	Nelson (1970)
Blue-footed Booby (*Sula nebouxii*)	1–3	41	18–0.75	161	105	161	—	105	56	161	Nelson (1970)
Peruvian Booby (*Sula variegata*)	3	42	0.3–0.5	167	105	167	—	105	62	167	Nelson (1970)
Phalacrocoracidae											
Great Cormorant (*Phalacrocorax carbo*)	3–4	28–31	0.3	84	50	100	5	50	50	100	Cramp and Simmons (1977); Palmer (1962)

(Continued)

Table II. (*Continued*)

Common name and scientific name	Clutch size	Incubation period	Incubation bouts	Time on nest (longest usual)	Age that chicks first fly	Age chicks leave colony	Period of intense care of young	Pre-fledging care	Post-fledging care	Maximum total days of care of young	Sources
Double-crested Cormorant (*Phalacrocorax auritus*)	2–7	25–29	0.1	42	42	70	5	42	28	70	Palmer (1962)
Shag (*Phalacrocorax aristotelis*)	1–6	30	0.3	42	53	110	16	53	62	115	Snow (1960, 1962a,b); Potts (1969); Cramp and Simmons (1977)
Anhingidae											
Anhinga (*Anhinga anhinga*)	5	25–28	0.2	42	42	56	14	42	14	56	Harriott (1970); Burger et al. (1978)
Fregatidae											
Magnificent Frigatebird (*Fregata magnificens*)	1	40–50	—	316	140–168	316	40	166	150	316	Diamond (1971, 1973); Cramp and Simmons (1977); Nelson (1965)
Great Frigatebird (*Fregata minor*)	1	55	6	310	120–160	310–548	31	160	150–428	588	Nelson (1967b, 1970); Diamond (1975); Schreiber and Ashmole (1970)
Ascension Frigatebird (*Fregata aquila*)	1	44	1–2	300	180–210	300	30	180	120	300	Stonehouse and Stonehouse (1963); Nelson (1975a)
Least Frigatebird (*Fregata ariel*)	1	45	3	260	140	260	35	140	120	260	Diamond (1975); Nelson (1975a)
Andrew's Frigatebird (*Fregata andrewsi*)	1	54	—	—	168	—	45	168	180–300	468	Nelson (1975a)

Species										References
Charadriiformes										
Alcidae										
Horned Puffin (*Fratercula corniculata*)	1	41	—	42	?	42	—	42	0	42	Sealy (1973a)
Southern Puffin (*Fratercula arctica*)	1	41	—	51	?	51	—	40	0	40	Witherby et al. (1941); Nettleship (1972)
Cassin's Auklet (*Ptychoramphus aleutica*)	1	37	—	50	40	50	4	50	0	50	Thoresen (1964)
Parakeet Auklet (*Cyclorrhynchus psittacula*)	1	35	—	37	35	37	—	37	0	37	Sealy and Bédard (1973)
Guillemot (Common Murre) (*Uria aalge*)	1	28–35	—	14–25	46	14–25	5	46	0	46	Greenwood (1964); Witherby et al. (1941); Tuck (1961)
Pigeon Guillemot (*Cepphus columba*)	2	—	—	35	35	35	—	35	0	35	Drent (1965); Drent et al. (1964)
Black Guillemot (*Cepphus grylle*)	2	26	—	36	36	36	—	36	0	36	Witherby et al. (1941); Bent (1919)
Northern Razorbill (*Alca torda*)	1	30	—	22	—	—	—	—	?	?	Witherby et al. (1941); Greenwood (1964); Bent (1919)
Brünnich's Murre (*Uria lomvia*)	1	28	—	20	?	20	—	—	—	?	Bent (1919); Birkhead (1977); Pennycuick (1956)
Stercorariidae										
South Polar Skua (*Catharacta maccormickii*)	2	28	0.1	27	53	59	3	53	6+	59+	Young (1963); Spellerberg (1971)
Long-tailed Jaeger (*Stercorarius longicaudus*)	1–2	24	—	45	21	45	2	21	24	45+	Witherby et al. (1941); Maher (1970)
Pomarine Jaeger (*Stercorarius pomarinus*)	2–3	24	—	46	31	46	2	31	15	46	Maher (1970); Bent (1921)

(*Continued*)

Table II. (*Continued*)

Common name and scientific name	Clutch size	Incubation period	Incubation bouts	Time on nest (longest usual)	Age that chicks first fly	Age chicks leave colony	Period of intense care of young	Pre-fledging care	Post-fledging care	Maximum total days of care of young	Sources
Rynchopidae											
Black Skimmer (*Rynchops niger*)	4	22	0.2	?	28	42	5	28	14+	42+	Erwin (1977a,b); Bent (1921); M. Gochfeld (personal communication)
Laridae											
Black-legged Kittiwake (*Rissa tridactyla*)	2	27	0.2–0.5	42	42	72	7	42	10	52	Dementev et al. (1951); Maunder and Threlfall (1972); Coulson and White (1958)
Glaucous Gull (*Larus hyperboreus*)	3	27	—	3	40	60	—	40	20+	60+	Strang (1973)
Glaucous-winged Gull (*Larus glaucescens*)	3	27	0.1	67	44	67	—	44	21	65	Vermeer (1963)
Silver Gull (*Larus novaehollandiae*)	3	24	—	28	30	42	—	30	14+	44+	Serventy et al. (1971)
Herring Gull (*Larus argentatus*)	3	28	0.5	90	42	90	3–4	45	45+	90+	J. Burger (unpublished data); Haycock and Threlfall (1975)
Southern Black-backed Gull[b] (*Larus dominicanus*)	3	29	0.3	79	49	79	3–4	49	41	90	Fordham (1964); Oliver (1955)
Kelp Gull[b] (*Larus dominicanus*)	3	29	—	47	36	90	—	36	54	90	Serventy et al. (1971)

Species											Reference
Lava Gull (*Larus fulginosus*)	2	32	—	74	60	—	74	60	28	88	Snow and Snow (1969)
Swallow-tailed Gull (*Creagrus furcatus*)	1	32	—	135	60	—	135	60	75	135	Harris (1970b); Snow and Snow (1967)
Laughing Gull (*Larus atricilla*)	3	24	0.3	61	32	2–3	61	32	29+	61+	J. Burger (unpublished data)
Great Black-backed Gull (*Larus marinus*)	3	28	0.3	65	45	—	75	45	40	85	Dementev et al. (1951); J. Burger (unpublished data)
Franklin's Gull (*Larus pipixcan*)	3	24	0.3	39	32	2–3	39	32	7	39	J. Burger (unpublished data)
Common Tern (*Sterna hirundo*)	3	23	0.05	20	30	—	49	30	21+	51+	Palmer (1941); Witherby et al. (1941); M. Gochfeld and I. Nisbet (personal communication)
Caspian Tern (*Sterna caspia*)	2–3	26	0.1	50	37	10	90+	37	90+	127+	Tomkins (1959); Bergman (1956); G. Shugart (personal communication)
Royal Tern (*Sterna maxima*)	1–2	30	0.5		26	—	90+	26	90+	116+	Tomkins (1959); Bent (1921); Buckley and Buckley (1972)
Least Tern (*Sterna albifrons*)	3–	22	—	+	20	2–3	+	20	26–56	76	Massey (1974); Tomkins (1959)
Sooty Tern (*Sterna fuscata*)	1	26	5.5	+	56	—	+	56	17	73	Brown (1957); Feare (1975, 1976); Lack (1968); Ashmole (1963)
Cayenne Tern (*Sterna sandvicensis*)		24	—	35	28	—	35	28	7	35	Ansingh et al. (1960)
Sandwich Tern (*Sterna sandvicensis*)	2	25		32	28	—	32	28	Few days	?	Smith (1975)

(Continued)

Table II. (*Continued*)

Common name and scientific name	Clutch size	Incubation period	Incubation bouts	Time on nest (longest usual)	Age that chicks first fly	Age chicks leave colony	Period of intense care of young	Pre-fledging care	Post-fledging care	Maximum total days of care of young	Sources
Arctic Tern (*Sterna paradisaea*)	3	21	—	28	21	28		21	7	28	Bent (1921); Witherby *et al.* (1941)
Black Tern (*Chlidonias niger*)	3	21	0.05	14	21	21+	2–7	21	40	61	Baggerman *et al* (1956)
Black Noddy (*Anous tenuirostris*)	1	35	1	+	42	+	—	42	Few days	?	Dorward (1963); Ashmole (1968)
Fairy Tern (*Gygis alba*)	1	36	0.04	95	60	95–120	Few days	60	35–60	95–120	Cullen and Ashmole (1963); Lack (1968)

[a] Means or ranges are given in days.
[b] Kelp Gull and Southern Black-backed Gull are races of *Larus dominicanus*, and are sometimes called Dominican Gull.

Fig. 1. Seabird families with no, some (<1 month), and extended (>1 month) postfledging parental care.

are particularly complete owing to the work of Nelson (1966, 1967a, 1969a, 1970, 1975b) and others. The difference is particularly noteworthy, as it should be easier to work in the dry, coastal colonies where terns and gulls breed than in the remote offshore islands. Field workers, however, may be more tempted to abandon coastal colonies late in the season when few birds are around. Flying young may return to such colonies for only a few minutes, and their movements may not be predictable. Under these conditions, colonies must be monitored continually during the critical transition stages.

This chapter is organized to include a description of foraging differences as a function of age, descriptions of parental care during the transition period for all seabird families, a discussion of comparisons among the families, and conclusions. Information on age-related foraging differences can be used to predict species in which it would be particularly adaptive to have extended parental care.

II. DIFFERENCES IN FORAGING BEHAVIOR AS A FUNCTION OF AGE IN SEABIRDS

Unlike many land birds, seabirds exhibit delayed maturity in that most species begin breeding when 3 years of older, and some species (for example albatrosses) do not breed until they are 8–11 years old (Lack, 1967). Presumably breeding activity is delayed until birds can obtain sufficient food to feed nestlings as well as themselves. Even so, young birds have lower breeding success than older birds, a subject discussed more fully in Chapter 5 (this volume). One reason often given for delayed breeding is that young have difficulty mastering foraging techniques and learning foraging areas (Lack, 1967, 1968). It takes young birds awhile to be able to obtain enough food consistently. Presumably the difficulty in learning these behaviors affects their early development as well. That is, it must be difficult to obtain enough food for themselves when they first begin to forage. Under these circumstances, one might expect young to be able to withstand long periods with little food, to be partially provisioned by their parents, or both. Thus age differences in foraging ability would predict some parental care during the transition period from dependence to independence. A review of studies examining age differences thus is useful in understanding the requirement for postfledging care.

Data on foraging ability as a function of age are lacking from several families including penguins, albatrosses, petrels, tropicbirds, and auks. The reasons include their use of offshore fishing areas and their method of foraging. It is difficult to gather foraging information on species, such as the auks, that forage for their food far offshore. Species that forage inshore, however, prove to be ideal subjects as they are easily observed, and often are not easily disturbed.

Orians (1969) first noted age differences in foraging ability in Brown Pelicans (*Pelecanus occidentalis*). He found that adults had higher success (69%) at diving for fish than did young (49%) for 8 days during the breeding and nonbreeding season. In the same year, Recher and Recher (1969) noted that a non-seabird, juvenile Little Blue Herons (*Florida caerulea*), also missed prey more frequently than adults. They further showed that juveniles captured less food per minute than did adults. Both studies concluded that the inefficiency of juveniles is an adequate explanation for the deferment of reproduction, and served as a basis for studies of several other species with deferred maturity.

Other than gulls and terns, foraging differences have been found for Adelie Penguins (scientific names for most species are in Table II; species not included are given in the text) and Olivaceous Cormorants (*Phalacrocorax olivaceus*). Morrison *et al*. (1978) found that foraging success of adults was higher than for immature cormorants, and the differences persisted as immatures entered their first winter. Immatures increased their food intake by feeding twice as often as adults. Although immatures acquired the diving ability soon after fledging, the lower capture success of juveniles (11%) compared to adults (18%) was a result of poor capture techniques, maneuverability and prey search image.

Most research on age differences in foraging ability has involved gulls and terns during the fall and winter months (Fig. 2). In a dump in England, Verbeek (1977a) found that first-year Herring Gulls moved fewer objects per minute than did adults, found fewer food items per minute, and attempted to steal food more often than did adults. Verbeek postulated that it takes time for immatures to learn to associate the removal of inedible objects with the food hidden beneath them. Such age differences in Herring Gulls were not limited to dumps, as Verbeek (1977b) also found that immatures were less efficient in capturing starfish by diving. Immatures stole much of their food; stealing gives way to independent methods of foraging as the birds mature (Verbeek, 1977c).

Ingolfsson and Estrella (1978) documented that first-year Herring Gulls were less successful (78%) than were adults (100%) at opening bay scallops (*Pecten irradians*) by dropping them. On the first drop, 65% of adults successfully cracked the shell, whereas only 43% of the first-year birds succeeded. The immature Herring Gulls dropped the scallops on soft rather than hard surfaces. From October to February the success of immatures improved as they learned to drop them on appropriate surfaces.

Similarly, immature Glaucous-winged Gulls had lower success than did adults in dropping clams (Barash *et al*., 1975). Again, young gulls dropped the clams on soft surfaces, whereas adults did not. Searcy (1978) found that this species exhibited age-related differences in fishing success with adults being more successful (67%) than yearlings (57%). On a salmon stream in Alaska, juvenile Glaucous-winged Gulls were often relegated to suboptimal foraging areas by adult aggression, and thus juveniles obtained less food (Moyle, 1966).

Social factors can affect foraging differences in some species, although this has not been examined in any detail. Laughing Gulls foraging on a dump in East Brunswick, New Jersey, did not show any age-related differences in the amount of food they located on the dump (J. Burger, unpublished data). However, some Laughing Gulls carried food away from the dump, and adults dropped less food than did first-year birds. The dump contained a large number of Herring Gulls that seemed to interfere with the

Fig. 2. (A) Laughing Gulls foraging at a garbage dump (Texas); (B) Western Gulls and pelicans foraging over California Sea lions on food scared up; (C) Laughing Gulls feeding on offal thrown to them by fishermen in Campeche, Mexico; (D) young Laughing Gull pecks at offal as adult (all-white tail) and subadult (partial tail band) attempt to get food.

Fig. 2. (*Continued*)

Laughing Gulls both at the dump and when they were carrying away food. Because the Herring Gulls did show age-related foraging differences, the lack of an age-related difference in Laughing Gulls is all the more noteworthy.

The foraging behavior of terns is less diversified than that of gulls, and they primarily forage by plunge diving for fish. In Sandwich Terns, the differences in the success of dives was slight but significant, with adults being more successful (Dunn, 1972). There were no major differences between the age groups in diving rates or in the size of fish captured, although adults tended to dive from higher locations than did first-winter birds. Buckley and Buckley (1974) found that in Royal Terns adults spent less time foraging over a given stretch of beach than young, adults made twice as many dives per minute as juveniles, and they had equivalent success rates per dive. Thus, although success was similar, adults found twice as much food because they dove twice as often.

Taken altogether, these data indicate that adults are more successful at foraging ability. Difficulties with foraging, alone, are not sufficient to explain than do immatures. They support the hypothesis that delayed breeding might be partially a result of a requirement for the young to improve their foraging ability. Difficulties with foraging alone are not sufficient to explain deferred maturity, because in most seabirds females breed at an earlier age than do males (Nelson, 1970, p. 560).

Several authors (Spaans, 1971; Schreiber, 1968; Davis, 1975) have reported that young gulls spend more time foraging than do adults and presumably make up for their inefficiencies by added time and energy expenditure during the rest of the day. Another method of increasing food intake during this period of learning is to have parents supplement the food that young find, and this is certainly the case in many species of seabirds. All the above seabird species that exhibit age-related foraging success differences have some postfledging parental care. A profitable area of future research would be to look for age-related foraging differences in species not having postfledging parental care (or postleaving the colony in penguins) and to look at postfledging care in those species showing pronounced age-related foraging differences.

III. THE TRANSITION AND POSTFLEDGING PERIOD IN SEABIRD FAMILIES

The major problems in the transition from dependence to independence in seabirds revolve around food acquisition. Young must obtain enough food to complete growth and feather development, yet still weigh enough when they leave the colony to survive the critical period of adapting to a hostile marine environment. Species solve this problem in many different ways. Seabirds either leave the colony partially dependent or completely independent. A critical role is played by parents who either desert the chicks prior to their departure, desert them at departure or shortly thereafter, or do not desert them for a considerable period of time after young leave the colony. In general, the solution to these two basic problems (colony or nest departure by the chicks, and parental desertion of the chicks) differs among avian families. Differences occur within families, but these differences seem to be less than among families. This section examines the transition and postfledging period in seabird families.

In this section I have compiled breeding behavior information on one-third of the seabird species to aid in understanding the transition to independence. Where complete, data are listed in Table II. Information in Table II includes clutch size, incubation period, incubation bouts, age at which the species leaves the nest, age the species first flies, age at which the species leaves the colony, period of intense care, prefledging care, postfledging care, and total period of parental care. Incubation bouts, given in days, refers to the average length of time one or the other parent incubates without being relieved by its mate. Authors often do not distinguish between the age at which chicks leave the nest and the colony, but these are sometimes distinguishable and are important. The period of intense care refers to the time the chicks are brooded or shaded fairly continuously by parents. Intense care information is often lacking in the literature. Age at first flight refers to first flights, and not to sustained flights. In most cases means are given. However, for some species means were not given in the literature and could not be calculated, and so ranges are given in Table II.

A. Sphenisciformes

1. Spheniscidae: Penguins

Penguins are flightless seabirds of the Southern Hemisphere whose feathers are uniformly distributed over their body (Thomson, 1964). Their wings, reduced to narrow flippers, allow them to swim rapidly through the water and feed by pursuing fish, crustaceans, and squids (Lack, 1968; Ashmole, 1971). There are 15 species of penguins.

In general, the parental care period can be divided into two phases—a guard phase and a postguard phase. During the guard phase the chick is continually guarded by one parent (usually the male), while the other parent brings back food (usually the female; Warham, 1975). In the Rockhopper Penguin, however, the female takes the first guard duty (Warham, 1963). The guard phase lasts 14–37 days in penguins (Table II). Following the guard phase, the chicks leave the nest site and wander about in creches, to return to the nest site to be fed by either parent when they arrive (Warham, 1975). At some point when their development and growth is partially complete, the young depart from the colony. The adults do not help the young during their departure and may attack their own or other chicks when they leave the colony. Departure from the colony usually occurs 56–390 days after hatching, depending on the species. As is clear from Table II, penguin parents provide no parental care once the young leave the colony. In fact, in Rockhopper, Adelie, and Yellow-eyed penguins, the adults continue to visit the nest site after the chicks leave.

For most penguins the length of the guard and postguard phases are known. However, the nature of parental care during the transition from postguard phase to colony departure (and independence) is unclear. In general, chicks leave the colony weighing from one-third to the eventual (nonbreeding season) adult weight (Lack, 1967). In King Penguins, however, the young temporarily increase above their future adult weight (Lack, 1968). They weigh the most at 4 months, and again at 10–11 months just prior to leaving the colony (Stonehouse, 1956).

The primary question surrounding the transition to independence for penguins is the interval between when the chicks are last fed by their parents and when they leave the colony. For many years researchers believed that in some species of seabirds the chicks were deserted by their parents and left to starve before they departed from the colony. Such starvation was thought to reduce the chick to flying weight (or leaving weight in penguins), and to force the chick to leave the colony. This starvation theory persists in the literature even today (see the Procellariidae, Diomedeidae, and Alcidae below). The length of the "starvation period" in penguins was often not defined, although its occurrence was often alluded to. Richdale (1957) reviewed this topic and found that five species of penguins are definitely fed until departure, and Lack (1968) added another. Information is simply lacking for the other nine penguin species. Furthermore, Richdale (1957) showed rather convincingly that such a starvation period would result in the production of inferior chicks at best, and dead chicks at the extreme. He concluded that normal chicks are, on the average, heaviest when about to leave the breeding grounds. It is clear, however, that adults may feed the chicks less often toward the end of the parental care period, as is the case in Adelie Penguins (Taylor, 1962).

Thus, in summary, penguins are guarded extensively for some time, and are then fed intermittently until the chicks leave the colony in the spring or early summer when they will have many months to learn to forage efficiently before winter comes. There is no evidence of parental care once the chicks leave the colonies, and adults actively attack their own and other chicks as they enter the icy waters.

B. Procellariiformes

1. Diomedeidae: Albatrosses

The 13 species of albatross or mollymauks are tubenosed pelagic species primarily of the Southern Hemisphere with wing spreads as great as 3.5 m. They are capable of extended flights, and wander throughout the oceans. They feed by seizing marine animals (especially squid) from the surface, although they follow ships and will scavenge (Thomson, 1964; Ashmole, 1971).

Upon hatching, albatross chicks are brooded intensively for 17–42 days, whereupon they are guarded for several more weeks before being left unattended (Harris, 1973). The parents then return to feed the chicks periodically for several months. The young fly 140–278 days after hatching, depending on the species and size (Table II). Two issues surround the transition period in albatrosses: (1) the existence of a starvation period, and (2) the existence of parental care after the chicks fly.

As with penguins, early investigators believed that albatross chicks were deserted for months before they left the colony (see Murphy, 1936;

Thomson, 1964). The starvation theory held that young Wandering Albatrosses, among others, were deserted throughout the Antarctic winter, subsisting on reserves of stored fat (Carrick *et al.*, 1960). The theory might have been initiated to explain the long periods without food, or the erratic weights of chicks. However, evidence for all the species listed in Table II indicates that none of these species is deserted prior to leaving the colonies (see Richdale, 1954). Several species, such as the Wandering Albatross, do lose weight (95–200 g/day) during their last 3 weeks in the colony. Nonetheless, there is no desertion period in this species (Serventy *et al.*, 1971) as they were fed regularly, although not as often as previously (Tickell, 1960). Richdale (1954) found that Royal Albatrosses were fed on 48% of the days up until 100 days of age, on 79% of the days from age 100–200 days, and on 55% of the days from 200–231 days of age. The chicks were fed 1–5 days ($\bar{x} = 2.7$) before they left (Richdale, 1954). Furthermore, their parents often returned a day or two after the chicks had departed. Richdale (1954) believed the starvation theory was a result of lack of study of albatross. I believe the prevalence of a starvation theory is also based on the occurrence of a period of time between the last feed and departure of the chick. This period, however, appears to be similar to that between the previous feed and the last feed.

Several authors believe that albatrosses have no postfledging parental care. However, the evidence is contradictory at best. For three of five species listed in Table II, some care occurred after the chicks could make short flights. Harris (1973) reported that Waved Albatross took short flights for some time before leaving the colony, and Rice and Kenyon (1962) found the same for Laysan Albatross. Black-footed Albatross are fed at the colony site for several weeks after they are able to fly (Rice and Kenyon, 1962). Thus it is clear that some parental care exists after the chicks can fly, and before they leave the colony.

Postcolony parental care in albatross, however, is believed to be nonexistent because of the long flights adults and young make upon leaving the colony (Lack, 1968; Thomson, 1964), and the relative difficulties of their maintaining contact (Ashmole, 1971). This latter difficulty may be more a problem to the investigators than to the birds, as some terns maintain such contact with their young far from breeding grounds (see Ashmole and Tovar, 1968, and below). One reference did suggest that this aspect of albatross breeding biology deserves scrutiny. Gibson (1963) reported that there was some evidence of a bond between Wandering Albatross far from the breeding grounds as he often observed first-year birds accompanied by an adult. Furthermore, he observed "family groups" of two adults and one young that would occasionally split up, but continued to come back together (Gibson, 1963, p. 219). Although the birds were not marked, and no feeding was observed, it suggests that this aspect should be further examined.

In summary, albatross chicks are brooded for several weeks, guarded for several more weeks, and then fed periodically for several months before they leave the colony. In more than one-half the species for which data are available, there is some parental feeding after the chicks can fly and before they leave the colony. Present evidence does not substantiate any parental care after the chicks leave the colony, but this aspect deserves further study.

2. Other Procellariids: Fulmars, Petrels, Shearwaters, Storm Petrels, Diving Petrels

There are 86 species of Procellariiformes (other than the Diomedeidae) in three families (here considered together). About one-half the species are restricted to the Southern Hemisphere, with progressively fewer species in the North Pacific, North Atlantic, Indian Ocean, and Mediterranean Sea (Thomson, 1964). They have similar life histories and are adapted to a uniform marine environment where they forage on the open seas by dipping, pattering, surface seizing, and pursuit plunging (Ashmole, 1971). They have a long period of immaturity, long breeding cycle, low reproductive rate, and long life expectancy.

Information on 20 members of this group (Table II) indicates that upon hatching, there is a short period (2–18 days) during which the chicks are guarded; thereafter, the parents are away gathering food, returning only to feed the chicks. Almost all members of this group nest in burrows, making the young relatively free from predators. Information on when chicks fly and leave the colony is precise; the period ranges from 46–110 days, a fairly extensive period considering their relatively small body size ($\bar{x}$ wt. for group = 703 g, see Table X).

Central to any understanding of the transition period in members of this group is the relationship between when the young leave the colony (by flying) and when the parents last feed them. Several authors note a "starvation" or "desertion period" for the eight species listed in Table III as well as Darkrumped Petrel (Harris, 1970a), Diving Petrel (Richdale, 1943a), and Bulwer's Petrel (*Bulweria bulwerii*; Cramp and Simmons, 1977). These authors list desertion periods of up to 23 days. For other species, authors clearly state that there is no desertion period (see Table III).

Contradictions exist even within species. Although Richdale (1963) reported no desertion period for Sooty Shearwater, others (Palmer, 1962; Serventy *et al.*, 1971) report a desertion period of up to 25 days. For Wilson's Petrel, Roberts (1934) listed a desertion period whereas Fisher and Lockley (1954) did not. Part of the difficulty comes from the assumption that whenever weights drop, the young have been deserted, as Allen (1962) did for the Madeiran Storm Petrel (*Oceanodroma castro*). In many petrels, the young gain weight rapidly, and often weigh 150% of normal adult

Table III. Relationship between Incubation Bouts, Feeding Intervals, and "Desertion Periods" for Selected Procellariids[a]

Common name	Incubation bouts	Feeding intervals	Desertion period	Source
Northern Fulmar	4–5	1–2	0	Palmer (1962)
Wilson's Petrel	1	1	0	Palmer (1962)
White-faced Storm Petrel	3–5	1 (1–2)	0	Palmer (1962)
Frigate Petrel	1–5	0 (0–1)	0	Palmer (1962)
Grey-faced Petrel	—	4 (0–10)	0	Palmer (1962)
Flesh-footed Shearwater	—	2	2	Harris (1969); Warham (1958)
Storm Petrel	1–5	2–3	2 (2–3)	Davis (1947); Fisher and Lockley (1954)
Audubon's Shearwater	—	1 (1–2)	1.5 (1–5)	Palmer (1962); Harris (1969)
Capped Petrel	8–14	(1–5)	(4–10)	Palmer (1962)
Manx Shearwater	1–6	2	8 (2–11)	Harris (1969); Cramp and Simmons (1977)
Little Shearwater	1	(8–13)	(8–11)	Fisher and Lockley (1954)
Sooty Shearwater	1–8	4 (1–25)	12 (0–27)	Palmer (1962); Serventy et al. (1971)
Short-tailed Shearwater	10–16	(1–16)	14 (1–23)	Palmer (1962); Serventy (1966)

[a] Means and ranges are given in days.

weight. Before flying, they lose the weight and usually fledge close to adult weight.

   Despite the disagreements over the desertion period in a few species, no one seems to question the occurrence of a desertion period in some species of this group. I would like to suggest that, like albatross, there is no desertion period in these Procellariids for four reasons:

   1. Food is scarce for these species, and they often fly great distances to forage (Lack, 1968). Low clutch size and long incubation bouts (up to 26 days—see Table II) substantiate these difficulties. The long incubation bouts correspond to long, and very variable feeding intervals (see Table III). That is, the time between when parents return to feed chicks in their nests can be three weeks in some species. I believe the so called "desertion period" simply reflects the difficulty of bringing back food. Table III shows the incubation bouts, feeding intervals, and desertion periods for five species that do not have desertion periods, and eight species that do. The species without desertion periods all have relatively short feeding intervals (mean of < 4 days), whereas those with a desertion period have long feeding intervals. Further, there seems to be a correlation between the length of the feed-

ing intervals and the length of the desertion period. The desertion period represents the final foraging interval, and chicks fledge before the parents return again with food.

2. In two species reported to have a desertion period, Cory's Shearwater (*Calonectres diomedea*: Zino, 1971) and Audubon's Shearwater (Snow, 1965), the parents are known to return to the nest site after the chicks departed, only to find the chicks gone. Parents are also known to return to the nest site after the chicks have departed in two species without desertion periods: White-faced Storm Petrel (Richdale, 1943*b*) and Grey-faced Petrel (Imber, 1976).

3. The assumption that a drop in the weight of the nestling reflects that it has been deserted by its parents is invalid. Figure 3 shows the mean weight curves for two species without desertion periods (after Beck and Brown, 1971; Imber, 1976) and four species with desertion periods (after Harris, 1966, 1969; Lack, 1968; Harris, 1969 and Harris, 1969; Richdale, 1963). As is clear, the young of species with and without desertion periods show a drop off in weight toward the end of their time in the nest. The Black-bellied Storm Petrel shows almost a 50% drop, equivalent to the Short-tailed Shearwater, yet the parents continue to return to the burrow until the chicks fledge. Furthermore, the drop in weight of Audubon's Shearwater is slight, yet it has a desertion period of 3–5 days (refer to Table III). In general, many species of seabirds experience a drop in weight prior to fledging when they are still being fed (see Lack, 1968; Ricklefs, 1973). Even Herring Gulls and Laughing Gulls have a drop in mean body weight at the time they learn to fly while they are being fed each day (J. Burger, unpublished data).

4. Observers often do not or cannot ascertain that parents have not returned, but simply assume it. Admittedly, it is difficult to determine if a parent has made a nocturnal visit to a burrow. Because the chicks often come out for several nights before departing, it is not possible to rely on a stick or string over the burrow entrance (Warham, 1956).

In view of the above-mentioned factors, it seems that the desertion period may be an artifact of their difficulty in obtaining food. Several factors indicate this difficulty: low clutches, long incubation bouts, long foraging intervals, and long developmental periods. The documented return of parents after the chicks have departed in two species with desertion periods corroborates this idea.

The concensus of the investigators working with these Procellariids is that there is no postfledging parental care. However, Palmer (1962) reported that in the Grey-backed Shearwater there was a fair amount of movement between the nesting island and the sea after the chicks began to fly. Palmer also reported that the chicks of Northern Fulmars spend several

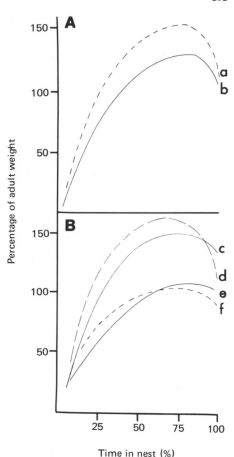

Fig. 3. Percent weight of Procellariid chicks, computed as a percentage of adult weight, as a function of the period of time spent in the nest by species reported to be (A) "without" desertion periods (a–b), and those (B) "with" desertion periods (c–f). a, Black-bellied Storm Petrel; b, Great-winged Petrel; c, Manx Shearwater; d, Short-tailed Shearwater; e, Audubon's Shearwater; f, Sooty Shearwater.

days afloat in compact groups on the water where they are visited by their parents. It was not known if the parents fed the chicks, but this possibility clearly needs investigating.

Thus, in summary, in the Procellarid species the chicks are guarded for a short period of time, and are then left unattended for periods up to three weeks while parents forage over long distances. The chicks of inshore species with comparatively short incubation bouts and feeding intervals are fed up until they depart from the nest. For species with long incubation bouts and feeding intervals, the interval between the last feed and the departure of the chick is long and variable, but similar in duration to the feeding intervals. There is no true desertion period as such, but merely long feeding intervals as parents sometimes return after the chick has departed. There is no evidence of parental care once the young leave the nesting ledges and colonies.

C. Pelecaniformes

1. Phaethontidae: Tropicbirds

There are three species of tropicbirds ranging over the tropical and sub-tropical oceans. They feed by plunging into the ocean for fish and cephalopods (Ashmole, 1971).

Upon hatching, tropicbird young are very shortly left alone in their nests while both parents forage (Table II, Thomson, 1964). The chicks remain in the nest for 85–105 days, then they fly and are independent. Red-tailed Tropicbirds make short flights at 75 days of age, and some return to their nest until they are 105 days old, presumably being fed (Fleet, 1974). Again, although the ages of their first flight and of departure from the nest are known, the role of their parents is less clear. For example, Gibson-Hill (1949) reported that Red-tailed Tropicbird young were abandoned 8–12 days before fledging, whereas Fleet (1974) reported that they were fed after flying while they were at the nest. Stonehouse (1962) suggests that young Red-billed Tropicbirds remain in the nest for ten days after they are no longer fed whereas Palmer (1962) suggests that fledglings may be tended by their parent. Thus, at present, the available evidence suggests that young Tropicbirds remain at their nests a few days after they first learn to fly, and they may be fed by their parents during this time. However, they quickly depart from the colony, and the evidence suggests that they are then completely independent.

2. Pelecanidae: Pelicans

There are 6–8 species of pelicans (depending on taxonomic opinion) in the genus *Pelecanus* distributed over the temperate and tropical zones of the world (Thomson, 1964). Although some are strictly marine, many species typically breed in inland lakes and lagoons. They are among the largest birds. They feed on fish and crustaceans by plunge diving and surface filtering (Ashmore, 1971).

Pelicans have a relatively short incubation period, considering their size (refer to Table II). Upon hatching the chicks are brooded for up to two weeks, guarded until about 1 month old (Schaller, 1964), and they then form groups or creches called pods. They begin making flights at 60–70 days old, but return to their nest sites to be fed by their parents. White Pelicans are fed for at least 20 days postfledging (Table II); Schaller (1964) mentioned that fully fledged young are fed away from the colony. Pink-backed Pelicans are also fed after they have learned to fly, but for at least 2 weeks they return to their nest to be fed (Din and Eltringham, 1974). Subsequently they were fed in trees hundreds of yards from their own nest tree

(Burke and Brown, 1970). Because Pink-backed Pelicans nest in trees, they do not form a creche, and so do not normally leave their nest tree until they can fly. Palmer (1962) mentions that Brown Pelicans (*Pelecanus occidentalis*) are fed by their parents for an unknown length of time after fledging.

Taken altogether, these data indicate that pelicans are fed by their parents for at least three weeks after fledging, initially they return to the nest site to be fed, subsequently they are fed at sites hundreds of yards away from their nest site, and still later they are fed away from the colony. During their first winter, the young are less efficient at foraging than adults (see above). Thus it seems likely that parental care might be more extended than the 20 days so far recorded, and that complete independence might be delayed for some weeks. That they are fed away from the colony indicates that parents and young can maintain contact and a bond, and suggests still further parental care.

3. Phalacrocoracidae: Cormorants

There are 35 species of cormorants inhabiting coastal and inland lakes and estuaries (Thomson, 1964). They feed on fish by plunging (Ashmole, 1971) and on fish, crustaceans, and amphibians by underwater pursuit and bottom feeding (Thomson, 1964; Ashmole, 1971). They forage underwater by swimming without the aid of their wings.

Cormorants usually nest in predator-free habitats, and so predator protection is minimal, but see Kury and Gochfeld (1975). However, the newly hatched young are brooded up to 2 weeks, and one parent often remains with the chicks until fledging. In some cormorants the young form creches called pods at about four weeks (Palmer, 1962). In my experience, pod formation in Double-crested Cormorants was a direct result of my disturbances at Agassiz Wildlife Refuge in 1970. When observed from a small plane and from a distance by boat they did not form pods.

Postfledging parental care occurs in all species examined (Table II). In the shag, the chicks can fly by about 50 days of age, but remain near the nest for another seven to nine weeks while they learn to swim, dive and fish (Potts, 1969). Similarly, in the Great Cormorant the chicks are fed for another seven weeks (Cramp and Simmons, 1977). Palmer (1962) reported that the young of this species are independent once they leave the colony, since the parents return to the nest site after the chicks have departed and call for them. In Double-crested Cormorants, chicks remain dependent on their parents for another five weeks after they can fly (Palmer, 1962). Thus, in cormorants the length of the postfledging care period is similar or longer than the prefledging phase. Chicks learn to fly, dive and swim before they learn to fish, and thus parents provision them during this time.

In the Shag, differences exist in the amount of postfledging care as a function of habitat (see Table IV). The normal period of postfledging care on Lundy in the Bristol Channel is 57 days (Snow, 1960), whereas it is generally only 22 days on the Farne Island (Potts, 1969). There is one record of a Shag being fed on the Farne Islands until it was 14 months old (Potts, 1969). This record, the occurrence of postfledging feeding for at least 5 weeks, and the difficulty of the foraging method (pursuing fish

Table IV. Habitat Differences in Length of Parental Care Periods (Days) in Selected Seabirds

Families and common names	Breeding location (source)	Prefledging period	Postfledging period
Sulidae			
Red-footed Booby	Galapagos (Nelson, 1970)	126	91
	British Honduras (Verner, 1961)	112	150
Blue-faced Booby	Galapagos (Nelson, 1970)	119	56
	Kure Atoll (Kepler, 1964)	119	30–60
Brown Booby	Ascension (Nelson, 1970)	119	63
	Christmas Island (Gibson-Hill, 1947)	98	42
	Ascension (Simmons, 1967)	119	377
Phalacrocoracidae			
Shag	Lundy, Bristol Channel (Snow, 1960)	53	57
	Farne Island (Potts, 1969)	55	22 (420)
Frigatidae			
Great Frigatebird	Aldabra Atoll, Indian Ocean (Diamond, 1971)	120	120
	Galapagos, Pacific Island (Nelson, 1967, 1975)	—	360
	Christmas Island (Schreiber and Ashmole, 1970)	—	420
Laridae			
Black-legged Kittiwake	Newfoundland (Mauder and Threlfall, 1972)	41	10
	Russia (Dement'ev et al., 1951)	45	30
Herring Gull	New Jersey, 1978 (J. Burger, unpublished data)	42	48
	New Jersey, 1977 (J. Burger, unpublished data)	42	21
	New England (Drury and Smith, 1968)	42	20
Fairy Tern	Ascension (Dorward and Ashmole, 1963)	60–75	35
	Cocos-Keeling (Gibson-Hill, 1951)	47–50	—

underwater) all suggest that parental care may well extend beyond the period now documented. It is possible that the young remain with adults when they leave the colony areas, but this needs to be further investigated.

4. Anhingidae: Darters

There are four species of darters in the genus *Anhinga* (Palmer, 1962), although only one species breeds in the new world. They are cormorantlike, but slimmer (Palmer, 1962). They feed on fish and crustaceans by swimming underwater and pursuing their prey, and are noted for inhabiting quiet inland waters and coastal bays.

Upon hatching, the young are brooded for about 2 weeks and are guarded until they can fly at 6 weeks of age (Harriott, 1970). By seven weeks they can swim submerged, but they are still entirely fed by their parents at the nest tree. According to Harriott (1970) they are independent by 8 weeks. Little else is known about postfledging parental care. However, the similarity of breeding and foraging behavior to cormorants suggests that parental care may extend beyond the period noted by Harriott (1970). Their foraging technique is equally as difficult as that of cormorants, which require at least 5 weeks of postfledging parental feeding while they learn to fish.

5. Sulidae: Gannets, Boobies

There are nine species of gannets and boobies in one genus inhabiting tropical and temperate marine environments. Gannet usually applies to the temperate species, whereas booby is used for the tropical species (Thomson, 1964). They are inshore foragers rather than pelagic, plunge diving for fish from as high as 30 m (Ashmole, 1971). They often, however, dive from the surface in search of squid (Thomson, 1964).

The breeding biology and behavior of the sulids has been extensively studied by Nelson (1966, 1967a, 1968, 1969a,b, 1970), Kepler (1969), Dorward (1962a,b) and others. In general, they lay clutches of one, although the more temperate species may have clutches of two (see Table II). However, within tropical areas, the location of the feeding site influences these behavior patterns. For example, the Red-footed Booby feeds further offshore and breeds in larger colonies, has a smaller clutch size, a longer incubation and fledging period, and a longer period of dependence on parents than does the inshore-feeding Blue-footed Booby (see Table II; Lack, 1968; Nelson, 1970). The Blue-faced Booby is intermediate in the location of its foraging and is intermediate with respect to these characteristics (Table II; Lack, 1968).

There are two general posthatching patterns in this family. In the Northern Gannet, the young are guarded for their entire developmental period (90 days), as the young do not defend their nests as do other sulids (Nelson, 1966). Upon fledging, there is no further parental care, and the newly fledged young gannets are attacked vigorously by adults in the vicinity (Nelson, 1966, 1970). The gannets' fledging period is shorter than that of the boobies, despite the fact that young are twice as heavy (Lack, 1968). Presumably the abundant food in the north temperate waters allows for rapid development of the young to feed independently. Indeed, the parents return and defend the nest sites for some time after the departure of the chicks (Nelson, 1966, 1970). I have seen the Northern Gannet still defending their nest sites on a rock off the Newfoundland coast in late October, several months after the chicks have departed.

Upon hatching, the young of boobies are brooded and guarded for about three weeks, after which they are generally left alone while the parents forage (see Table II). The young are able to defend themselves and their nest sites thereafter (Nelson, 1970). The development period of the boobies is slower than that of gannets. The long prefledging period seems a consequence of food scarcity (Nelson, 1969a,b), and as such, there are variations even within species that seem attributable to habitat differences (Table IV). Even the prefledging period can vary by 2–3 weeks. Some apparent habitat differences may relate to differing food availabilities in different years. Only longer studies can distinguish between yearly and habitat variations.

Postfledging care is particularly well developed in the tropical boobies (see Table II), species that apparently time their breeding activities so that they can maintain themselves with adequate food supplies during the incubation period (Simmons, 1967). Given the generally low levels of food in tropical waters, and its unpredictability, the outcome of their reproductive effort is a "matter of chance" (Simmons, 1967). As a consequence, the timing of the breeding season of the boobies, such as the Red-footed Booby, is erratic (Nelson, 1969a,b).

In boobies, the cessation of postfledging parental care seems to be a result of the chicks behavior, not of the parents. Nelson (1968) and Gibson-Hill (1949) both mention that the young Red-footed Booby and Brown Booby, respectively, begin to take increasingly longer flights from the breeding colony before they finally depart. Eventually the young Brown Boobies are away most of the day, and the adults apparently lose interest (Gibson-Hill, 1949). In Red-footed Boobies, the young make exploratory flights but return each day, until they suddenly cease returning to the colony. The adults experience increased sexual activity, and vigorously defend nest sites for a few days, and then they leave also (Nelson, 1968). Thus booby juveniles seem to maintain contact with their parents at the breeding colony while capable of flying, and when they leave the colony for good they become independent. This aspect requires further examination on marked birds as the possibility exists that such care occurs away from the colony as well. The length of the postfledging care period varies from 56–377 days, depending on the species (Table II) and on habitat variables (Table IV). Although care is sometimes extended for these long periods, the mean period of postfledging parental care is 2 months (Abbott's, Blue-footed, Brown, and Blue-faced Boobies) to 3–4 months (Red-footed Booby, Nelson, 1970, p. 548). Nelson (1969a,b, 1970) and others have proposed that the long postfledging development period is a consequence of food scarcity. That food is scarce is shown by twinning experiments in Red-footed Boobies, where they could not successfully raise additional young (Nelson, 1970). Nelson (1969a,b, 1970) proposed that the ancestral tropical sulid evolved postfledging care in order to cope with the limited, unpredictable food source. When gannet became established in temperate waters where food was plentiful and predictable it evolved a shortened, well-synchronized breeding season. To make use of the plentiful, but seasonal, food supply, the development period of juveniles was shortened so the young gannet became independent before food decreased. Thus gannets produce fat young with sufficient food reserves (50% above adult weight, Nelson, 1970, p. 557) to get them over the difficult period when they leave the colony and begin to forage on their own. Most boobies, on the contrary, have remained in tropical waters where the food supply is not predictable and is never abundant. Under these conditions, the postfledging period is highly variable within and among species as a consequence of habitat location and year (Table IV; Nelson, 1970).

The longest postfledging care periods occur in the Brown and Abbott's Boobies which have highly specialized feeding techniques and foraging areas (Nelson, 1970). For example, Abbott's Booby, which maintains the longest mean postfledging care period (Nelson, 1970, p. 555), apparently uses a very specific feeding area northwest of Christmas Island. In combination with a highly specialized feeding method, the young require extensive provisioning beyond fledging.

In summary, the temperate gannets have seasonal breeding periods, short developmental periods, and no postfledging care. The boobies, inhabiting tropical waters, generally have erratic timing of breeding, long prefledging care, and extended postfledging care averaging 2–4 months, although it may extend up to a year in some species. The shortened developmental period, heavy fledging weight, and lack of postfledging care of gannets are attributable to the dependable, seasonal, and abundant food reserves of their temperate breeding grounds. The long development period, and the variable and long postfledging parental care period in the tropical boobies are attributable to the erratic, ephemeral, and scarce food reserves in the tropical oceans. The variability in length of the postfledging care period depends on species, habitat, and season. The length of the period reflects particularly scarce food reserves, whereas the frequency of feeds within this period is correlated with the abundance of food (Simmons, 1967).

5. Fregatidae: Frigatebirds

There are five species of frigatebirds (or man-of-war birds) in a single genus (*Fregata*) that inhabit tropical and subtropical waters (Thomson, 1964). Compared to some tropical seabirds, they remain relatively near their breeding colonies throughout the year. They feed by dipping for fish and squid, catching flying fish from above the waters surface, and pirating food from other species (Ashmole, 1971; Nelson, 1975a). All five species have been studied in some detail (see review in Nelson, 1975a), and so a clear picture of parental care emerges.

Like all pelecaniformes other than tropicbirds, frigatebird nestlings are altricial, and so are brooded (shaded from the sun) fairly continuously for about 1 month (Table II). Their developmental period prior to fledging is fairly long ($\bar{x}$ of 5.4 months, Table II), apparently because of the difficulty parents have in obtaining sufficient food. Because the parents catch prey only by surface dipping or aerial snatching, they cannot exploit the deeper layers available to boobies, tropicbirds, cormorants, and pelicans (Nelson, 1975a). Their specialized feeding techniques and the general paucity of food in tropical waters results in long intervals between feedings. Even so, some young starve prior to fledging (Nelson, 1975a). The evolutionary answer to the feeding constraints of frigatebirds is to have the longest postfledging parental care period of any bird (Nelson, 1976). In general, the usual postfledging period is 4–9 or 10 months (Nelson, 1975a), however, the record of extreme periods of postfledging care by parents (given in Table II) is as long as 15.5 months in Andrew's Frigatebird. There is some evidence for Magnificent Frigatebirds that the females do most of the postfledging feeding allowing the males to breed every year, whereas the females can breed only every other year (Diamond, 1973). This difference in ability to breed every year versus every other year corresponds to an excess of females at fledging (Diamond, 1973, 1975). Nelson (1975) did not find this sex difference in Andrew's Frigatebird, as males fed 8-month-old juveniles.

Within species, variations in the apparent postfledging period exist as a function of the location of the breeding colony (see Table IV). Presumably, the difference represents differences in available food, as the foraging techniques are similar within a species. During the first four months after fledging in the Great Frigatebird, the young spend 26–50% of their time in or flying over the colony (Diamond, 1975). Thereafter, their time at the colony decreases, but they still return.

Even with extended postfledging feeding by their parents, many juveniles lose weight when parents decrease their feeding rate (Nelson, 1967b). For example, year-old Great Frigatebirds lose weight from an average of 1083 g in April to 983 g in June, even though their parents had been feeding them for the preceding 6 months postflying. Several died at 600 g when their parents ceased feeding them (Nelson, 1968, 1967). Juvenile Great Frigatebirds characteristically do not attempt to pirate food from boobies with any regularity (Nelson, 1967), presumably because the task is exceedingly difficult. However, juvenile Magnificent Frigatebirds in Yucatan in January did pirate from Laughing Gulls. Indeed, one-half the frigatebirds foraging by this method were juveniles (M. Gochfeld and J. Burger, unpublished results).

In summary, frigatebirds are relatively sedentary, inhabiting tropical and subtropical islands. They have a long prefledging period and the longest postfledging periods, with juveniles being partially provisioned for a year or longer. Their restrictive, specialized feeding method, along with a general paucity of food in tropical waters, makes the long parental care period essential. Even so, many young starve to death when parents finally cease to feed them. Thus frigatebirds represent an extreme of parental care for the Pelicaniformes. Tropicbirds have almost no care; pelicans, anhingas, and cormorants have postfledging care of about 1 month; boobies (but not gannets) generally have parental care of 4–6 months; and frigatebirds have parental care for more than 6 months beyond fledging. Within the order, these differences correlate with feeding locations and techniques, i.e., tropicbirds feed onshore by diving for fish, whereas pelicans and cormorants feed inshore and in estuaries, respectively. Although the food for cormorants and anhingas is abundant, the technique of swimming underwater in pursuit of prey requires a period of learning. Boobies feed offshore by plunging for fish; therefore the technique is difficult, but the food source is larger and more diverse than that of frigatebirds. Frigatebirds are limited to dipping for food, piracy, or snatching fish from the air above the water's surface. With respect to postfledging parental care, there is more quantitative data for this order than for the other orders, although I feel that tropicbirds and cormorants bear further examination.

D. Charadriiformes

1. Alcidae: Auks

The 22 species of auks primarily inhabit the colder parts of the northern oceans although they do reach the warm waters of the Gulf of California (Thomson, 1964; Drent and Guiguet, 1961). The family is most numerous in the North Pacific, where 16 species occur. They are relatively pelagic and remain at sea except during the breeding season. They generally feed solitarily by pursuit diving for fish and crustaceans, although some species feed on the bottom and others seize prey from the surface waters (Ashmole, 1971). They are the only Northern Hemisphere group occupying the subsurface waters of the ocean (Bédard, 1969), and grade from plankton feeders to fish feeders.

There are three basic patterns of development of young alcids: four species are precocial, 15 species are semiprecocial, and three show an intermediate pattern (Sealy, 1968, 1973, 1974c). The distinction between these groups seems to lie in the timing of their departure from the nest. Unfortunately, most investigators working with alcids use the term "fledging" to mean when the chicks leave the nest (see Greenwood, 1964). Thus the precocial species would "fledge" at 2 or 3 days. Because I believe the term fledging should be restricted to when the birds are free flying, I will clearly distinguish leaving the colony from flying. Thus, in the three patterns (see Table V), the precocial species leave the nest at 2 days, swim off in search of foraging areas with their parents, and are fed until they are able to fly (age unknown) and are then independent. Presumably, the parents do *not* return to the nest site after departure.

In the three species showing an intermediate pattern, the young leave the nest with their parents when one-third to one-half adult weight, remain with their parents until they can fly (age unknown) and are then presumably independent. For two of these species, one parent returns to the nest after the chick has departed to defend the nest site (Birkhead, 1978). In the case of the semiprecocial species (auklets and puffins), the young remain at the nest until fully grown and fully feathered, and depart fully independent (refer to Table V). There is no evidence that the parents of any of these species continue to feed the young once they leave. For two species (Southern Puffin, Parakeet Auklet), the adults appear to desert the young before they depart (Witherby *et al.*, 1941; Sealy and Bédard, 1973, respectively). Again, I believe this point deserves substantiation. For at least two species (Pigeon Guillemot, Black Guillemot) the parents return to the nest, sometimes with fish, after the chicks have departed (Drent, 1964, 1965).

Thus, using my definition of fledging (first flying), there is no postfledging parental care in any alcid. There is, however, considerable care after leaving the colony for those species that leave below the adult weight. This has often been referred to in the literature as postfledging care. This is the only seabird group in which very young chicks leave the natal colony and travel far out to sea. Some members leave when half-grown (Witherby *et al.*, 1941) and travel to sea with their parents. Penguins also often leave the colony at one-half the winter-adult weight, but they are *not* accompanied by their parents. The ecological advantage of taking the young to the food source is obvious in species not well adapted to long distance flights

(Sealy, 1972). Sealy (1973b,c) believes that young murres and Razorbill Auks leave for the sea when only one-third grown because both adults and young can forage over greater areas offshore than would be possible if the adults had to keep returning to the nest sites. In general, the growth rates of alcids are more rapid than those of many other seabirds (Lack, 1968; Ricklefs, 1968), and there is a slight tendency for a prefledging weight loss in the semiprecocial young (Sealy, 1973b). Their rapid growth rates are a result of daily feeds, unusual for many offshore pelagic species.

Given that it is efficient for adults to take chicks to the food source rather than bring food, it is relevant to ask why more alcid species are not precocial or intermediate. Birkhead (1977) has proposed that the small chicks of cliff nesting species might not be able to safely descend the 300-m cliffs. Furthermore, the chicks may need to develop a suitable weight–wing-area ratio, which would allow them to glide down to the sea. Such wing loading would allow chicks to immediately dive upon hitting the water to prevent falling prey to gulls (Birkhead, 1977). The precocial species all nest in crevices, and chicks can easily scramble to the sea (Brent, 1919; Sealy, 1972, 1974).

The descent to the ocean is certainly critical in all alcids since the precocial and intermediate forms must avoid predation and locate their parents, and the semiprecocial (independent and capable of flying when they leave) must avoid predation. Tuck (1961) found that in murres the parents call the chicks from the cliffs. In the excitement of parent calls, some chicks leave before they are homeothermic, and up to 10% of the chicks leaving may drown or be picked off by gulls. Similarly, in Brunnich's Murre, Pennycuick (1956) found that the adult first flutters to the sea, the chick follows, and both then swim out to sea. In the Guillemot (Common Murre) and Razorbill, the descent to the sea is also cooperative (Greenwood, 1964). There is much bobbing and bowing as the chick and parent make their way to the cliff edge, with the adult defending the chick. The adult descends, followed by the chick, which calls loudly until joined by its parent, which leads it out to sea. The adult must also protect the chick on the water from the mass of other parents and gulls. Predation seems largely a case of parental neglect (Greenwood, 1964). Gulls do not prey on chicks attended by parents. Greenwood (1964) observed deserted chicks trying to link up with strange adults, but in no case was a permanent bond established.

In summary, chick development in alcids can be categorized into three types: (1) precocial young that leave their nests at two days of age, follow their parents to sea, and are fed until they can fly and are independent; (2) intermediate forms in which the young leave the ledges when they are a half to a third adult weight, and are fed by parents until they fly and can feed on their own; and (3) semiprecocial young that leave when fully grown and are independent. In some species of the last two groups, one parent returns to

Table V. Development in Selected Alcids

Alcid patterns	Age chick leaves nest	Parental care away from nest	Parent at nest after chick leaves	Age of flight	Source
Precocial					
Ancient Murrelet (*Synthliboramphus antiquis*)	2	Yes	No	?	Sealy (1973) Thomas (1964)
Craveri's Murrelet (*Endomychura craveri*)	2	Yes	No	?	Thomas (1964)
Xantus's Murrelet (*E. hypoleucus*)	2	Yes	?	?	Thomas (1964)
Partial development (Intermediate)					
Guillemot (*Uria aalge*)	14–25	Yes	Yes	?	Birkhead (1978) Greenwood (1964)
Brunnich's Murre (*U. lomvia*)	20	Yes	No	?	Witherby *et al.* (1941) Bent (1919); Tuck (1961)

Species					References
Razorbill (Alca torda)	20	Yes	Yes		Birkhead (1977) Birkhead (1947, 1978)
Leave when able to fly (Semiprecocial)					
Pigeon Guillemot (Cepphus columba)	35–39	No	Yes	35–39	Drent (1964, 1965)
Black Guillemot (C. grylle)	34–37	No	Yes	34–37	Drent (1965)
Horned Puffin (Fratercula corniculata)	42	No	?	42	Nettleship (1972) Drent (1964)
Southern Puffin (F. arctica)	51	No	?	51	Drent (1964) Witherby et al. (1941)
Cassin's Auklet (Ptychoramphus aleutica)	50	No	?	44	Thorensen (1964); Drent (1964)
Parakeet Auklet (Cyclorrhynchus psittacula)	37	No	?	35	Sealy and Bédard (1973)
Crested Auklet (Aethia cristatella)		No			Sealy and Bédard (1973)
Least Auklet (Aethia pusilla)		No			Sealy and Bédard (1973)

defend the nest site after the young have departed. Thus, using a definition of fledging as equal to flying, there is no evidence for postfledging care in this group. The crucial factor in the transition to independence is leaving the colony site successfully and finding their parents (in the first two cases), and avoiding predation and eventually learning to forage on their own. The precocial and intermediate forms have parental aid in the descent from the nesting colonies, whereas the semiprecocial species are on their own.

2. Stercorariidae: Skuas

There are 4–6 species of skuas in two genera (*Catharacta* and *Stercorarius*). In general, they are limited to boreal and arctic regions as well as Antarctic and sub-Antarctic regions. Skuas are related to gulls and are considered members of the Laridae by some authors (Thomson, 1964). Both genera generally specialize in kleptoparasitism. On the ocean, *Catharacta* forage by piracy, dipping, and some plunging, catching fish, cephalopods, and some crustaceans, whereas *Stercorarius* feed similarly, but do not take squid (Ashmole, 1971). In the breeding colonies both genera feed on young gulls and small mammals in the Arctic, and young seabirds in the Antarctic.

Being primarily high-latitude species, the skuas have shortened developmental periods (see Table II). *Stercorarius* young fly in 3–4 weeks, and *Catharacta* young fly in 7 weeks. Despite the difference in fledging time of the Long-tailed and Pomarine Jaegers, the total period of parental care is similar. The short development time is an adaptation to the short Arctic season (Maher, 1970). In Churchill, Manitoba, their breeding areas are not clear of snow and ice until June, yet jaegers can lay eggs by late June and produce young by the end of August (J. Burger, unpublished data). Both Arctic jaegers listed in Table II (Long-tailed and Pomarine) feed the young for 2–3 weeks after the young can fly and before they migrate. Young are fed at the nest site. By the end of August, however, birds begin to leave the Arctic, and no evidence exists to suggest that the young are fed once they leave the nesting territory. Nonetheless, in view of the difficulty of their foraging techniques (piracy, fishing), it seems worthwhile to follow parents and young to ascertain the degree of such feeding.

Prefledging development is slower in the larger South Polar Skua, although the period of postfledging parental care is considerably less (Table II). Chicks are always fed on their territory before and after fledging, and are usually accompanied by a parent to protect them from the attacks of other skuas (Spellerberg, 1971). The relatively short postfledging period reflects the short season.

In summary, the shortened breeding cycle of the skuas reflects their nesting at high latitudes where conditions are only favorable for a short time of year. Their postfledging parental care period is shorter than gulls (see Table II), some of which are smaller in size. The short period of feeding following fledging is an adaptation to encroaching winter conditions.

3. Rynchopidae: Skimmers

There are three species of skimmers living in tropical and low temperate regions of the world (Lockley, 1954; Thomson, 1964). They forage on small fish and other aquatic life taken from just below the surface of the water (Ashmole, 1971; Wolk, 1959). Their feeding method is unique in that with open bill, the mandible ploughs through the water, skimming the surface.

Little information is available on the transition to independence or postfledging care in the skimmers. M. Gochfeld (personal communication) has observed Black Skimmer young still being fed at the colony site at least 2 weeks after fledging, and it is probable that the period of partial parental care is longer, because young remain in flocks with parents for at least 2 months. Furthermore, the young may simply migrate with their parents and continue to be fed elsewhere. Erwin (1977a,b) reported that in July all foraging juveniles were in groups with adults, whereas by mid-August only 19% were in such groups. This suggests either imitative learning (Erwin's suggestion, 1977a) or that the young are still partially dependent on parents. In any case, because of the highly specialized nature of the foraging behavior, this species bears further study, for some postfledging provisioning would be adaptive.

4. Laridae

Larinae: Gulls. Gulls are a cosmopolitan subfamily generally distributed along coasts with some species nesting in the interior. There are 44 species, ranging from the Arctic regions to the tropics (Moynihan, 1959; Burger, 1974). Over the oceans they feed by surface dipping and plunging for fish, crustaceans, and other invertebrates (Ashmole, 1971). They also scavenge along seashores, behind boats, and at fish docks, feed on clams and other shellfish on mudflats, and scavenge at garbage dumps (see Harris, 1965; Spaans, 1971; Hunt and Hunt, 1973 for review).

Upon hatching, gull chicks are brooded intensively for up to a week (see Table II) depending on the species, weather, and hatching order of the chicks. For example, it is still cold in Minnesota marshes when Franklin's Gull chicks hatch, and brooding may be extended when it is particularly cold (J. Burger, unpublished data). Similarly, parents often stop brooding at a particular point, so that within a family, some chicks may be brooded for a shorter period of time than older chicks. Generally young gulls begin to fly when 5 weeks old in the smaller hooded species, and seven weeks old in the larger white-headed species (refer to Table II). Swallow-tailed Gull and Lava Gull both have fledging periods of almost 9 weeks. They nest in tropical islands with less abundant food supplies and no seasonal constraints. Presumably, the longer development period as well as low clutch sizes are adaptations to a low, erratic food supply.

For the 12 species of gulls for which there are data, the young of all species are fed for some time at the nest site following their first flights. This postfledging care period ranges from about a week in Franklin's Gull to over 10 weeks in the Swallow-tailed Gull. The longest periods of being fed after fledging occur in the two tropical species, and are presumably a result of the limited and erratic food supply. The relationship of postfledging care to prefledging care is shown in Fig. 4. Using a Spearmann rank (r_s = 0.63; n = 10; p < 0.05), there is a positive correlation between prefledging and postfledging care. The opposite trend occurs in the closely related skuas, no doubt because the short arctic seasons require short total developmental periods, and if the prefledging period is long, the postfledging period is short.

For most of the species listed in Table II, the investigators merely state that young are fed until a certain age at the colony. No one has carefully examined how frequent such feedings are, what percentage of young are fed at what age, and the rate of feeding during this transition period. Thus there are no behavioral data in the literature on the nature of the transition period. It is not even clear if young obtain most of their food from their parents up until one age, and then are completely independent, or are fed decreasing amounts until they are independent.

I have examined the transition period and postfledging parental care in three species of gulls. Franklin's Gulls were observed from 1969 to 1971 at Agassiz National Wildlife Refuge in Minnesota, where a colony of 15,000 pairs nested in a marsh (Burger, 1974). Laughing Gulls were examined on Clam Island, New Jersey, in 1976, where 4000 pairs nested in a *Spartina* salt marsh (Burger and Shisler, 1978). Herring Gulls were examined from 1976 to 1978 at Clam Island, where 800 pairs nested in the high areas of a salt marsh among the small bushes (Burger, in press). In all cases the adults were sexed and color marked, and the young were color banded for identification. Observations were made 6–12 hr a day during these periods.

The percentage of young returning to the nest at least once a day to be fed by parents is shown in Fig. 5 for the three species. For Franklin's Gull, the percentage of chicks (n = 28) fed at the nest dropped off rapidly after 38 days, 1 week after they began flying. The rapid drop was partly due to a general desertion of the colony by most pairs. Franklin's Gulls are very synchronous, and by late July the colony is entirely deserted over a 2-week period (see Burger, 1974). This desertion may be attributable to a lack of

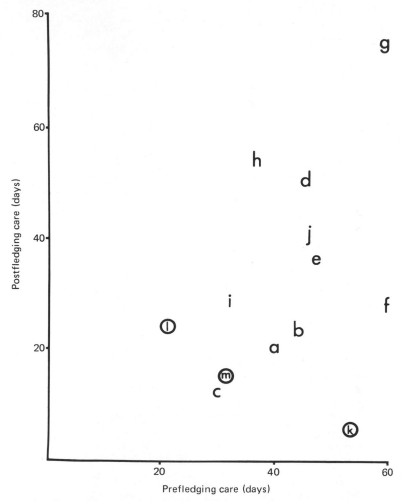

Fig. 4. Relationship of postfledging care (in days) to prefledging care (in days) for gulls, and skuas. a, Glaucous Gull; b, Glaucous-winged Gull; c, Silver Gull; d, Herring Gull; e, Southern Black-backed Gull; f, Lava Gull; g, Swallow-tailed Gull; h, Kelp Gull; i, Laughing Gull; j, Great Black-backed Gull; k, South Polar Skua; l, Long-tailed Jaeger; m, Pomarine Jaeger. Kelp Gull and Southern Black-backed Gull are races of *Larus dominicanus*.

food in the area, as the sudden influx of new recruits puts a tremendous burden on the surrounding farming areas. Franklin's Gulls primarily feed on earthworms obtained from the farmlands, and on midges and other insects from the marshes. Therefore it is clearly adaptive to disperse outward from the colony, and this happens rapidly in late July and August (see Burger, 1972). It is possible that chicks disperse with their parents and are fed by them, but this has not been investigated. In one case, two wing-tag-

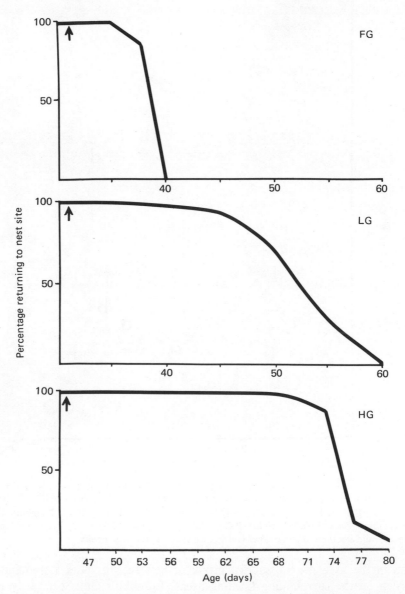

Fig. 5. Percentage of chicks returning to the nest site to be fed by parents in Franklin's Gulls (FG), Laughing Gulls (LG), and Herring Gulls (HG) as a function of age (in days). The arrow indicates the age of fledging.

ged siblings repeatedly occurred together with a color-dyed adult in two different locations in North Dakota. However, as I did not see the adult, I could not positively identify it as the chicks' parent.

In Laughing Gulls, the percentage of chicks (n = 36) returning to be fed decreased gradually (Fig. 5). Most chicks were fed until 45 days of age (12 days postfledging), while some chicks were fed until 60 days of age. Laughing Gull colonies are less synchronous than Franklin's Gull colonies, but instead have synchronous subgroups. The colonies are not as large as Franklin's Gull colonies, and there are more feeding areas in the surrounding marshlands and estuaries. Laughing Gulls also feed on a higher diversity of food items than Franklin's Gulls (J. Burger, unpublished data), so the pressure for dispersal would be less.

In Herring Gulls, most chicks (n = 34) were fed at least once a day until 75 days of age (32 days post hatch), and some were fed until 90 days. Again, the drop in the percentage of chicks being fed was steep, and occurred in mid-August. It was not a result of parental activity since in all nests both parents returned to the colony for 2–7 days after the chicks no longer returned to their territories. As I was present during this critical period from 0630 to 2000 hr, I feel certain that chicks did not return.

In Herring Gulls, the percentage of total time chicks were on their territory began to decrease at 50 days of age (Fig. 6), and was similar to the time spent on territory by their parents. When not on territory, chicks frequently stayed in flocks on the nearby Bay. Returning adults that landed on territory Long-Called until chicks returned to be fed. When in the bay with chicks, I could hear Long Calls, but could not identify individuals. Members of particular broods, however, frequently got up together and flew directly to their territory where there were waiting adults. Chicks *always* landed on their territory except when they first learned to fly and couldn't accurately land on their own area (Burger, in press).

The mean number of feedings per brood per hour of my observation periods is shown for Herring Gulls in Fig. 6. The number increased at about 48 days of age, remained high until 70 days when it again dropped. The initial increase may represent the increased food needs of young that are flying about and using more energy. Prior to 45 days, chicks make only short flights around their territory, or circle in the air, but they return immediately to the territory. The decrease at 70 days may represent a real decrease in food provided to the chicks, or merely a change in the location of feedings. Chicks may begin to accompany parents on their foraging trips, and may be fed during these trips. Unfortunately, it is difficult to follow chicks during this stage, but such data need to be gathered to better understand the transition period.

Although Trivers (1972) and others have argued about the nature of the parent–young conflict during the transition period, my data on Herring

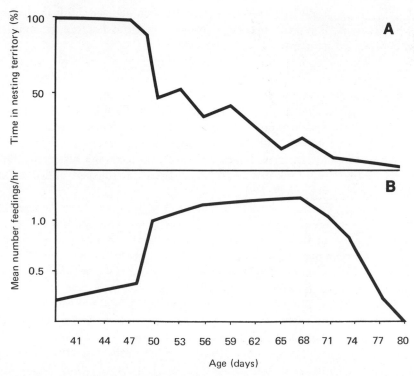

Fig. 6. Percentage of time Herring Gull chicks remained in their nesting territory as a function of age (A), and the mean number of feedings per hour for each brood as a function of age (B).

Gulls showed that adults Long-Called to bring young back to the colony. This certainly suggests that the adults encouraged such care. As Fig. 7 and 8 show, adults frequently stood very close to full-grown, fully fledged, 70-day-old chicks showing no indication of a desire to escape, and they fed chicks whenever they begged. Similarly, there was no apparent conflict between siblings, and they usually left and returned to the colony together. The desire to escape shown by parents of prefledging chicks may reflect their desire to escape the badgering of their chicks, and not a desire to stop feeding chicks. When both parents and young must actively seek contact for feeding to continue, they did so, and conflict seemed to disappear. When parents returned to find chicks missing from their territory, they always Long Called to call in the chicks (Fig. 9). In 60% of the cases, chicks immediately returned from the surrounding bay areas and were always immediately fed.

　　　　Several authors (Drury and Smith, 1968; Haycock and Threlfall, 1975) have suggested that Herring Gull young remain near the colony for a few days postfledging, but this is the first report of such an extended period of

Fig. 7. Young Herring Gulls (age: 75 days). (A) Waiting for parent to return; (B) upon return, young hunch (a submissive posture).

Fig. 8. Young Herring Gulls. (A) Standing by parents waiting to be fed; (B) adult beginning to regurgitate food.

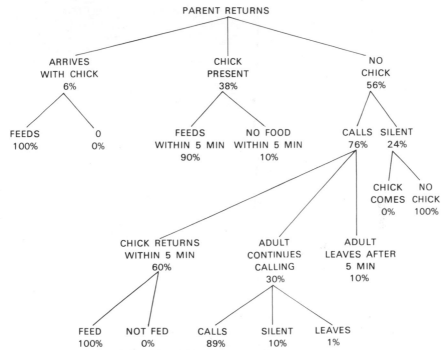

Fig. 9. Behavior of parent Herring Gull upon returning to its nesting territory. Each level shows parental behavior toward chicks, or in calling the chicks back to the nest.

postfledging care for all chicks. Either this behavior is variable depending on such factors as colony location (see Table IV), food supply, and disturbance, or researchers have overlooked it. Certainly human disturbance would influence this behavior pattern. Food supply may also influence the length of the care period at the colony since presumably low food supplies might stimulate adults and chicks to disperse to more productive areas. The birds at Clam Island were not food stressed, as both parents usually stood idly around the territories. Frequently they flew out and returned within 2 min with a crab. They fed their young only on natural foods despite the close proximity of a garbage dump. That the pattern of postfledging parental care has been missed in the past seems likely since the Clam Island colony looked deserted with a casual walk through it. The timing of returns of the chicks to their territories was not constant, and they often remained for only 3–15 min before leaving. The possiblity that chicks are fed once they leave the colony bears further examination. Herring Gulls have been seen feeding young away from the colony in October (see Table VI), and Glaucous-winged Gull chicks were fed at a dump by adults when they were 8 months old (Vermeer, 1963).

Table VI. Evidence of Postfledging Feeding Away from Breeding Colonies by
Presumed Parents

Species	Source	Age of young[a]
Royal Tern	Escalante (1968)	225
	Ashmole and Tovar (1968)	210
Elegant Tern	Ashmole and Tovar (1968)	127
(*Thalasseus elegans*)		
	Monroe (1956)	127
Crested Tern	Feare (1975)	120
(*T. bergii*)		
Sooty Tern	Feare (1975)	60
Caspian Tern	Tomkins (1959)	127
Sandwich Tern	D. Lack cited in Ashmole and Tovar (1968)	67
Glaucous-winged Gull	Vermeer (1963)	240
Herring Gull	R. Schreiber cited in Ashmole and Tovar (1968)	115
	Lack cited in Ashmole and Tovar (1968)	115
Shag	Potts (1969)	420
Brown Booby	Simmons (1967)	377

[a] In days from hatching.

Another way to examine the possiblity that postfledging care has been missed in gulls is to approach it from another direction. That is, it is generally assumed that the transition period to independence, and the period immediately following independence is the most critical (Drury and Smith, 1968). It therefore seems reasonable to assume that an increase in death rate during a particular time might pinpoint this difficult period. I therefore examined published banding data for species where data are available for juveniles in the 6 months following fledging (Table VII). In this table I have taken all the deaths for the 6-month period, and have computed the percentage occurring in each month. Where the percentage is higher than expected (17%) for 1–2 months, I am assuming this is the period immediately following independence.

For Herring Gulls, data from five colonies in Europe and North America indicate that usually a half or more deaths (69%) occur in September and October. In New Jersey, where the young were fed at the colony until late August, most deaths occurred in September. Since the birds had been flying in the area throughout most of July and August, this high death rate in September is not a result of being in unfamiliar surroundings and seems likely to be due to difficulties of finding food. The results are fairly consistant over several geographic areas.

For other species, the time of peak deaths varied (Table VII) from July to December. For the small hooded gulls, high death rates occurred immediately upon leaving the colonies, suggesting that parental care might be minimal during this period. For the Dominican Gull (*Larus dominicanus*), in the Southern Hemisphere, the highest death rates occur in April, or 4 months after fledging. California Gulls (*L. californicus*) have high death rates immediately after departing from the colony, whereas Western Gulls (*L. occidentalis*) have high death rates from October to December. Glaucous-winged Gulls experience high death rates in November, and Vermeer's (1963) data from a dump clearly indicate that some parental feeding is occurring at this time (see Table VI). His data indicate that some parents and young leave the colony area together, and migrate south feeding at dumps.

One objection that could be raised to using banding data in this way is that gulls migrate and have a different probability of being found in different geographical areas. This is certainly true. However, the consistency of the Herring Gull data from vastly different areas suggest that the technique is useful. Second, I looked at these data by using only returns within a 200-mile radius of the colony to eliminate this variable (Table VIII). For these species, the results are similar to the pooled data. Despite the difficulties with the technique, it does suggest extended parental care in some species after they leave the breeding colonies. From these data I would look for extensive and extended post-fledging care in Glaucous-winged, Dominican, and Western Gulls. The low death rates in the months immediately following fledging despite the difficulties of mastering flying and being in new habitats suggests that they are being partially provisioned by their parents.

In summary, gull chicks are fed at the colony site for an average of over a month after fledging. The length of postfledging care is related to the prefledging period. In Herring Gulls, the young spend a decreasing percentage of their time on their territories after 50 days, but when undisturbed, nearly all return to be fed at least once a day until they are over 75 days old. This suggests that the transition period is one of decreasing attachment to the colony while still receiving significant quantities of food from their parents. Banding data and casual observations indicate that some young of some species are fed away from the breeding colony for some time.

Sterninae: Terns. There are 39 species of terns nesting from the Arctic to the tropics, most of which nest along coasts, although some species nest in the interior. Related to the gulls, they are generally a more southern latitude group than gulls. They primarily feed by plunge diving for fish, although they occasionally dip for fish and crustaceans (Ashmole, 1971), and by feeding on large insects (Thomson, 1964). Nesting colonies may vary from only a few individuals (see Burger and Lesser, 1978) to hundreds of thousands (see Schreiber and Ashmole, 1970).

Upon hatching, tern chicks are brooded for up to 10 days (see Table II) depending on species, weather conditions, and hatch order of the chicks. Most terns can fly when 3–5 weeks old, although the tropical species (Sooty

Table VII. Death of Gulls (%) That Died from

	Herring Gull				
Month	Study 1[a]	Study 2[b]	Study 3[c]	Study 4[d]	Study 5[e]
July	4	8	0	8	1
August	25	16	26	14	7
September	37	46	38	17	32
October	16	23	20	27	37
November	13	5	8	20	15
December	5	2	8	14	8
No. Returns:	241	42	117	36	406
Location:	Britain	New Jersey	Newfoundland	Minnesota	Netherlands
Fledging date:	Early July	Early July	Early July	Early July	Early July

[a] Parsons and Duncan (1978).
[b] Burger, Clam Island, 1976–1978 (unpublished data).
[c] Threlfall (1978).
[d] Hofslund (1959).
[e] Spaans (1971).

and Fairy Terns, and Black Noddy) require longer periods (Table II). Within this range, the larger species (e.g., Caspian Tern) take longer to fly than do the smaller species (e.g., Black Tern).

In all tern species examined there was some parental feeding following fledging (see Table II). However, authors often only mentioned that terns were fed for a few days (see Massey, 1974; Cullen and Ashmole, 1963; and Brown, 1976). Data on the postfledging care of terns are less precise despite the frequent mention of its existence. Seemingly, tern adults and young are not marked so that these relationships can be stated with certainty. Partly the problem lies in the mobility of the tern chicks. That is, when disturbed, broods often leave the colony or move elsewhere (M. Gochfeld and G. Shugart, personal communication) making such observations difficult. As this group clearly demonstrates extensive postfledging care, it seems essential to examine this aspect of their breeding biology.

For some terns, the flying chicks seem only to be fed at their nest sites (Black Noddy: Cullen and Ashmole 1963), whereas others are fed on the ground beneath their nest tree [Brown Noody (*Anous stolidus*): Murphy, 1936] as well as at their nest sites (Brown Noddy: Dorward and Ashmole, 1963). Brown (1975) found that 30+ day-old Sooty Tern chicks were fed at night, and each chick was given 25.1 g of food per feeding, whereas 30+ day-old Brown Noddy chicks were fed only half as often. The Sooty chicks were fed half fish and half squid, while the Noddies received more fish. Although useful, Brown did not know the precise ages of the chicks,

July to December of Year of Birth

Laughing Gull[f]	Franklin's Gull[g]	Black-headed Gull[h]	Glaucous-winged Gull[i]	California Gull[i]	Western Gull[i]		Dominican Gull[j]
2	22	25	7	26	4	Jan.	0
54	45	34	5	30	8	Feb.	21
23	28	20	16	18	16	Mar.	21
9	4	9	15	4	23	April	27
0	—	6	49	12	26	May	20
12	—	6	8	10	23	June	11
95	108	256	74	84	74		1733
New Jersey	Minnesota	Belgium	British Columbia	California	Oregon		New Zealand
Early July	Mid-July	Mid-July	Mid-July	Mid-July	Mid-July		January

[f] Burger, Clam Island, 1976–1978 (unpublished data).
[g] Burger, Agassiz National Wildlife Refuge, calculated for July to October only, as they began to leave in mid-November (unpublished data).
[h] Roggeman (1970).
[i] Woodbury and Knight (1951).
[j] Fordham (1968).

nor did he quantify age-related differences during the transition period. Nonetheless, these are the *only* quantitative data on postfledging parental care in terns.

For others terns, the chicks are fed in a number of locations. Once they can fly, Common Tern chicks fly to the beach with their parents. Initially they simply wait on the beach for the adults to bring back food (Palmer, 1962). After 4–5 days, however, they begin to follow the parents on fishing trips. Bent (1921) recorded three methods of feeding juvenile Common Terns: in the air, on the sand, and on the water. In the air they transfer the fish from bill to bill. Feare (1975) also saw the aerial transfer of fish between adult and young Sooty Terns. On the sand, the Common Tern young take the fish in their bills, and either swallow it immediately or fly and swallow it while in the air. While the young sits on the water, the adult delivers the fish aerially; the chick then flies and swallows it. Similar behavior has been reported for Least Tern chicks (Tomkins, 1959).

Bergman (1956) proposed that young Caspian Terns (58–60 days old) do not fish because their primaries are too weak to allow the sudden movements required for accurate diving. Soon after, their fishing ability develops rapidly. G. Shugart (personal communication) found that Caspian Tern young are fed at the nest site for a least 2 weeks after fledging, if left undisturbed.

The role of practice, and of being taught by parents with respect to

Table VIII. An Examination of Death Rates for Selected Gulls by
Eliminating Distance from the Colony[a]

Month	Glaucous-winged[b] Gull	California[b] Gull	Black-headed[c] Gull
July	8	40	32
August	6	15	38
September	16	40	9
October	14	0	9
November	48	5	9
December	8	0	3
Distance from colony (miles):	200	200	200
Number:	63	20	50

[a] Percentage that died each month are given.
[b] Woodbury and Knight (1951).
[c] Roggeman (1970).

plunge diving seems unclear and contradictory. Several authors have noted
young following parents on fishing trips (see above). LeCroy (1972) noted
adults passing fish underwater to young Common Terns, and suggested they
were being taught to fish. Similarly, she reported young closely following
the fishing maneuvers of adults. I. Nisbet (personal communication),
however, has frequently observed young Common Terns first fishing
without adults near their breeding colony and believes they are not taught
nor do they learn by observing their parents. Smith (1975) believed that
young Sandwich Terns imitated the hunting technique of their parents.
Furthermore, he observed them "practicing" by dipping at small sticks,
flotsam, and seaweed. M. Gochfeld (personal communication) stated that it
is unlikely that young skimmers learn from their parents, since he has
observed them skimming in flocks of only young at 4 days following fledging. Parents provisioning young after fledging does not necessarily indicate
that they must learn from them. It is possible that they learn independently
as their coordination develops, and parental feeding is essential to tide them
over this period.

Prolonged feeding of young by presumed parents certainly occurs in
some tern species (see Table VI). The differences between species, however,
may reflect differences in our knowledge rather than real differences among
species. It is clear that terns are fed on or near the colony for some period
of time (1 week to 1 month). Thereafter, adults and their chicks move to
nearby beaches. As the chicks learn to forage they are partially provisioned
by their parents. Eventually they migrate, but young and adults apparently
stay together as young are fed by parents thousands of miles from their

breeding colonies (see Buckley and Buckley, 1972; Ashmole and Tovar, 1968; Escalante, 1968).

In reviewing parental care within the Charadriiformes, there is a progression from no postfledging care (alcids) to prolonged parental care (terns). Problems in the transition to independence in the Alcidae are limited to the relationship between colony abandonment and flying, where the young of some species leave when 2 days old to swim to foraging areas with their parents, and the young of others leave the colony when completely independent. Skuas have limited postfledging parental care, a result of their nesting in Arctic and Antarctic habitats requiring shortened reproductive cycles. Information on skimmers is scanty but suggests prolonged parental care for several weeks or months as they migrate in groups with adults. In gulls and terns, there is extensive parental care beyond the fledging stage. Depending on disturbance and food scarcity, family groups abandon the colony soon after fledging or use the territories for another month or more. Once they leave the natal colonies, some parental feeding occurs in some individuals for extended periods of time. It seems clear that terns may have longer periods of postfledging parental care than do gulls. This is imposed on them by their more specialized feeding methods. Gulls are generalists, foraging on different foods using several techniques (including foraging on dumps), whereas terns primarily feed by plunge diving for fish.

IV. COMPARISONS AMONG FAMILIES OF SEABIRDS

In all species of birds, some parents must successfully raise young to contribute to the next generation. From a parental point of view, the energy adults expend in a reproductive effort should be a compromise between increasing their offsprings' chances of survival and decreasing their own chances of mortality. The energy expended should result in the maintenance of the adult at least until the production of replacement young (averaged over their entire lifespan). In seabirds, this compromise is met by a number of patterns. In some families, there is no parental care beyond the point when the chicks leave the colony (penguins) or when they fledge (Procellariids, Alcids). In a few families, the length of the postfledging parental care period is very short (albatrosses, tropicbirds). In these latter cases the young may simply be learning to fly, and once this is accomplished they are independent. In pelicans, anhingas, and skuas, parental care after fledging is less than 1 month. The care seems to be a result of learning a specialized foraging technique (commorants, anhingas) or constrained by weather pressures which necessitate a short breeding season (skuas). Gulls and terns have extended care for a month or more following flying, and then exhibit partial parental care for some months. Postfledging care is highly developed and prolonged in the boobies and frigatebirds.

In this section I will develop some generalization among the families, and examine the variables leading to extended parental care.

A. The Starvation Theory in Seabirds

Very early in the study of seabirds, investigators noticed that in some species the young seemed to be deserted by parents and left to starve for some time prior to their departure from the nesting colonies. The period of time between the last visit of the parents and the leaving of the nest site by the young is referred to as a desertion period. Initially, the starvation theory, with its associated desertion period, was assumed to apply to penguins, some alcids, albatrosses, and other Procellariiformes. Presumably, the desertion period resulted in a decrease in weight of the young allowing them to fly and forcing young to leave their nest site in search of food. In general, albatross and petrel young exceed the mean adult weight by up to 50% in the months or weeks before fledging. Flight would be difficult for such over-weight, inexperienced young.

The observations of the early investigators were correct in that some species of seabird young are not visited for some period prior to fledging. However, to name this period as a desertion period and invoke a starvation theory is both unfortunate and misleading. Richdale (1957) and others have successfully refuted the existence of a starvation theory in penguins. Richdale (1957) showed that young penguins are heaviest when about to leave the breeding grounds, although they may be fed less often toward the end of the parental care period. Similarly, in alcids, parents may feed chicks less often near their departure from the nesting islands. However, chicks are not deserted, and parents often return to defend their nesting burrows after chicks have departed.

Early investigators believed that Wandering Albatross chicks were deserted for months and left to starve during the Antarctic winter. Again, Richdale (1954) has shown that although the frequency of feeding drops toward the end of the prefledging period, there is no desertion period. Young albatross do lose weight prior to fledging, but this does not indicate that chicks are not fed during this period. Thus, for alcids, penguins, and albatrosses, seabird biologists have acknowledged the untenability of the starvation theory.

The starvation theory, however, still persists in the petrel literature. Even within the last year, investigators have noted that some species have desertion periods (see Procellariids Section III.B.3). Contradictions exist even within species with some authors still maintaining that there is a desertion period. As mentioned above, I suggest that there is no desertion period for petrels. Parents merely have difficulty in finding sufficient food for their offspring, and may not return with food until after the chicks have departed. Those species that seem to have long desertion periods (refer to Table III) also have long intervals between feeds and long incubation bouts, which suggests that parents have difficulty foraging for themselves and their offspring. Furthermore, in some species claimed to have desertion periods (Cory's and Audubon's Shearwater) parents are known to return to their nest sites after the chicks have departed (Zino, 1971; Snow, 1965). This does not suggest desertion, but simply an inability to return earlier as a result of foraging difficulties. The assumption that a drop in weight of nestling petrels indicates that they have been deserted is invalid, since species lacking such a desertion period also lose weight (see Fig. 3). I propose that young petrels may experience prolonged periods without parental feeding throughout their prefledging periods, but that they are not deserted by parents. Young petrels simply leave when fully developed, and their parents may return to find them departed.

The starvation theory thus was developed to explain the long interval between the last feed and the chicks departure in some penguins, alcids, albatrosses and petrels. It implies that parents desert chicks so that chicks will lose weight and eventually fledge, as well as allowing parents to migrate early. The evidence for all species suggests that parents do not desert young. However, many of these species have difficulty obtaining food, resulting in long interfood intervals, interpreted as desertion periods.

B. Factors Affecting Postfledging Care

1. Prefledging and Postfledging Care

Lack (1967, 1968) first noted that there was a linear relationship between clutch size, the incubation period and the fledging period. Ricklefs (1973) developed this further and suggested that growth rates have been driven to their physiological maximum, and egg and chick development rates are correlated because of these physiological constraints. This suggests that, within limits, postfledging periods of parental care may relate to prefledging periods when both relate to food constraints.

Figure 10 shows the mean length of the post fledging period for individual species (from Table II only) as a function of their prefledging

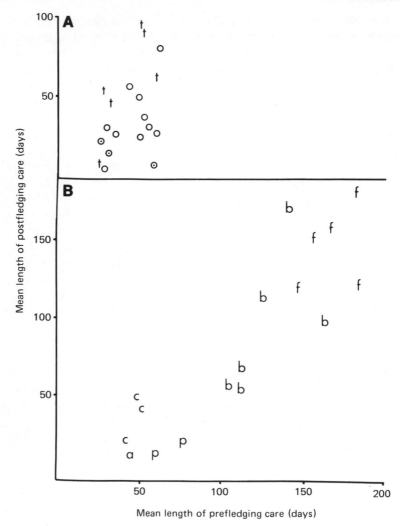

Fig. 10. Mean length of postfledging care (in days) compared to prefledging care for selected seabirds. Shown are Laridae (A) and all other seabird families (B). a, anhinga; b, booby; c, cormorant; f, frigatebird; p, pelican; t, tern; O, gull; ⊙, Jaeger.

periods. Lumping all the data (from the top and bottom groups), the length of the postfledging period is directly related to the length of the postfledging parental care period ($r = 0.78$; $df = 35$; $p < 0.001$), and prefledging care accounts for 61% of the variability in the length of the postfledging care period ($F = 48.4$; $df = 1,35$; $p < 0.001$). I then separated the Charadriiformes from all other species. In the other species, primarily tropical and subtropical in breeding range, the relationship is even stronger

$(r = 0.86; df = 15; p < 0.001)$. Thus development rates prior to fledging parallel the time required by young to become independent after fledging, suggesting both are caused by the same constraints (low food supplies).

For the gulls and terns, the relationship is not significant, although it is when the gulls are considered alone (see above). I believe this is a result of three factors: incomplete data, migration, and foraging technique difficulties. First, information on postfledging care in this group is spotty and imprecise. Whereas the length of the postfledging care period is known for some species, the information on other species is lacking beyond the time they use the colony. In the non-Charadriiformes discussed above, parental care occurs primarily at the colony site. Second, gulls and terns migrate, and presumably the young of some species accompany adults and are sometimes fed. Information, however, is difficult to obtain for this period. Third, in the non-Charadriiformes, food may be difficult to obtain because it is erratic and in scarce supply. Thus the same constraints that cause slow prefledging development might cause slow postfledging independence. For terns and gulls, the difficulties may relate to the difficulty of learning to plunge dive and not to scarcity of food. Since most terns plunge dive, differences in the duration of postfledging care among species might not be expected.

Table IX. Relationship of Postfledging Parental Care, Incubation, and Fledging Periods in Marine Birds[a]

Family	Number of species[b]	Incubation period[c]	Fledging (fly or leave)[c]	Postfledging period[d]
Diomedeidae	5	70	196	10 (0–40)
Procellariidae	20	44	62	0
Phaethontidae	3	42	84	10 (0–30)
Pelecanidae	2	30[d]	60	20 (20–21+)
Sulidae	7	53	112	100 (0–294)
Phalacrocoracidae	3	28	53	40 (28–50)
Anhingidae	1	28	30	14+
Fregatidae	5	49	182	168 (120–428)
Spheniscidae	5	39	77	0
Alcidae	9	32	39	0
Stercorariidae	3	28	42	15 (6–24)
Laridae				
Larinae	12	28	45	38 (0–75)
Sterninae	10	28	32	33 (0–90)
Rynchopidae	1	28	28[e]	14+[e]

[a] Means are given in days.
[b] Number of species used to calculate postfledging period.
[c] From Lack (1968, 1967); leave applies to penguins only.
[d] Calculated from summary sheets discussed in text.
[e] From M. Gochfeld (personal communication).

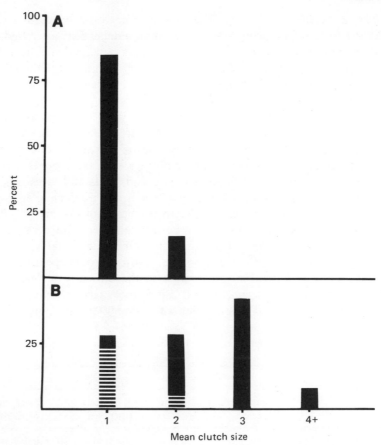

Fig. 11. Mean clutch size of species without postfledging care (A) and with (B) such care. For species with extended parental care, the hatched line indicates the percentage due to those with care periods of well over a month.

In examining mean postfledging care periods for families (Table IX) having some care, there is a significant correlation ($r = 0.96$, $df = 6$; $p < 0.001$) between prefledging and postfledging care. The length of the prefledging care period accounts for 82% of the variability in the length of the postfledging parental care periods ($F = 61.8$; $df = 1,6$; $p < 0.001$). For this calculation, I used only families with care of more than 10 days, as some authors defined flying differently. I wanted to eliminate care excluded by other authors who had defined fledging as sustained flight.

2. Clutch Size and Incubation Bouts

Just as the length of the prefledging care period is a measure of physiological development rates with respect to food constraints, clutch size and the length of incubation bouts are also indicators (see Lack, 1968). Clutch size reflects both ultimate as well as proximate factors, whereas the length of incubation bouts clearly reflect proximate food constraints. Figure 11 shows the percent of species having particular clutch sizes for species with and without postfledging parental care. In general, species without such care (alcids, petrels) have low clutch sizes ($\bar{x}$ = 1.15 ± 0.37). Species with postfledging parental care have significantly larger clutch sizes ($\bar{x}$ = 2.37 ± 1.2) than do those without care (t = 6.61; df = 73; p < 0.001). However, comparing species with long parental care (boobies, frigatebirds) with those with shorter postfledging parental care periods (all others with care) reveals that those with the longest care periods have lower mean clutch sizes (t = 4.7; df = 38; p < 0.001), although there was no difference in mean clutch size of those with no care and those with the most extended care (t = 0.48; df = 42). Hence it appears that with no parental care, clutch sizes are small reflecting food constraints, whereas long periods of extended parental care also reflects food constraints. Species having larger clutches provide limited

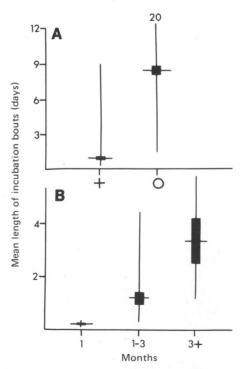

Fig. 12. (A) Mean length of incubation bouts (in days) in seabird species with (+) and without (O) postfledging parental care. (B) Length of incubation bouts (in days) of seabirds as related to the length of the postfledging parental care period. Shown are means (horizontal line), range (vertical line), and standard deviation (bar).

parental care to get chicks over the difficulties of learning new foraging techniques.

Just as clutch size is partially a function of food constraints in an evolutionary sense, the length of incubation bouts reflects proximate as well as ultimate food constraints. A bird usually will not return to begin its incubation bout unless it has found enough food for itself. Thus a species that can return several times a day to take an incubation bout has less difficulty finding food than one that comes back only after several days. The mean lengths of incubation bouts (computed from Table II) vary by family. The longest mean bouts occur in albatrosses ($\bar{x}$ = 14.5 days), penguins ($\bar{x}$ = 10.7 days), and petrels ($\bar{x}$ = 7.7 days). Intermediate incubation bouts occur in tropicbirds ($\bar{x}$ = 5.3 days), frigatebirds ($\bar{x}$ = 3.3 days) and boobies ($\bar{x}$ = 1.4 days). Short bouts of less than 1 day occur in terns, gulls, pelicans, skuas, and skimmers. Information is not available for alcids. Figure 12A shows the relationship between the mean length of incubation bouts and the occurrence of a postfledging parental care period for seabirds. In general, species with parental care have shorter incubation bouts ($\bar{x}$, 1.07 ± 2.0 days) than do those without care ($\bar{x}$ = 8.8 ± 5.3 days; t = 7.4; df = 53; p < 0.001). In comparing the length of bouts with the length of the postfledging parental care period (Fig. 12B), it is clear that species with longer mean

Table X. Postfledging Care, Weights, and Feeding Locations of Seabird Families

Family	Foraging location	Adult weight (g)		Mean number of days of Postfledging care[b]
		$\bar{x}$ ± SD	(Range)[a]	
Spheniscidae	Both	6353 ± 7274	1100–30,000	0
Diomedeidae	Offshore	4974 ± 2924	2450–8700	10
Procellariidae	Offshore	703 ± 972	130–4000	0
Phaethontidae	Offshore	571 ± 239	300–750	0
Pelecanidae	Inshore	3500	—	20
Sulidae	Both	1874 ± 868	715–3300	100
Phalacrocoracidae	Inshore	2086 ± 349	1750–2500	40
Anhingidae	Inshore	1326[c]		14
Fregatidae	Offshore	1337 ± 123	1250–1425	168
Stercorariidae	Inshore	500	—	15
Laridae				
Larinae	Inshore	736 ± 414	250–1600	38
Sterninae	Inshore	124 ± 64	40–175	33
Rynchopidae	Inshore	310[d]		14
Alcidae	Offshore	605 ± 333	125–1,000	0

[a] Lack (1968); J. Burger (unpublished data).
[b] From Table II.
[c] Palmer (1962).
[d] From M. Gochfeld (unpublished data).

bout lengths have longer postfledging parental care. Species with less than 1 month of care have shorter incubation bouts than do species with 1–3 months of postfledging care ($t = 2.17$; $df = 31$; $p < 0.05$), which in turn have shorter incubation bouts than do species with more than 3 months of parental care ($t = 3.17$; $df = 10$; $p < 0.01$).

Taken together, the clutch size and incubation bout information indicates that species without parental care have low clutches and long bouts, indicating food stress. Species with care have larger clutches and shorter bouts indicating less food strees. Nonetheless, the species with the longest postfledging parental care periods have low clutch sizes and long incubation bouts.

3. Location of Feeding

I then examined the mean length of the postfledging parental care period with respect to body weight and the location of foraging. There seems to be no relationship between either factor (Table X). Both large (penguins) and small (petrels) species exhibit no postfledging parental care, and large (boobies, cormorants) and small species (terns) exhibit extended postfledging care. Postfledging parental care is lacking in some inshore feeders (penguins) and is present in others (cormorants). Such care is lacking in some offshore feeders (petrels) and present in others (boobies).

Even within a family such as the Sulidae there seems to be no relationship. Gannets have no parental care and feed offshore, whereas the Blue-faced ($\bar{x} = 56$ days), Red-footed ($\bar{x} = 180$), and Abbotts' Boobies ($\bar{x} = 180$) have extended care and also feed offshore.

4. Latitude

Table XI shows the breeding location of some species of marine birds as a function of having ($+$) or not having ($-$) parental care. The percentage of species having parental care increases toward the equator (Fig. 13A). No Arctic species have postfledging parental care.

The mean length of the postfledging parental care period varies significantly among the latitude categories ($F = 15.71$; $df = 3,47$; LSI = 12 days; $p < 0.001$) (Fig. 13B). There were no differences between the Arctic–temperate and the temperate, nor between the temperate and the temperate–tropical; but there were significant differences between the Arctic–temperate and the temperate–tropical, and between the tropical and all other latitude categories. The incidence of postfledging parental care increases toward the tropics.

Table XI. Postfledging Care and Breeding Locations in Seabird Families[a]

Family	Arctic[b]		Arctic-temperate		Temperate		Temperate-tropical		Tropical	
	+	0	+	0	+	0	+	0	+	0
Diomedeidae					1	3			1	
Procellariidae		1		1		9		6		
Phaethontidae							1	1		
Pelecanidae					2					
Sulidae						1			5	
Phalacrocoracidae					1		2			
Anhingidae									1	
Fregatidae									5	
Spheniscidae		2		3						
Alcidae				9						
Stercorariidae			3							
Laridae			4	1	6		5		5	
Rynchopidae					1					
Total	0	3	7	14	11	13	8	7	17	0
Percent having postfledging care:	0		33		46		53		100	

[a] +, Postfledging care; 0, no postfledging care.
[b] Arctic and antartic.

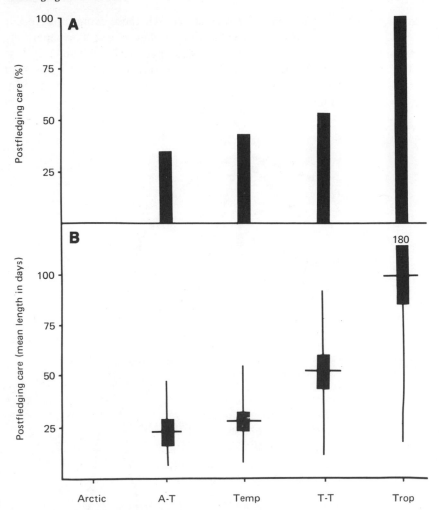

Fig. 13. (A) Percentage of seabird species having postfledging parental care as a function of latitude. A–T, Arctic–temperate species; Temp, temperate; T–T, temperate–tropical; Trop, tropical. (B) Mean length of the postfledging care period (in days) for seabirds as a function of latitude.

V. SYNTHESIS AND CONCLUSIONS

There is a great deal of variation in the nature and amount of postfledging parental care in seabirds. The important questions relative to the transition period from dependence to independence for the young deals with learning to fly, leaving the colony site, being finally abandoned by parents, and learning to forage by themselves. All these events may occur together (as in most alcids and tropicbirds), or they may occur separately (as in gulls and terns). The patterns exhibited in seabirds are shown in Fig. 14. The diversity eventually results in some recruitment into the next generation though recruitment may be very low (as in frigatebirds) or relatively high (as in sulids).

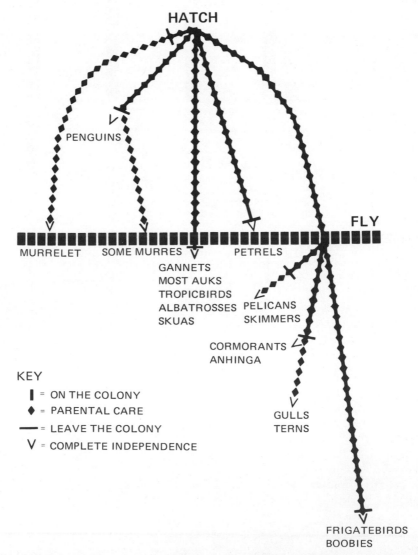

Fig. 14. Patterns of parental care in seabirds as a function of major developmental stages such as hatching, leaving the colony, fledging (first flight), and becoming independent of parents.

The pattern shown by each family and species represents a compromise imposed by its environment. From the generalizations discussed above, it is clear that postfledging parental care is present in species with large clutches and short incubation bouts, unrelated to foraging location (onshore, offshore), is directly related to the length of the prefledging parental care period, and increases in frequency and duration from the Arctic to the equator. These differences in postfledging parental care seem to relate directly to seasonal constraints and food constraints (e.g., short Arctic breeding season and nutrient poor tropical waters).

Lack (1954, 1966, 1968), Nelson (1970), and Ashmole (1971) have suggested that food is erratic and limited in the tropics. Data on food availability are generally lacking, although Jespersen (in Lack, 1968) did sample macroplankton density. Figure 15 shows macroplankton density (after Lack, 1968) for the Atlantic Ocean. Generally, the lowest food levels were present in the tropics. Tropical seabirds attempt to combat the unpredictability of the food supply by ranging over large distances, but this places constraints on incubation bout length and chick-feeding intervals. Given this constraint, families nesting in the tropics have long developmental periods and prolonged postfledging parental care periods. Even so, young frigatebirds frequently starve when completely independent. Species belonging to families that are generally more northern often have the longest postfledging care periods in the tropics. For example, in gulls and terns, the longest postfledging parental care periods occur in Lava and Swallow-tailed Gulls, Fairy and Sooty Terns, and the Black Noddy. Similarly, the gannets (temperate regions) have no postfledging care, whereas the boobies (tropical) have extended care. Tropicbirds do not fit the general pattern in that they are tropical yet have limited postfledging parental care. It is unclear why they have not developed such care, except that the prefledging period is comparatively long. These birds require further study before conclusions can be drawn.

In temperate regions food is more abundant, but seasonal constraints require specific breeding seasons. Temperate seabirds thus have larger clutches, small incubation bouts, and moderate, but variable postfledging parental care periods. The variation in parental care seems partially the result of the difficulty of the foraging technique. Thus gulls have shorter postfledging dependency periods than do terns that must learn to plunge dive, and feed on a low diversity of food types.

Going still further North, food supplies increase, but so do seasonal constraints. Most Arctic–temperate and Arctic species do not exhibit postfledging parental care. Only the gull and terns breeding in these zones exhibit such care, and the care is generally of short duration. Kittiwakes and skuas have relatively short postfledging parental care periods. The abundant food supply results in rapid development of chicks and a shortened

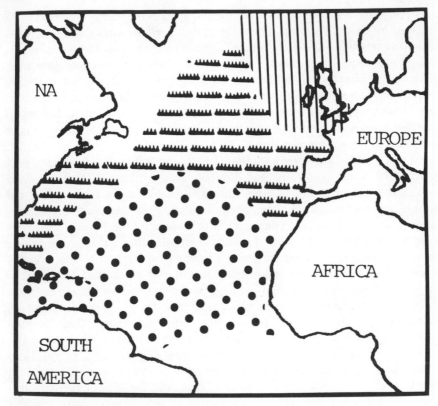

Fig. 15. Density of macroplankton in areas of the Atlantic Ocean. (Vertical lines) 7500 cm³; (horizontal marks) 1000–5000 cm³; (dots) <1000 cm³. See text for further explanation.

reproductive cycle in most species. Impending winter requires that adults lay down fat stores for the upcoming winter or for migration.

Taken together, these data suggest that seasonal and food constraints interact to determine the nature and extent of postfledging parental care in seabirds. In general, these two factors vary predictably as a function of latitude (Fig. 16). In northern regions, food is abundant in the cold waters. The long daylight hours provide sufficient opportunity for foraging. Toward the tropics, food constraints increase as food is less abundant and unpredictable. On the contrary, seasonal constraints are high in the Arctic, as breeding colony sites are often covered with ice and snow until early June, making the nesting season short. No such clear-cut seasonal constraints exist for tropical species, although food availability may be seasonal in many locations.

Postfledging parental care increases toward the tropics with respect to the percentage of species nesting in these areas having such care, and with respect to the length of that care. Summing the food and seasonal constraints produces a predictive model for the relative length of the postfledging parental care period (Fig. 16). The model predicts that the length of this period should be longer where the combined constraints are greatest.

In Fig. 16, species having postfledging parental care are averaged according to their primary breeding locations with respect to latitude. Most of these seabirds fit the model in that the length of the postfledging parental care period is longest in gulls and terns nesting in northern regions, and in frigatebirds and boobies in the tropics. Species nesting in low temperate and high tropical regions, such as cormorants, pelicans, and albatrosses, have shorter postfledging parental care periods.

Toward the ends (tropics and tundra), the constraints seem to either produce extended parental care (frigatebirds), or no parental care (gannets). Seasonal constraints can be so high as to preclude postfledging parental care, or to make it unnecessary where food is superabundant (gannets). With no seasonal constraints, the high food constraints in the tropics can preclude extended postfledging care, suggesting that parents must concentrate on foraging for themselves (tropicbirds). Tropicbirds, however,

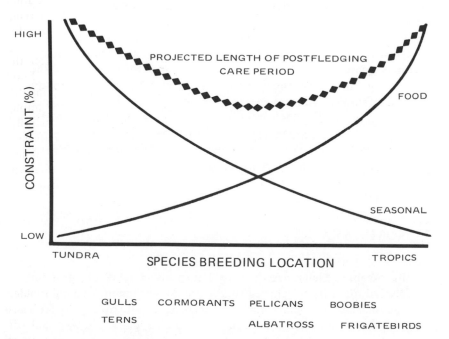

Fig. 16. Model for predicting length of postfledging care periods as a function of seasonal and food constraints. (Solid line) Relative level of constraint; (triangle line) predicted length of postfledging care period. See text for further explanation.

have long prefledging periods in which chicks may be exposed to periods without food. Presumably, this indicates that the parents are having difficulty finding enough food for themselves and their young. Nonetheless, tropicbirds have not been studied as extensively as have boobies and frigatebirds, and further study may reveal some limited postfledging parental care. Such care is easier to assess in boobies and frigatebirds because the young remain at the colony site long after they can fly. Tropicbird young, however, leave the colony site soon after they fledge. Extended postfledging care would require that parents and young maintain contact away from the colonies. Such contact would not be impossible to maintain as tern young frequently remain with parents while migrating hundreds of miles.

Tropicbirds are the only tropical-nesting seabirds that lack long postfledging care periods. This in itself suggests either intense food pressure on parents or a lack of information on the part of investigators.

Alcids, petrels, and penguins are not shown on the postfledging parental care model shown in Fig. 16, since they lack this care. Alcids and penguins generally experience seasonal constraints with relatively low food constraints. Petrels solve the problem of intense food constraints by having relatively long prefledging developmental periods with slow growth rates for their body weight (see Ricklefs, 1973).

Thus, in summary, the presence of parental care dispensed after the chicks learn to fly is a result of food and seasonal constraints. Food constraints relate both to the quantity of food (as in the tropics) and the difficulty of the foraging technique (plunge diving). Variations exist among and within families with respect to the amount of such care. Generally, petrels, penguins, and alcids have no postfledging parental care, albatrosses and tropicbirds have limited care, pelicans, cormorants, anhingas, skuas, gulls, terns, and skimmers have extensive care, and boobies and frigatebirds regularly have postfledging parental care periods of 4 months to a year.

VI. SUMMARY

Most seabirds defer breeding until they are 3 years of age or older. This deferred maturity is believed to allow sufficient time for young to learn to forage successfully before they are themselves attempting to feed offspring. Studies with pelicans, penguins, cormorants, gulls, and terns have all indicated that young are less successful using most foraging methods when compared to adults. Factors contributing to their lower success include their failure to locate food sources, lower attempt rates, and lower success rates. Presumably, delayed maturity provides young with sufficient time to learn or discover foraging areas and to perfect hunting techniques. That young forage less successfully predicts that the period from their complete dependence on parents until independence would be particularly critical in that they are vulnerable to starvation while they learn to forage. The transition to independence and the postfledging period were discussed in this chapter by families and comparisons were made among families.

The transition period from complete dependence to independence involves protection and provisioning by parents and learning of foraging areas and techniques by young. Major developmental stages for the young involve leaving the nest, vacating the colony, learning to fly, and being left by parents. These stages may occur together or separately, and great variations exist within and among families with respect to these developmental stages.

In this chapter fledging is defined as the first flight of the young. Postfledging parental care thus refers to any provisioning by parents after the chicks have learned to fly, regardless of the location of that care. Presumably, such care is a compromise between increasing the survival prospects of the young and not decreasing the longevity of the parents. In general, postfledging parental care is lacking in penguins, petrels, and alcids; occurs for up to about 1 month in pelicans, skuas, tropicbirds, and albatrosses; and is extended for more than 1 month after fledging in cormorants, terns, gulls, boobies, and frigatebirds.

Penguin (Spheniscidae) young are fed at the nest site by their parents until they both leave the colony area. Departure and independence occur at the same time, although the young may be only one-third the adult weight. There is no evidence of either a desertion period or postfledging care by parents.

Albatross (Diomedeidae) chicks are brooded for several weeks, guarded for a few more weeks, and then fed only periodically for several more weeks before they leave the colony. In half the species examined there was some parental feeding after the chicks learned to fly (up to 40 days) before they departed from the colony. Present evidence does not substantiate postfledging care after the chicks leave the colony site.

In the other Procellariids (e.g., petrels) there is a relatively long prefledging period when the young are fed at intervals of up to 3 weeks. Food is apparently limiting, and parents have difficulty foraging. Although several authors believe that some petrel young experience desertion periods prior to fledging when the parents abandon them to starve, evidence is presented that no desertion periods exist. The desertion period is merely a long interfood period, chicks simply fledge before parents return, and parents often return with food to find that the chicks have departed. There is no evidence of postfledging care in this group.

Tropicbird (Phaethontidae) young have long prefledging periods with long intervals between feedings by their parents. Once the young learn to fly, they are fed for a few days by parents at the nest site, but thereafter there is no evidence of postfledging parental care.

Pelican (Pelecanidae) young are fed for at least 3 weeks after fledging. Initially they return to the nest site to be fed, but subsequently they are fed away from the colony. Evidence is lacking on the total time young are fed after they leave the colony.

Cormorant (Phalacrocoracidae) young are fed near the colony for at least 1 month after they learn to fly. One record of a young shag being fed by its parents at 14 months of age suggests that extended parental care may exist in this family, but it requires further study. Anhinga (Anhingidae) young are known to be fed for at least 2 weeks after they can fly. However, the difficulty of their foraging method suggests that further study will show that Anhinga young are provisioned by parents for several weeks after they learn to fly.

In the Sulidae, the gannets experience no postfledging parental care, whereas the boobies have extended care of up to 1 year. The shortened developmental period, heavy fledging weight, and lack of postfledging care in gannets are attributable to the dependable, abundant food reserves of temperate waters. The long pre- and postfledging care periods of the tropical boobies are attributable to the erratic, ephemeral, and scarce food reserves of the tropics.

Frigatebird (Fregatidae) young exhibit the longest postfledging care periods of any seabird. The mean postfledging care period of the five frigatebirds is over 5 months. Such long periods of dependence relate to the general paucity of food in tropical waters where they nest.

There are three developmental patterns in alcids (Alcidae): four species are precocial and young leave the nest site with their parents when only 2 or 3 days old to travel to foraging grounds; three semiprecocial species leave the nest site at about 2 weeks when less than half-grown to travel with their parents to foraging areas; and 15 species leave the nest site independent of parents when fully grown and able to fly. There is no evidence of any postfledging care in any alcids.

The six species of skuas (Stercorariidae) have short postfledging care periods of less than 1 month. Presumably nesting at high latitudes precludes extended postfledging parental care at the colony site, and there is no evidence of such care away from the colonies.

Little information is available on skimmer young (Rynchopidae), but the evidence suggests that young are fed for at least 2 weeks. Perhaps they are fed for considerably longer periods, as young and adults migrate together.

In the Laridae, gull and tern chicks are fed for at least 1 month in and around the colony site. The length of the postfledging care period relates directly to the prefledging care period in gulls, and such data are lacking for terns. Evidence away from the colony sites for gulls and terns indicates that gulls may be fed for a few months, and terns may be fed for several months. Presumably these differences are attributable to the difficulty of the plunge-diving technique of terns.

There are some generalizations that emerge from comparisions within and among families. The length of the postfledging care period, for species having such care, directly relates to the length of the prefledging care period. Presumably, the factors (such as low food supply) that influence development rates before fledging also influence postfledging parental care. Species without parental care have low clutch sizes and long incubation bouts, indicating food stress. Species with care have larger clutches and shorter incubation bouts, indicating less food stress. However, of those with such postfledging care, the species with the longest periods of care have low clutches and long incubation bouts. There is no relationship between postfledging parental care and either body weight or location of foraging areas (inshore versus offshore). The percentage of nesting species having postfledging parental care increases from the north to the tropics. All Arctic-nesting species lack postfledging care, and all strictly tropical species have postfledging parental care.

A model is presented that predicts the length of the postfledging parental care period as a function of seasonal and food constraints. Such care is expected to be the longest where the constraints are highest. In general the presence and amount of parental care relates to the quantity of food, the difficulty of foraging techniques, and to seasonal constraints.

ACKNOWLEDGMENTS

I especially wish to thank M. Gochfeld for valuable discussions and arguments during the development and writing of this chapter; without his help, attention, and love, this chapter would have been written much earlier. Several people discussed various aspects of the topic with me over the years and I am grateful: C. Beer, F. Buckley, P. Buckley, M. Fitch, F. Lesser, B. Murray, J. Ryder, J. Shisler, G. Shugart, and W. Southern. The research on Herring and Laughing Gulls was partially funded by the Penrose Fund of the American Philosophical Society, The Research Council of Rutgers University, and the New Jersey State Mosquito Commission. Logistical support and encouragement was provided by the Ocean County Mosquito Commission. The research on Franklin's Gulls was funded by the Frank M. Chapman Fund of the American Museum of Natural History and Sigma Xi.

REFERENCES

Allan, R. G., 1962, The Madeiran Storm Petrel *Oceandroma castro*, *Ibis* **103b**:274–296.

Ansingh, F. H., Kuelen, H. J., Van der Werf, P. S., and Voous, K. H., 1960, The breeding of the Cayenna or Yellow-billed Sandwich Tern in Curacao in 1958, *Ardea* **48**:51–65.

Ashmole, M. J., and Ashmole, N. P., 1968, The use of food samples from sea birds in the study of seasonal variation in the surface fauna of tropical oceanic areas, *Pac. Sci.* **22**:1–10.

Ashmole, N. P., 1962, The Black Noody *Anous tenuirostris* on Ascension Island. Part I. General biology, *Ibis* **103b**:235–273.

Ashmole, N. P., 1963, The biology of the Wideawake or Sooty Tern *Sterna fuscata* on Ascension Island, *Ibis* **103b**:197–364.

Ashmole, N. P., 1968, Breeding and molt in the White Tern (*Gygis alba*) on Christmas Island, Pacific Ocean, *Condor* **70**:35–55.

Ashmole, N. P., 1971, Seabird ecology and the marine environment, in: *Avian Biology* (D. S. Farner and J. R. King, eds.), Vol. 1, pp. 224–286, Academic Press, New York.

Ashmole, N. P., and Tovar S. H., 1968, Prolonged parental care in Royal Terns and other birds, *Auk* **85**:90–100.

Austin, O. L., Jr., 1961, *Birds of the World*, Golden Press, New York, N.Y.

Baggerman, G., Baerends, G. P., Heikens, H. S., and Mook, J. H., 1956, Observations on the behavior of the Black Tern, *Chlidonias n. niger* (L.), in the breeding area, *Ardea* **44**:1–71.

Barash, D. P., Donovan, P., and Myrick, R., 1975, Clam dropping behavior of the Glaucous-winged Gull (*Larus glaucescens*), *Auk* **87**:60–64.

Beck, J. R., and Brown, D. W., 1971, The breeding biology of the Black-bellied Storm-petrel *Fregatta tropica*, *Ibis* **113**:73–89.

Bédard, J., 1969, Adaptive radiation in Alcidae, *Ibis* **111**:189–197.

Bent, A. C., 1919, *Life Histories of North American Diving Birds*, U. S. National Museum Bulletin, Smithsonian Institution, Washington, D.C.

Bent, A. C., 1921, *Life Histories of North American Gulls and Terns*, U.S. National Museum Bulletin, Smithsonian Institution, Washington, D.C.

Bent, A. C., 1922, *Life Histories of North American Petrels, Pelicans and Their Allies*, U.S. National Museum Bulletin, Smithsonian Institution, Washington, D.C.

Bergman, G., 1956, Beteendestudier över ungar av skrantarna (*Hydro-progne tschegrava*), *Var Fagelvarld* **15**:223–245.

Birkhead, T. R., 1977, Adaptive significance of the nesting period of Guillemots *Uria aalge*, *Ibis* **119**:544–549.

Birkhead, T. R., 1978, Behavioral adaptations to high density nesting in the Common Guillemot *Uria aalge*, *Anim. Behav.* **36**:321–331.

Boersma, P. D., 1977, An ecological and behavioral study of the Galapagos Penguins, *Living Bird* **15**:43–93.

Belopol'skii, L. O., 1957, *Ecology of Sea Colony Birds of the Barents Sea*, Israel Program for Scientific Translation, Jerusalem.

Buckley, F. G., and Buckley, P., 1972, The breeding ecology of Royal Terns *Sterna* (*Thalasseus*) *maxima maxima*, *Ibis* **114**:344–359.

Buckley, F. G., and Buckley, P., 1974, Comparative feeding ecology of wintering adult and juvenile Royal Terns (Aves: Laridae, Sterninae), *Ecology* **55**:1033–1063.

Burger, J., 1972, Dispersal and post-fledging survival of Franklin's Gull, *Bird-Banding.* **43**:267–275.

Burger, J., 1974, Breeding adaptations of Franklin's Gull (*Larus pipixcan*) to a marsh habitat, *Anim. Behav.* **22**:521–567.

Burger, J., 1978, Competition between Cattle Egrets and native North American herons, egrets, and ibises, *Condor* **80**:15–23.

Burger, J., 1979, Herring Gull versus Laughing Gull: Competition and predation, *Condor* **81**:1269–1277.

Burger, J., in press, On becoming independent in Herring Gulls: Parent–young conflict, *Am. Nat.*

Burger, J., and Lesser, F., 1978, Selection of colony sites and nest sites by Common Terns *Sterna hirundo* in Ocean County, New Jersey, *Ibis* **120**:443–449.

Burger, J., and Shisler, J., 1978, Nest site selection and competitive interactions of Herring and Laughing Gulls in New Jersey, *Auk* **95**:252–266.

Burger, J., Miller, L., and Hahn, D.C., 1978, Behavior and sex roles of nesting Anhingas at San Blas, Mexico, *Wilson Bull.* **90**:359–375.

Burke, V. E. M., and Brown, L. J., 1970, Observations on the breeding of the Pink-backed Pelican *Pelecanus rufescens*, *Ibis* **112**:499–512.

Brown, W. Y., 1975, Parental feeding of young Sooty Terns [*Sterna fuscata* (L)] and Brown Noddies [*Anous stolidus* (L)] in Hawaii, *J. Anim. Ecol.* **44**:731–742.

Brown, W. Y., 1976, Growth and fledging age of Sooty Tern chicks, *Auk* **93**:179–183.

Carrick, R., Keith, K., and Gwynn, A. M., 1960, Fact and fiction on the breeding of the Wandering Albatross, *Nature (London)* **185**:112–114.

Coulson, J. C., and White, E., 1958, Observations on the breeding of the Kittiwake, *Bird Study* **5**:74–83.

Cramp. S., and Simmons, K. E. L., 1977, *Handbook of the Birds of Europe, The Middle East and North Africa.* Vol. I. *Ostrich to Ducks*, Oxford University Press, Oxford.

Cromwell, T., 1953, Circulation in a meridional plan in the central equatorial Pacific, *J. Mar. Res.* **12**:196–213.

Cullen, J. M., and Ashmole, N. P., 1963, The Black Noddy *Anous tenuirostris* on Ascension Island. Part 2. Behaviour, *Ibis* **103b**:423–457.

Davis, P., 1947, The breeding of the Storm Petrel, *Br. Birds* **50**:371–384.

Davis, J. W. F., 1975, Specialization in feeding location by Herring Gulls, *J. Anim. Ecol.* **44**:795–804.

Dement'ev, G. P., Gladkov, N. A., and Spangenberg, E. P., 1951, *Ptitsy Sovetskogo Soyuza, Moskua [Birds of the Soviet Union]*, Israel Program of Scientific Translation, Jerusalem, transl.: 1969).

Diamond, A. W., 1971, The ecology of seabirds breeding at Aldabra Atoll, Indian Ocean, Unpublished Ph.D. thesis, Aberdeen University, Scotland.

Diamond, A. W., 1973, Notes on the breeding biology and behavior of Magnificent Frigatebird, *Condor* **75**:200–209.

Diamond, A. W., 1975, Biology and behavior of frigatebirds *Fregate* spp. on Aldabra Atoll, *Ibis* **117**:302–323.

Din, H. A., and Eltringham, S. K., 1974, Breeding of the Pink-backed Pelican *Pelecanus rufescens* in Rwenzori National Park, Uganda, *Ibis* **116**:477–493.

Dorward, D. F., 1962a, Comparative biology of the White Booby and the Brown Booby *Sula* spp. at Ascension, *Ibis* **103b**:174–220.

Dorward, D. F., 1962b, Behavior of boobies *Sula* spp., *Ibis* **103**:221–234.

Dorward, D. F., 1963, The Fairy Tern *Gygis alba* on Ascension Island, *Ibis* **103b**:365–378.

Dorward, D. F., and Ashmole, N. P., 1963, Notes on the biology of the Brown Noddy *Anous stolidus* on Ascension Island, *Ibis* **103b**:447–457.

Drent, R. H., and Viguet, C. J. G., 1961, A catalogue of British Columbia seabird colonies, *Occ. Pap. B.C. Prov. Mus.* **12**:173 pp.

Drent, R. H., 1965, Breeding biology of the Pigeon Guillemot, *Cepphus columba*, *Ardea* **93**:99–159.

Drent, R. H., VanTets, G. F., Tompa, F., and Vermeer, K., 1964, The breeding birds of Mandarte Island, British Columbia, *Can. Field-Nat.* **78**:208–263.

Drury, W. H., Jr., and Smith, W. J., 1968, Defense of feeding areas by adult Herring Gulls and intrusion by young, *Evolution* **22**:193–201.

Dunn, E. K., 1972, Effect of age on the fishing ability of Sandwich Terns *Sterna sandvicensis*, *Ibis* **114**:360–366.

Erwin, R. M., 1977*a*, Black Skimmer breeding ecology and behavior, *Auk* **94**:709–717.

Erwin, R. M., 1977*b*, Foraging and breeding adaptations to different food regimes in three seabirds: The Common Tern, *Sterna hirundo*, Royal Tern *Sterna maxima*, and Black Skimmer, *Rynchops niger*, *Ecology* **58**:389–397.

Escalante, R., 1968, Notes on the Royal Tern in Uruguay, *Condor* **70**:243–247.

Feare, C. J., 1975, Post-fledging parental care in Crested and Sooty Terns, *Condor* **77**:368–370.

Feare, C. J., 1976, The breeding of the Sooty Tern *Sterna fuscata* in the Seychelles and the effects of experimental removal of eggs, *J. Zool. Lond.* **179**:317–360.

Fisher, J., and Lockley, R. M., 1954, *Sea-birds*, Houghton-Mifflin, Boston.

Fleet, R. R., 1974, The Red-tailed Tropicbird on Kare Atoll, *AOU Monogr.* **16**:1–64.

Fordham, R. A., 1964, Breeding biology of the Southern Black-backed Gull: 1: Pre-egg stage and egg stage, and 2: Chick stage, *Notornis* **11**:3–34, 110–126.

Fordham, R. A., 1968, Dispersion and dispersal of Domincan Gull in Wellington, New Zealand, *Proc. N.Z. Ecol. Soc.* **15**:40–50.

Gison, J. D., 1963, Third report of the New South Wales Albatross Study Group (1962) summarizing activities to date, *Emu* **63**:215–223.

Gibson-Hill, G. A., 1949, Notes on the nesting habits of seven representative tropical sea birds, *J. Bombay Nat. Hist. Soc.* **48**:214–235.

Glauert, L., 1946, The Little Shearwater's year, *Emu* **46**:187–192.

Greenwood, J., 1964, The fledging of the Guillemot *Uria aalge* with notes on the Razorbill *Alca torda*, *Ibis* **106**:469–481.

Hall, E. R., 1925, Pelicans versus fishes in Pyramid Lake, *Condor* **27**:147–160.

Harriott, M. C.. 1970, Breeding behavior of the Anhinga, *Fl. Nat.* **43**:138–142.

Harris, M. P., 1965, The food of some *Larus* Gulls, *Ibis* **107**:43–51.

Harris, M. P., 1966, Breeding biology of the Manx Shearwater *Puffinus puffinus*, *Ibis* **108**:17–33.

Harris, M. P., 1969, Food as a factor controlling the breeding of *Puffinus lherminieri*, *Ibis* **111**:139–156.

Harris, M. P., 1970*a*, The biology on endangered species, the Dark-rumped Petrel (*Pterodroma phaeopygia*), in the Galapagos Islands, *Condor* **72**:76–84.

Harris, M. P., 1970*b*, Breeding ecology of the Swallow-tailed Gull, *Creagrus furcatus*, *Auk* **87**:215–243.

Harris, M. P., 1973, The biology of the Waved Albatross *Diomedea irrorata* of Hood Island, Galapagos, *Ibis* **115**:483–510.

Harris, M. P., 1978, Supplementary feeding of young puffins, *Fratercula arctica*, *J. Anim. Ecol.* **47**:15–23.

Haycock, K. A., and Threlfall, W., 1975, The breeding biology of the Herring Gull in Newfoundland, *Auk* **92**:678–697.

Hofslund, P. B., 1959, Fall migration of Herring Gulls from Knife Island, Minnesota, *Bird-Banding* **30**:104–114.

Hunt, G. L., Jr., and Hunt, M. W., 1973, Habitat partitioning by foraging Gulls in Maine and Northwestern Europe, *Auk* **90**:827–839.

Imber, M. J., 1976, The breeding biology of the Grey-faced Petrel *Pterodroma macroptera gouldi*, *Ibis* **118**:51–64.

Ingolfsson, A., and Estrella, J. T., 1978, The development of shellcracking behavior in Herring Gulls, *Auk* **95**:577–579.

Jarvis, M. J. E., 1974, The ecological significance of clutch size in the South African Gannet (*Sula capensis* Lichtenstein), *J. Anim. Ecol.* **43**:1–17.

Kepler, C. B., 1969, *Breeding Biology of the Blue-faced Booby Sula dactylatra personata on Green Island, Kure Atoll*, Nuttall Pub. No. 8 (R. Paynter, Jr., ed), Cambridge, Mass.

Kury, C. R., and Gochfeld, M., 1975, Human interference and gull predation in cormorant colonies, *Biol. Conserv.* **8**:23–34.

Lack, D., 1954, *The Natural Regulation of Animal Numbers*, Clarendon Press, Oxford.

Lack. D., 1966, *Population Studies of Birds*, Clarendon Press, Oxford.

Lack, D., 1967, Interrelationships in breeding adaptation as shown by marine birds, *Proc. 14th Int. Ornith. Congr.* **1967**:3–42.

Lack, D., 1968, *Ecological Adaptations for Breeding in Birds*, Methuen, London.

LeCroy, M., 1972, Young Common and Roseate Terns learning to fish, *Wilson Bull.* **84**:201–202.

Lockley, R. M., 1974, *Ocean-Wanderers*, Stackpole Books, Harrisburg, Pa., 168 pp.

Maher, W. J., 1970, Ecology of the Long-tailed Jaeger at Lake Hazen, Ellesmere Island, *Arctic* **23**:112–129.

Marshall, A. J., and Serventy, D. L., 1956, The breeding cycle of the Short-tailed Shearwater, *Puffinus tenuirostris* (Temminck) in relation to transequatorial migration and its environment, *Proc. Zool. Soc. London* **127**:489–510.

Massey, B. W., 1974, Breeding biology of the California Least Tern, *Proc. Linn. Soc. N.Y.* **72**:1–24.

Maunder, J. E., and Threlfall, W., 1972, The breeding biology of the Black-legged Kittiwake in Newfoundland, *Auk* **89**:789–816.

Miller, L., 1940, Observations on the Black-footed Albatross, *Condor* **42**:227–238.

Monroe, B. L., Jr., 1956, Observations of Elegant Terns at San Diego, California, *Wilson Bull.* **68**:239–244.

Morrison, M. L., Slack, R. D., and Shanley, E., Jr., 1978, Age and foraging ability relationships of Olivaceous Cormorants, *Wilson Bull.* **90**:414–422.

Moyle, P., 1966, Feeding behavior of Glaucous-winged Gull on an Alaskan salmon stream, *Wilson Bull.* **78**:175–190.

Moynihan, M., 1959, A revision of the family Laridae (*Aves*), *Am. Mus. Novit.* **1928**:1–42.

Murphy, E. C., 1979, Seasonal variation in reproductive output of House Sparrows: The Determination of clutch size, *Ecology* **59**:1189–1199.

Murphy, R. C., 1936, *Oceanic Birds of South America*, Vol. II, MacMillan, New York.

Nelson, J. B., 1966, The breeding biology of the Gannet *Sula bassana* on the Bass Rock, Scotland, *Ibis* **108**:584–626.

Nelson, J. B., 1967a, The breeding behaviour of the White-Footed Booby *Sula dactylatra*, *Ibis* **109**:194–231.

Nelson, J. B., 1967b, Etho-ecological adaptations in the Great Frigatebird, *Nature (London)* **214**:318.

Nelson, J. B., 1968, *Galapagos: Islands of Birds*, Morrow, New York, 335 pp.

Nelson, J. B., 1969a, The breeding behavior of the Red-footed Booby *Sula sula*, *Ibis* **111**:357–385.

Nelson, J. B., 1969b. The breeding ecology of the Red-Footed Booby in the Galapagos, *J. Anim. Ecol.* **38**:181–198.

Nelson, J. B., 1970, The relationship between behavior and ecology in the Sulidae with reference to other sea birds, *Oceanogr. Mar. Biol. Annu. Rev.* **8**:501–574.

Nelson, J. B., 1975a. The breeding biology of frigatebirds—A comparative review, *Living Bird* **14**:113–156.

Nelson, J. B., 1975b, Functional aspects of behavior in the Sulidae, in: *Function and Evolution in Behavior* (G. Baerends, C. Beer, and A. Manning, eds.), pp. 313–330, Clarendon Press, Oxford.

Nettleship, D. N., 1972, Breeding success of the Common Puffin (*Fratercula arctic* L.) on different habitats at Great Island, Newfoundland, *Ecol. Monogr.* **42**:239–268.

Nisbet, I. C. T., 1973, Courtship feeding, egg-size and breeding success in Common Terns (*Sterna hirundo*), *Nature (London)* **241**:141–142.

Oliver, W. P. B., 1955, *New Zealand Birds*, A. H. and A. W. Reed, Wellington, Australia.

Orians, G. H., 1969, Age and hunting success in the Brown Pelican (*Pelecanus occidentalis*), *Anim. Behav.* **17**:316–319.

Palmer, R. S., 1941, A behavior study of the Common Tern (*Sterna hirundo hirundo* L.), *Proc. Boston Soc. Nat. Hist.* **42**:1–119.

Palmer, R. S., 1962, *Handbook of North American Birds*, Vol I, Yale University Press, New Haven, Conn. 567 pp.

Parker, G. A., and MacNair, M. R., 1978, Models of parent–offspring conflict. I. Monogamy, *Anim. Behav.* **26**:97–110.

Parsons, F., and Duncan, N., 1978, Recoveries and dispersal of Herring Gulls from the Isle of May, *J. Anim. Ecol.* **47**:993–1005.

Pennycuick, C. J., 1956, Observations on a colony of a Brunnich's Guillemot *Uria lomvia* in Spitsbergen, *Ibis* **78**:80–99.

Potts, G. R., 1969, The influence of eruptive movements, age, population size and other factors on the survival of the Shag (*Phalacrocorax aristotelis* L.), *J. Anim. Ecol.* **38**:53–102.

Recher, H. F., and Recher, J. A., 1969, Comparative foraging efficiency of adult and immature Little Blue Herons (*Floride caerulea*), *Anim. Behav.* **17**:320–322.

Reilly, P. N., and Balmford, P., 1975, A breeding study of the Little Penguin *Eudyptula minor* in Australia, in: *The Biology of Penguins* (B. Stonehouse, ed.), pp. 161–187, University Park Press, London.

Richdale, L. E., 1943a, The Kuaka or Diving Petrel, *Pelecanoides urinatrix* (Gmelin), *Emu* **43**:24–48.

Richdale, L. E., 1943b, The White-faced Storm Petrel. Part 1, *Trans. R. Soc. N.Z.* **73**:97–115.

Richdale, L. E., 1943c, The White-faced Storm Petrel. Part II, *Trans, R. Soc. N.Z.* **73**:335–350. **73**:335–350.

Richdale, L. E., 1945, Supplementary notes on the Diving Petrel, *Trans, R. Soc. N.Z.* **75**:42–53.

Richdale, L. E., 1954, The starvation theory in Albatrosses, *Auk* **71**:239–252.

Richdale, L. E., 1957, *A Population Study of Penguins*, Claredon Press, Oxford.

Richdale, L. E., 1963, *Biology of the Sooty Shearwater*, *Proc. Zool. Soc. London* **141**:1–117.

Rice, D. W., and Kenyon, K. W., 1962, Breeding cycles and behavior of Laysan and Black-footed Albatrosses, *Auk* **79**:517–567.

Ricklefs, R. E., 1968, Patterns of growth in birds, *Ibis* **110**:419–541.

Ricklefs, R. E., 1973, Patterns of growth in birds. II. Growth rate and mode of development, *Ibis* **115**:177–201.

Ricklefs, R. E., 1977, On the evolution of reproductive strategies in birds: reproductive effort, *Am. Nat.* **111**:453–478.

Roberts, B., 1934–1937, The life cycle of Wilson's Petrel *Oceanites oceanicus* (Kuhl), *Br. Graham Land Exped. Sci. Rep.* **1**:141–194.

Roggerman, W., 1970, The migration of *Larus ridibundus* ringed as chicks in the north of Belgium, *Die Giervalk* **60**:301–321.

Rowan, M. K., 1952, The Greater Shearwater *Puffinus gravis* at its breeding grounds, *Ibis* 94:97–121.

Schaller, G. B., 1964, Breeding behavior of the White Pelican at Yellowstone Lake, Wyoming, *Condor* 66:3–23.

Schreiber, R. W., 1968, Seasonal population fluctuations of Herring Gulls in Central Maine, *Bird-Banding* 39:81–106.

Schreiber, R. W., and Ashmole, N. P., 1970, Sea-bird breeding seasons on Christmas Island, Pacific Ocean, *Ibis* 112:361–394.

Schwartz, L. G., 1966, Seacliff Birds of Cape Thompson, in: *Environment of Cape Thompson Region, Alaska* (N. J. Wilimorsky and J. N. Wolfe, eds.), U.S. Atomic Energy Commission, Oak Ridge.

Sealy, S. G., 1968, A comparative study of breeding ecology and timing in plankton-feeding alcids (*Cyclorrhynchus* and *Aethia* spp) St. Lawrence Island, Alaska, M.Sc. thesis, University of British Columbia, Vancouver, 193 pp.

Sealy, S. G., 1972, Adaptive differences in breeding biology in the marine bird family Alcidae, Unpublished Ph.D. thesis, University of Michigan, Ann Arbor.

Sealy, S. G., 1973a, Breeding biology of the Horned Puffin on St. Lawrence Island, Bering Sea, with zoogeographical notes on the North Pacific Puffins, *Pac. Sci.* 27:99–119.

Sealy, S. G., 1973b, Adaptive significance of post-fledging developmental patterns in growth rates in the Alcidae, *Ornis Scand.* 4:113–121.

Sealy, S. G., 1973c, Interspecific feeding assemblages of marine birds off British Columbia, *Auk* 90:796–800.

Sealy, S. G., 1974, Breeding phenology and clutch size in the Marbled Murrelet, *Auk* 91:10–23.

Sealy, S. G., and Bédard, J., 1973, Breeding biology of the Parakeet Auklet (*Cyclorrhynchus psittacula*) on St. Lawrence Island, Alaska, *Astarte* 6:59–68.

Searcy, W. A., 1978, Foraging success in three age classes of Glaucous-winged Gulls, *Auk* 95:587–588.

Serventy, D. L., 1966, Aspects of the population ecology of the Short-tailed Shearwater *Puffinus tenuirostris*, *Proc. 14th Int. Ornithol. Congr.* (D. W. Snow, ed.), pp. 165–190, Blackwell, Oxford.

Serventy, D. L., Serventy, V., and Warham, J., 1971, *The Handbook of Australian Sea-birds*, A. H. and A. W. Reed, Sydney, Australia, 254 pp.

Simmons, K. E. L., 1967, Ecological adaptations in the life history of the Brown Booby at Ascension Island, *Living Bird* 6:187–212.

Smith, J. M., 1975, Studies of breeding Sandwich Terns, *Br. Birds* 68:142–156.

Snow, B. K., 1960, The breeding biology of the Shag *Phalacrocorax aristotelis* on the island of Lundy, Bristol Channel, *Ibis* 102:554–575.

Snow, B. K., 1963a, The behavior of the Shag, *Br. Birds* 56:77–103.

Snow, B. K., 1963b, The behavior of the Shag, *Br. Birds* 56:164–186.

Snow, B. K., and Snow, D. W., 1967, Behavior of the Swallow-tailed Gull on the Galapagos, *Condor* 70:252–264.

Snow, B. K., and Snow, D. W., 1969, Observations on the Lava Gull *Larus fuliginosus*, *Ibis* 111:30–35.

Snow, D. W., 1965, The breeding of Audubon's Shearwater (*Puffinus lherminieri*) in the Galapagos, *Auk* 82:591–597.

Snow, D. W., and Snow, B. K. 1967. The breeding cycle of the Swallow-tailed Gull *Greagrus furcatus*, *ibis* 109:14–24.

Spaans, A. L., 1971, On the feeding ecology of the Herring Gull *Larus argentatus* Pont. in the northern part of the Netherlands, *Ardea* 59:1–188.

Spellerberg, I. F., 1971, Breeding behavior of the McCormick Skua *Catharacta maccormicki* in Antarctica, *Ardea* **59**:3–5, 180–230.

Stonehouse, B., 1956, The King Penguin of South Georgia, *Nature* (*London*) **178**:1424–1426.

Stonehouse, B., 1962, The tropic birds (genus Phaethon) of Ascension Island, *Ibis* **103b**:124–161.

Stonehouse, B., and Stonehouse, S., 1963, The frigatebird *Fregata aquila* of Ascension Island, *Ibis* **103b**:409–422.

Strang, C. A., 1973, The Alaskan Glaucous Gull (*Larus hyperboreus barrovianus* Ridgway): Autecology, Taxonomy, Behavior, Department of Forestry and Conservation, Purdue University, W. Lafayette, Indiana, 20 pp.

Taylor, R. H., 1962, The Adelie Penguin *Pygoscelis adeliae* at Cape Royds, *Ibis* **104**:176–204.

Thomson, A. L., 1964, *A New Dictionary of Birds*, McGraw-Hill, New York, 928 pp.

Thorensen, A. C., 1964, The breeding behavior of the Cassin's Auklet, *Condor* **66**:456–476.

Threlfall, W., 1978, Dispersal of Herring Gulls from the Witless Bay Sea Bird Sanctuary, Newfoundland, *Bird-Banding*, **49**:116–124.

Tickell, W. L. N., 1960, Chick feeding in the Wandering Albatross *Diomedea exulans* Linnaeus, *Nature* (*London*) **185**:116–117.

Tickell, W. L. N., 1968, The Biology of the Great Albatross *Diomedea exulans* and *Diomedea epomophora*, in: *Antarctic Bird Studies* (O. L. Austin, Jr., ed.), National Academy of Sciences, Washington, D.C.

Tomkins, I. R., 1959, Life history notes on the Least Tern, *Wilson Bull.* **71**:314–322.

Tremblay, J., and Ellison, L. N., 1979, Effects of human disturbance on breeding of Black-crowned Night Herons, *Auk* **96**:364–369.

Trivers, R. L., 1972, Parental investment and sexual selection, in: *Sexual Selection and the Descent of Man* (B. Campbell, ed.), pp. 136–179, Aldine, Chicago.

Tuck, L. M., 1961, *The Murres*, Queen's Printer, Ottawa.

Verbeek, N. A. M., 1977a, Age differences in the digging frequency of Herring Gulls on a dump, *Condor* **79**:123–125.

Verbeek, N. A. M., 1977b Comparative feeding behavior of immature and adult Herring Gulls, *Wilson Bull.* **89**:415–421.

Verbeek, N. A. M., 1977c, Comparative feeding ecology of Herring Gulls *Larus argentatus* and Lesser Black-backed Gulls *Larus fuscus*, *Ardea* **65**:25–42.

Vermeer, K., 1963, The breeding ecology of the Glaucous-winged Gull (*Larus glaucescens*) on Mandarte Island, B.C., *Occ. Papers B.C. Prov. Mus.* **13**:1–104.

Verner, J., 1961, Nesting activities of the Red-footed Booby in British Honduras, *Auk* **78**:573–594.

Warham, J., 1956, The breeding of the Great-winged Petrel *Pterodroma macroptera*, *Ibis* **98**:171–185.

Warham, J., 1958, The nesting of the Shearwater *Puffinus carneipes*, *Auk* **75**:1–14.

Warham, J., 1962, The biology of the Giant Petrel, *Macronectes giganteus*, *Auk* **79**:139–160.

Warham, J., 1963, The Rockhopper Penguin, *Eudyptes chrysocome* at Macquarie Island, *Auk* **80**:229–256.

Warham, J., 1966, Giant Petrels: The birds called "Stinkers," *Animal Kingdom* (*Bronx Zoo*) Feb. 1966.

Warham, J., 1974, The Fiordland Crested Penguin *Eudyptes pachyrhynchus*, *Ibis* **116**:1–27.

Warham, J., 1975, The Crested Penguin, in: *The Biology of Penguins* (B. Stonehouse, ed.), pp. 189–270, University Park Press, London.

Wilbur, H. M., 1969, The breeding biology of Leach's Petrel, *Oceanodroma leucorhoa*, *Auk* **86**:433–442.

Williams, G. C., 1966, Natural selection, the cost of reproduction, and a refinement of Lack's principle, *Am. Nat.* **100**:687–692.

Witherby, H. F., Jourdain, F. C. R., Ticehurst, N. F., and Tucker, B. W., 1941, *The Hand-book of British Birds*, Vol. 5, H. F. and G. Witherby, London.

Wolk, R. G., 1959, Some reproductive behavior patterns of the Black Skimmer, Unpublished Ph.D. thesis, Cornell University, Ithaca, N.Y., 78 pp.

Woodbury, A. M., and Knight, H., 1951, Results of the Pacific Gull colorbanding project, *Condor* **53**:57–77.

Young, E. C., 1963, The breeding behavior of the South Polar Skua *Catharacta maccormicki*, *Ibis* **105**:203–233.

Zino, P. A., 1971, The breeding of Cory's Shearwater *Calonectris diomedea* on the Salvage Islands, *Ibis* **113**:212–217.

Chapter 11

COMPARATIVE DISTRIBUTION AND ORIENTATION OF NORTH AMERICAN GULLS

William E. Southern

Department of Biological Sciences
Northern Illinois University
DeKalb, Illinois 60115

I. INTRODUCTION

The abundance of many gull species and consequently their distributional patterns have shown considerable variation during the last century and a half. By the mid-19th century human exploitation resulted in the populations of many seabirds, including gulls, declining to dangerously low levels. The birds and their eggs were used for food by the growing human population, feathers were processed by the millinery trade, and many islands and coastal areas were usurped for human endeavors thereby destroying nesting habitats.

Legislation passed early in the 20th century marked the beginning of a recovery period for some of the impacted species. The New England Herring Gull population responded positively to protection and the abundant food provided as waste products of an urbanizing society. Population increases were such that between 1940 and 1967 attempts were made by the U.S. Fish and Wildlife Service to control its growth (Kadlec and Drury, 1968; Kadlec, 1971). Gull populations in other parts of the continent responded similarly, although the timetable was slightly different for each (e.g., Ludwig, 1974). By the mid-1960s most gull species apparently had returned to levels of abundance similar to, or possibly even exceeding, those believed to have existed prior to human decimation. Under environmental conditions existing in the late 1970s, most gulls appear capable of maintaining their populations and a few, such as the Herring Gull and Ring-billed Gull, are still expanding their ranges.

Human modification of the landscape and the by-products of our urban society are still having an influence on gull populations and also on the seasonal distribution of most species. It is likely that the present seasonal ranges and migration routes of most gulls have been modified, and will be modified further as a consequence of man's activities. Factors envisioned as having played particularly important roles in molding the present distribution patterns of gulls include (1) proliferation of artificial ponds, lakes, and reservoirs in historically dry areas; (2) heated effluents that retard icing in winter and provide water for use by gulls; (3) agricultural practices that expose bare ground, thereby making terrestrial invertebrates available periodically to foraging gulls; (4) creation of dredge spoil islands where nesting substrate was not previously available; and (5) provision of dependable year-round food sources, such as garbage at landfills. Because of the influence of these factors, the distributional trends of most gull species are expected to differ from those that existed in the past.

Drury and Nisbet (1972) reported that the center of Herring Gull abundance in New England shifted in the 1940s from Maine to Massachusetts. They also recognized other population shifts that had occurred over 80 years in response to changes in food supply and human disturbance. Besides responses to environmental changes by local or large-scale shifts in breeding distribution, there have been at least two demonstrable changes in migration patterns (Drury and Nisbet, 1972). They concluded that young Herring Gulls now migrate, on average, less far in the first winter than they were inclined to travel 30–40 years ago. Nevertheless, the young birds on the average still travel farther than do adults of most species. Evidence provided by Drury and Nisbet (1972) for Herring Gulls indicates that young birds may remain through the summer at localities up to 350 km from their natal colony. This represents an apparent shift in behavior, as in the late 19th century, Herring Gulls were unknown in summer at this distance from the colonies that persisted in Maine (Drury and Nisbet, 1972).

We should expect further modifications in seasonal ranges as responses to changing patterns of human activity, such as changes in our waste-disposal practices. Most gull species appear to be opportunistic, quickly taking advantage of environmental changes. The published accounts of such responses are most detailed for the Herring Gull, but similarities in behavior have been documented for other species. For example, most North American gulls take advantage of landfills as a dependable food source. In the East, Glaucous Gulls, Great Black-backed Gulls, Herring Gulls, Ring-billed Gulls, and Laughing Gulls habitually visit landfills (W. E. Southern, personal observations). In the West, Glaucous-winged Gulls, Western Gulls, California Gulls, and Mew Gulls, in addition to the eastern forms that overlap with them, all visit operational landfills (W. E. Southern, personal observations). The growing dependence of many gull species on this particular foraging strategy tends to concentrate the birds during winter, results in more gulls being able to forage successfully at northern latitudes, and undoubtedly intensifies competitive interactions between species that may ultimately influence future distributional patterns.

The seasonal distributions of the various gull species across North America pose a variety of stimulating ecological questions as a result of their responses to a changing environment. Obvious questions arise about the ways in which the various species partition resources. The spatial distribution of the respective species influences the extent to which they may compete for colony sites, seasonal food resources, and roosting areas. In order to fully appreciate the extent to which the various species might interact, it is essential to conceive their continentwide distribution during all seasons of the year. Data presented in this chapter represent the initial phase of such an effort.

This chapter compares the seasonal distribution of eight species of North American gulls. Seven of these were selected because sufficient band-recovery data were recovered for denoting seasonal trends in their ranges. More than 1100 recoveries were available for seven of the eight species: Glaucous-winged Gull, Great Black-backed Gull, Western Gull, Herring Gull, California Gull, Ring-billed Gull, and Laughing Gull (Table I). The eighth one, Franklin's Gull, for which there were but 244 recoveries, was included in the discussion of breeding and postbreeding ranges because of its prevalence at inland locations. It was excluded from the monthly maps depicting seasonal patterns (Fig. 2) because the percentages were biased by

Table I. Number of Band Recoveries Available for Each Gull Species—Through August 1978

Common name	Scientific name	AOU No.	Recoveries (*n*)
Black-legged Kittiwake	*Rissa tridactyla*	0400	24
Glaucous Gull	*Larus hyperboreus*	0420	3
Glaucous-winged Gull	*L. glaucescens*	0440	17,226
Great Black-backed Gull	*L. marinus*	0470	1,156
Western Gull	*L. occidentalis*	0490	1,516
Lesser Black-backed Gull	*L. fuscus*	0500	2
Herring Gull	*L. argentatus*	0510	32,240
California Gull	*L. californicus*	0530	2,734
Ring-billed Gull	*L. delawarensis*	0540	21,458
Mew Gull	*L. canus*	0550	18
Laughing Gull	*L. atricilla*	0580	1,521
Franklin's Gull	*L. pipixcan*	0590	244
Bonaparte's Gull	*L. philadelphia*	0600	1

the low sample sizes. No attempt was made on this occasion to discern the seasonal distribution for each age class of the various species. The primary purpose of this chapter is to compare the annual ranges of the eight species. More detailed analysis of each species' distribution is left up to the investigators specifically working on each species.

Although various investigators have analyzed the band-recovery data for regional populations of several of the species discussed in this chapter, I believe this is the first time that anyone has managed to analyze all the available recoveries for the entire continent. The feasibility of undertaking such a task was enhanced by a computerized sorting and mapping procedure developed in cooperation with Jerrold H. Zar at Northern Illinois University. Without this advancement, the task would not have been undertaken.

II. DATA SOURCES AND PROCEDURES

The banding data available for all North American species of gulls were obtained from the U.S. Fish and Wildlife Service and the Canadian Wildlife Service, following approval by the various bird banders. The data set was complete through August 1978. The data were statistically summarized, and geographic locations for recoveries were mapped by a computer program that accepts data in the format of the recovery data card used by the U.S. Fish and Wildlife Service. The input data can then be sorted according to one or more of the record card characteristics—in this case, month of the year—prior to submitting them to the computer program. The details pertaining to this program have been summarized by Zar and Southern (1977).

The computer plots each recovery location on a map of North America (from 19° to 59°N latitude and from 52° to 125°W longitude). The number of gulls recovered at each set of coordinates is printed on a map. If any data are for individuals outside the range of the map, they are counted and listed, but are not plotted. The program also provides for great-circle calculations of the distance and direction between banding and recovery sites.

The maps provided in this chapter are summaries of those produced by the computer. The basic map format is quantitative. The map area is divided into zones, most of which are 6° square. The exceptions are on the edges of the map. For reference purposes, the zones are assigned letters along each axis, which allows each to be designated by a set of letters (e.g., A'A). For the species and month in question, each map shows the proportion of the month's total band recoveries for the species that occurred in each zone of the map. The total number of recoveries actually plotted on each map does not agree with the total number of band recoveries analyzed because some had to be culled for various reasons (e.g., lack of complete coordinates).

Various biases are inherent to the use of banding data, and it is impossible to compensate for all of these. For example: (1) some species have been banded more intensively than others; (2) particular studies have concentrated on the live capture of large numbers of gulls, which has increased the number of recoveries; and (3) people report bands, and hence the frequency of bird recoveries is often highest near human population centers. The combined effects of these factors make it impossible to determine the exact proportion of a gull population that occurs at any particular region solely on the basis of band-recovery data. Nevertheless, this approach represents the best procedure we presently have for sampling the populations of a variety of species across the entire continent. The resulting figures probably provide reasonably accurate indicators of trends in distribution, even though the numbers generated cannot be considered absolute.

The maps (Fig. 2) delineating the general breeding range of each species have been prepared on the basis of descriptions provided in the *AOU Check-list of North American Birds* (1957). No attempt has been made to update this information, although several studies have produced supplemental information for some species and localities.

The seasonal distributions of several North American gull species have been examined in some detail but other species have received little attention. Most investigators have addressed the problem of gull migration on a single species basis or with reference to populations of specific geographic origin (e.g., a specific colony). Both types of studies have resulted in useful information, but particularly interesting questions require an overview of the seasonal distribution of the various gull species across the continent. No attempt has been made to gleen information from all the regional studies, as this would detract from the main goal of this chapter—to provide a survey of the continental patterns of gull distribution.

III. BREEDING RANGES

The populations of the various gull species are most concentrated during the breeding season. At this time a large proportion of the adults and subadults are either at the breeding sites or have moved closer to the breeding range than they were during the postbreeding period. Even birds that are not going to breed seem to express tnedencies to move closer to the breeding range during the spring months. This group is composed of subadults, particularly individuals reared the preceding year, that may have moved farther from the breeding sites during the postbreeding period than do most adults.

Because representatives of all age classes show tendencies of migrating back to, or toward, the breeding sites, this period in the annual cycle depicts the most compacted distributional pattern. It stands out as the logical starting point for a general comparison of the seasonal distribution of the eight species under consideration.

A variety of factors differentially affect the dates on which representatives of the various gull species arrive on their respective breeding grounds. The most important ones are, or include (1) the phenology of the postbreeding range; (2) the physiological condition of the migrants upon depature; (3) the age and amount of prior breeding experience; (4) the onset of favorable environmental conditions on the breeding range; and (5) the distance separating the postbreeding and breeding ranges. Belopol'skii (1957) emphasized the importance of this last factor by citing Altman's rule which states that the shorter the distance between the winter and breeding grounds, the earlier and more protracted the arrival; conversely, the more distant the wintering grounds, the later the arrival, although the dates are more regular and concentrated within a shorter period. Considerable variation exists between the arrival times of species that breed sympatrically. For instance, in the Great Lakes region Herring Gulls are often on territories about a month before Ring-billed Gulls arrive at the colonies. At this latitude, even though Herring Gulls characteristically are on territory much earlier than are Ring-bills, the peak of hatching in adjacent colonies is frequently only a few days apart. Relatively little research effort has been devoted to this phase of gull reproductive biology and so very little information is available on arrival times for the various species.

More up-to-date information is available on the breeding ranges of some of the species listed in Fig. 1. Some of this material has been eluded to, but an attempt was not made to provide a catalog of known breeding sites. Various agencies are in the process of preparing such documents, and some of these are cited in this chapter. The information provided on breeding ranges indicates the amount of spatial separation that exists between populations of some species, the extent of overlap between others, and the tendencies for coastal as opposed to inland nesting. Several other gull species have breeding ranges in North America that bring them into contact with one or more of the species under consideration. These forms are not considered in detail, as insufficient banding data are available, but those known to possibly influence distributional tendencies of the others are acknowledged in the appropriate sections.

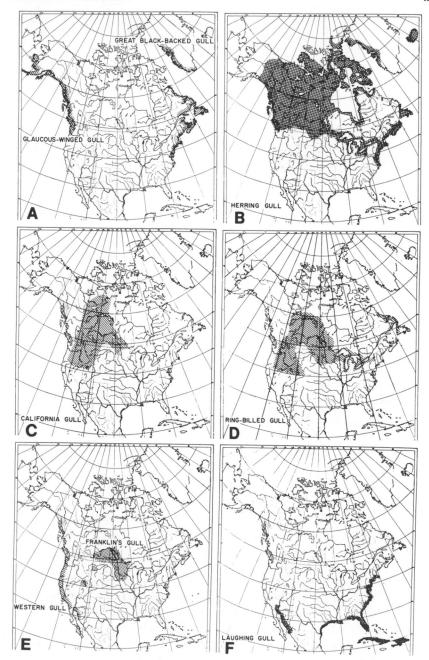

Fig.1. Breeding ranges of North American gulls emphasized in the text: (A) Glaucous-winged Gull and Great Black-backed Gull; (B) Herring Gull; (C) California Gull; (D) Ring-billed Gull; (E) Western Gull and Franklin's Gull; (F) Laughing Gull.

The general breeding range of each of the gull species emphasized in this chapter is depicted in Fig. 1. A narrative follows that elaborates on pertinent aspects of each species' breeding range.

A. Glaucous-winged Gull

In North America this species is restricted to the northwestern coast of the continent. Its breeding range extends from near 45°N latitude to about 62°N and between 120° and 150°W longitude (Fig. 1A). The Alaska sea-bird inventory has identified 547 coastal breeding sites with an estimated population of 500,000 birds. In much of this region it is the most abundant large gull. Only the Black-legged Kittiwake and the Red-legged Kittiwake (*Rissa brevirostris*) outnumber it along the Alaska coast (Sowls *et al.*, 1978). These pelagic foragers and cliff nesters do not appear to compete directly with species such as the Glaucous-winged Gull for food or nesting space. Hence, the distributions of the species emphasized in this chapter are not envisioned as being affected by these two smaller forms.

The Glaucous-winged Gull prefers nearshore habitats for foraging and nesting. It expresses a preference for sandbar islands, the flat tops of rugged islands, or beaches as nest sites. Their omnivorous and opportunistic tendencies have brought them into close contact with human settlements where they forage on refuse. As a result, their population appears to be enlarging in response to the food resources provided by human wastes. This phenomenon was eluded to earlier in the case of the Herring Gull.

Some overlap occurs in the southern portion of the breeding range between the Glaucous-winged Gull and the Western Gull, with relatively small numbers of Herring Gulls that are scattered along coastal areas of western Canada and Alaska, and with the Glaucous Gull in sections of northern Alaska. The center of abundance of the Glaucous-winged Gull is south of the primary breeding range of the Glaucous Gull. It is possible that this large species serves as a limiting factor to further northward advancement by Glaucous-winged Gulls.

Peters *et al.* (1978) reported Glaucous-winged Gulls at eight coastal breeding sites along the Oregon and Washington coasts. At six of these, it was concluded that the birds present were hybrids with the Western Gull. In the northern part of its range where it overlaps with the Herring Gull, hybrids have also been reported (Sowls *et al.*, 1978).

The Mew Gull also occurs over portions of the Glaucous-winged Gull's range. The coastal population of Mew Gulls, however, appears to be small in Alaska (Sowls *et al.*, 1978), and so the amount of actual contact between the two species during the breeding season may be minimal. In addition,

Mew Gulls tend to be rather dispersed nesters, as opposed to being highly colonial. This undoubtedly reduces further the likelihood of the two species having significant contact while nesting. From all indications, the largest proportion of the Mew Gull population probably occurs at inland rather than coastal localities in Alaska (Sowls *et al.*, 1978).

B. Great Black-backed Gull

The breeding range of this large gull is restricted to the northeastern seaboard of the United States, eastern Canada and the western coast of Greenland (Fig. 1A). The Great Black-backed Gull typically breeds in the Atlantic Boreal and Low Arctic regions (Brown *et al.*, 1975). In the northern part of its range it overlaps with the Glaucous Gull, whereas in the south its primary congener is the Herring Gull. Recently occasional pairs of Great Black-backed Gulls have been reported breeding south of New York in the United States, but the center of abundance during the breeding season remains north of 45°N. It is primarily a coastal species, although occasional breeding records have been reported from inland localities, such as the Great Lakes. It is possible that we are witnessing the early stages of this species' attempts to invade the Great Lakes region as a breeding species. If it is successful in doing so, its impact on existing larid populations should be significant.

C. Western Gull

The range of this species during the breeding season is restricted to the Pacific coast of northwestern Mexico, the United States and possibly the extreme southwestern portion of Canada (Fig. 1E). Its breeding range overlaps slightly with that of the Glaucous-winged Gull to the north, but over most of its range it is the only breeding gull present. It is, in fact, the only gull species that breeds along the California coast (Schreiber, 1970). To the south, colonies of Heermann's Gull (*L. heermanni*) along the Baja California peninsula and on the San Benito Islands (28°20'N) are within the breeding range of the Western Gull. Jehl (1976) reported these as being the northernmost colonies of Heermann's Gull. A limited breeding population of Laughing Gulls in northwestern Mexico and southern California is referred to in the *AOU Check-list* (Fig. 1F), but no recent information has been located about it. Some overlap may occur therefore in the breeding ranges of Western, Heermann's, and Laughing Gulls in the vicinity of the Gulf of California.

Western Gulls are largely coastal in all their activities. The primary breeding range is between 30° and 48°N latitude and 114° and 125°W longitude. Although the Western Gull is the only species breeding over much of this area, an assortment of other gull species co-inhabit the area during the nonbreeding season or as subadults during the breeding season.

Schrieber (1970) estimated, from the state of nest construction and courtship behavior observed during his California study, that the first eggs would have been laid in about mid-May. It appears that this species initiates breeding considerably later in these southern localities than do larids nesting at more northern latitudes. For example, in northern Michigan the peak of hatching in Herring Gull and Ring-billed Gull colonies frequently occurs during the 3rd week in May. This means that the peak of egg laying would occur about 26–28 days earlier.

D. Herring Gull

This is a widespread circumpolar species during both the breeding and nonbreeding seasons. Its breeding range covers a larger geographical area than that of any of the other North American gulls. Population density, however, varies greatly across this range. In eastern North America, most Herring Gulls breed north of 35°N, whereas in the west few breed farther south than 48°N (Fig. 1B). Herring Gulls are concentrated along the Atlantic seaboard of the United States and southern Canada and in the Great Lakes region between 36° and 48°N. In addition the species breeds in lower desities at numerous coastal and inland localities across northern North America that provide suitable habitat. They breed as far north as 80°N latitude but at these northern latitudes they are generally replaced by other gull species. They also occur along the Pacific coast northward to southeastern Alaska but in much lower population densities than in eastern North America.

Northward in the Boreal and Low Arctic zones of North America the Herring Gull is replaced by the Iceland Gull (*L. glaucoides glaucoides*) in west Greenland, by Kumlien's Gull (*L. g. kumlieni*) on southwest Baffin Island, and by Thayer's Gull (*L. thayeri*) farther north in the Canadian Arctic (Brown *et al.*, 1975). Glaucous Gulls breed through most of the High Arctic of the continent and extend southward to about 55°N on the Labrador coast. The presence of this large species undoubtedly influences the distribution and density of Herring Gulls.

Great Black-backed Gulls are also Boreal and Low Arctic in distribution in eastern North America. Herring Gull abundance also appears to taper off as its range extends into regions of higher population density of Great Black-backed Gulls.

Herring Gulls are not particularly oceanic and the species is widely distributed in coastal habitats and across inland regions having suitable aquatic habitats. Colonies are located on the margins of inland lakes and streams as well as on islands. The birds may breed as isolated pairs on small bodies of water or in colonies composed of hundreds of pairs. Its versatility with respect to nesting conditions as well as diet largely account for its widespread distribution. In Alaska, the Herring Gull is more common in the interior than along the coast. It is known to interbreed with the Glaucous-winged Gull, and therefore a wide array of hybrid plumages occur (Sowls *et al.*, 1978).

Only 12 coastal breeding sites have been reported in Alaska and the estimated coastal breeding population is about 300 pairs. No estimate is available for the more dispersed inland population. In much of coastal Alaska the Herring Gull appears to be replaced by the Glaucous-winged Gull or the Glaucous Gull. In portions of its Alaskan range it also hybridizes with the latter species (Strang, 1977).

The Herring Gull has shown significant population increases in recent decades. Several factors have contributed to its success along the Atlantic Coast of the United States, but of particular importance is the contribution made by landfills to the energetics of this and other larid species. The availability of garbage, particularly during the most stressful periods of the year, as a supplemental food source has enabled larger numbers of individuals to survive to breeding age. Prior to the availability of large landfills, novice juveniles in particular were undoubtedly hard-pressed to secure food during winter conditions. As a result, many more probably perished from such causes than do now. Several other species of gulls exploit landfills to some extent, particularly during the nonbreeding season. This tendency has greatly influenced their seasonal and daily distributional patterns.

E. California Gull

Seasonally this species occurs sympatrically with the Glaucous-winged Gull and the Western Gull, but there are noticeable differences in its distributional tendencies by the onset of breeding. The California Gull has an extensive breeding range in the western United States and Canada that is generally inland as opposed to coastal (Fig. 1C). It breeds from central California (about 35°N) northward to approximately 65°N. Colonies are on islands in man-made reservoirs, inland lakes at wildlife refuges and on river shingles. This results in a spotty distribution across the outlined breeding range. It appears that the present breeding success of the species is closely linked with human activities that affect water levels.

Because of its preference for inland breeding sites, the California Gull has essentially no contact with the Glaucous-winged or Western Gull while on the breeding range. Instead, it encounters the Herring Gull to a minor extent, the Ring-billed Gull in many localities, and possibly the Franklin's Gull in the eastern portions of its range.

F. Ring-billed Gull

In eastern North America this species breeds in southern and southeastern Labrador, the northern Gulf of St. Lawrence, southern Quebec, northeastern New Brunswick and Newfoundland southward to Lake Champlain in New York, and then westward to the Great Lakes (Fig. 1D). In recent decades, the species has prospered in this latter region with the population increasing many times over (Ludwig, 1974). This has occurred in response to fluctuating water levels on the Great Lakes that have periodically increased the amount of nesting space and the availability of introduced forage fishes. The increase in fish abundance and the occurrence of lows in the water cycle appear to have coincided on two occasions, and each resulted in significant increases in Ring-billed Gulls.

The breeding range of this species has expanded in the Great Lakes region as its population size has enlarged. It is currently more widespread than at any other time in its recorded history. Its rate of dispersal into new breeding areas, such as western Lake Superior, intensified during the early 1970s as rising Great Lakes water levels inundated previously available colony sites. Some traditional colony sites were completely covered by rising water, and others were severely reduced in size. As a result, the large gull population that resulted from optimal breeding conditions during low water levels of the early 1960s was severely stressed by the lack of traditional nesting sites in the 1970s. Gradually many of these birds moved to other colony sites where space was still available or they dispersed to new areas and established colonies where none previously existed.

In the Great Lakes region, the Ring-billed Gull far outnumbers the Herring Gull, its frequent congener throughout the eastern part of its range.

The Ring-billed Gull also ranges over a large portion of central and western North America (Fig. 1D). The western segment of the population appears to be distinct from the eastern component. There is little evidence of mixing of individuals from the two regions during any season of the year. The western population appears smaller than that of the Great Lakes region, although little population data exist.

The Ring-billed Gull is primarily an inland nester that expresses preference for islands in freshwater lakes. A few exceptions to this pattern exist in northeastern Canada, where small numbers of Ring-bills have nested on oceanic islands. This population appears to be a remnant of a much larger Ring-billed Gull population that inhabited the Atlantic Coast prior to the human-induced decimation that occurred during the 19th century.

In the East the Ring-billed Gull characteristically breeds sympatrically with the Herring Gull, whereas in the West its breeding range is commonly shared with the California Gull. The overall impact of these species on one another has not been studied, but at present it appears that the smaller Ring-billed Gull is not severely limited in either case.

G. Laughing Gull

This is the only North American gull that characteristically breeds in the southeastern United States. The primary breeding range extends from the Greater and Lesser Antilles and the Gulf Coast northward along the New England coast (Fig. 1F). A small breeding component has also been reported from northwestern Mexico and southern California, but little information exists about this population.

Throughout its breeding range in the southern United States, the Laughing Gull is the only breeding gull in the region. In the northeastern section of its range, however, it overlaps with the Herring Gull, and to a lesser extent with the Great Black-backed Gull. These two larger species may exert sufficient competitive pressure on the Laughing Gull to denote the northern limit of its range. In the West it may overlap with the Western Gull on part of its range.

H. Franklin's Gull

This inland marsh nesting species has a restricted breeding range in North America. Essentially it is limited to suitable wetland areas in a narrow segment of the continent between about 43° and 54°N latitude and 90° and 110°W longitude (Fig. 1F). Its dependence on marshes for nesting cover apparently contributes to its range being restricted to this region. From all indications, however, the species is experiencing adequate reproductive success to sustain the population.

In portions of its breeding range, it may encounter nesting Herring Gulls, Ring-billed Gulls, or California Gulls. Its nesting habitat is unique enough from that of these other species, however, that little spatial competition is likely.

This summary of breeding ranges of the various species under consideration indicates that the primary breeding ranges of most species are spatially segregated from most of the others. When overlap does occur, it is seldom with more than one other gull species that is prevalent in the area.

IV. POSTBREEDING DISTRIBUTION

Gulls as a group express considerable variations as to the extent of their seasonal movements or migrations. A few species are quite sedentary, with most of the population remaining within perhaps a few hundred kilometers of their breeding sites and possibly not leaving the breeding range of the species. Others travel moderate distances of up to 1000 km or so outside their breeding range. The Laughing Gull and Franklin's Gull, however, frequently travel several thousand kilometers in one direction. The general tendency for some southward movement on the part of at least a portion of the postbreeding population of each larid covered by this chapter results in species associations not encountered during the breeding season. In some instances species composition may remain constant but the relative abundance of the species involved may change.

Most North American gulls tend to be littoral and remain in reasonably close proximity to coastal areas during migration. A few species, however, commonly range inland from the oceans where they frequent freshwater lakes and marshes. Some, such as the Franklin's Gull, appear to follow inland routes characteristically, but others, such as Ring-billed Gulls and California Gulls, simply have segments of their populations that follow routes distinct from those of most of the species. As a result, Ring-billed Gulls may be widely distributed across the United States during migration and during much of the nonbreeding period.

Frequently juveniles range far outside the usual postbreeding range of adults. The proportion of the various subadult age classes involved in such long-distance movements is unknown. By ranging southward large northern species may end up wintering with smaller species of gulls in many cases. This possibly provides novice foragers, such as most juveniles, a competitive advantage because of their size. In the southeastern United States, for example, limited numbers of juvenile Herring Gulls and Great Black-backed Gulls are scattered through the range predominated by Ring-billed and Laughing Gulls. Their relative abundance is generally low in comparison with the smaller species, but such associations may have sufficient survival value to result in selection pressures favoring long-distance travel by juveniles.

Some adults remain closer to the breeding range than do others. The assumption is sometimes made that those doing so have a selective advantage in that they can return to the breeding range more promptly and obtain choice sites. Possibly those staying closest to a colony site are the most dominant individuals and the ones most capable of defending the limited winter food supply against competitors. This may be true in some instances, but it does not provide a complete explanation for intraspecific and interspecific differences in postbreeding travel distances by adults. In Ring-billed Gulls, successful breeders with several years of experience migrate to the southernmost part of the primary postbreeding range. Wing-marked individuals from a Lake Huron colony have repeatedly traveled between Michigan and Florida. All these individuals return in time to secure what appear to be optimal locations in the central part of the colony. They also appear to be in breeding synchrony with the major part of the colony (Southern, manuscript in preparation). Thus conspecifics that winter in Florida do not appear to be at a disadvantage over those staying closer to Michigan. In the case of Ring-billed Gulls, it is likely that the prominence of Herring Gulls may result in it being advantageous for most of the population to winter south of the primary range of the larger competitor. Thus, in spite of the possible selective value of being close to colony sites, some species may not have that option open to them. Obviously the abundance and distribution of food items in a locality will directly influence the intensity of competition between gulls for these resources. Thus in richer environments, the diversity and density of gulls will be greater during the nonbreeding season.

In most species, it is a portion of the juvenile population that normally ranges farthest from the breeding sites each winter. Because adults, or experienced migrants, of at least some gull species appear to be goal oriented during both spring and fall migration, an explanation for the exceptionally long travels of some juveniles in fall might be that they lack a recognizable goal to stop them sooner. Some juvenile Ring-bills, for example, depart from the breeding range almost immediately after leaving the colony site. As a result, they appear to be traveling alone or a least without the company of adults. Banding data indicate that some of these birds arrive along the Gulf Coast within a few weeks. It is possible that without adult Ring-bills being present some of these birds might not recognize the southeastern United States as being the southern terminus of the postbreeding range for most the eastern population. Because of this, they continue farther south into Mexico and Central America.

Figure 2A–L compares the proportion of the total band recoveries available for each species of gull recorded in each 6° zone designated on the maps. A brief discussion of the seasonal trends expressed by each species follows.

A. Glaucous-winged Gull

The Glaucous-winged Gull is one of the two essentially sedentary gull species discussed in this chapter, the other being the Western Gull. During all seasons a significant proportion of the Glaucous-winged Gull population appears to remain on or near the breeding range. The species seems to be largely coastal during the breeding and nonbreeding seasons alike. In February, Glaucous-winged Gulls are as widely dispersed across their annual range as they will be at any time (Fig. 2B). By this time, the mean distance between breeding (i.e., colony sites) and recovery locations peaks at about 232 km (Fig. 3). Presumably juveniles wander farther than do the adults or subadults during their first winter and this probably accounts for the gradual increase in dispersal distance after fledging. Although the available data have not been analyzed as yet on the basis of age, intuitive evidence supports the contention that juveniles travel farther. For example, the mean distance between banding and recovery sites for every month but July is more than 100 km. It is likely that many of these nonbreeders are juveniles.

The monthly distribution of Glaucous-winged Gull band recoveries is provided in Fig. 2A–L. Although these maps do not cover the entire range of the species, the data plotted account for all but 1.74% of all the available band recoveries (n = 17,226). A majority of those outside the map range (86%) were recovered along the Canadian coast and in southeastern Alaska. The exceptions (42 recoveries) were scattered along the Alaskan coast and westward to Japan. In this species, there appears to be a tendency for juveniles inclined to wander great distances to disperse westward rather than southward, as is exemplified by species such as the Herring Gull.

Without exception, the largest proportion of each month's recoveries came from Zone B'A (Fig. 2A–L). During November–February the frequency of recoveries for this Zone is comparatively low. At this time, the species is most prevalent in the various parts of its nonbreeding range. But even during these months, only a small proportion of the population ranges as far south as central California. Very few inland recoveries have been obtained at any time of the year.

Adults appear to begin their return to the breeding range in March, and a significant proportion of the birds remain in B'A until late October, when migration apparently begins.

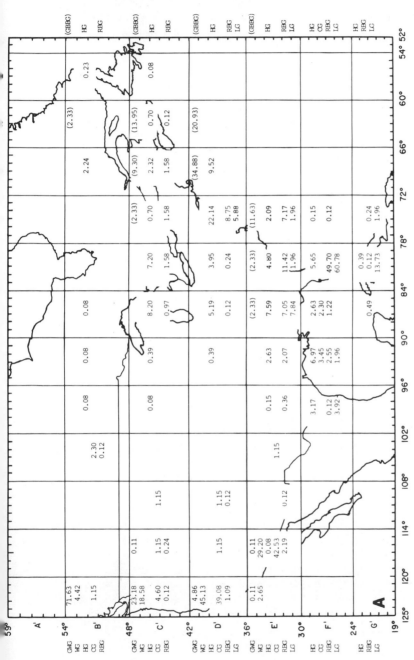

Fig. 2. Monthly distribution of band recoveries for each species for which more than 1100 total recoveries exist. In each zone designated on the map by heavy vertical and horizontal lines and letters (e.g., B'A), the percentage of that month's total band recoveries is indicated for each species. The species are indicated by initials listed on the margin: GWG, Glaucous-winged Gull; GBBG, Great Black-backed Gull; WG, Western Gull; HG, Herring Gull; CG, California Gull; RBG, Ring-billed Gull; LG, Laughing Gull. (A) January: Sample sizes for each species are: GWG, 906 recoveries; GBBG, 41; WG, 113; HG, 1292; CG, 87; RBG, 823; LG, 51. (Recoveries outside the range of the map are not included in these totals.)

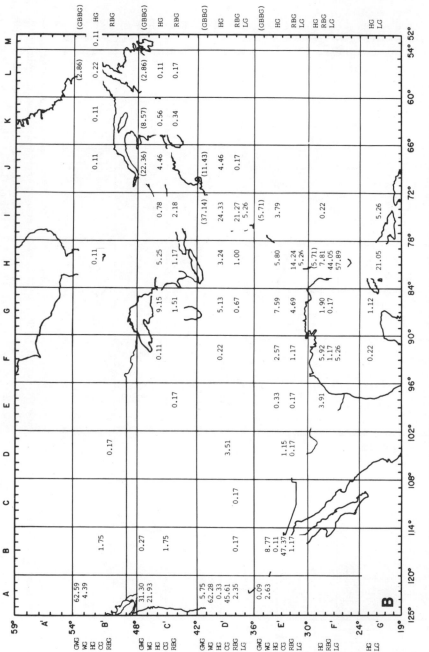

Fig. 2. (B) February: GWG, 1096 recoveries; GBBG, 35; WG, 114; HG, 896; CG, 57; RBG, 597; LG, 19.

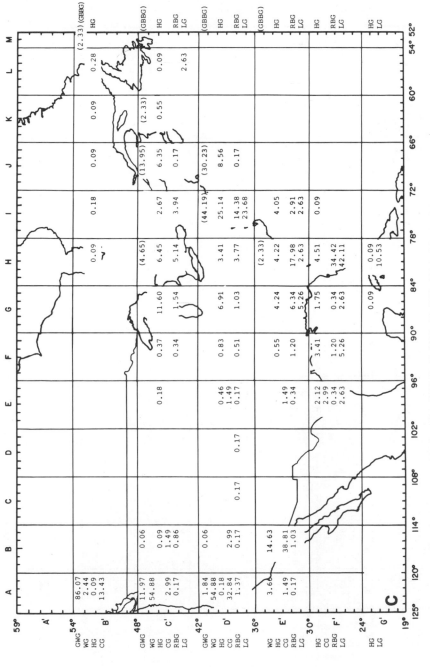

Fig. 2. (C) March: GWG, 1579; GBBG, 43; WG, 82; HG, 1086; CG, 67; RBG, 584; LG, 38.

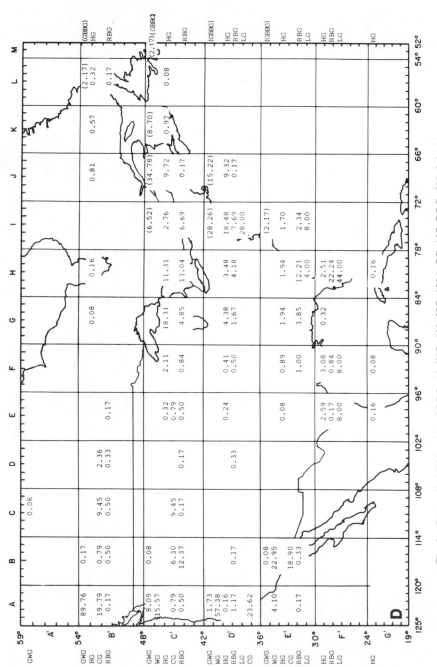

Fig. 2. (D) April: GWG, 1211; GBBG, 46; WG, 122; HG, 1234; CG, 127; RBG, 598; LG, 25.

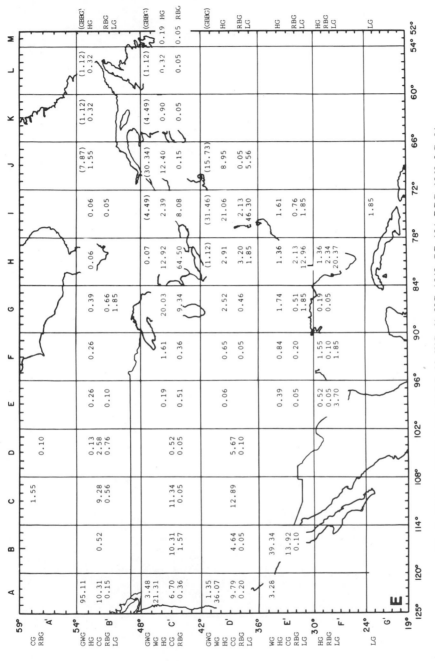

Fig. 2. (E) May: GWG, 1410; GBBG, 89; WG, 61; HG, 1548; CG, 194; RBG, 1969; LG, 54.

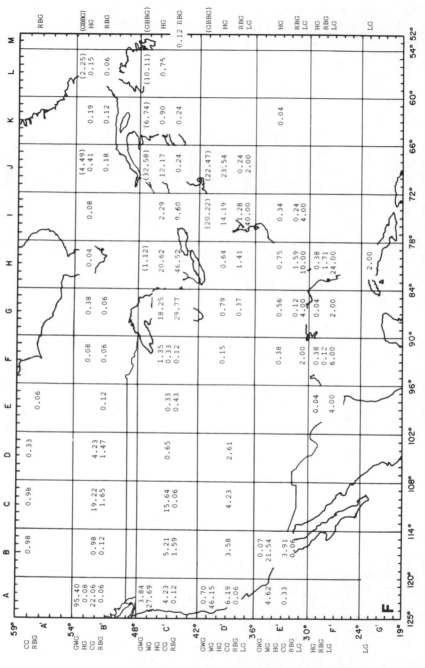

Fig. 2. (F) June: GWG, 1434; GBBG, 89; WG, 65; HG, 2663; CG, 307; RBG, 1636; LG, 50.

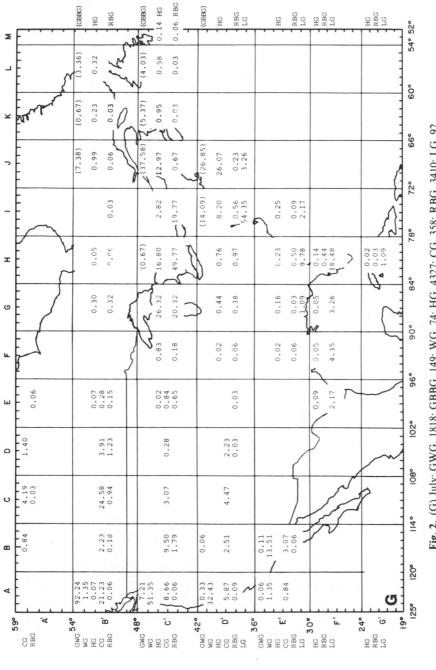

Fig. 2. (G) July: GWG, 1818; GBBG, 149; WG, 74; HG, 4327; CG, 358; RBG, 3410; LG, 92.

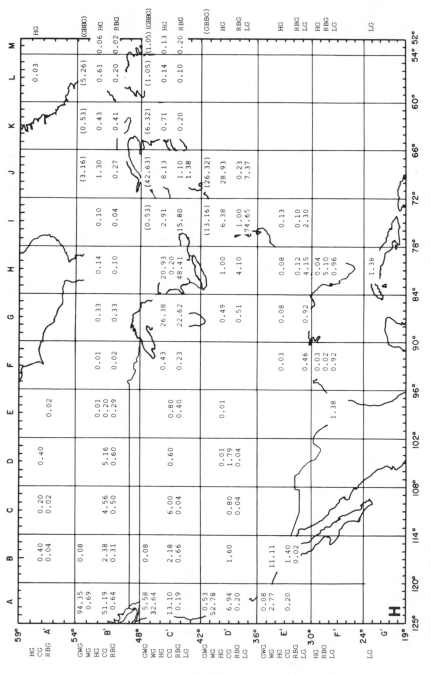

Fig. 2. (H) August: GWG, 1309; GBBG, 190; WG, 74; HG, 7206; CG, 504; RBG, 5141; LG, 217.

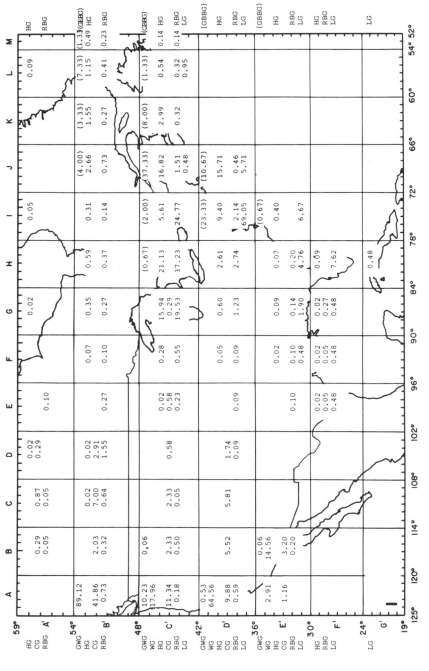

Fig. 2. (I) September: GWG, 1691; GBBG, 150; WG, 206; HG, 4246; CG, 344; RBG, 2192; LG, 210.

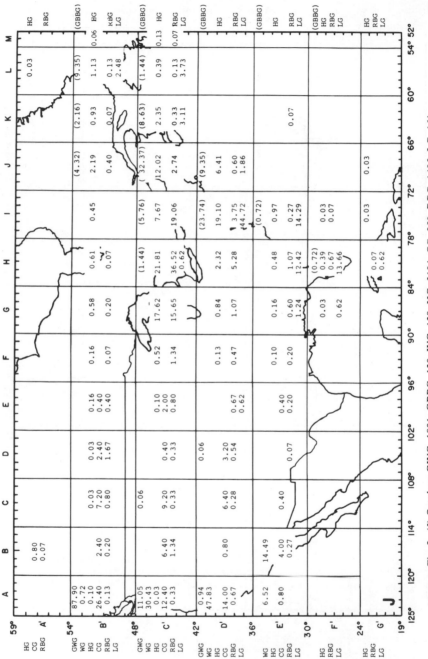

Fig. 2. (J) October: GWG, 1801; GBBG, 139; WG, 138; HG, 3104; CG, 250; RBG, 1495; LG, 161.

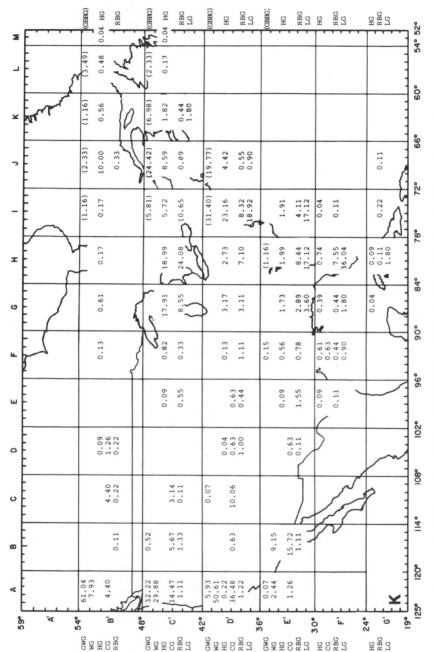

Fig. 2. (K) November: GWG, 1350; GBBG, 86; WG, 164; HG, 2306; CG, 159; RBG, 901; LG, 111.

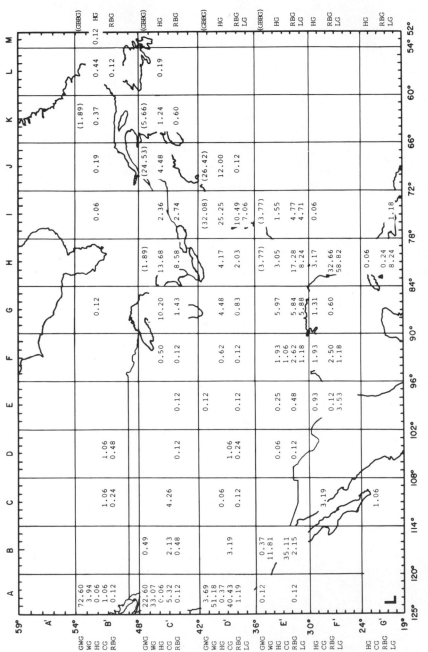

Fig. 2. (L) December: GWG, 814; GBBG, 53; WG, 127; HG, 1608; CG, 94; RBG, 839; LG, 85.

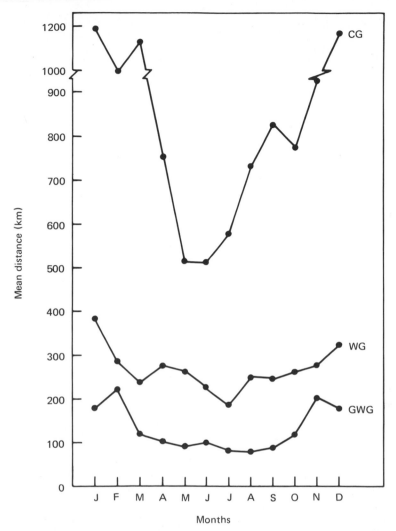

Fig. 3. Mean distances between banding and recovery sites during each month of the year for Glaucous-winged Gulls (GWG), Western Gulls (WG), and California Gulls (CG).

As is true of most other gull species, migration is not an all-or-none phenomenon. Some individuals, particularly juveniles, may depart earlier than do most of the potential migrants. A portion of the population remain on or near the breeding range, but population density is generally much lower than it was during the breeding season. Hence the July, August, and September reports from localities far to the south of the primary breeding range are from such individuals. For March–October, the mean distance at which recoveries were made does not exceed 130 km in any month (Fig. 3).

B. Great Black-backed Gull

During most of the year adult members of this eastern species are rather sedentary, as reflected by the mean distance between banding and recovery sites being less than 300 km from April to September. During June–August, the mean distance is 200 km or less (Fig. 4). During the post-breeding season, however, some individuals disperse rather widely. Adults and subadults range westward to the Great Lakes region, and some travel as far as Florida. In February the greatest distance between banding and recovery sites is achieved (583 km). No individuals have been recorded outside the range of the map. Most of the population appears to remain north of 36°N throughout the year (Fig. 2A–L).

During essentially all months, the majority of this species is found in Zone D'J and northward along the Atlantic Coast. Over the year, however, it is fairly widespread along the East Coast of North America, as the juveniles disperse quite far to the south. Typically, however, it is a northern species with comparatively few individuals being found as far south as Chesapeake Bay.

The mean distance between the place of banding and the recovery site is the greatest (583 km) in February, suggesting that the proportion of the population away from the breeding range is greatest by this time (Fig. 4). Thereafter, there is a progressive decrease in the distance between banding and recovery sites until July, when the largest proportion of the population (i.e., all age classes) appear to be nearest the breeding sites. The mean distance between banding and recovery locations for 189 birds recovered during July was 164 km. Following this, there is a gradual increase in the distances (186 to 436 km) at which Great Black-backed Gulls were recovered away from the colonies (Fig. 2A–L).

Some individuals, primarily juveniles, range farther south to Florida and Georgia. In this region they have been recovered in small numbers from September to April. During the summer months no recoveries of Great Black-backed Gulls have been reported south of Zone D'I (36°N). Breeding colonies occur in the northern part of this Zone, in northwestern D'J and on northward along the coast.

C. Western Gull

This is a very sedentary species when compared with most of the others discussed in this chapter. In fact, it may be the most sedentary of those species found in the United States. The majority of all age classes remain on or close to the same range during the breeding and nonbreeding seasons (Fig. 2A–L). Only four (0.26%) of the 1515 band recoveries for this species

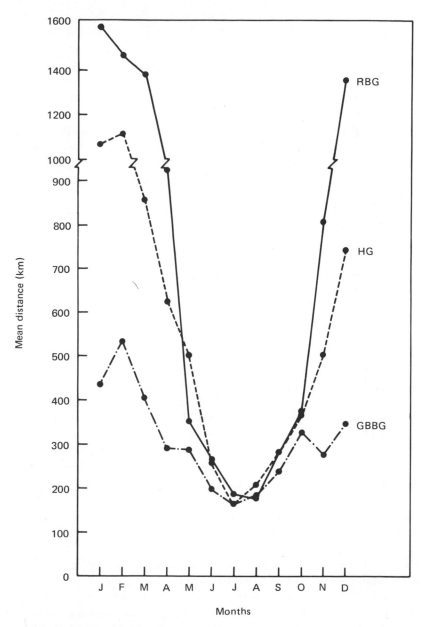

Fig. 4. Mean distances between banding and recovery sites during each month of the year for Great Black-backed Gulls (GBBG), Herring Gulls (HG), and Ring-billed Gulls (RBG).

were from outside the range of the computer maps. The monthly mean distances are very similar (Fig. 3). Postbreeding dispersal begins in late July and there is a tendency for some birds, presumably juveniles, to move northward into Zone B′Z (Fig. 2F). Each of the four recoveries from outside the range of the map were also from areas north of the breeding range. Three of the four came from coastal areas immediately northwest of the map boundary of Zone B′A, whereas the remaining recovery came from southeastern Alaska.

The highest mean distance between banding and recovery sites, 384 km, occurred in January (Fig. 3). Two of the reports from outside the map range were obtained during this month, which contributes to the higher mean. It appears, therefore, that essentially no migration occurs in this species. It is sedentary on its breeding range with juveniles and subadults dispersing short distances from their natal colonies, but generally remaining within the breeding range of the species.

D. Herring Gull

Various investigators (e.g., Moore, 1976; Southern, 1968; Threlfall, 1978) have examined the seasonal distribution of Herring Gulls banded at specific colonies or in a particular region (e.g., Great Lakes). This is the first time, however, that the available band recovery data from across the continent have been examined as a single group.

Herring Gulls remain widespread throughout the year. Essentially the entire breeding range is occupied during each month but there are regional shifts in population density. Many Herring Gulls remain in the Atlantic Boreal and Low Arctic waters throughout the winter (Brown et al., 1975). Drury and Nisbet (1972) pointed out that marked differences exist in the migratory behavior of different populations of this species. They found that most birds from Cape Ann and Boston Harbor in New England stayed near their breeding islands throughout the winter. Birds from Maine, on the other hand, moved southward. Such differences are probably related to the availability of food during winter.

Many adults from the northern colonies in Canada migrate southward to winter in the northeastern United States. As a rule, young birds migrate farther than do adults. Sixty-two (0.19%) of the 32,240 recoveries analyzed for this study were from outside the range of the computer map (Fig. 2). Most (88.7%) of these birds were recovered in Central America or Cuba. These figures suggest that a majority of the North American Herring Gull population remains within the confines of the continent throughout the year.

In June the largest proportion of all age classes of Herring Gulls appears to be closest to the breeding range. About 96% of the recoveries (n = 2663) are from Long Island and northward on the East Coast. For this month, the mean distance between banding and recovery sites was 258 km (Fig. 4). This distance decreased further during July and August, but this is probably an artifact resulting from young-of-the-year being recovered at the colony sites.

Dispersal from breeding colonies may begin in late July, but few individuals appear to leave the breeding range (Fig. 2A–L). In the Great Lakes region, many Herring Gulls remain on territory well into August, even though most juveniles may be capable of flight. During September–November, the mean distance between banding and recovery sites gradually increases from 283 to 502 km (Fig. 4). During these months there is a gradual increase in the proportion of recoveries from localities south of Chesapeake Bay and from the Gulf Coast. Moore (1976) noted that most adult Herring Gulls remain in close proximity to their breeding grounds throughout the year. He found that the monthly proportion of adult recoveries from outside the Great Lakes region ranged between about 2 and 12% of the total. By contrast, the monthly proportion of juveniles recovered outside the breeding range increased from about 2 to 72% for October–February. A similiar pattern of age-related migration is expected from populations of this species in other parts of its range.

During December–February (Fig. 2A–B, L), a larger proportion of the recoveries were from outside the breeding range and these were primarily subadults. In December, the average distance between banding and recovery sites was 746 km, in January 1070 km, and in February 1116 km. By January and February, Herring Gulls have moved as far south of the breeding range as they normally will, with many having reached the Gulf Coast of the United States, Mexico, and Central America. There is a tendency for subadult Herring Gulls to concentrate along the southern Atlantic seaboard and Gulf Coast during the winter season (Moore, 1976). In contrast, less than 1% of the adults are recorded in this area during winter.

There is an apparent correlation between winter concentrations of band recoveries and dense urban centers. Such tendencies may be associated with the abundant food resources generated by such human population centers.

The spring return of young to the breeding range probably commences in late February and early March (Fig. 2B, C). The mean distance of recoveries from the colonies decreased in March (Fig. 2C) to 857 km, continues to do so in April as well as during the subsequent months, representing the early stages of the reproductive cycle (Fig. 4).

E. California Gull

Throughout much of the year, this species is widespread in the western United States and parts of Canada (Fig. 2). Zones B′A, B′C, C′C, and D′A stand out as having particularly high proportions of the breeding season recoveries (Fig. 2F). Over this range, habitat availability probably determines population density.

The most obvious difference between the distributional trends of this species and all the others covered by this chapter is the high mean distance between recovery and banding sites during the summer months. In both May (n = 194 recoveries) and June (n = 307) the mean is greater than 500 km (Fig. 3). This suggests that a larger proportion of the subadults, and possibly nonbreeding adults, remain farther outside the breeding range during summer than is the case in the other species. The mean distance between banding and recovery sites is strikingly different during all months of the year from those for either the Western Gull or Glaucous-winged Gull (Fig. 3). The monthly distance curve for the California Gull (Fig. 3) bears most resemblance to that for the Herring Gull (Fig. 4). Fewer Herring Gulls, however, appear to be recovered far from the breeding range during May and June.

During December–March, the mean distance between banding and recovery sites is 1000 km or higher (Fig. 3). A substantial decrease occurs in April and May, indicating that many adults are returning to the colony sites. Postbreeding dispersal begins again in late July and the birds range widely along the Pacific Coast during winter (Fig. 2G).

Although this species is widely dispersed across the western United States and Canada during most months, very few individuals appear to travel outside the range of the computer map (Fig. 2). Only 14 (0.51%) of 2734 recoveries were from localities outside the map range. All but one of these were from localities between 49° and 55°N latitude and 125.2° and 129.2°W longitude. The exception was reported from farther northwest (60.3°N 148.7°W).

Of the distinctly western gulls, this species is the most migratory, with a sizeable proportion of the population moving upward to about 1200 km from the breeding sites. The California Gull differs from the typically eastern species by confining most of its long-distance movements within the continent (Fig. 2). Thus birds breeding northward in central Canada migrate to southern California, where they winter along the coast and take advantage of landfills.

F. Ring-billed Gull

Elsewhere (Southern, 1974a,b) I described in detail the seasonal distribution of Ring-billed Gulls from the Great Lakes population. This is the first time I have examined the banding data stemming from the western part of the species' range. The sample sizes for this portion of the population are small which makes it difficult to compare the chronology of migration in the West and East. It appears that the Ring-billed Gulls coming from breeding colonies in the western states or provinces migrate south or southwest in the fall. This results in a postbreeding population along the Pacific Coast. This is in contrast to representatives of the eastern population that express a tendency to migrate south–southeast in fall. As a result, a large part of this population winters along the Atlantic Coast southward to the Florida peninsula.

The latitudinal distribution of Ring-bills along the Pacific Coast resembles that for the species on the Atlantic Coast (Fig. 2). It appears, however, that a larger proportion of the eastern population migrates south of 30°N latitude than is the case in the West (Fig. 2A, B). This may indicate that better foraging conditions exist in Florida than in Baja California.

In both populations there are some individuals that migrate along the coast and others that follow assorted inland routes. The inland migrants frequently follow river systems to their terminus, where they reside for the winter. This results in Ring-billed Gulls being widely distributed during migration. Usually, however, the highest concentrations of individuals are along the coasts, bordering major rivers, such as the Mississippi, or along the Great Lakes. It is not uncommon for migrants to diverge from their route and spend one or more days foraging in farm fields or other suitable areas away from water courses. They return to the water at dusk for roosting purposes. In fall, migration is more casual than in spring, and birds destined to winter farther south will spend time en route at suitable foraging areas. In spring, less time is spent in transit and so groups of birds may not pause as long at stopping places.

Data from wing-marked adults show that many individuals are migrating toward specifc goals in both spring and fall. Some evidence also exists to indicate that particular birds follow the same or similar migration route during successive trips (Southern, manuscript in preparation).

Mid-July to early August marks the onset of juvenile and adult dispersal from the nesting colonies. Beginning in August, the mean distance between banding and recovery sites increases, and it continues to do so every month through January (Fig. 4). Initially upon departing from the colonies, Ring-billed Gulls appear to scatter in a variety of directions, but as fall approaches a southward trend becomes increasingly apparent. Very few individuals travel as far south as the southern states during July–September (Fig. 2G–I).

During October, southward migration is slightly more apparent with the southern portions of the range showing a subtle population increase (Fig. 2J). The most extensive changes in distributional pattern occur during November and December. By January the mean distance between banding and recovery sites is the highest, at 1587 km (Fig. 4). Return toward the breeding colonies begins in late February and intensifies during March and

April. In the Great Lakes Region, Ring-billed Gulls often begin arriving at colony sites in late March.

G. Laughing Gull

This species has an elongated coastal breeding range in the eastern United State (Fig. 1F). Individuals breeding at the more northern latitudes essentially vacate the breeding range during the fall months. Many Laughing Gulls, however, remain on the southern part of the range throughout the postbreeding period. Presumably many of these are resident birds, but data are not available to sbustantiate this possibility.

This black-headed species has a significantly different pattern of migration when compared with the white-headed forms discussed previously. For instance, from December to April the mean distance between banding and recovery sites is more than 1500 km, and in February it peaks at almost 3000 km (Fig. 5). Thus for five months sufficient representatives of this species are far enough from the breeding range to boost the mean distance of travel higher than that for any of the other species.

A large proportion (21.70%) of the total recoveries for this species were from localities south of the limits of the computer map. Fifteen (4.55%) of these were from close to, or south of, the Equator. The majority were from Central America and northwestern South America. Because the probability of bands being reported from this part of the world is probably even less than in the United States and Canada, these data probably indicate that a significant proportion of the North American Laughing Gull population winters in this general area.

Of those remaining within the range of the maps in Fig. 2, the largest proportion is consistently reported from Florida, the next highest from Cuba and vicinity, and the remainder are from the Gulf Coast and the Carolinas.

As spring advances in March (Fig. 2C) there is a decrease in the average distance from colony sites, indicating that northward migration is under way (Fig. 5). An increase also occurs in the proportion of the recoveries from the United States, and this trend continues into April and May (Fig. 2D, E). In June and July most adults are on the breeding grounds, but some subadults remain scattered at more distant localities.

In late July and August, dispersal from the colonies again occurs. It is during August and September (Fig. 2H, I) that the largest number of recoveries is obtained from localities between North Carolina and Cape Cod, Massachusetts. In October, the mean distance between recovery and

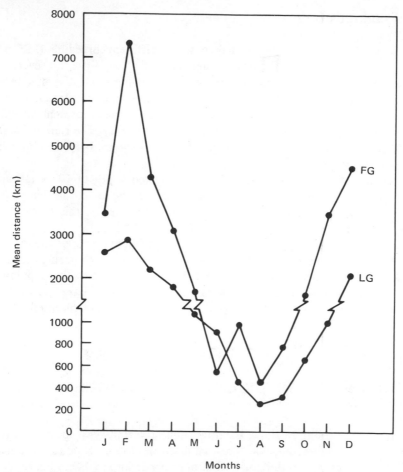

Fig. 5. Mean distances between banding and recovery sites during each month of the year for Laughing Gulls (LG) and Franklin's Gulls (FG).

banding locations (Fig. 5) begins to increase at a more rapid rate thereby indicating that southward migration is well underway.

On their postbreeding range in the southeastern United States, Laughing Gulls coexist with large numbers of Ring-billed Gulls, a few subadult Herring Gulls and occasional juvenile Great Black-backed Gulls. The extent to which these species might interact is influenced by all four foraging at landfills. Without landfills, it is conceivable that less overlap might occur within local areas and the various species might exhibit some sort of geogrphical separation along the Gulf and Atlantic Coasts.

H. Franklin's Gull

This black-headed species has notably different breeding (Fig. 1F) and postbreeding ranges than the Laughing Gull. Very little, however, can be said about its distribution on the basis of banding data. The number of recoveries per month within the map coordinates ranges between zero in February and 41 in October. Many of these recoveries represent juveniles recovered on or near the breeding range (Burger, 1972). The remainder fail to provide definitive information about the routes traveled between breeding and postbreeding ranges.

Of the 244 recoveries, 16.39% were reported south of 19°N latitude. Many (60%) of these birds traveled south of the equator. The mean distances between banding and recovery sites are higher for this species than for any of the others, particularly during February (Fig. 5).

Personal observations indicate that a significant proportion of this population migrates through the Great Plains during both spring and fall. Some individuals scatter as far as both coasts but most of the population seems to be restricted to the central part of the continent. Large numbers of Franklin's Gulls move through central and eastern Texas and then on southward into Mexico.

V. MIGRATIONAL ORIENTATION BY GULLS

In the preceding sections, the migration of various gull species between breeding and nonbreeding localities has been described. In many instances, a sizable proportion of the population may travel hundreds or thousands of kilometers between these sites. The ability to travel between specific geographic regions with reasonable accuracy and speed requires the species to possess a system for orientation or direction finding. The greater the distance traveled and the more precise the goals, the greater the need for a sophisticated orientation system.

Various hypotheses have been presented to explain the ability of birds to orient during migration. Although we are still far from solving all the problems associated with this ability, important advances have occurred in recent years.

Over the years I have conducted a series of studies dealing with the orientational ability of Ring-billed Gulls. As these results are pertinent to the subject of this chapter, they are presented for review.

I have used Ring-billed Gulls in three general types of experiments (Southern, 1969, 1972, 1974c), each involving a distinct age class of birds.

These are (1) classical homing experiments with adults, (2) free-flight trials with juveniles not previously out of the colony, and (3) orientation cage trials with chicks of various ages, usually 3–10 days old. By using birds of different ages, and therefore different amounts of experience with environmental cues, it has been possible to obtain some information about which cues appear to be of primary importance.

All the gulls used in experiments discussed in this chapter were from the Calcite Colony near Rogers City, Presque Isle County, Michigan. A thorough long-term banding program at this colony has provided considerable information about the seasonal distribution of gulls from the precise population being studied (Southern, 1974a). The results from band-recovery analysis and orientation studies have been compared in an attempt to verify that the responses of experimental subjects are biologically relevant (Southern, 1976). My findings indicate that essentially all ages of Ring-billed Gulls are capable of expressing an oriented response when provided with adequate environmental information. The mean initial bearings selected by displaced adult subjects usually can be correlated with the direction of travel required for reaching the home colony, whereas that taken by juveniles and chicks is a bearing appropriate for fall migration.

I have used a variety of procedures to monitor for the possible effect of several environmental cues. In general there has been an acceptable level of consistency in my data, but on occasion the results have been inconclusive, or contradictory, and most certainly perplexing. The amount of ambiguity in my gull data, however, does not appear to be any greater than that encountered in the data for some other species, such as Homing Pigeons. In spite of the analytical problems, some patterns are beginning to emerge from these data. Recently I (Southern, 1978) re-analyzed most of the gull orientation data I had collected between 1963 and 1977. The results are interesting and have caused me to modify some of my earlier conclusions.

In earlier publications, I presented the following working hypothesis: Juvenile Ring-billed Gulls select an appropriate bearing for their first fall migration on the basis of geomagnetic cues. Such directional preferences apparently are innate, as they are expressed by chicks tested soon after hatching (about 2 days old). During, and subsequent to, their first migration the young birds gain experience with a variety of other cues which are then used to supplement their magnetic compass. Eventually this series of cues is integrated into an orientational system that permits the gulls to migrate under a variety of environmental conditions. The multiplicity of cues available to adult gulls provide sufficient redundancy so that equivocal information from one cue may not cause disorientation.

It now appears that Ring-billed Gull chicks may develop an ability to use a sun compass at a much earlier age that I previously proposed. My data from experiments with adults and chicks also indicate that there may

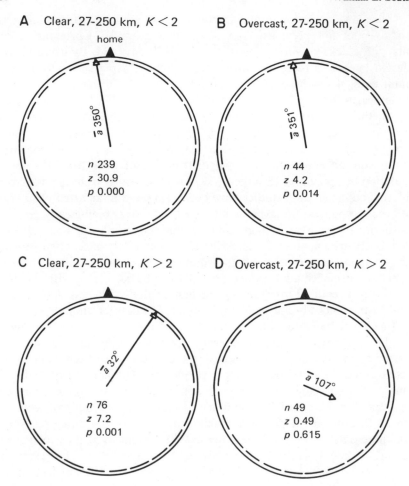

A Clear, 27-250 km, $K < 2$

home

$\bar{a}\ 350°$

n 239
z 30.9
p 0.000

B Overcast, 27-250 km, $K < 2$

$\bar{a}\ 351°$

n 44
z 4.2
p 0.014

C Clear, 27-250 km, $K > 2$

$\bar{a}\ 32°$

n 76
z 7.2
p 0.001

D Overcast, 27-250 km, $K > 2$

$\bar{a}\ 107°$

n 49
z 0.49
p 0.615

Fig. 6. Summary of the directional responses of 408 adult Ring-billed Gulls during homing trials under clear and overcast conditions and K values of 0–4.

be a link of some sort between the magnetic compass of a gull and the solar compass. The link relating these cues may exist in the bird's orientation mechanism, or it may be an interrelationship of some type between the natural occurrence of the respective cues.

In this section I have discussed three topics: (1) the data indicating a link between geomagnetic and solar cues, (2) a little information about the potential influence of other cues, and (3) a model suggesting ways in which the components of a Ring-billed Gull's orientation system may be arranged.

A. Solar and Magnetic Cues

A total of 408 adult Ring-billed Gulls (3-year-old or older) were used in homing trials during clear and overcast skies when levels of geomagnetic disturbances ranged between $K0$ and 4 ($K0-2$: quiet conditions; $K3$: unsettled; $K4$: active). The birds were released at distances of 27–250 km and at

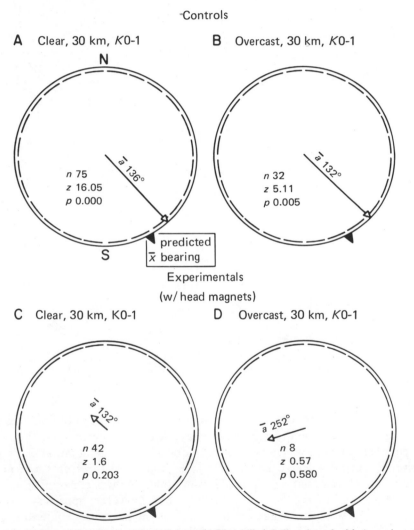

Fig. 7. Results of free-flight trials with juvenile Ring-billed Gulls released with (experimental group) and without (controls) magnets glued to their heads.

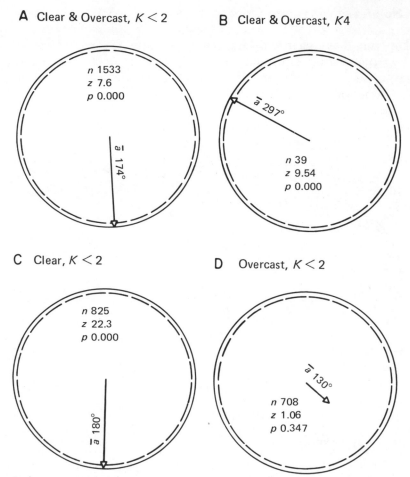

Fig. 8. Summary of the directional responses of Ring-billed Gull chicks tested in Southern-type standard orientation cages during clear and overcast sky conditions and varying levels of geomagnetic disturbance.

various directions from home. When K was 2 or less, the initial bearing of gulls released under both clear and overcast skies was toward home (Fig. 6). In contrast, when K was greater than 2, a notable change in initial bearings occurred. Adults released under clear skies continued to express a significant mean bearing, but an easterly shift of about 40° was apparent (Fig. 6). The departure bearings for the overcast group, however, approached randomness. These results suggest that so long as the birds have either a near-normal geomagnetic field or a view of the sun, a significant proportion of the sample will be able to select similar bearings. Usually the heading selected is homeward oriented. Disturbance to the magnetic field, however, apparently affects the orientation mechanism and even though the sun is visible a directional shift, but not disorientation, may occur. When detection of both cues is interfered with, disorientation results.

Juvenile Ring-bills ($n = 157$) were released at a site 30 km from the colony. None of these birds had flown previously outside the colony. The mean departure bearings for gulls released under clear and overcast skies were similar, 136° and 132°, respectively (Fig. 7). Birds in both groups were released during periods of low geomagnetic disturbance. In order to test the effect of an altered magnetic field, a ceramic magnet was glued to the head of each experimental Ring-bill, whereas controls wore brass disks. The magnetic field produced by the magnet at the level of a bird's head was about 0.5 Gauss. Gulls bearing magnets dispersed randomly when released under clear or overcast conditions (Fig. 7), whereas the controls headed southeast. These results are consistent with those obtained from homing trials. It seems that the initial heading, relative to home or to the fall migration route, is selected most accurately when unequivocal information is available to the birds from both magnetic and solar sources.

Several thousand orientation trials with Ring-billed Gull chicks have been conducted in standard orientation cages (Southern, 1974c). Of these, 1533 were tested during clear and overcast skies, and K values were less than 2. The combined sample and the clear sky trials have significant mean bearings (Fig. 8), whereas those conducted under overcast skies do not. The significance of this group of trials is not readily apparent, but it is possible that it indicates that the link between magnetic and solar cues is particularly important to chicks. This view is reinforced when the sample is divided into subgroups based on K value and sky condition (Table II). For higher K values there is a directional shift rather than disorientation (Fig. 8). When the $K4$ samples are split into those conducted under clear and overcast skies, a significant bearing persists for the clear group, but the headings are random under overcast skies (Table II). This dichotomy was evident in all but one K category (Table II).

It appears that geomagnetic and solar cues are of primary importance in the orientation process of Ring-billed Gulls of all ages. Some kind of link appears to exist between the two cues, and severe disruption of one may alter a bird's ability to effectively use the other. During high-intensity magnetic disturbances, such as those produced by superimposed magnetic fields and possibly high-intensity magnetic storms, the chicks dispersed randomly regardless of sky conditions (Table III). However, during low-level geomagnetic disturbances, an oriented response occurred under clear skies (Table II). Just how these two cues interact to provide a gull with directional information is unknown at this time.

Attempts to describe more precisely the characteristics of the cues to which gull chicks are responding have complicated the matter even further.

Table II. Directional Responses of Gull Chicks[a] Tested Under Various Sky and Geomagnetic Conditions

K value	Sky	n	$\bar{a}$	z	Probability of z
0	Clear	126	109	4.12	0.398
0	Overcast	136	96	1.56	0.210
1	Clear	103	95	3.05	0.047
1	Overcast	108	196	6.80	0.001
2	Clear	156	191	5.94	0.003
2	Overcast	162	7	0.16	0.853
3	Clear	162	215	4.23	0.014
3	Overcast	67	46	0.40	0.672
4	Clear	33	292	10.57	0.000
4	Overcast	17	46	0.63	0.539

[a] Only 1070 trials are included here, as the K value during some trials was not recorded.

Cage trials were conducted in which a bird's view of the sun was altered by a translucent fiberglass top placed above the orientation cage. In another set of trials, a 2.4-m-high stockade encircled the test cage, thereby preventing chicks from seeing the natural horizon, but the natural sky was visible overhead. Chicks in the covered cage expressed statistically significant mean bearings, although these were shifted far to the north of the predicted southeasterly bearing (Table IV). Birds wearing magnets when tested under these same circumstances dispersed randomly. It appears that a view of the sun or horizon is not always necessary for an oriented response, but an absence of these cues may cause noteworthy directional shifts away from the bearing taken by controls and predicted for the population on the basis of banding data. Magnetic anomalies beyond those produced by $K2$–3 levels of natural disturbance apparently disrupt the tendency for chicks to express any directional preference and near random dispersal occurred in the test apparatus.

Table III. Directional Responses of Gull Chicks Tested Under the Influence of Sky Conditions and Superimposed Magnetic Fields

Sky condition	Source of field	n	$\bar{a}$	z	Probability of z
Clear and overcast	Magnets on head, back or cage floor	1333	344	1.03	0.357
Clear	As above	723	33	1.62	0.198
Overcast	As above	610	273	0.35	0.705

In the stockade, both groups (i.e., clear and overcast skies) had a significant mean bearing toward east–southeast and southeast (Table IV). Only low levels of magnetic disturbance ($K0$–3) occurred during these trials, and this in some way may account for the ability of this group of chicks to express a consistent preference for southeast during both clear and overcast conditions. The results from this set of trials are not consistent with those for my other data sets, and I lack an explanation for this fact. Inconsistencies of this type are not uncommon in orientation research, and our inability to explain such occurrences has greatly impeded our progress.

B. Influence of Other Cues

During homing experiements with adult Ring-bills an attempt was made to record their responses to various landscape features. There was a general tendency to follow shorelines upon release and to head toward large bodies of water almost anywhere along the flight path. Often these reactions resulted in the direction of travel being other than homeward oriented. Occasionally several hours were spent bathing and apparently loafing on bodies of water near release sites. Subsequent to this the birds that were radio-tracked usually made a direct nonstop flight to the home colony. These observations indicate that homing Ring-billed Gulls are aware of, and respond to, particular types of landmarks. There was no indication, however, that directional information relative to the homeward route was obtained from this source.

Migrating Ring-billed Gulls express similar tendencies and follow water courses during migration. Hundreds of adults follow the Mississippi River northward each spring. Others follow the shoreline of Lake Michigan to the north. Usually these migrants are flying at altitudes of less than

Table IV. Results of Trials in Which Vision of the Natural Sky or Horizon Was Distorted

Cage type	K value	Magnets or brass	n	$\bar{a}$	z	Probability of z
Translucent top	0–3	Neither	90	1	3.89	0.20
	2–3	Brass	80	8	5.17	0.005
	2–3	Magnets	79	61	2.44	0.087
Stockade						
Clear	0–3	Neither	67	165	10.85	0.000
Overcast	0–3	Neither	65	123	11.05	0.000

400 m, traveling in loose flocks of 6–10 individuals up to several hundred birds. When such landmarks are aligned in the appropriate direction of migration, they appear to be followed by migrants. Whether they function as a source of directional information is unknown. At points along the Mississippi River, groups of Ring-bills may leave the river and follow a diagonal inland course to the northeast in spring that takes them to the southern end of Lake Michigan. In the fall, the reverse pattern is followed by an unknown part of the population, including flocks of juveniles unaccompanied by adults. During the inland migratory flights, the birds usually fly at altitudes above 400 m, and many are high enough that they are usually missed by observers. During these portions of the journey cues other than landmarks probably influence flight direction.

The results from first-flight trials with juvenile Ring-billed Gulls reinforce my contention that geomagnetic and solar cues take priority over any information provided by landmarks. These young birds, which were positively unfamiliar with terrain around the release site and which had no prior experience with any land features outside the colony, had highly significant southeasterly mean bearings (Fig. 7). As juveniles are not dependent on landmarks for initial orientation, it seems unlikely that more experienced Ring-bills would be either. It is possible that landmarks serve as a source of local or site-specific information that is particularly useful during foraging forays. During migration certain landmarks probably serve as supplemental cues for experienced birds.

Chicks used in a series of orientation cage trials were permitted a view of the natural landscape to ascertain what effect this had on their directional response. When the landscape was visible throughout the 2-min trial (i.e., cages had clear plastic walls or no walls), the gull chicks in three of the data subsets expressed a statistically significant mean bearing (Table V). In two instances, the mean angle for the sample approximated that predicted for the population on the basis of banding data. These data are not interpreted as indicating that gull chicks are using landmarks in their selection of a course to follow within the test apparatus. Such a possibility seems unlikely, as none of the test subjects had prior experience at, or near, the test site. It seems reasonable to suspect, however, that the more natural surroundings

Table V. Results for Trials Conducted in Open Cages

Sky condition	K value	Magnets?	n	$\bar{a}$	z	Probability of z
Clear	1–2	No	96	40	5.90	0.003
Overcast	1	No	34	181	6.40	0.001
Clear	0–4	Yes	102	160	4.76	0.008

resulted in a higher response rate (100% in most cases) and possibly in an increased tendency for chicks to give an orientation response rather than some other type of reaction (e.g., fear) to experimental procedures.

It is interesting, as well as perplexing, to note that the test group for superimposed fields (i.e., involving magnets) also had a statistically significant mean heading. In this case, the 160° bearing closely resembles that predicted for the population on the basis of banding data. This is the only occasion on which magnets failed to produce disorientation. As these trials were conducted under clear skies, it is possible that the sun compass served as a primary cue. As birds tested in open-walled cages were provided with a view of the natural horizon, the accuracy of their sun compass may have been better than that of chicks tested in the standard apparatus that involved an artificial horizon.

Presently I possess no evidence to indicate that Ring-billed Gulls are using any of the other forms of environmental information that have been proposed as orientational cues. Wind direction was considered during homing trials with adult gulls and for each release I plotted the initial heading against wind direction at release time. A total of 446 adults were released when the wind was blowing from eight different directions. Although half these groups had a significant mean bearing, there was no consistent correlation between wind direction and directional headings. During these trials, I measured surface wind, which tends to be highly variable with respect to local direction and velocity. It seems unlikely that dependence on such a variable source of information would be favored by natural selection as an orientation cue. Winds aloft, however, are more dependable with respect to direction, and it is conceivable that they serve as a directional cue for high-flying gulls during migration. This latter possibility needs to be addressed by investigators.

C. A Preliminary Orientation Model for Ring-billed Gulls

Although we are far from understanding the complexities and interrelationships that exist in the orientation system of any avian species, we possess bits and pieces of information about the cues responded to by Ringbills, the relative importance of particular cues, the factors that disrupt cue availability, and the orientation ability of various age classes of gulls.

I have attempted to organize the available information into a model (Fig. 9) depicting how these various components may interact, the possible effects of maturation and experience, and the environmental conditions that could alter the efficiency of the system. In this model I have speculated that complexity, or the degree of sophistication, increases through experience, and hence, time. I have not attempted to determine if the magnetic compass

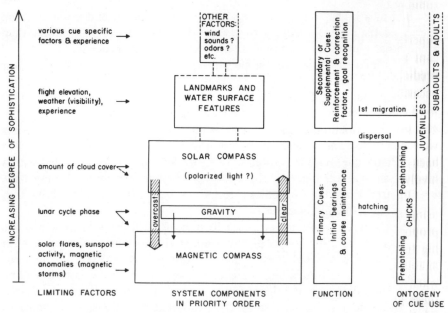

Fig. 9. A preliminary model of the orientation system of Ring-billed Gulls that is probably applicable to other gulls as well. Refer to the text for details.

is more complex than the sun compass; instead, I have indicated when I think each becomes functional. This may occur almost simultaneously, rather than sequentially, as suggested by the model. As the number of functional components in the system increases, the flexibility of the system also expands as the birds have more options open to them with respect to the environmental information that can be used for orientation. An ability to selectively use one or more cues from among those available in an integrated system of multiple cues probably places greater demands on the nervous system than would a system dependent on only one cue. For this reason, I have postulated that the orientation system of an adult Ring-bill is more complex, or sophisticated, than that of a chick, juvenile, or subadult (Fig. 9).

On the basis of my present data, it appears that the magnetic compass is the basic component of the system. It exists first in young gulls and plays an important role in selection of an initial bearing. After hatching, the young develop a solar compass that becomes closely linked with the magnetic compass. These two components are the primary ones in a young gull's orientation system and interference with one may alter the effectiveness of the other. After dispersal from the colony in late July or early August, experience with other sources of environmental information (e.g., landmarks) provide the birds with a set of supplemental cues. Added experience with this assortment of cues results in adults having an integrated system that is functional under a variety of environmental conditions and provides greater accuracy than the system possessed by gull chicks.

This model is expected to change as new information is acquired. Until then, it will serve as a summary of our present knowledge about the orientation behavior of Ring-billed Gulls.

REFERENCES

American Ornithologists' Union, 1957, *Check-list of North American Birds*, 5th ed., Port City Press, Baltimore, Md, 691 pp.

Belopol'skii, L. O., 1957, *Ecology of Sea Colony Birds of the Barents Sea*, Academy of Science of USSR, Karelian Branch.

Brown, R. G. B., Nettleship, D. N., Germain, P., Tull, C. E., and Davis, T., 1975, *Atlas of Eastern Canadian Seabirds*, Canadian Wildlife Service CW66-44, Ottawa, 220 pp.

Burger, J., 1972, Dispersal and post-fledging survival of Franklin's Gulls, *Bird-Banding* **43**:267–275.

Drury, W. H., and Nisbet, I. C. T., 1972, The Importance of Movements in the Biology of Herring Gulls in New England. *Population Ecology of Migratory Birds: A Symposium*, pp. 173–212, U.S. Department of the Interior, Wildlife Research Report No. 2.

Forsythe, D. M., 1974, *An Ecological Study of Gull Populations to Reduce the Bird–Aircraft Strike Hazard at Charleston Air Force Base*, USAF Weapons Lab, AFWL-TR-73-42.

Jehl, J. R., Jr., 1976, The northernmost colony of Heermann's Gull, *West. Birds* **7**:25–26.

Kadlec, J. A., 1971, Effect of introducing foxes and raccoons on Herring Gull colonies, *J. Wild. Mgmt.* **35**:625–636.

Ludwig, J. P., 1974, Recent changes in the Ring-billed Gull population and biology in the Laurentian Great Lakes, *Auk* **91**:575–594.

Moore, F. R., 1976, The dynamics of seasonal distribution of Great Lakes Herring Gulls, *Bird-Banding* **47**:141–159.

Peters, C. F., Richter, K. O., Manuwal, D. A., and Hermani, S. G., 1978, *Colonial Nesting Sea and Wading Bird Use of Estuarine Islands in the Pacific Northwest*, U.S. Army Corps of Engineers, Dredged Material Research Program, Technical Rep. D-78-17.

Schreiber, R. W., 1970, Breeding biology of Western Gulls (*Larus occidentalis*) on San Nicolas Island, California, 1968, *Condor* **72**:133–140.

Southern, W. E., 1968, Dispersal patterns of subadult Herring Gulls from Rogers City, Michigan, *Jack-Pine Warbler* **46**:2–6.

Southern, W. E., 1969, Sky conditions in relation to Ring-billed and Herring Gull orientation, *Trans. Ill. State Acad. Sci.*, **62**:342–349.

Southern, W. E., 1972, Magnets disrupt the orientation of juvenile Ring-billed Gulls, *BioScience* **22**:476–479.

Southern, W. E., 1974*a*, Seasonal distribution of Great Lakes region Ring-billed Gulls, *Jack-Pine Warbler* **52**:154–179.

Southern, W. E., 1974*b*, Florida distribution of Ring-billed Gulls from the Great Lakes, *Bird-Banding* **45**:341–352.

Southern, W. E., 1974*c*, The effects of superimposed magnetic fields on gull orientation, *Wilson Bull.* **86**:256–271.

Southern, W. E., 1976, Migrational orientation in Ring-billed Gull chicks, *Auk* **93**:78–85.

Southern, W. E., 1978, Orientational responses of Ring-billed Gull chicks: A re-evaluation, in: *Proceedings in Life Sciences, Animal Migration, Navigation and Homing* (K. Schmidt-Koenig and W. T. Keeton, eds.), pp. 311–317, Springer-Verlag, Berlin.

Sowls, A. L., Hatch, S. A., and Lensink, C. J., 1978, *Catalog of Alaska Seabird Colonies*, U.S. Fish and Wildlife Service, Biol. Services Program FWS/OBS-78-78.

Strang, C. A., 1977, Variation and distribution of Glaucous Gulls in western Alaska, *Condor* **79**:170–175.

Threlfall, W., 1978, Dispersal of Herring Gulls form the Witless Bay Sea Bird Sanctuary, Newfoundland, *Bird-Banding* **49**:116–124.

Zar, J. H., and Southern, W. E., 1977, Computer determination of flight distance and direction with statistical analysis and mapping, *Proc. Colonial Waterbird Gp.* **1977**:140–153.

SPECIES INDEX

SUBJECT INDEX